Bacterial Physiology and Metabolism

Recent determination of the genome sequences for a wide range of bacteria has made an in-depth knowledge of prokaryotic metabolic function even more essential in order to give biochemical, physiological and ecological meaning to the genomic information. Clearly describing the important metabolic processes that occur in prokaryotes under different conditions and in different environments, this advanced text provides an overview of the key cellular processes that determine bacterial roles in the environment, biotechnology and human health. Prokaryotic structure and composition are described as well as the means by which nutrients are transported into cells across membranes. Discussion of biosynthesis and growth is followed by detailed accounts of glucose metabolism through glycolysis, the TCA cycle, electron transport and oxidative phosphorylation, as well as other trophic variations found in prokaryotes including the use of organic compounds other than glucose, anaerobic fermentation, anaerobic respiration, chemolithotrophy and photosynthesis. The regulation of metabolism through control of gene expression and enzyme activity is also covered, as well as the survival mechanisms used under starvation conditions.

Professor Byung Hong Kim is an expert on anaerobic metabolism, organic degradation and bioelectrochemistry. He graduated from Kyungpook National University, Korea, and obtained a Ph.D. from University College Cardiff. He has carried out research at several universities around the world, with an established career in the Korea Institute of Science and Technology. He has been honoured by the Korean Government, which designated his research group a National Research Laboratory, the Bioelectricity Laboratory, and has served as President of the Korean Society for Microbiology and Biotechnology. Professor Kim wrote the classic Korean microbiology text on *Microbial Physiology* and has published over 100 refereed papers and reviews, and holds over 20 patents relating to applications of his research in environmental biotechnology.

Professor Geoffrey Michael Gadd is an authority on microbial interactions with metals, minerals and radionuclides and their applications in environmental biotechnology. He holds a personal Chair in Microbiology at the University of Dundee and is the Head of the Division of Molecular and Environmental Microbiology in the College of Life Sciences. He has published over 200 refereed scientific papers, books and reviews and has received invitations to speak at international conferences in over 20 countries. Professor Gadd has served as President of the British Mycological Society and is an elected Fellow of the Institute of Biology, the American Academy of Microbiology, the Linnean Society, and the Royal Society of Edinburgh. He has received the Berkeley Award from the British Mycological Society and the Charles Thom award from the Society for Industrial Microbiology for his outstanding research contributions to the microbiological sciences.

Bacterial Physiology and Metabolism

Byung Hong Kim
Korea Institute of Science and Technology

Geoffrey Michael Gadd
University of Dundee

CAMBRIDGE
UNIVERSITY PRESS

CAMBRIDGE UNIVERSITY PRESS
Cambridge, New York, Melbourne, Madrid, Cape Town, Singapore, São Paulo
Delhi, Tokyo, Mexico City

Cambridge University Press
The Edinburgh Building, Cambridge CB2 8RU, UK

Published in the United States of America by Cambridge University Press, New York

www.cambridge.org
Information on this title: www.cambridge.org/9780521846363

First published 2008
3rd printing 2011

Printed in the United Kingdom at the University Press, Cambridge

A catalogue record for this publication is available from the British Library

ISBN 978-0-521-84636-3 Hardback
ISBN 978-0-521-71230-9 Paperback

To our families

Hyungock Hong, Kyoungha Kim and Youngha Kim
and
Julia, Katie and Richard Gadd

Contents in brief

Contents

Preface

Knowledge of the physiology and metabolism of prokaryotes under-pins our understanding of the roles and activities of these organisms in the environment, including pathogenic and symbiotic relation-ships, as well as their exploitation in biotechnology. Prokaryotic organisms include bacteria and archaea and, although remaining relatively small and simple in structure throughout their evolutionary history, exhibit incredible diversity regarding their metabolism and physiology. Such metabolic diversity is reflective of the wide range of habitats where prokaryotes can thrive and in many cases dominate the biota, and is a distinguishing contrast with eukaryotes that exhibit a more restricted metabolic versatility. Thus, prokaryotes can be found almost everywhere under a wide range of physical and chemical conditions, including aerobic to anaerobic, light and dark, low to high pressure, low to high salt concentrations, extremes of acidity and alkalinity, and extremes of nutrient availability. Some physiolo-gies, e.g. lithotrophy and nitrogen fixation, are only found in certain groups of prokaryotes, while the use of inorganic compounds, such as nitrate and sulfate, as electron acceptors in respiration is another prokaryotic ability. The explosion of knowledge resulting from the development and application of molecular biology to microbial sys-tems has perhaps led to a reduced emphasis on their physiology and biochemistry, yet paradoxically has enabled further detailed analysis and understanding of metabolic processes. Almost in a reflection of the bacterial growth pattern, the number of scientific papers has grown at an exponential rate, while the number of prokaryotic genome sequences determined is also increasing rapidly. This pro-duction of genome sequences for a wide range of organisms has made an in-depth knowledge of prokaryotic metabolic function even more essential in order to give biochemical, physiological and ecological meaning to the genomic information. Our objective in writing this new textbook was to provide a thorough survey of the prokaryotic metabolic diversity that occurs under different conditions and in different environments, emphasizing the key biochemical mechan-isms involved. We believe that this approach provides a useful over-view of the key cellular processes that determine bacterial and archaeal roles in the environment, biotechnology and human health. We concentrate on bacteria and archaea but, where appropriate, also provide comparisons with eukaryotic organisms. It should be noted that many important metabolic pathways found in prokaryotes also occur in eukaryotes further emphasizing prokaryotic importance as research models in providing knowledge of relevance to eukaryotic processes.

This book can be considered in three main parts. In the first part, prokaryotic structure and composition is described as well as the means by which nutrients are transported into cells across membranes. Discussion of biosynthesis and growth is followed by detailed accounts of glucose metabolism through glycolysis, the TCA cycle, electron transport and oxidative phosphorylation, largely based on the model bacterium *Escherichia coli*. In the second part, the trophic variations found in prokaryotes are described, including the use of organic compounds other than glucose, anaerobic fermentation, anaerobic respiration, chemolithotrophy and photosynthesis. In the third part, the regulation of metabolism through control of gene expression and enzyme activity is covered, as well as the survival mechanisms used by prokaryotes under starvation conditions. This text is relevant to advanced undergraduate and postgraduate courses, as well as being of use to teachers and researchers in microbiology, molecular biology, biotechnology, biochemistry and related disciplines.

We would like to express our thanks to all those who helped and made this book possible. We appreciate the staff of Academy Publisher (Seoul, Korea) who re-drew the figures for the book, and those at Cambridge University Press involved at various stages of the publication process, including Katrina Halliday, Clare Georgy, Dawn Preston, Alison Evans and Janice Robertson. Special thanks also go to Diane Purves in Dundee, who greatly assisted correction, collation, editing and formatting of chapters, and production of the index, and Dr Nicola Stanley-Wall, also in Dundee, for the cover illustration images. Thanks also to all those teachers and researchers in microbiology around the world who have helped and stimulated us throughout our careers. Our families deserve special thanks for their support and patience.

Byung Hong Kim
Geoffrey Michael Gadd

Introduction to bacterial physiology and metabolism

The biosphere has been shaped both by physical events and by interactions with the organisms that occupy it. Among living organisms, prokaryotes are much more metabolically diverse than eukaryotes and can also thrive under a variety of extreme conditions where eukaryotes cannot. This is possible because of the wealth of genes, metabolic pathways and molecular processes that are unique to prokaryotic cells. For this reason, prokaryotes are very important in the cycling of elements, including carbon, nitrogen, sulfur and phosphorus, as well as metals and metalloids such as copper, mercury, selenium, arsenic and chromium. A full understanding of the complex biological phenomena that occur in the biosphere therefore requires a deep knowledge of the unique biological processes that occur in this vast prokaryotic world.

After publication in 1995 of the first full DNA sequence of a free-living bacterium, *Haemophilus influenzae*, whole genome sequences of hundreds of prokaryotes have now been determined and many others are currently being sequenced (www.genomesonline.org/). Our knowledge of the whole genome profoundly influences all aspects of microbiology. Determination of entire genome sequences, however, is only a first step in fully understanding the properties of an organism and the environment in which the organism lives. The functions encoded by these sequences need to be elucidated to give biochemical, physiological and ecological meaning to the information. Furthermore, sequence analysis indicates that the biological functions of substantial portions of complete genomes are so far unknown. Defining the role of each gene in the complex cellular metabolic network is a formidable task. In addition, genomes contain hundreds to thousands of genes, many of which encode multiple proteins that interact and function together as multicomponent systems for accomplishing specific cellular processes. The products of many genes are often co-regulated in complex signal transduction networks, and understanding how the genome functions as a whole presents an even greater challenge. It is also known that for a significant proportion of metabolic activities, no representative genes have been identified across all organisms, such activities being termed 'orphan' to indicate they are not currently assigned to any gene. This also represents a major future challenge and will require both computational and experimental approaches.

It is widely accepted that less than 1% of prokaryotes have been cultivated in pure culture under laboratory conditions. Development of new sequencing techniques has allowed us to obtain genomic information from the multitudes of unculturable prokaryotic species and complex microbial populations that exist in nature. Such information might provide a basis for the development of new cultivation techniques. Elucidation of the function of unknown genes through a better understanding of biochemistry and physiology could ultimately result in a fuller understanding of the complex biological phenomena occurring in the biosphere.

Unlike multicellular eukaryotes, individual cells of unicellular prokaryotes are more exposed to the continuously changing environment, and have evolved unique structures to survive under such conditions. Chapter 2 describes the main aspects of the composition and structure of prokaryotic cells.

Life can be defined as a reproduction process using materials available from the environment according to the genetic information possessed by the organism. Utilization of the materials available in the environment necessitates transport into cells that are separated from the environment by a membrane. Chapter 3 outlines transport mechanisms, not only for intracellular entry of nutrients, but also for excretion of materials including extracellular enzymes and materials that form cell surface structures.

Many prokaryotes, including *Escherichia coli*, can grow in a simple mineral salts medium containing glucose as the sole organic compound. Glucose is metabolized through glycolytic pathways and the tricarboxylic acid (TCA) cycle, supplying all carbon skeletons, energy in the form of ATP and reducing equivalents in the form of NADPH for growth and reproduction. Glycolysis is described in Chapter 4 with emphasis on the reverse reactions of the EMP pathway and on prokaryote-specific metabolic pathways. When substrates other than glucose are used, parts of the metabolic pathways are employed in either forward or reverse directions. Chapter 5 describes the TCA cycle and related metabolic pathways, and energy transduction mechanisms. Chapter 6 describes the biosynthetic metabolic processes that utilize carbon skeletons, ATP and NADPH, the production of which is discussed in the previous chapters. These chapters summarize the biochemistry of central metabolism that is employed by prokaryotes to enable growth on a glucose–mineral salts medium.

The next five chapters describe metabolism in some of the various trophic variations found in prokaryotes. These are the use of organic compounds other than glucose as carbon and energy sources (Chapter 7), anaerobic fermentation (Chapter 8), anaerobic respiratory processes (Chapter 9), chemolithotrophy (Chapter 10) and photosynthesis (Chapter 11). Some of these metabolic processes are prokaryote specific, while others are found in both prokaryotes and eukaryotes.

Prokaryotes only express a proportion of their genes at any given time, just like eukaryotes. This enables them to grow in the most efficient way under any given conditions. Metabolism is regulated not only through control of gene expression but also by controlling the activity of enzymes. These regulatory mechanisms are discussed in Chapter 12. Finally, the survival of prokaryotic organisms under starvation conditions is discussed in terms of storage materials and resting cell structures in Chapter 13.

This book has been written as a text for senior students at undergraduate level and postgraduates in microbiology and related subjects. A major proportion of the book has been based on review papers published in various scientific journals including those listed below:

Annual Review of Microbiology
Annual Review of Biochemistry
Current Opinion in Microbiology
FEMS Microbiology Reviews
Journal of Bacteriology
Microbiology and Molecular Biology Reviews (formerly *Microbiology Reviews*)
Nature Reviews Microbiology
Trends in Microbiology.

The authors would also like to acknowledge the authors of the books listed below that have been consulted during the preparation of this book.

Caldwell, D. R. (2000). *Microbial Physiology and Metabolism*, 2nd edn. Belm, CA: Star Publishing Co.

Dawes, D. A. (1986). *Microbial Energetics*. Glasgow: Blackie.

Dawes, I. W. & Sutherland, I. W. (1992). *Microbial Physiology*, 2nd edn. Basic Microbiology Series, 4. Oxford: Blackwell.

Gottschalk, G. (1986). *Bacterial Metabolism*, 2nd edn. New York: Springer-Verlag.

Ingraham, J. L., Maaloe, O. & Neidhardt, F. C. (1983). *Growth of the Bacterial Cell*. Sunderland, MA: Sinauer Associates Inc.

Mandelstam, J., McQuillin, K. & Dawes, I. (1982). *Biochemistry of Bacterial Growth*, 3rd edn. Oxford: Blackwell.

Moat, A. G., Foster, J. W. & Spector, M. P. (2002). *Microbial Physiology*, 4th edn. New York: Wiley.

Neidhardt, F. C. & Curtiss, R. (eds.) (1996). *Escherichia coli and Salmonella: Cellular and Molecular Biology*, 2nd edn. Washington, DC: ASM Press.

Neidhardt, F. C., Ingraham, J. L. & Schaechter, M. (1990). *Physiology of the Bacterial Cell: A Molecular Approach*. Sunderland, MA: Sinauer Associates Inc.

Stanier, R. J., Ingraham, J. L., Wheelis, M. K. & Painter, P. R. (1986). *The Microbial World*, 5th edn. Upper Saddle River, NJ: Prentice-Hall.

White, D. (2000). *The Physiology and Biochemistry of Prokaryotes*, 2nd edn. Oxford: Oxford University Press.

FURTHER READING

General

Downs, D. M. (2006). Understanding microbial metabolism. *Annual Review of Microbiology* **60**, 533–559.

Galperin, M. Y. (2004). All bugs, big and small. *Environmental Microbiology* **6**, 435–437.

Klamt, S. & Stelling, J. (2003). Two approaches for metabolic pathway analysis? *Trends in Biotechnology* **21**, 64–69.

Papin, J. A., Price, N. D., Wiback, S. J., Fell, D. A. & Palsson, B. O. (2003). Metabolic pathways in the post-genome era. *Trends in Biochemical Sciences* **28**, 250–258.

Park, S., Lee, S., Cho, J., Kim, T., Lee, J., Park, J. & Han, M. J. (2005). Global physiological understanding and metabolic engineering of microorganisms based on omics studies. *Applied Microbiology and Biotechnology* **68**, 567–579.

Postgate, J. R. (1992). *Microbes and Man*, 3rd edn. Cambridge: Cambridge University Press.

Diversity

Crawford, R. L. (2005). Microbial diversity and its relationship to planetary protection. *Applied and Environmental Microbiology* **71**, 4163–4168.

DeLong, E. F. (2001). Microbial seascapes revisited. *Current Opinion in Microbiology* **4**, 290–295.

Fernandez, L. A. (2005). Exploring prokaryotic diversity: there are other molecular worlds. *Molecular Microbiology* **55**, 5–15.

Fredrickson, J. & Balkwill, D. (2006). Geomicrobial processes and biodiversity in the deep terrestrial subsurface. *Geomicrobiology Journal* **23**, 345–356.

Pedros-Alio, C. (2006). Marine microbial diversity: can it be determined? *Trends in Microbiology* **14**, 257–263.

Rappe, M. S. & Giovannoni, S. J. (2003). The uncultured microbial majority. *Annual Review of Microbiology* **57**, 369–394.

Ecology

Gadd, G. M., Semple, K. T. & Lappin-Scott, H. M. (2005). *Micro-organisms and Earth Systems: Advances in Geomicrobiology*. Cambridge: Cambridge University Press.

Galperin, M. Y. (2004). Metagenomics: from acid mine to shining sea. *Environmental Microbiology* **6**, 543–545.

Geesey, G. G. (2001). Bacterial behavior at surfaces. *Current Opinion in Microbiology* **4**, 296–300.

Ivanov, M. V. & Karavaiko, G. I. (2004). Geological microbiology. *Microbiology-Moscow* **73**, 493–508.

Johnston, A. W. B., Li, Y. & Ogilvie, L. (2005). Metagenomic marine nitrogen fixation – feast or famine? *Trends in Microbiology* **13**, 416–420.

Karl, D. (2002). Nutrient dynamics in the deep blue sea. *Trends in Microbiology* **10**, 410–418.

Riesenfeld, C. S., Schloss, P. D. & Handelsman, J. (2004). Metagenomics: genomic analysis of microbial communities. *Annual Review of Genetics* **38**, 525–552.

Shively, J. M., English, R. S., Baker, S. H. & Cannon, G. C. (2001). Carbon cycling: the prokaryotic contribution. *Current Opinion in Microbiology* **4**, 301–306.

Tyson, G. W. & Banfield, J. F. (2005). Cultivating the uncultivated: a community genomics perspective. *Trends in Microbiology* **13**, 411–415.

Evolution

Altermann, W. & Kazmierczak, J. (2003). Archean microfossils: a reappraisal of early life on Earth. *Research in Microbiology* **154**, 611–617.

Arber, W. (2000). Genetic variation: molecular mechanisms and impact on microbial evolution. *FEMS Microbiology Reviews* **24**, 1–7.

Boucher, Y., Douady, C. J., Papke, R. T., Walsh, D. A., Boudreau, M. E., Nesbo, C. L., Case, R. J. & Doolittle, W. F. (2003). Lateral gene transfer and the origins of prokaryotic groups. *Annual Review of Genetics* **37**, 283–328.

Groisman, E. A. & Casadesus, J. (2005). The origin and evolution of human pathogens. *Molecular Microbiology* **56**, 1–7.

Koch, A. L. (2003). Were Gram-positive rods the first bacteria? *Trends in Microbiology* **11**, 166–170.

Koch, A. L. & Silver, S. (2005). The first cell. *Advances in Microbial Physiology* **50**, 227–259.

Moran, N. (2003). Tracing the evolution of gene loss in obligate bacterial symbionts. *Current Opinion in Microbiology* **6**, 512–518.

Orgel, L. E. (1998). The origin of life – a review of facts and speculations. *Trends in Biochemical Sciences* **23**, 491–495.

Ouzounis, C. A., Kunin, V., Darzentas, N. & Goldovsky, L. (2006). A minimal estimate for the gene content of the last universal common ancestor – exobiology from a terrestrial perspective. *Research in Microbiology* **157**, 57–68.

Rainey, P. B. & Cooper, T. F. (2004). Evolution of bacterial diversity and the origins of modularity. *Research in Microbiology* **155**, 370–375.

Sallstrom, B. & Andersson, S. G. E. (2005). Genome reduction in the α-proteobacteria. *Current Opinion in Microbiology* **8**, 579–585.

Trevors, J. T. (1997). Bacterial evolution and metabolism. *Antonie van Leeuwenhoek* **71**, 257–263.

Trevors, J. T. (2003). Origin of the first cells on Earth: a possible scenario. *Geomicrobiology Journal* **20**, 175–183.

van der Meer, J. R. & Sentchilo, V. (2003). Genomic islands and the evolution of catabolic pathways in bacteria. *Current Opinion in Biotechnology* **14**, 248–254.

Weinbauer, M. G. & Rassoulzadegan, F. (2004). Are viruses driving microbial diversification and diversity? *Environmental Microbiology* **6**, 1–11.

Genomics

Boucher, Y., Nesbo, C. L. & Doolittle, W. F. (2001). Microbial genomes: dealing with diversity. *Current Opinion in Microbiology* **4**, 285–289.

Clayton, R. A., White, O. & Fraser, C. M. (1998). Findings emerging from complete microbial genome sequences. *Current Opinion in Microbiology* **1**, 562–566.

Conway, T. & Schoolnik, G. K. (2003). Microarray expression profiling: capturing a genome-wide portrait of the transcriptome. *Molecular Microbiology* **47**, 879–889.

Doolittle, R. F. (2005). Evolutionary aspects of whole-genome biology. *Current Opinion in Structural Biology* **15**, 248–253.

Francke, C., Siezen, R. J. & Teusink, B. (2005). Reconstructing the metabolic network of a bacterium from its genome. *Trends in Microbiology* **13**, 550–558.

Glaser, P. & Boone, C. (2004). Beyond the genome: from genomics to systems biology. *Current Opinion in Microbiology* **7**, 489–491.

Groisman, E. A. & Ehrlich, S. D. (2003). Genomics: a global view of gene gain, loss, regulation and function. *Current Opinion in Microbiology* **6**, 479–481.

Koonin, E. V. (2004). Comparative genomics, minimal gene-sets and the last universal common ancestor. *Nature Reviews Microbiology* **1**, 127–136.

Nelson, K. E. (2003). The future of microbial genomics. *Environmental Microbiology* **5**, 1223–1225.

Puhler, A. & Selbitschka, W. (2003). Genome research on bacteria relevant for agriculture, environment and biotechnology. *Journal of Biotechnology* **106**, 119–120.

Ward, N. & Fraser, C. M. (2005). How genomics has affected the concept of microbiology. *Current Opinion in Microbiology* **8**, 564–571.

Extreme environments

Cowan, D. A. (2004). The upper temperature for life – where do we draw the line? *Trends in Microbiology* **12**, 58–60.

Deming, J. (2002). Psychrophiles and polar regions. *Current Opinion in Microbiology* **5**, 301–309.

Javaux, E. J. (2006). Extreme life on Earth: past, present and possibly beyond. *Research in Microbiology* **157**, 37–48.

Mock, T. & Thomas, D. N. (2005). Recent advances in sea-ice microbiology. *Environmental Microbiology* **7**, 605–619.

Simonato, F., Campanaro, S., Lauro, F. M., Vezzi, A., D'Angelo, M., Vitulo, N., Valle, G. & Bartlett, D. H. (2006). Piezophilic adaptation: a genomic point of view. *Journal of Biotechnology* **126**, 11–25.

Steven, B., Leveille, R., Pollard, W. H. & Whyte, L. G. (2006). Microbial ecology and biodiversity in permafrost. *Extremophiles* **10**, 259–267.

New areas

Anderson, N. L., Matheson, A. D. & Steiner, S. (2000). Proteomics: applications in basic and applied biology. *Current Opinion in Biotechnology* **11**, 408–412.

Chen, L. & Vitkup, D. (2007). Distribution of orphan metabolic activities. *Trends in Biotechnology* **25**, 343–348.

Dufrene, Y. F. (2002). Atomic force microscopy, a powerful tool in microbiology. *Journal of Bacteriology* **184**, 5205–5213.

Whitfield, E. J., Pruess, M. & Apweiler, R. (2006). Bioinformatics database infrastructure for biotechnology research. *Journal of Biotechnology* **124**, 629–639.

Composition and structure
of prokaryotic cells

Like all organisms, microorganisms grow, metabolize and replicate utilizing materials available from the environment. Such materials include those chemical elements required for structural aspects of cellular composition and metabolic activities such as enzyme regulation and redox processes. To understand bacterial metabolism, it is therefore helpful to know the chemical composition of the cell and component structures. This chapter describes the elemental composition and structure of prokaryotic cells, and the kinds of nutrients needed for biosynthesis and energy-yielding metabolism.

2.1 | Elemental composition

From over 100 natural elements, microbial cells generally only contain 12 in significant quantities. These are known as major elements, and are listed in Table 2.1 together with some of their major functions and predominant chemical forms used by microorganisms.

They include elements such as carbon (C), oxygen (O) and hydrogen (H) constituting organic compounds like carbohydrates. Nitrogen (N) is found in microbial cells in proteins, nucleic acids and coenzymes. Sulfur (S) is needed for S-containing amino acids such as methionine and cysteine and for various coenzymes. Phosphorus (P) is present in nucleic acids, phospholipids, teichoic acid and nucleotides including NAD(P) and ATP. Potassium is the major inorganic cation (K^+), while chloride (Cl^-) is the major inorganic anion. K^+ is required as a cofactor for certain enzymes, e.g. pyruvate kinase. Chloride is involved in the energy conservation process operated by halophilic archaea (Section 11.6). Sodium (Na^+) participates in several transport and energy transduction processes, and plays a crucial role in microbial growth under alkaline conditions (Section 5.7.4). Magnesium (Mg^{2+}) forms complexes with phosphate groups including those found in nucleic acids, ATP, phospholipids and lipopolysaccharides. Several microbial intracellular enzymes, e.g. monomeric alkaline phosphatase, are calcium dependent. Ferrous and ferric ions play a

Table 2.1. *Major elements found in microbial cells with their functions and predominant chemical forms used by microorganisms*

Element	Atomic number	Chemical forms used by microbes	Function
C	6	organic compounds, CO, CO_2	major constituents of cell material in proteins, nucleic acids, lipids, carbohydrates and others
O	8	organic compounds, CO_2, H_2O, O_2	
H	1	organic compounds, H_2O, H_2	
N	6	organic compounds, NH_4^+, NO_3^-, N_2	
S	16	organic sulfur compounds, SO_4^{2-}, HS^-, S^0, $S_2O_3^{2-}$	proteins, coenzymes
P	15	HPO_4^{2-}	nucleic acids, phospholipids, teichoic acid, coenzymes
K	19	K^+	major inorganic cation, compatible solute, enzyme cofactor
Mg	12	Mg^{2+}	enzyme cofactor, bound to cell wall, membrane and phosphate esters including nucleic acids and ATP
Ca	20	Ca^{2+}	enzyme cofactor, bound to cell wall
Fe	26	Fe^{2+}, Fe^{3+}	cytochromes, ferredoxin, Fe-S proteins, enzyme cofactor
Na	11	Na^+	involved in transport and energy transduction
Cl	17	Cl^-	major inorganic anion

crucial role in oxidation–reduction reactions as components of electron carriers such as Fe-S proteins and cytochromes.

In addition to these 12 major elements, others are also found in microbial cells as minor elements (Table 2.2). All the metals listed in Table 2.2 are required for specific enzymes. It is interesting to note that the atomic number of tungsten is far higher than that of the other elements and that this element is only required in rare cases.

2.2 | Importance of chemical form

2.2.1 Five major elements

The elements listed in Tables 2.1 and 2.2 need to be supplied or be present in the chemical forms that the organisms can use. Carbon is the most abundant element in all living organisms. Prokaryotes are broadly classified according to the carbon sources they use: organotrophs (heterotrophs) use organic compounds as their carbon source while CO_2 is used by lithotrophs (autotrophs). These groups

Table 2.2. | *Minor elements found in microbial cells with their functions and predominant chemical form used by microorganisms*

Element	Atomic number	Chemical form used by microbes	Function
Mn	23	Mn^{2+}	superoxide dismutase, photosystem II
Co	27	Co^{2+}	coenzyme B_{12}
Ni	28	Ni^+	hydrogenase, urease
Cu	29	Cu^{2+}	cytochrome oxidase, oxygenase
Zn	30	Zn^{2+}	alcohol dehydrogenase, aldolase, alkaline phosphatase, RNA and DNA polymerase, arsenate reductase
Se	34	SeO_3^{2-}	formate dehydrogenase, glycine reductase
Mo	42	MoO_4^{2-}	nitrogenase, nitrate reductase, formate dehydrogenase, arsenate reductase
W	74	WO_4^{2-}	formate dehydrogenase, aldehyde oxidoreductase

are divided further according to the form of energy they use: chemotrophs (chemoorganotrophs and chemolithotrophs) depend on chemical forms for energy while phototrophs (photoorganotrophs and photolithotrophs) utilize light energy ('organo' refers to an organic substance while 'litho' refers to an inorganic substance).

Nitrogen sources commonly used by microbes include organic nitrogenous compounds such as amino acids, and inorganic forms such as ammonium and nitrate. Gaseous N_2 can serve as a nitrogen source for a limited number of nitrogen-fixing prokaryotes. Nitrogen fixation is not known in eukaryotes. Some chemolithotrophs can use ammonium as their energy source (electron donor, Section 10.2) while nitrate can be used as an electron acceptor by denitrifiers (Section 9.1).

Sulfate is the most commonly used sulfur source, while other sulfur sources used include organic sulfur compounds, sulfide, elemental sulfur and thiosulfate. Sulfide and sulfur can serve as electron donors in certain chemolithotrophs (Section 10.3), and sulfate and elemental sulfur are used as electron acceptors and reduced to sulfide by sulfidogens (Section 9.3).

2.2.2 Oxygen

Oxygen in cells originates mainly from organic compounds, water or CO_2. Molecular oxygen (O_2) is seldom used in biosynthetic processes. Some prokaryotes use O_2 as the electron acceptor, but some cannot grow in its presence. Thus, organisms can be grouped according to their reaction with O_2 into aerobes that require O_2, facultative anaerobes that use O_2 when it is available but can also grow in its absence, and obligate anaerobes that do not use O_2. Some obligate anaerobes cannot grow and/or lose their viability in the presence of O_2 while others can tolerate it. The former are termed strict anaerobes and the latter aerotolerant anaerobes.

Table 2.3.	*Common growth factors required by prokaryotes and their major function*
Growth factor	Function
p-aminobenzoate	part of tetrahydrofolate, a one-carbon unit carrier
Biotin	prosthetic group of carboxylase and mutase
Coenzyme M	methyl carrier in methanogenic archaea
Folate	part of tetrahydrofolate
Hemin	precursor of cytochromes and hemoproteins
Lipoate	prosthetic group of 2-keto acid decarboxylase
Nicotinate	precursor of pyridine nucleotides (NAD^+, $NADP^+$)
Pantothenate	precursor of coenzyme A and acyl carrier protein
Pyridoxine	precursor of pyridoxal phosphate
Riboflavin	precursor of flavins (FAD, FMN)
Thiamine	precursor of thiamine pyrophosphate
Vitamin B_{12}	precursor of coenzyme B_{12}
Vitamin K	precursor of menaquinone

2.2.3 Growth factors

Some organotrophs such as *Escherichia coli* can grow in simple media containing glucose and mineral salts, while others, like lactic acid bacteria, require complex media containing various vitamins, amino acids and nucleic acid bases. This is because the latter organisms cannot synthesize certain essential cellular materials from only glucose and mineral salts. These required compounds should therefore be supplied in the growth media: such compounds are known as growth factors. Growth factor requirements differ between organisms with vitamins being the most commonly required growth factors (Table 2.3).

2.3 | Structure of microbial cells

Microorganisms are grouped into either prokaryotes or eukaryotes according to their cellular structure. With only a few exceptions, prokaryotic cells do not have subcellular organelles separated from the cytoplasm by phospholipid membranes such as the nuclear and mitochondrial membranes. Organelles like the nucleus, mitochondria and endoplasmic reticulum are only found in eukaryotic cells. The detailed structure of prokaryotic cells is described below.

2.3.1 Flagella and pili

Motile prokaryotic cells have an appendage called a flagellum (plural, flagella) involved in motility, and a similar but smaller structure, the fimbria (plural, fimbriae). Fimbriae are not involved in motility and are composed of proteins.

The bacterial flagellum consists of three parts. These are a basal body, a hook and a filament (Figure 2.1). The basal body is embedded in the cytoplasmic membrane and cell surface structure and

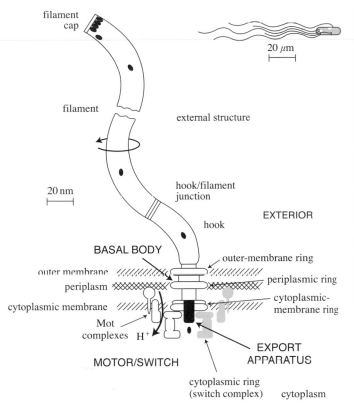

filament cap

filament

external structure

20 μm

20 nm

hook/filament junction

hook

EXTERIOR

BASAL BODY

outer-membrane ring

outer membrane

periplasm

periplasmic ring

cytoplasmic membrane

cytoplasmic-membrane ring

Mot complexes H⁺

EXPORT APPARATUS

MOTOR/SWITCH

cytoplasmic ring (switch complex)   cytoplasm

Figure 2.1 **The structure of the flagellum in Gram-negative bacteria.**

(*J. Bacteriol.* 180:1009–1022, 1998)

Three rings of the basal body are embedded in the cytoplasmic membrane, murein layer and outer membrane. The outer filament is connected to the basal body through the hook. The cytoplasmic membrane ring of the basal body is associated with the Mot complex. This complex functions like a motor rotating the flagellum, thus rendering the cell motile. Energy for this rotation is provided in the form of the proton motive (or sodium motive) force.

connected to the filament through the hook. In Gram-negative bacteria the basal body consists of a cytoplasmic membrane ring, a periplasmic ring and an outer-membrane ring through which the central rod passes. The diameter of the rings can be 20–50 nm depending on the species. The cytoplasmic ring of the basal body is associated with additional proteins known as the Mot complex. The Mot complex rotates the basal body with the entire flagellum consuming a proton motive force (or sodium motive force). The cytoplasmic membrane ring is therefore believed to function as a motor with the Mot complex. A more detailed description of motility is given in Section 12.2.11. In addition to the Mot complex, the basal body is associated with an export apparatus through which the building blocks of the filament are transported.

The hook connects the central rod of the basal body to the filament and is composed of a single protein called the hook protein. The filament, with a diameter of 10–20 nm, can be dissolved at pH 3–4 with surfactants to a single protein solution of flagellin. The molecular weight of flagellin varies from 20 to 65 kD depending on the bacterial species. The hook and the filament are tube-shaped and the flagellin moves through the tube to the growing tip of the filament. The tip of the filament is covered with filament cap protein. Flagellin can be exported to the medium in mutants defective in expression of this protein.

The number and location of flagella vary depending on the bacterial species. In some prokaryotes they are located at one or both poles, while the entire cell surface may be covered with flagella in others.

The fimbria, also known as the pilus (plural, pili), is observed in many Gram-negative bacteria but rarely in Gram-positive bacteria. Fimbriae have been proposed as the fibrils that mediate attachment to surfaces. For this reason, the term pilus should be used only to describe the F-pilus, the structure that mediates conjugation. Fimbriae are generally smaller in length (0.2–20 μm) and width (3–14 nm) than flagella. Fimbriae help the organism to stick to surfaces of other bacteria, to host cells of animals and plants, and to solid surfaces. Different kinds of fimbriae are known which depend on the species as well as the growth conditions for a given organism. Fimbriae consist of a major protein with minor proteins called adhesins that facilitate bacterial attachment to surfaces by recognizing the appropriate receptor molecules. They are classified as type I through to type IV according to such receptor recognition properties. Adhesive properties are inhibited by sugars such as mannose, galactose and their oligomers, suggesting that the receptors are carbohydrate in nature.

A fibril bigger in size than fimbriae occurs in many Gram-negative bacteria that harbour the conjugative F-plasmid. This is called the F-pilus or sex pilus and mediates attachment between mating cells for the purpose of transmitting DNA from the donor cell by means of the F-pilus to a recipient cell. The F-pilus recognizes a receptor molecule on the surface of the recipient cell and after attachment, the F-pilus is depolymerized so that there is direct contact between the cells for DNA transmission.

2.3.2 Capsules and slime layers

Many prokaryotic cells are covered with polysaccharides. In some cases the polymers are tightly integrated with the cell while in others they are loosely associated. The former is called a capsule, and the latter a slime layer (Figure 2.2). Slime layer materials can diffuse into the medium with their structure and composition being dependent on growth conditions. An important role for these structures is adhesion to host cells for invasion or to a solid surface to initiate and stabilize biofilm formation. These structures are also responsible for resistance to phagocytosis, thereby increasing virulence. In some bacteria the capsule functions as a receptor for phage. Since the polysaccharides are hydrophilic, they can also protect cells from desiccation.

The term glycocalyx can be used to describe extracellular structures including the capsule and S-layer, the latter being described below.

2.3.3 S-layer, outer membrane and cell wall

Unicellular prokaryotes have elaborate surface structures. These include the S-layer, outer membrane and cell wall. The cell wall

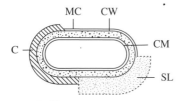

Figure 2.2 Diagram of the surface structure of a prokaryotic cell showing the capsule and slime layer.

C, capsule; CM, cytoplasmic membrane; CW, cell wall; MC, micro-capsule; SL, slime layer.

determines the physical shape of prokaryotic cells in most cases. Prokaryotes can be classified into four groups according to their cell wall structure. These are mycoplasmas, Gram-negative bacteria, Gram-positive bacteria and archaea. Cell walls are not found in mycoplasmas which are obligate intracellular pathogens. Murein is the sole building block of the cell wall in Gram-negative bacteria which also have an outer membrane. Murein and teichoic acid constitute the cell wall in Gram-positive bacteria which do not possess an outer membrane. Some archaea have cell walls that do not contain murein, while others are devoid of cell walls.

2.3.3.1   S-layer

A protein or glycoprotein layer is found on the surface of all prokaryotic cells except mycoplasmas. This is called an S-layer (Figure 2.3). All prokaryotic cells studied are surrounded by this layer, but this property can be lost in some laboratory strains. This suggests that this layer is indispensable in natural environments. The proposed functions of the S-layer are (1) protection from toxic compounds, (2) adhesion to solid surfaces, (3) a phage receptor, (4) a physical structure to maintain cell morphology, and (5) a binding site for certain extracellular enzymes.

Strains devoid of an S-layer are less resistant to murein- and protein-hydrolyzing enzymes, and more prone to release from biofilms with environmental changes such as pH. The S-layer serves as a cell wall in some archaea. Amylase is bound to the S-layer in *Bacillus stearothermophilus*.

The protein forming the S-layer is one of the most abundant proteins in bacterial cells, comprising 5–10% of total cellular protein. In a fast-growing organism, this protein needs to be synthesized and exported very efficiently. The promoter and signal sequence of this protein has therefore been studied for use in foreign protein production using bacteria.

2.3.3.2   Outer membrane

Gram-negative bacteria are more resistant to lysozyme, hydrolytic enzymes, surfactants, bile salts and hydrophobic antibiotics than Gram-positive bacteria. These properties are due to the presence of the outer membrane in Gram-negative bacteria (Figure 2.3). The outer membrane (OM) is different in structure from the cytoplasmic membrane (CM). The CM consists of phospholipids while lipopolysaccharide (LPS) forms the outer leaflet of the OM with the inner leaflet composed of phospholipids. LPS provides a permeability barrier against the hydrophobic compounds listed above. In addition to these lipids, the OM contains protein and lipoprotein (Table 2.4).

Lipopolysaccharide (LPS) consists of three components: lipid A, core polysaccharide and repeating polysaccharide. The repeating polysaccharide is referred to as O-antigen. Lipid A is embedded in the membrane to form the lipid layer, and the sugar moieties extend into the surface. The sugar moieties of LPS consist of hexoses, hexosamines,

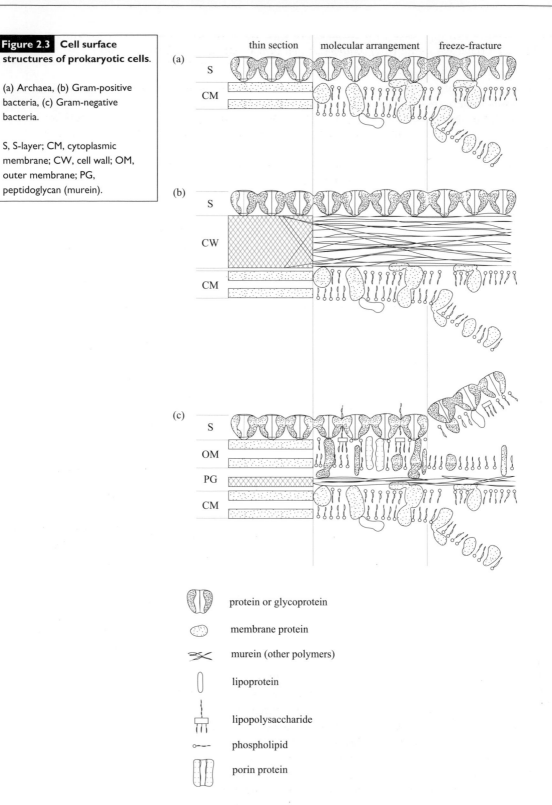

Figure 2.3 **Cell surface structures of prokaryotic cells**.

(a) Archaea, (b) Gram-positive bacteria, (c) Gram-negative bacteria.

S, S-layer; CM, cytoplasmic membrane; CW, cell wall; OM, outer membrane; PG, peptidoglycan (murein).

thin section   molecular arrangement   freeze-fracture

(a) S
CM

(b) S
CW
CM

(c) S
OM
PG
CM

protein or glycoprotein

membrane protein

murein (other polymers)

lipoprotein

lipopolysaccharide

phospholipid

porin protein

Table 2.4. | *Outer membrane (OM) components and their functions in* Escherichia coli

Component	Function
Phospholipid	inner leaflet
Lipopolysaccharide	outer leaflet, hydrophilic in nature providing a barrier against hydrophobic compounds. Stabilization of the surface structure by bonding with metal ions such as Mg^{2+}
Lipoprotein	lipid part is embedded in the OM hydrophobic region, and the sugar part is covalently bound to murein which stabilizes the OM
OM protein A	maintenance of OM stability, receptor for amino acids and peptides, F-pilus in recipient cell for conjugation
Porin	three different porins, OmpC, OmpF and PhoE, each consisting of three peptides, act as specific and non-specific channels for hydrophilic solutes
Receptor proteins	for sugars, amino acids, vitamins, etc.
Other proteins	enzymes such as phospholipase, protease, etc., extracellular protein export machinery

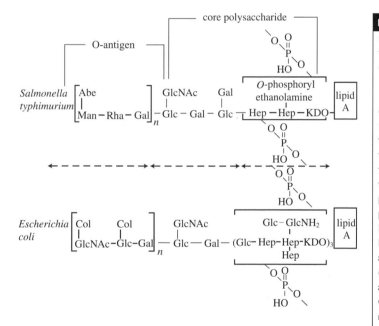

Figure 2.4 **Lipopolysaccharide structure in *Salmonella typhimurium* and *Escherichia coli*.**

Lipopolysaccharide (LPS) consists of lipid A, core polysaccharide and repeating polysaccharide. Lipid A forms a bilayer membrane with the lipid part of phospholipid and the sugar part extending into the surface. The repeating polysaccharide is involved in pathogenesis and is called O-antigen. LPS contains unique sugars L-glycero-D-mannoheptose (Hep) and 2-keto-3-deoxyoctonate (KDO), and rare sugars such as abequose (Abe) and colitose (Col). Galactose (Gal), glucose (Glc), mannose (Man) and rhamnose (Rha) can also be present.

deoxyhexoses and keto-sugars with different structures depending on the species and on the culture conditions. Figure 2.4 shows the LPS structures found in *Salmonella typhimurium* and *Escherichia coli*.

O-antigen mutants of pathogens are less virulent and show a different phage susceptibility, suggesting that the O-antigen is involved in pathogenesis and phage recognition processes. The structure of O-antigen varies with changes in growth conditions such as temperature, and this polysaccharide may therefore protect the cell during such changes. Lipid A consists of 6–7 molecules of 3-hydroxy fatty acids bound to phosphorylated glucosamine (Figure 2.5).

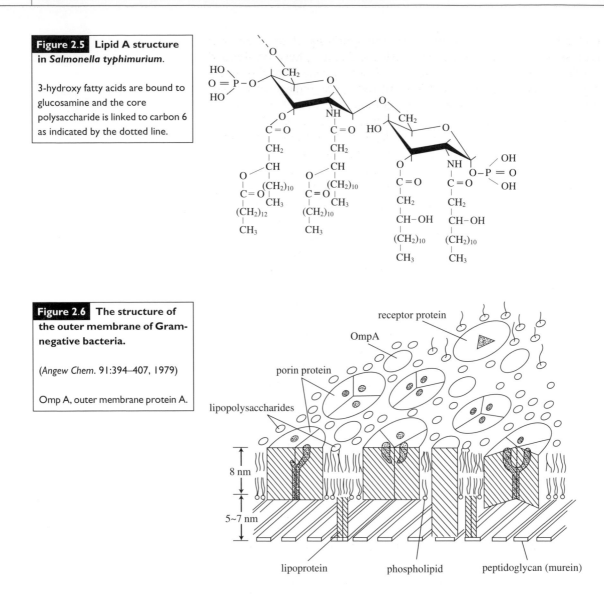

Figure 2.5 Lipid A structure in *Salmonella typhimurium*.

3-hydroxy fatty acids are bound to glucosamine and the core polysaccharide is linked to carbon 6 as indicated by the dotted line.

Figure 2.6 The structure of the outer membrane of Gram-negative bacteria.

(*Angew Chem.* 91:394–407, 1979)

Omp A, outer membrane protein A.

Due to phosphate and carboxylic moieties in LPS, the Gram-negative bacterial cell surface is negatively charged. Divalent cations such as Mg^{2+} cross-link LPS, thus stabilizing OM structure. The OM becomes more permeable to hydrophobic compounds when the cations are removed using chelating agents such as EDTA. In addition to LPS and phospholipids, the OM contains various proteins which have various functions including lipoprotein (Figure 2.6).

Active transport of nutrients takes place across the cytoplasmic membrane. Nutrients therefore need to cross the OM, and for this the OM has porins and receptors. Porins are hydrophilic channels formed by three peptides spanning the OM, allowing the diffusion of solutes of molecular weight less than 600 Da. *Escherichia coli* has three porins, OmpC, OmpF and PhoE – OmpC and OmpF being non-specific. OmpF is slightly bigger than OmpC, the synthesis of which is stimulated under high osmotic pressure when the synthesis of the former is repressed (Section 12.2.9). PhoE is synthesized under phosphate-limited conditions and is specific for anions. The pore size of porins varies depending on the species, and is regulated by the molecular arrangement around it. Nutrients with a higher molecular weight cross the OM through receptor proteins. The speed of diffusion across the OM through non-specific porins and receptors is dependent on the concentration gradient, but the speed increases only up to the saturation concentration through the porins and receptors specific for that nutrient.

OmpA functions as a receptor for various amino acids and peptides and for F-pili. Various other receptors are identified for nucleosides, vitamins and other nutrients. Lipoprotein stabilizes OM structure. The lipid end of the molecule is embedded in the lipid area of OM, while the protein end is covalently bound to murein. A lipoprotein in a plant pathogenic bacterium, *Erwinia chrysanthemi*, has pectin methylesterase activity.

Like other microbes, many Gram-negative bacteria can use polymers as their carbon and energy sources (Section 7.1) and excrete hydrolyzing enzymes into their external environment. The OM has proteins responsible for the translocation of extracellular proteins (Section 3.8.2). Cytochromes and ferric reductase are located in the OM in Fe(III)-reducing bacteria such as *Shewanella putrefaciens* and *Geobacter sulfurreducens* (Section 9.2.1).

2.3.3.3 Cell wall and periplasm

With a few exceptions, prokaryotic cells have a cell wall that provides the physical strength to maintain their shapes. Murein is the main component of the cell wall of bacteria. The cell wall in Gram-negative bacteria is much thinner than in Gram-positive bacteria, which have a complex cell wall with other polymers and do not possess an outer membrane (Figure 2.3).

Murein (peptidoglycan) is a polymer with a backbone of β-1,4-linked N-acetylglucosamine and N-acetylmuramate (Figure 2.7), the lactyl group of which is cross-linked through tetrapeptides (Figure 2.8). Some Gram-positive bacteria, including *Lactobacillus acidophilus*, are resistant to lysozyme since they have murein O-acetylated at the 6-OH of N-acetylmuramate.

D-glutamate and D-alanine occupy the 2nd and 4th positions of the tetrapeptide in most cases, and L-serine or glycine the 1st position depending on the species. Various amino acids are found in the 3rd position including L-ornithine, L-lysine, L,L-diaminopimelate, *meso*-diaminopimelate and rarely L-homoserine. The 3rd amino acid in the murein tetrapeptide is a key criterion used for bacterial classification. The degree of cross-linking differs depending on the species, and is low at the growing tip of a single bacterial cell. Murein with reduced cross-linking is more susceptible to hydrolyzing enzymes.

Gram-positive bacteria do not have an outer membrane but have a much thicker cell wall containing teichoic acid, lipoteichoic acid and lipoglycan in addition to murein.

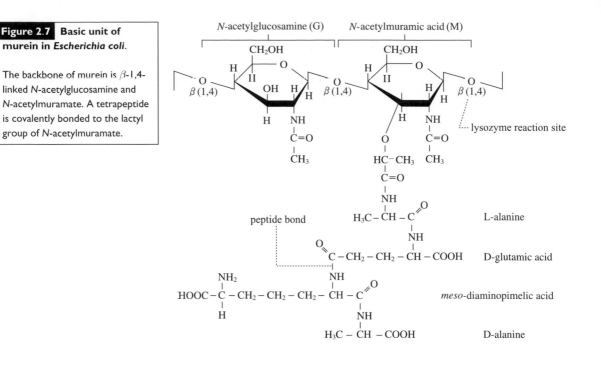

Figure 2.7 Basic unit of murein in *Escherichia coli*.

The backbone of murein is β-1,4-linked N-acetylglucosamine and N-acetylmuramate. A tetrapeptide is covalently bonded to the lactyl group of N-acetylmuramate.

The structure of teichoic acid differs depending on the bacterial species. Teichoic acids are polymers of ribitol phosphate, glycerol phosphate and their derivatives. In some cases, the hydroxyl groups of the poly alcohols are bonded with amino acids or sugars. Teichoic acid is linked to murein through a linkage unit (Figure 2.9). While murein is a structure rendering physical strength, teichoic acid has some physiological functions. Due to the phosphate in teichoic acids, the surface of Gram-positive bacteria is negatively charged, attracting cations including Mg^{2+} which stabilize cell wall structure. An autolytic enzyme is necessary to hydrolyze murein at the growing tip for growth of a cell. This enzyme is attached to teichoic acid, ensuring it is not released into the medium and enabling control of activity. Teichoic acid is also a receptor for certain phages. Gram-positive bacteria synthesize teichuronic acid in place of teichoic acid under phosphate-limited conditions.

Lipoteichoic acid has a similar structure to teichoic acid. A diacylglycerol-containing sugar is attached to the terminal polyalcohol phosphate. This lipid part of lipoteichoic acid is embedded in the cytoplasmic membrane, while the polyalcohol phosphate partly constitutes the cell wall.

The Gram-positive bacterial cell wall contains various proteins, including the autolytic enzymes mentioned above. They are attached to the cell wall through the action of an enzyme called sortase (Section 6.9.2.4) after they are excreted through the cytoplasmic membrane. Their functions include (1) metabolism of cell surface structures, (2) invasion into the host, (3) hydrolysis of polymers such as proteins and polysaccharides and (4) adhesion to solid surfaces.

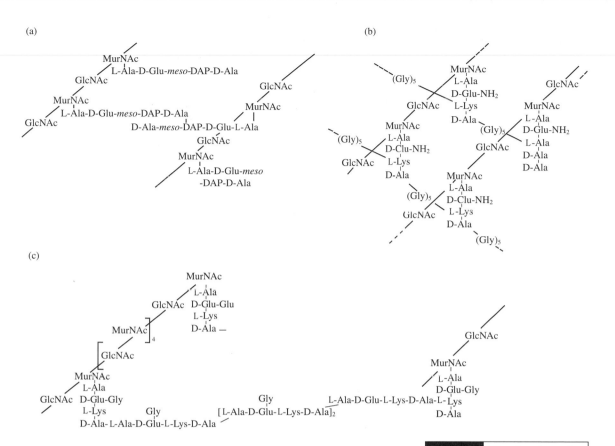

Figure 2.8 **The structure of murein: the main component of the eubacterial cell wall**.

The backbone of murein consists of a repeat of β-1,4-linked N-acetylglucosamine and N-acetylmuramate that is bonded with a tetrapeptide. The tetrapeptides of neighbouring backbones are cross-linked to make a net-like structure.

(a) *Escherichia coli*; (b) *Staphylococcus aureus*; (c) *Micrococcus luteus* (*Micrococcus lysodeikticus*).

Mycobacterium species possess a distinctive staining property called acid-alcohol fastness. This property is due to the presence of a high concentration of mycolic acid in their cell wall. This cell wall has some similar properties to the outer membrane of Gram-negative bacteria and has a porin-like structure.

The term periplasm is used to describe a separate compartment between the outer membrane and the cytoplasmic membrane in Gram-negative bacteria. Murein (cell wall) is contained within this compartment. The periplasm is in a gel state and contains proteins and oligosaccharides. Under high osmotic pressure conditions, Gram-negative bacteria accumulate an oligosaccharide known as osmoregulated periplasmic glucan, which buffers the osmolarity.

A variety of proteins are found in the periplasm including sensor proteins (Section 12.2.10), enzymes for the synthesis of cell surface components (Section 6.9), transporters (Section 3.8), solute-binding proteins, regulatory proteins (Section 12.2.9), part of the electron transport system (Sections 5.8.2, 5.8.3, 9.1.1), and hydrolytic enzymes such as β-lactamase, amylase and alkaline phosphatase. *Pseudomonas aeruginosa* releases outer membrane vesicles containing autolytic enzymes into the medium, causing the lysis of other bacteria.

Members of the family *Mycoplasmataceae* and the class *Chlamydiae* are obligate intracellular pathogens of humans and other animals and do not have cell wall structures. However, the genome of

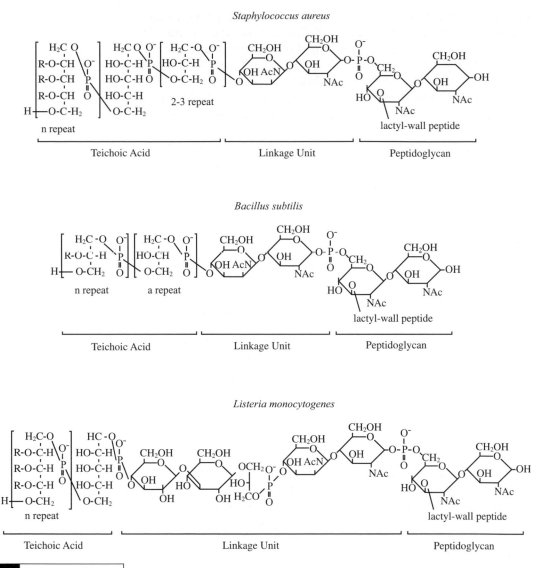

Figure 2.9 **Structure of teichoic acid.**

(*Microbiol. Mol. Biol. Rev.* 63:174–229, 1999)

Teichoic acid is a polymer of ribitol phosphate or glycerol phosphate. In some Gram-positive bacteria, the poly-alcohols are bonded with alanine, galactose and N-acetylglucosamine. Linkage units link teichoic acid to murein.

Chlamydiae includes a complete set of genes for the synthesis of murein. They are also susceptible to antibiotics such as β-lactams and D-cycloserine which inhibit murein synthesis. It is not known how these biosynthetic genes are repressed or how the antibiotics inhibit growth. These organisms have cysteine-rich proteins in their outer membrane. Disulfide bonds in the proteins render the cells resistant to osmotic pressure.

The phylum *Planctomycetes* is one of the major divisions of the domain bacteria and is phylogenetically deep-rooted. These bacteria lack murein in their cell wall, which mainly comprises protein, and is probably an S-layer.

The archaeal cell wall is different from the bacterial cell wall. The main components of the archaeal cell wall include pseudomurein (pseudopeptidoglycan), sulfonated polysaccharide and glycoprotein.

Pseudomurein is found in the cell wall of *Methanobacterium* species. The structure of pseudomurein is different from that of murein. *N*-acetyltalosaminuronate is linked to *N*-acetylglucosamine through a β-1,3-linkage in the place of β-1,4-linked *N*-acetylmuramate. Consequently lysozyme does not hydrolyze pseudomurein. D-amino acids are not found in pseudomurein. Members of the genus *Methanosarcina* have a cell wall consisting of polysaccharide, and the halophilic archaeon *Halococcus* has a cell wall of sulfonated polysaccharide. Members of the genus *Halobacterium* and hyperthermophilic archaea have glycoprotein as their cell wall, probably as an S-layer.

2.3.4 Cytoplasmic membrane

2.3.4.1 Properties and functions

The cell contents (cytoplasm) need to be isolated from the external environment, but at the same time nutrients must be transported into the cell. The cytoplasmic membrane mediates not only these functions but also other important physiological activities. These include solute transport (Chapter 3), oxidative phosphorylation through electron transport (Section 5.8), photosynthetic electron transport in photosynthetic prokaryotes (Section 11.4), maintenance of electrochemical gradients and ATP synthesis (Section 5.8.4), motility (Section 12.2.11), signal transduction (Section 12.2), synthesis of cell surface structures and protein secretion (Section 3.8). The cytoplasmic membrane consists of phospholipid (35–50%) and protein (50–65%). The phospholipid is responsible for the isolation property of the membrane with the various proteins being involved in the rest of the membrane functions.

Phospholipids have unique properties which determine the isolation of the cytoplasm. They are amphipathic compounds having hydrophobic hydrocarbon chains (tails) and a hydrophilic head group in a single molecule. When an amphipathic compound is mixed with water, the hydrophilic heads and water interact with each other and the hydrophobic tails do so in such a way that the molecules form micelles or liposomes (Figure 2.10). Because of this property, a damaged cytoplasmic membrane can restore its structure spontaneously, and when the cells are broken the membrane forms smaller vesicles (Figure 2.11).

When cells are broken with physical methods such as sonication or the French press, inside-out vesicles are formed, while right-side-out vesicles are formed when the protoplast is osmotically lyzed after the cell wall is removed using enzymes. Vesicles are a useful tool to study membranes without the interference of cytosolic activities.

Phospholipid forms both inner and outer leaflets of the cytoplasmic membrane, but the membrane is asymmetrical due to proteins present in the membrane. The phospholipid bilayer membrane is permeable to hydrophobic solutes and water but not to charged solutes and polymers. Membrane proteins transport these in and out of the cell. Though water can diffuse through the membrane, the diffusion rate is too low

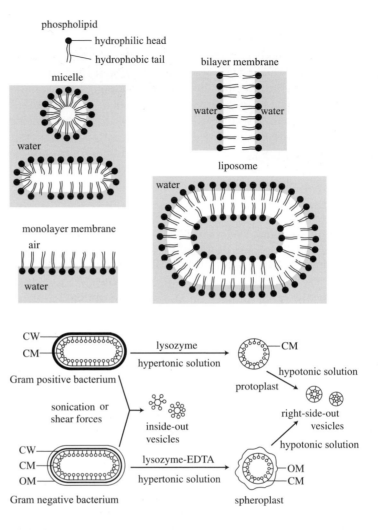

Figure 2.10 Diagram of a micelle and liposome formed when amphipathic phospholipid is mixed with water.

Figure 2.11 Membrane vesicle formation from broken cells.

(Dawes, E. A. 1986, *Microbial Energetics*, Figure 1.1. Blackie & Son, Glasgow)

Right-side-out vesicles are obtained when enzymically prepared protoplasts are burst by osmotic shock, while physical methods which break cells result in inside-out vesicles.

CM, cytoplasmic membrane; CW, cell wall; OM, outer membrane.

to explain the rapid water flux which counters osmotic shock. This rapid flux is mediated by a protein, aquaporin.

2.3.4.2 Membrane structure

The hydrocarbon part of the cytoplasmic membrane phospholipids is in a semi-solid state at physiological temperature. The proteins in the membrane are clustered in functional aggregates, with the aggregates floating around within the membrane. This membrane structure is called a fluid mosaic model (Figure 2.12). Thin sections of the cytoplasmic membrane stained with metal compounds show two lines of strong electron absorption separated by a less electron dense zone (Figure 2.3). The electron dense area is the metal-binding head group separated by the hydrocarbon part which binds less metal. This structure is consistent with the behaviour of an amphipathic phospholipid.

The fluidity of the membrane is determined by the melting point of the hydrocarbon part of the phospholipid. Phospholipid

(a)

(b)

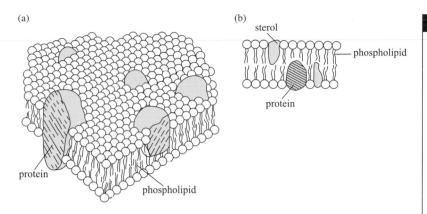

sterol

phospholipid

protein

protein

phospholipid

Figure 2.12 Schematic diagram of the cytoplasmic membrane.

Protein comprises about 50–65% of the membrane. The proteins in the membrane are clustered in functional aggregates, and the aggregates float around within the membrane. This membrane structure is called a fluid mosaic model. Sterols are found only in obligate intracellular pathogenic mycoplasmas among prokaryotes as in eukaryotic cells (b), but not in most prokaryotes (a). Hopanoids, a member of the isoprenoid family, are widespread in many bacteria.

containing more unsaturated fatty acid has a low melting point and is found in the cytoplasmic membrane of psychrophiles. Thermophiles have a high degree of saturation in their membrane that manifests itself as impaired transport function at low temperature. Freeze-fractured membranes show integral as well as peripheral proteins (Figure 2.3).

2.3.4.3 Phospholipids

Lipids found in the bacterial cytoplasmic membrane are shown in Figure 2.13. Phospholipids are the major lipid component in the cytoplasmic membrane. Sterols are universal lipids in eukaryotic cells, but are found only in obligate intracellular pathogenic mycoplasmas among prokaryotic cells. Hopanoids, members of the isoprenoid family, are widespread in many bacteria (Figure 2.13g).

Glycerophospholipids are dominant phospholipids in bacteria, and sphingolipids are found in certain bacterial cytoplasmic membranes. Sphingolipids (Figure 2.13f) comprise more than 50% of cytoplasmic phospholipids in the obligate anaerobe *Prevotella melaninogenicus* (formerly *Bacteroides melaninogenicus*). Other bacteria known to have sphingolipids include *Prevotella ruminicola*, *Bdellovibrio bacterivorus*, *Flectobacillus major*, *Bacteroides fragilis*, *Sphingobacterium spiritivorum* and *Porphyromonas gingivalis* (formerly *Bacteroides gingivalis*).

As shown in Figure 2.13, membrane phospholipids contain fatty acids with a carbon number of 10–20, but mainly 16–20. The obligately anaerobic fermentative bacterium *Sarcina ventriculi* contains fatty acids with carbon numbers of 14–18 in its cytoplasmic membrane, but α,ω-dicarboxylic acids with carbon numbers of 28–36 become the major fatty acids in its membrane in hostile environments such as those of low pH, high ethanol concentration, etc. Since the dicarboxylic acids form diglycerol tetraesters, this molecule spans the thickness of the membrane and reduces its fluidity. Diglycerol tetraesters are found in other anaerobes including members of *Desulfotomaculum*, *Butyrivibrio*, and thermophilic *Clostridium* species. Diglycerol tetraethers are also found in archaea, as discussed later.

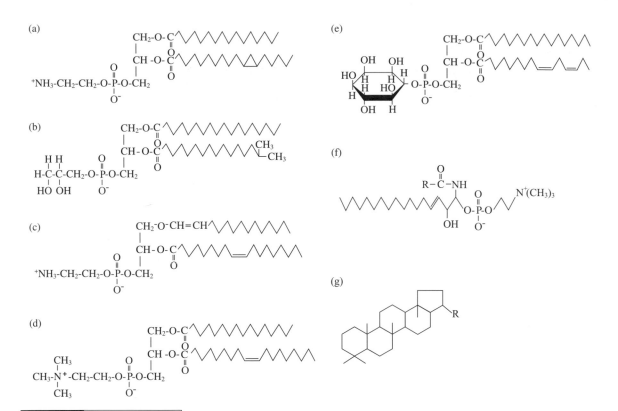

Figure 2.13 **Phospholipids in bacteria.**

Hydrophobic tails of phospholipids in the bacterial cytoplasmic membrane include saturated, unsaturated, branched and cyclopropane fatty acids. Phosphatidyls are the most common lipid in the bacterial cytoplasmic membrane. In most phospholipids, fatty acids form an ester linkage with glycerol, and in *Clostridium butyricum* ether forms are found as plasmalogens. Sphingolipids are the main phospholipids in certain anaerobes and hopanoids are found in many bacteria.

(a) Phosphatidylethanolamine;
(b) phosphatidylglycerol;
(c) phosphatidylethanolamine plasmalogen;
(d) phosphatidylcholine;
(e) phosphatidylinositol;
(f) sphingomyelin; (g) hopanoids.

Plasmalogens are found in strictly anaerobic bacteria (Figure 2.13c). The C-1 of glycerol in this phospholipid has an ether linkage to a long chain alcohol. Plasmalogens are not found in aerobic, facultative anaerobic and microaerophilic bacteria. They are known in species of the genera *Bacteroides*, *Clostridium*, *Desulfovibrio*, *Peptostreptococcus*, *Propionibacterium*, *Ruminococcus*, *Selenomonas* and *Veillonella* among others.

To maintain membrane fluidity, more saturated fatty acids are found in cells growing at higher temperature. In addition to temperature, fatty acid composition is influenced by other environmental factors such as pH, osmotic pressure and organic solvents. Fatty acids containing a cyclopropane ring are found in many bacteria. These are known as lactobacillic acids and were first identified in *Lactobacillus arabinosus*. They are formed under certain growth conditions or a certain growth phase. They are not present in vegetative cells of *Azotobacter vinelandii*, but cysts of this organism possess them as major fatty acids.

Archaea possess different phospholipids from bacteria. Archaea comprise three physiologically distinct groups of organisms. These are halophiles and methanogens (both belong to *Euryarchaeota*), and hyperthermophiles (*Crenarchaeota*). Though they are distinct in their physiological characteristics and habitats, they share important phylogenetic characters different from bacteria, including unique ether-linked phospholipids, ribosomal structure, and the lack of murein in their cell walls.

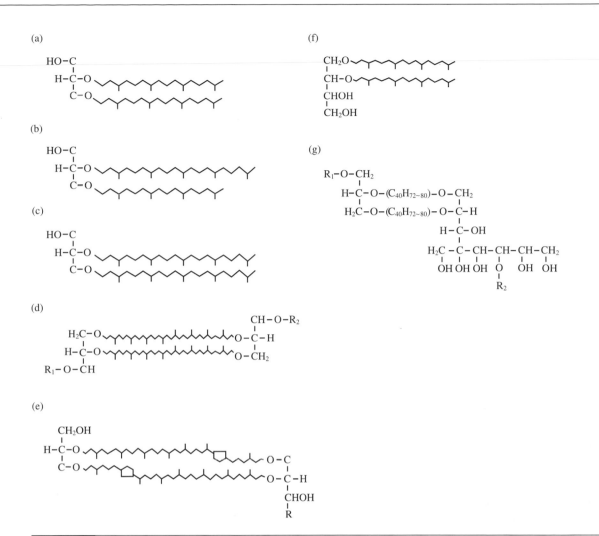

Figure 2.14 **Phospholipids in archaea.**

The hydrophobic tails of the phospholipids in archaea are isoprenoid alcohols, either 20 (phytanyl) or 25 (sesterterpanyl) carbons long and fully saturated. They are ether-linked to polyalcohols, mainly glycerol (a, b, c), and tetritol (f) and nonitol (g) are also found in archaeal phospholipids.

Phospholipids in halophilic archaea: complex lipids consisting of (a) 2,3-di-O-phytanyl-sn-glycerol (b), 2-O-sesterterpanyl-3-O-phytanyl-sn-glycerol, and (c) 2,3-di-O-sesterterpanyl-sn-glycerol, the C1 of which is linked to phosphate compounds, oligosaccharides or sulfonated oligosaccharides.

The common phospholipids in methanogenic archaea are (a) 2,3-di-O-phytanyl-sn-glycerol and (d) di-diphytanyl-diglycerol-tetraether. The C40 molecule (diphytanol) comprises two phytanols linked head-to-head as in (d). Oligosaccharides and phosphate compounds are linked to the C1 of glycerol. In some methanogenic archaeal phospholipids, tetritol replaces glycerol.

Di-diphytanyl-diglycerol-tetraether is the major phospholipid in hyperthermophilic archaea. Cyclopentanes are found in their hydrophobic tails (e). Glycerol is the common polyalcohol in the tetraether (R = H in (e)), and rarely nonitol (g) occupies one end of the tetraether compounds (R = $C_6H_{13}O_6$ in (e)).

The hydrophobic tails of the phospholipids in archaea are isoprenoid alcohols ether-linked to glycerophosphate to form monoglycerol-diether or diglycerol-tetraether (Figure 2.14). The alcohols are either 20 or 25 carbons long in monoglycerol-diethers. The di-diphytanyl-diglycerol-tetraether contains a C40 dialcohol (diphytanol) which is fully saturated and comprises two phytanols linked head-to-head.

Di-diphytanyl-diglycerol-tetraether is found in methanogenic archaea and is a major component of hyperthermophilic archaea. This molecule spans the whole thickness of their cytoplasmic membrane as in some other bacteria, discussed previously. Cyclopentanes are found in the hydrophobic tails of hyperthermophilic archaeal membranes. In addition to glycerol, tetritol and nonitol constitute methanogenic archaeal phospholipids. Unlike bacteria, the hydrophilic head groups in archaea are oligosaccharides or sulfonated oligosaccharides in addition to phosphate compounds.

In bacteria and eukaryotes, membrane fluidity is determined by the degree of saturation of the phospholipid fatty acid according to the temperature at which they grow. Little is known about how membrane fluidity is controlled in archaea, which can have optimum temperatures that span from ambient to over $100\,^{\circ}$C.

2.3.4.4 Proteins

In addition to an isolation function, the cytoplasmic membrane has many physiologically important functions carried out by the proteins which constitute 50–65% of the membrane. Membrane proteins are divided into two classes according to their location: integral and peripheral proteins. The surface of the integral proteins is hydrophobic and they are embedded in the membrane through hydrophobic interactions with the hydrophobic tails of the phospholipids. They can be removed from the membrane with detergents. Peripheral proteins can be removed by washing using salt solutions since they are attached to the membrane by ionic interactions.

Many of the membrane proteins mediate solute transport and protein secretion. Water molecules are diffusible through the lipid bilayer but the diffusion rate is too low to explain the rapid water flux that occurs to counter osmotic shock. A water channel known as aquaporin has been identified in the cytoplasmic membrane of many organisms including bacteria. The aquaporin gene (aqpZ) in *Escherichia coli* is growth phase and osmotically regulated. AqpZ has a role in both short- and long-term osmoregulatory responses, and is required for rapid cell growth. Aquaporin is known to be necessary for expression of virulence in pathogenic bacteria, and for sporulation and germination.

Enzymes responsible for synthesis and turnover of surface structures, such as phospholipase, protease and peptidase, are also associated with the cytoplasmic membrane. The bacterial chromosome is attached to the membrane, and also some of the enzymes involved in DNA replication. Bacterial DNA is attached to the cytoplasmic membrane via a protein that is believed to be responsible for chromosome segregation into daughter cells during cell division.

The ATP synthase (ATPase) enzyme complex is a membrane protein that couples ATP synthesis and hydrolysis to transmembrane proton transfer, thereby governing the proton motive force, Δp (Section 5.6.4), and the energy status of the cell (Section 5.6.2).

The cytoplasmic membrane is the site of oxidative phosphorylation. Proteins of the electron transport system are arranged in the cytoplasmic membrane in such a way that protons are expelled into the periplasmic region with the free energy available from the electron transport process (Section 5.8). Photosynthetic proteins are localized in the cytoplasmic membrane of photosynthetic bacteria (Section 11.3). Rhodopsins are found in halophilic archaeal membranes when grown under oxygen-limited conditions. They are responsible for phototaxis and for proton export utilizing light energy (Section 11.6).

Prokaryotes employ two-component systems to regulate metabolism in response to environmental change. The sensor proteins of such two-component systems are localized within the cytoplasmic membrane (Section 12.2.10).

2.3.5 Cytoplasm

The cytoplasm refers to everything inside the cytoplasmic membrane. Cells are classified as prokaryotes or eukaryotes depending on the possession of a nucleus. Eukaryotic cells have well-developed intracellular organelles such as mitochondria, chloroplasts and endoplasmic reticulum in addition to the nucleus. With only a few exceptions, prokaryotic cells do not have subcellular organelles within the cytoplasm. Prokaryotic cytoplasm contains DNA, ribosomes, proteins, RNA, salts and metabolites and is viscous due to the high concentration of macromolecules. Some of these macromolecules form aggregates, while others are soluble. The soluble part is called the cytosol. Though the cytoplasm is not compartmentalized, the cytoplasm is not a random mixture of its components. Proteins in the cytoplasm are in high concentration and can interact with each other to form a kind of network. The enzymes involved in a particular metabolic process are adjacent for their required interaction. The term 'metabolon' has been proposed to describe such a set of enzymes and their cofactors involved in such a fashion. A bacterium from the surgeonfish, *Epulopiscium fishelsoni*, has a cell size ($80 \times 600\,\mu\text{m}$) that can be seen with the naked eye. It was predicted that a cell of this size could not function properly without possessing a metabolic network such as the metabolon.

There are a few examples of intracytoplasmic membrane structures in bacteria. Some of them are continuous with the cytoplasmic membrane and are therefore believed to be derived from invaginations of chemically modified areas of the cytoplasmic membrane. Others are independent of the cytoplasmic membrane.

The thylakoids of cyanobacteria are the best known example of an intracellular organelle among prokaryotes. This organelle contains photosynthetic pigments (Section 11.3.1). Other examples are in the members of the phylogenetically deep-rooted bacterial family, the *Planctomycetaceae*, which have a compartmentalized cytoplasm (Figure 2.15). They have a bilayered intracytoplasmic membrane (ICM) that is discontinuous with the cytoplasmic membrane. The

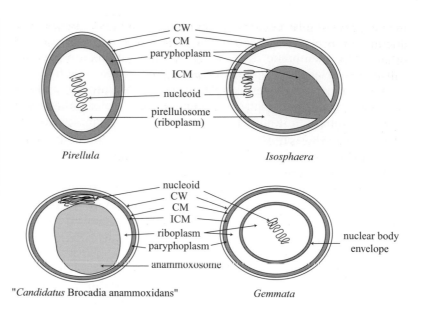

Figure 2.15 Cell organization and compartmentalization in (upper) *Pirellula marina* and *Isosphaera pallida* (also applies to *Planctomyces maris*), and (lower) '*Candidatus* Brocadia anammoxidans' and *Gemmata obscuriglobus.*

(*Arch. Microbiol.* 175:413–429, 2001)

CW, cell wall; CM, cytoplasmic membrane; ICM, intracytoplasmic membrane.

compartment enclosed by the ICM is called the riboplasm and contains the nucleoid and ribosomes. Little is known of the function of the paryphoplasm that occupies the region between the ICM and the cytoplasmic membrane. *Gemmata obscuriglobus*, a planctomycete originally isolated from fresh water, has been found to possess another unique feature not found in other bacteria. *Gemmata obscuriglobus* has a membrane-bound nuclear body consisting of two membranes surrounding the fibrillar DNA-containing nucleoid.

Many bacteria have intracytoplasmic membrane structures continuous with the cytoplasmic membrane. These include nitrogen-fixers (Section 6.2.1), methanotrophs (Section 7.9), nitrifiers (Section 10.2), and photosynthetic bacteria (Section 11.3). The intracytoplasmic membrane structures in these bacteria contain proteins that determine the specific physiological properties of these bacteria.

The chromosome is not separated by a membrane within the cytoplasm in most prokaryotes. DNA is highly coiled in the nucleoid and bound to several proteins and nascent RNA chains. Proteins bound to DNA include those involved in replication and transcription, and those regulating these processes (Section 12.1.1). In most cases, the prokaryotic chromosome is in a circular form. Some bacteria, e.g. *Streptomyces lividans*, have linear DNA. Since transcription and translation takes place in the same compartment and an intervening sequence is absent in prokaryotes, ribosomes bind to the nascent mRNA for translation (Section 6.12). For this reason, certain genes in prokaryotes can be controlled by attenuation (Section 12.1.4) and this regulatory mechanism is not known in eukaryotes. Intervening sequences have, however, been identified in some prokaryotes, including archaea. In addition to the chromosome, smaller DNA molecules are common in prokaryotes. They are called plasmids. Cells harbouring plasmids show different phenotypic traits, including resistance to particular antibiotics and toxic metals, production of bacteriocins and toxins, virulence, conjugation and others. Some large-sized plasmids carry essential genes. These are called megaplasmids or second chromosomes.

A major proportion of the RNAs in the cytoplasm is present as ribosomes. The ribosome is a complex consisting of 50 different proteins and three different kinds of RNA (i.e. 23S, 16S and 5S rRNA). A ribosome is the site of protein synthesis. Multiple ribosomes bind mRNA, and this structure is called a polysome.

Multienzyme aggregates are found in some bacteria. Obligate lithotrophs use CO_2 as their carbon source. Ribulose-1,5-bisphosphate carboxylase (Section 10.8.1.1) fixes CO_2 and this enzyme is included in an enzyme complex called a carboxysome. It is interesting to note that some cellulolytic bacteria produce a similar but extracellular complex known as a cellulosome.

Many bacteria accumulate intracellular storage materials. They include poly-β-hydroxyalkanoic acid, glycogen and polyphosphate (Section 13.2), all found as granules in the bacterial cytoplasm.

The liquid part of the cytoplasm, the cytosol, contains many enzymes including those mediating central metabolism, metabolites and salts. The cytosol contains small molecular weight organic compounds such as betaine, amino acids and sugars to balance the external osmotic pressure in addition to inorganic salts. These are called compatible solutes. A hyperthermophilic methanogenic archaeon, *Methanopyrus kandleri*, produces 2,3-diphosphoglycerate (cDPG) as its compatible solute. Di-myo-inositol-1,1'-phosphate is used for this purpose in the hyperthermophilic archaea *Pyrococcus woesei* and *Pyrococcus furiosus*.

2.3.6 Resting cells

Many bacteria differentiate into resting cells when the growth environment becomes unfavourable, such as by depletion of nutrients (Section 13.3). The best known resting cells are spores as found in the Gram-positive aerobic *Bacillus* and anaerobic *Clostridium* genera. Cysts are another form of resting cells. These are resistant against physical and chemical stresses such as desiccation and ionizing radiation. Spores are resistant to high temperatures, but cysts are not. Spores can remain viable for several decades under dry conditions.

Spore-specific structures include the outer coat, inner coat and cortex. The coats mainly comprise protein and the cortex murein. The cortex, occupying the region between the spore wall and the coats, renders the resistance property of the spores. The structure of cysts is similar to vegetative cells except for the exine and intine. The exine is the outer cyst wall consisting of mainly alginate, protein and lipid, and the intine is the inner cyst wall comprising polymannuronic acid. The exine is stabilized with divalent metal ions such as Ca^{2+} which bridge the carboxyl groups of alginate. A cyst central body can be prepared by treating cysts with chelating agents such as EDTA and citrate. The cyst central body transforms to a vegetative cell under favourable growth conditions but is not resistant to external stresses, showing that the exine is responsible for resistance. Resting cells are discussed in detail later (Section 13.3).

FURTHER READING

Cellular elements

Beinert, H. (2000). A tribute to sulfur. *European Journal of Biochemistry* **267**, 5657–5664.

Dosanjh, N. S. & Michel, S. L. J. (2006). Microbial nickel metalloregulation: NikRs for nickel ions. *Current Opinion in Chemical Biology* **10**, 123–130.

Hille, R. (2002). Molybdenum and tungsten in biology. *Trends in Biochemical Sciences* **27**, 360–367.

Itoh, S. (2006). Mononuclear copper active-oxygen complexes. *Current Opinion in Chemical Biology* **10**, 115–122.

Jakubovics, N. S. & Jenkinson, H. F. (2001). Out of the iron age: new insights into the critical role of manganese homeostasis in bacteria. *Microbiology–UK* **147**, 1709–1718.

Kobayashi, M. & Shimizu, S. (1999). Cobalt proteins. *European Journal of Biochemistry* **261**, 1–9.

Maroney, M. J. (1999). Structure/function relationships in nickel metallo-biochemistry. *Current Opinion in Chemical Biology* **3**, 188–199.

Stadtman, T. C. (2002). Discoveries of vitamin B_{12} and selenium enzymes. *Annual Review of Biochemistry* **71**, 1–16.

Cell surface appendages

Beatson, S. A., Minamino, T. & Pallen, M. J. (2006). Variation in bacterial flagellins: from sequence to structure. *Trends in Microbiology* **14**, 151–155.

Berry, R. M. & Armitage, J. P. (1999). The bacterial flagella motor. *Advances in Microbial Physiology* **41**, 291–337.

Fraser, G. M. & Hughes, C. (1999). Swarming motility. *Current Opinion in Microbiology* **2**, 630–635.

He, S. Y. & Jin, Q. (2003). The Hrp pilus: learning from flagella. *Current Opinion in Microbiology* **6**, 15–19.

Kalkum, M., Eisenbrandt, R., Lurz, R. & Lanka, E. (2002). Tying rings for sex. *Trends in Microbiology* **10**, 382–387.

Pallen, M. J., Penn, C. W. & Chaudhuri, R. R. (2005). Bacterial flagellar diversity in the post-genomic era. *Trends in Microbiology* **13**, 143–149.

Scott, J. R. & Zahner, D. (2006). Pili with strong attachments: Gram-positive bacteria do it differently. *Molecular Microbiology* **62**, 320–330.

S-layer and other surface structures

Desvaux, M., Dumas, E., Chafsey, I. & Hebraud, M. (2006). Protein cell surface display in Gram-positive bacteria: from single protein to macromolecular protein structure. *FEMS Microbiology Letters* **256**, 1–15.

Lasa, I. & Penades, J. R. (2006). Bap: a family of surface proteins involved in biofilm formation. *Research in Microbiology* **157**, 99–107.

Moens, S. & Vanderleyden, J. (1997). Glycoproteins in prokaryotes. *Archives of Microbiology* **168**, 169–175.

Navarre, W. W. & Schneewind, O. (1999). Surface proteins of Gram-positive bacteria and mechanisms of their targeting to the cell wall envelope. *Microbiology and Molecular Biology Reviews* **63**, 174–229.

Ron, E. & Rosenberg, E. (2002). Biosurfactants and oil bioremediation. *Current Opinion in Biotechnology* **13**, 249–252.

Sara, M. (2001). Conserved anchoring mechanisms between crystalline cell surface S-layer proteins and secondary cell wall polymers in Gram-positive bacteria? *Trends in Microbiology* **9**, 47–49.

Sleytr, U. B. & Beveridge, T. J. (1999). Bacterial S-layers. *Trends in Microbiology* **7**, 253–260.

Outer membrane in Gram-negative bacteria

Achouak, W., Heulin, T. & Pages, J. M. (2001). Multiple facets of bacterial porins. *FEMS Microbiology Letters* **199**, 1–7.

Begley, M., Gahan, C. G. M. & Hill, C. (2005). The interaction between bacteria and bile. *FEMS Microbiology Reviews* **29**, 625–651.

Bitter, W. (2003). Secretins of *Pseudomonas aeruginosa*: large holes in the outer membrane. *Archives of Microbiology* **179**, 307–314.

Chatterjee, S. N. & Chaudhuri, K. (2003). Lipopolysaccharides of *Vibrio cholerae*: I. Physical and chemical characterization. *Biochimica et Biophysica Acta – Molecular Basis of Disease* **1639**, 65–79.

Cullen, P. A., Haake, D. A. & Adler, B. (2004). Outer membrane proteins of pathogenic spirochetes. *FEMS Microbiology Reviews* **28**, 291–318.

Helander, I. M., Haikara, A., Sadovskaya, I., Vinogradov, E. & Salkinoja-Salonen, M. S. (2004). Lipopolysaccharides of anaerobic beer spoilage bacteria of the genus *Pectinatus* – lipopolysaccharides of a Gram-positive genus. *FEMS Microbiology Reviews* **28**, 543–552.

Koebnik, R., Locher, K. P. & Van Gelder, P. (2000). Structure and function of bacterial outer membrane proteins: barrels in a nutshell. *Molecular Microbiology* **37**, 239–253.

Lloubes, R., Cascales, E., Walburger, A., Bouveret, E., Lazdunski, C., Bernadac, A. & Journet, L. (2001). The Tol-Pal proteins of the *Escherichia coli* cell envelope: an energized system required for outer membrane integrity? *Research in Microbiology* **152**, 523–529.

Mogensen, J. E. & Otzen, D. E. (2005). Interactions between folding factors and bacterial outer membrane proteins. *Molecular Microbiology* **57**, 326–346.

Nikaido, H. (2003). Molecular basis of bacterial outer membrane permeability revisited. *Microbiology and Molecular Biology Reviews* **67**, 593–656.

Schulz, G. E. (2002). The structure of bacterial outer membrane proteins. *Biochimica et Biophysica Acta – Biomembranes* **1565**, 308–317.

Cell wall

Beveridge, T. J. (1999). Structures of Gram-negative cell walls and their derived membrane vesicles. *Journal of Bacteriology* **181**, 4725–4733.

Bhavsar, A. P. & Brown, E. D. (2006). Cell wall assembly in *Bacillus subtilis*: how spirals and spaces challenge paradigms. *Molecular Microbiology* **60**, 1077–1090.

Cabeen, M. T. & Jacobs-Wagner, C. (2005). Bacterial cell shape. *Nature Reviews Microbiology* **3**, 601–610.

Dmitriev, B., Toukach, F. & Ehlers, S. (2005). Towards a comprehensive view of the bacterial cell wall. *Trends in Microbiology* **13**, 569–574.

Hatch, T. P. (1996). Disulfide cross-linked envelope proteins: the functional equivalent of peptidoglycan in chlamydiae? *Journal of Bacteriology* **178**, 1–5.

Keep, N. H., Ward, J. M., Cohen-Gonsaud, M. & Henderson, B. (2006). Wake up! Peptidoglycan lysis and bacterial non-growth states. *Trends in Microbiology* **14**, 271–276.

Koch, A. L. (1998). Orientation of the peptidoglycan chains in the sacculus of *Escherichia coli*. *Research in Microbiology* **149**, 689–701.

Koch, A. L. (2002). Why are rod-shaped bacteria rod shaped? *Trends in Microbiology* **10**, 452–455.

McCoy, A. J. & Maurelli, A. T. (2006). Building the invisible wall: updating the chlamydial peptidoglycan anomaly. *Trends in Microbiology* **14**, 70–77.

Neuhaus, F. C. & Baddiley, J. (2003). A continuum of anionic charge: structures and functions of D-alanyl-teichoic acids in Gram-positive bacteria. *Microbiology and Molecular Biology Reviews* **67**, 686–723.

Schaffer, C. & Messner, P. (2005). The structure of secondary cell wall polymers: how Gram-positive bacteria stick their cell walls together. *Microbiology–UK* **151**, 643–651.

Smith, T. J., Blackman, S. A. & Foster, S. J. (2000). Autolysins of *Bacillus subtilis*: multiple enzymes with multiple functions. *Microbiology–UK* **146**, 249–262.

Periplasm

Bohin, J. P. (2000). Osmoregulated periplasmic glucans in Proteobacteria. *FEMS Microbiology Letters* **186**, 11–19.

Koch, A. L. (1998). The biophysics of the Gram-negative periplasmic space. *Critical Reviews in Microbiology* **24**, 23–59.

Cytoplasmic membrane

Bernstein, H. D. (2000). The biogenesis and assembly of bacterial membrane proteins. *Current Opinion in Microbiology* **3**, 203–209.

Cronan, J. E. (2006). A bacterium that has three pathways to regulate membrane lipid fluidity. *Molecular Microbiology* **60**, 256–259.

Driessen, A. J. M., Vandevossenberg, J. L. C. M. & Konings, W. N. (1996). Membrane composition and ion-permeability in extremophiles. *FEMS Microbiology Reviews* **18**, 139–148.

Edwards, M. D., Booth, I. R. & Miller, S. (2004). Gating the bacterial mechanosensitive channels: MscS a new paradigm? *Current Opinion in Microbiology* **7**, 163–167.

Engelman, D. M. (2005). Membranes are more mosaic than fluid. *Nature* **438**, 578–580.

Fleming, K. G. (2000). Riding the wave: structural and energetic principles of helical membrane proteins. *Current Opinion in Biotechnology* **11**, 67–71.

Flores, E., Herrero, A., Wolk, C. P. & Maldener, I. (2006). Is the periplasm continuous in filamentous multicellular cyanobacteria? *Trends in Microbiology* **14**, 439–443.

Gumbart, J., Wang, Y., Aksimentiev, A., Tajkhorshid, E. & Schulten, K. (2005). Molecular dynamics simulations of proteins in lipid bilayers. *Current Opinion in Structural Biology* **15**, 423–431.

Hanford, M. J. & Peeples, T. L. (2002). Archaeal tetraether lipids: unique structures and applications. *Applied Biochemistry and Biotechnology* **97**, 45–62.

Hedfalk, K., Tornroth-Horsefield, S., Nyblom, M., Johanson, U., Kjellbom, P. & Neutze, R. (2006). Aquaporin gating. *Current Opinion in Structural Biology* **16**, 447–456.

Kung, C. & Blount, P. (2004). Channels in microbes: so many holes to fill. *Molecular Microbiology* **53**, 373–380.

Mansilla, M. C., Cybulski, L. E., Albanesi, D. & de Mendoza, D. (2004). Control of membrane lipid fluidity by molecular thermosensors. *Journal of Bacteriology* **186**, 6681–6688.

Martinac, B. (2001). Mechanosensitive channels in prokaryotes. *Cellular Physiology and Biochemistry* **11**, 61–76.

Matsumoto, K., Kusaka, J., Nishibori, A. & Hara, H. (2006). Lipid domains in bacterial membranes. *Molecular Microbiology* **61**, 1110–1117.

Olsen, I. & Jantzen, E. (2001). Sphingolipids in bacteria and fungi. *Anaerobe* **7**, 103–112.

Pivetti, C. D., Yen, M. R., Miller, S., Busch, W., Tseng, Y. H., Booth, I. R. & Saier, M. H., Jr. (2003). Two families of mechanosensitive channel proteins. *Microbiology and Molecular Biology Reviews* **67**, 66–85.

Porat, A., Cho, S. H. & Beckwith, J. (2004). The unusual transmembrane electron transporter DsbD and its homologues: a bacterial family of disulfide reductases. *Research in Microbiology* **155**, 617–622.

Sajbidor, J. (1997). Effect of some environmental factors on the content and composition of microbial membrane lipids. *Critical Reviews in Biotechnology* **17**, 87–103.

Cytoplasm

Boccard, F., Esnault, E. & Valens, M. (2005). Spatial arrangement and macro-domain organization of bacterial chromosomes. *Molecular Microbiology* **57**, 9–16.

Candela, T. & Fouet, A. (2006). Poly-gamma-glutamate in bacteria. *Molecular Microbiology* **60**, 1091–1098.

Dame, R. T. (2005). The role of nucleoid-associated proteins in the organization and compaction of bacterial chromatin. *Molecular Microbiology* **56**, 858–870.

Dennis, P. P., Omer, A. & Lowe, T. (2001). A guided tour: small RNA function in Archaea. *Molecular Microbiology* **40**, 509–519.

Docampo, R., de Souza, W., Miranda, K., Rohloff, P. & Moreno, S. N. J. (2005). Acidocalcisome – conserved from bacteria to man. *Nature Reviews Microbiology* **3**, 251–261.

Frankel, R. B. & Bazylinski, D. A. (2006). How magnetotactic bacteria make magnetosomes queue up. *Trends in Microbiology* **14**, 329–331.

Graumann, P. L. (2004). Cytoskeletal elements in bacteria. *Current Opinion in Microbiology* **7**, 565–571.

Janakiraman, A. & Goldberg, M. B. (2004). Recent advances on the development of bacterial poles. *Trends in Microbiology* **12**, 518–525.

Lowe, J., van den Ent, F. & Amos, L. A. (2004). Molecules of the bacterial cytoskeleton. *Annual Review of Biophysics and Biomolecular Structure* **33**, 177–198.

Lybarger, S. R. & Maddock, J. R. (2001). Polarity in action: asymmetric protein localization in bacteria. *Journal of Bacteriology* **183**, 3261–3267.

Mathews, C. K. (1993). The cell – bag of enzymes or network of channels? *Journal of Bacteriology* **175**, 6377–6381.

Noirot, P. & Noirot-Gros, M. F. (2004). Protein interaction networks in bacteria. *Current Opinion in Microbiology* **7**, 505–512.

Shih, Y. L. & Rothfield, L. (2006). The bacterial cytoskeleton. *Microbiology and Molecular Biology Reviews* **70**, 729–754.

Sleator, R. D. & Hill, C. (2002). Bacterial osmoadaptation: the role of osmolytes in bacterial stress and virulence. *FEMS Microbiology Reviews* **26**, 49–71.

Spitzer, J. J. & Poolman, B. (2005). Electrochemical structure of the crowded cytoplasm. *Trends in Biochemical Sciences* **30**, 536–541.

Prokaryotic intracellular organelles

Bobik, T. (2006). Polyhedral organelles compartmenting bacterial metabolic processes. *Applied Microbiology and Biotechnology* **70**, 517–525.

Bohm, A. & Boos, W. (2004). Gene regulation in prokaryotes by subcellular relocalization of transcription factors. *Current Opinion in Microbiology* **7**, 151–156.

Bravo, A., Serrano-Heras, G. & Salas, M. (2005). Compartmentalization of prokaryotic DNA replication. *FEMS Microbiology Letters* **29**, 25–47.

Cannon, G. C., Bradburne, C. E., Aldrich, H. C., Baker, S. H., Heinhorst, S. & Shively, J. M. (2001). Microcompartments in prokaryotes: carboxysomes and related polyhedra. *Applied and Environmental Microbiology* **67**, 5351–5361.

Lewis, P. J. (2004). Bacterial subcellular architecture: recent advances and future prospects. *Molecular Microbiology* **54**, 1135–1150.

Resting cells

Driks, A. (1999). *Bacillus subtilis* spore coat. *Microbiology and Molecular Biology Reviews* **63**, 1–20.

Membrane transport – nutrient uptake and protein excretion

Microbes import the materials needed for growth and survival from their environment and export metabolites. As described in the previous chapter, the cytoplasm is separated from the environment by the hydrophobic cytoplasmic membrane, which is impermeable to hydrophilic solutes. Because of this permeability barrier exerted by the phospholipid component, almost all hydrophilic compounds can only pass through the membrane by means of integral membrane proteins. These are called carrier proteins, transporters or permeases (a website devoted entirely to transport can be found at www-biology.ucsd.edu/~msaier/transport/).

Solute transport can be classified as diffusion, active transport or group translocation according to the mechanisms involved. Diffusion does not require energy; energy is invested for active transport; and solutes transported by group translocation are chemically modified during this process. Some solutes are accumulated in the cell against a concentration gradient of several orders of magnitude, and energy needs to be invested for such accumulation.

3.1 | Ionophores: models of carrier proteins

There are two models which explain solute transport mediated by carrier proteins: the mobile carrier model and the pore model. The solute binds the carrier at one side of the membrane and dissociates at the other side according to the mobile carrier model, while the pore model proposes that the carrier protein forms a pore across the membrane through which the solute passes. A certain group of antibiotics can make the membrane permeable to ions. These are called ionophores and are useful compounds to assist the study of membrane transport.

One ionophore, valinomycin, transports ions according to the carrier model, while gramicidin A, another ionophore, makes a

Figure 3.1 **Structure of valinomycin, a model mobile carrier.**

Cations bind to the hydrophilic interior of the molecule at the side of the membrane with the higher concentration of the cation, and the complex moves through the hydrophilic membrane to the other side where the cation concentration is lower. The cations dissociate to equilibrate with the low concentration.

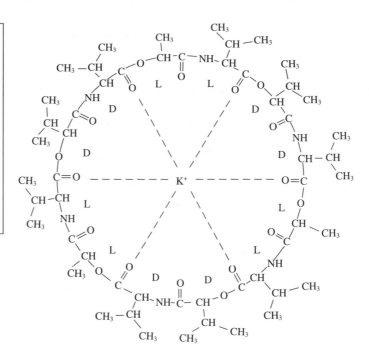

Gramicidin A

Figure 3.2 **Gramicidin A, a model of a pore-forming carrier protein.**

Two molecules of this peptide with 15 amino acid residues form a pore across the membrane through a hydrophobic interaction between the side chains of the amino acids and the membrane lipid. Solutes can move through the pore.

pore across the membrane. Valinomycin is a circular molecule consisting of valine, lactate and hydroxy isovalerate (Figure 3.1). Hydrophobic methyl and isopropyl groups form the surface of the circular molecule while hydrophilic carbonyl groups are arranged inside. Cations such as K^+ bind to the hydrophilic interior of the molecule. The complex moves through the hydrophilic membrane to the other side of the membrane where the cation dissociates. The efficiency of a mobile carrier is temperature dependent and becomes less efficient at low temperature due to decreased membrane fluidity. Uncouplers (Section 5.8.5) are mobile carriers of H^+.

Gramicidin A is a peptide consisting of 15 amino acid residues (Figure 3.2). This linear peptide has one hydrophilic side and one hydrophobic side. Two molecules of this compound form a hydrophilic pore across the membrane with the interaction between the hydrophobic side of the compound and the membrane lipid. Various cations can move through the pore thus created, the efficiency being temperature independent. It is believed that some carrier proteins are mobile carriers while others form pores.

3.2 | Diffusion

Some solutes enter cells by diffusion according to the concentration gradient without energy expenditure. Hydrophobic solutes diffuse through the lipid part of the membrane, and others diffuse through carrier proteins. The former is called simple diffusion and the latter facilitated diffusion. Facilitated diffusion shows different kinetics from that of simple diffusion. The initial diffusion rate is proportional to the concentration gradient in simple diffusion, while facilitated diffusion shows a relationship between the diffusion rate and the concentration of the solute similar to the Michaelis–Menten kinetics known in enzyme catalysis. Solutes transported through facilitated diffusion do not passively leak into the cell to any significant extent, and the rate of transport is directly proportional to the fraction of carrier proteins associated with them. When the carrier protein is fully saturated with the solute, the rate of transport reaches a maximum, and the rate does not increase further with any further increase in solute concentration. By definition, charged solutes are not transported through diffusion, since the transport of charged solutes changes the membrane potential (Section 5.7.1).

3.3 | Active transport and role of electrochemical gradients

Solute transport coupled to energy transduction is divided into primary and secondary transport according to the energy source. Primary transport systems are driven by energy-generating metabolism. Primary transport includes proton export driven by electron transport in respiration (Section 5.8) and photosynthesis (Section 11.4.3), by ATP hydrolysis (Section 5.8.4), and by light in halophilic archaea (Section 11.6). Also included in primary transport are chloride ion import in halophilic archaea driven by light (Section 11.6), sodium ion export coupled to decarboxylation reactions, proton export coupled to fumarate reduction and fermentation product excretion (Section 5.8.6), and import of sugars through group translocation (Section 3.5). These ion transport mechanisms are energy-conserving processes, except for sugar transport by group translocation, and will be discussed in the appropriate sections. This section is devoted to energy-dependent transport of materials needed for growth and survival and includes the secondary transport and group translocation of sugars.

Energy needed for secondary transport is supplied as an electrochemical gradient (a proton motive or sodium motive force) or from high energy phosphate bonds, as the results of primary transport (the proton (acidic internal pH) gradient and membrane potential) are established (Section 5.8). Since sodium ions are exported due to their coupling with

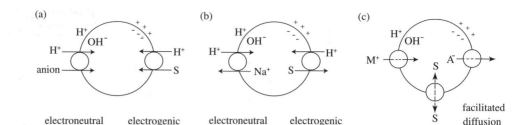

electroneutral   electrogenic   electroneutral   electrogenic   facilitated
                                                                                   diffusion

Figure 3.3. Electrochemical gradient dependent active transport.

(Dawes, E. A. 1986, *Microbial Energetics*, Figure 6.1. Blackie & Son, Glasgow)

Some solutes move across the membrane together with protons (or sodium ions) in the same direction (a, symport), while others move in opposite directions (b, antiport). Some ions cross the membrane along the electrochemical gradient (c, uniport). Transport involving the membrane potential is called electrogenic transport. Uniport and symport of uncharged solutes are electrogenic. When anions with a total negative charge of n are symported with m protons, the transport becomes electroneutral in the case of $n = m$ and electrogenic in the case of $n \# m$. In antiport, where cations with a total positive charge of n are exchanged with m protons it becomes electroneutral in the case of $n = m$ and electrogenic in the case of $n \# m$.

some energy-yielding reactions such as decarboxylation and anaerobic respiration (Chapter 9), a sodium gradient across the prokaryotic membrane is also established. Proton and sodium gradients are collectively termed electrochemical gradients, and they are used as energy for many secondary transport processes. All the carrier proteins studied have 12 helices spanning the membrane, some of which function as binding sites for solutes and others for protons (or sodium ions).

The proton motive force consists of a proton gradient (ΔpH) and membrane potential ($\Delta\psi$) (Section 5.7). Depending on the nature of the solute, transport requires energy in the form of either ΔpH or $\Delta\psi$, or both. According to the carrier proteins involved, electrochemical gradient-dependent solute transport can be classified as symport, antiport and uniport mechanisms (Figure 3.3). A solute crosses the membrane in the same direction with protons (or sodium ions) in the symport mechanism but in the opposite direction in an antiport system. Uniporters transport ions along the electrochemical gradient without involving protons or sodium ions. Uniporters consume only the $\Delta\psi$ part of the proton motive (or sodium motive) force. This is called electrogenic transport. When a monovalent anion is symported (cotransported) with a proton, the ΔpH is reduced without any change in the $\Delta\psi$. This is called electroneutral transport. When an uncharged solute is symported with protons, the ΔpH as well as the $\Delta\psi$ supply the energy needed for the accumulation of the solute. Since $\Delta\psi$ is reduced this becomes an electrogenic transport system.

3.4 | ATP-dependent transport: ATP-binding cassette (ABC) pathway

Solute transport can be driven not only by the electrochemical gradient but also by ATP hydrolysis. An *unc* (ATPase) mutant of *Escherichia coli* was unable to take up maltose under conditions where a large proton motive force was established, but the disaccharide was transported when the mutant was supplied with substrates that can produce ATP through substrate-level phosphorylation (Section 5.6.5) such as 1,3-diphosphoglycerate and phosphoenolpyruvate. Maltose transport requires not only ATP but also a binding protein in the periplasm.

Gram-negative bacteria have solute-binding proteins in the periplasm that are released when cells are subjected to osmotic shock with EDTA and MgCl$_2$ (cold osmotic shock). For this reason,

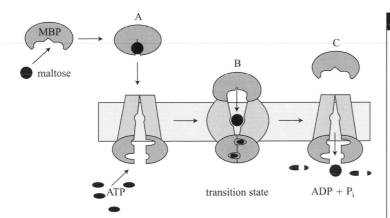

Figure 3.4 **Maltose transport through the ATP-binding cassette (ABC) transporter.**

(*J. Bacteriol.* 184:1225–1233, 2002)

(A) Maltose-binding protein (MBP) binds maltose, undergoing a change from an open to a closed conformation, generating a high-affinity sugar-binding site. In the closed conformation, MBP binds the maltose ATP-binding cassette (ABC) to initiate transport and ATP hydrolysis. (B) In the transition state for ATP hydrolysis, the MBP becomes tightly bound to the maltose ABC to transfer the sugar. (C) Maltose is transported, and MBP is released.

solute-binding protein-dependent transport is called a shock-sensitive transport system. A variety of nutrients including sugars, amino acids and ions can be transported through the shock-sensitive transport system. Solutes cross the outer membrane through porins and bind specific binding proteins before being transported through the cytoplasmic membrane by a membrane-bound protein complex. This protein complex is a member of a large superfamily of proteins that import nutrients or export cell surface constituents and extracellular proteins (Section 3.8.2). They have an ATP-binding motif and hydrolyze ATP to supply energy for the transport. They are called ATP-binding cassette (ABC) transporters and are known in all organisms (Figure 3.4).

A glutamate-binding protein is known in *Rhodobacter sphaeroides*, but glutamate transport is driven by the electrochemical gradient in this bacterium. This system is known as the TRAP (tripartite ATP-independent periplasmic) transporter. TRAP transporters are shock insensitive.

Through active transport, nutrients are accumulated in the cell with the expenditure of a large amount of energy in the form of the electrochemical gradient or ATP. Microbes can grow efficiently in environments with low nutrient concentrations due to active transport systems. Active transport can be summarized as follows:

(1) Carrier proteins have solute specificity as in the enzyme–substrate relationship.
(2) Energy is needed to change the affinity of the transporter for the transported solute at the other side of the membrane.
(3) The transported solute can be accumulated against a concentration gradient.
(4) The structure of the solute does not change during active transport.

3.5 | Group translocation

Sugars are phosphorylated during their transport in many bacteria, especially in anaerobes. The phosphate donor in these transport

Figure 3.5 Group translocation of sugars mediated by the phosphotransferase system in anaerobic bacteria.

(*Microbiol. Rev.* **57**:543–594, 1993)

Sugars are transported into the cell as phosphorylated forms mediated by the phosphotransferase (PT) system. Phosphoenolpyruvate (PEP) serves as the phosphate donor.

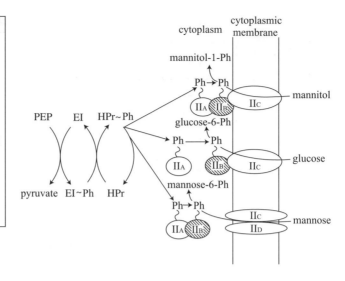

systems is phosphoenolpyruvate (PEP), a glycolytic intermediate. Since the solute is phosphorylated, this transport is referred to as group translocation. This system is not known in eukaryotes.

A group of proteins known as the phosphotransferase (PT) system transports and phosphorylates sugars (Figure 3.5). They are cytoplasmic enzyme I and HPr (histidine-containing protein), and membrane-bound enzymes II_A, II_B and II_C. Enzyme I transfers phosphate from PEP to HPr. Phosphorylated HPr transfers phosphate to enzyme II_A. The solute passes through the membrane-embedded enzyme II_C and enzyme II_B transfers phosphate from enzyme II_A to the solute. The cytolasmic proteins HPr and enzyme I do not have any specificity towards the sugar and are common components of PT systems. On the other hand, the membrane-bound proteins are specific for each sugar transported. The specific enzymes are described as enzyme $II_A{}^{Glu}$ etc. As shown in Figure 3.5, there are variations in the proteins involved in the transport of different sugars. Enzyme $II_A{}^{Glu}$ is a soluble protein, and enzyme II_A and II_B involved in mannose transport are also soluble. In addition, the mannose PT system has an extra membrane protein, enzyme II_D.

3.6 | Precursor/product antiport

Some lactic acid bacteria utilize the potential energy developed by the accumulation of fermentation products in the cell to drive nutrient transport in a similar manner to antiport. Instead of H^+ or Na^+, a fermentation product is exchanged with its precursor and this system is therefore referred to as the precursor/product antiport (Figure 3.6).

A malate/lactate antiport system is known in species of *Lactobacillus*, *Streptococcus*, *Leuconostoc* and *Pediococcus*. They generate a proton motive force fermenting malate to lactate through the well-documented malolactic fermentation pathway (Section 8.4.6).

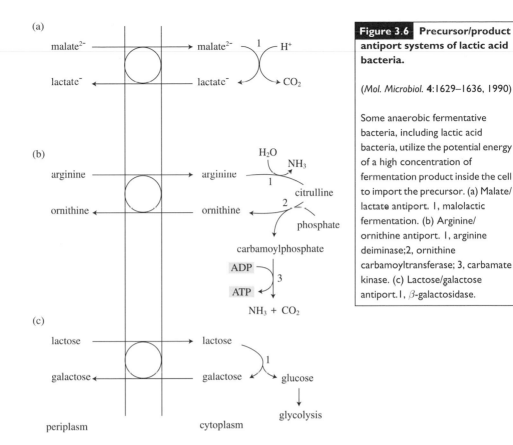

Figure 3.6 **Precursor/product antiport systems of lactic acid bacteria.**

(Mol. Microbiol. **4**:1629–1636, 1990)

Some anaerobic fermentative bacteria, including lactic acid bacteria, utilize the potential energy of a high concentration of fermentation product inside the cell to import the precursor. (a) Malate/lactate antiport. 1, malolactic fermentation. (b) Arginine/ornithine antiport. 1, arginine deiminase;2, ornithine carbamoyltransferase; 3, carbamate kinase. (c) Lactose/galactose antiport.1, β-galactosidase.

Species of *Streptococcus* ferment arginine to ornithine through citrulline to produce ATP. Ornithine is exchanged with arginine in a 1:1 ratio.

Lactose is imported through a H^+-symport system in *Streptococcus thermophilus* and *Lactobacillus bulgaricus*. Lactose can be transported by a lactose/galactose antiport system in these bacteria when the lactose concentration is high. Lactose is hydrolyzed to glucose and galactose by β-galactosidase. Glucose is metabolized through glycolysis and galactose is exchanged with lactose. Excreted galactose is then utilized after all the lactose is consumed.

In addition to lactic acid bacteria, precursor/product antiport systems are known in other anaerobic fermentative bacteria such as the oxalate/formate antiporter in *Oxalobacter formigenes* and the betaine/*N,N*-dimethyl glycine antiporter in *Eubacterium limosum*. Figure 3.7 summarizes some of the nutrient uptake pathways known in prokaryotes.

3.7 | Ferric ion (Fe(III)) uptake

In natural aerobic ecosystems, almost all iron is present as the ferric ion (Fe(III)) since ferrous iron (Fe(II)) is auto-oxidized with molecular

Figure 3.7 Nutrient import systems in prokaryotes.

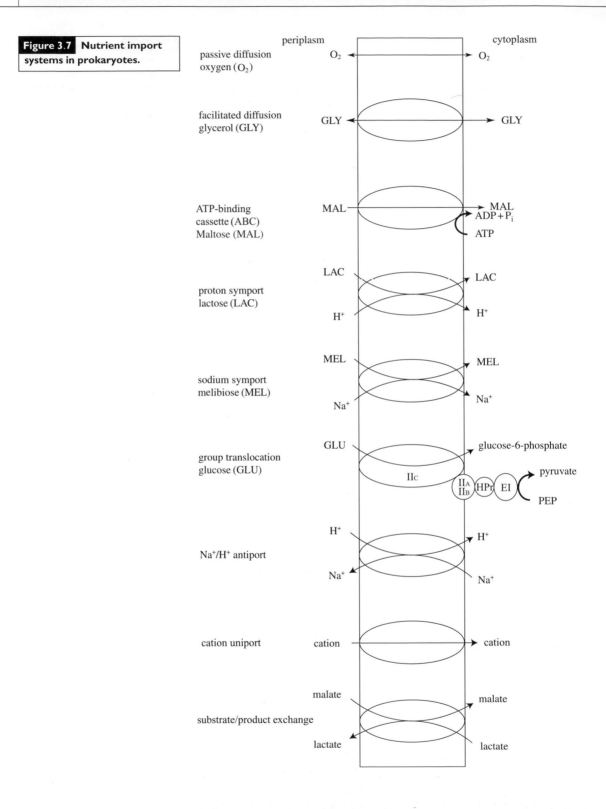

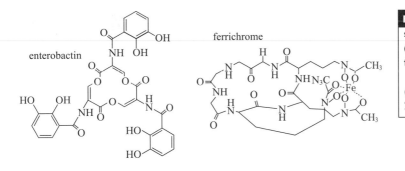

Figure 3.8 Structures of the siderophores enterobactin (a catecholamide) and ferrichrome (a hydroxamate).

(*FEMS Microbiol. Rev.* **27**:215–237, 2003)

oxygen at neutral pH. Ferric iron is virtually insoluble in water with a solubility of around 10^{-20} M and this is much lower than the 10^{-6} M necessary to supply adequate iron for most microbes. To overcome this problem, many microbes synthesize and excrete low molecular weight ferric iron chelating compounds known as siderophores for the sequestration and uptake of iron. Siderophores form complexes with ferric iron that are imported into the cell by an ABC transport system. Siderophores are a collection of compounds with a variety of chemical structures. Two main structural classes of siderophores are the catecholamides and the hydroxamates (Figure 3.8). A given organism may produce siderophores of one or both classes.

3.8 | Export of cell surface structural components

Large numbers of proteins and polysaccharides constitute the cytoplasmic membrane, periplasm (including the cell wall), and outer membrane. Many prokaryotes also secrete extracellular enzymes and toxins. These must be translocated through the cytoplasmic membrane after their synthesis in the cytoplasm. Translocation into and through the cytoplasmic membrane is referred to as protein transport. The term 'secretion' is used to refer to protein translocation further away from the cytoplasmic membrane to the cell surface and to the extracellular medium. Proteins are transported through one of three mechanisms. These are the general secretory pathway (GSP), the ABC pathway and the twin-arginine translocation (TAT) pathway.

The outer membrane is another barrier for protein secretion in Gram-negative bacteria and several different mechanisms have been identified as protein secretion pathways (see Section 3.8.2).

3.8.1 Protein transport

3.8.1.1 General secretory pathway (GSP)

The general secretory pathway (GSP) is also known as the Sec system. Without exception, proteins transported through the GSP have a unique *N*-terminal sequence. This sequence is called the signal sequence (or signal peptide) and consists of three regions: a basic

Figure 3.9 **Protein translocation through the general secretory pathway (GSP).**

(*Curr. Opin. Microbiol.* **3**:203–209, 2000)

The GSP is classified according to the mechanisms by which the nascent peptides are prevented from folding before excretion. The molecular chaperone SecB binds peptides in the SecB pathway (a) and a signal recognition particle (SRP) plays a similar role in the SRP pathway. (b) An ATPase, SecA, provides the energy needed for translocation with the hydrolysis of ATP. During the translocation process the membrane enzyme, signal peptidase, cleaves the *N*-terminal signal peptide. In Gram-negative bacteria the outer membrane proteins (OMP) are excreted exclusively through the SecB pathway, and the cytoplasmic membrane proteins (CMP) are embedded in the membrane by the SRP pathway. Proteins are excreted by Gram-positive bacteria in a similar GSP mechanism, but SecB has not been identified.

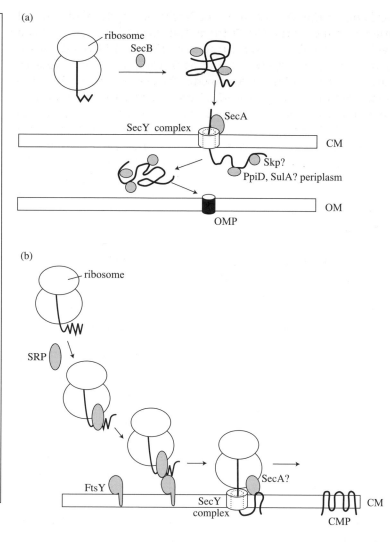

region at the *N*-terminal end, a central hydrophobic region and a recognition site for a peptidase. The basic end is positively charged at physiological pH values and is believed to attach to the negatively charged membrane phospholipid at the beginning of the transport process. The central hydrophobic region inserts itself into the cytoplasmic membrane, facilitating the transport of the main peptide, and the recognition site is cleaved by a signal peptidase at the membrane during or after transport.

Folded proteins cannot be transported through the GSP. The GSP can be classified into a SRP (signal recognition particle) pathway and a SecB pathway according to the mechanism of how the nascent peptides are kept unfolded in a transport compatible state (Figure 3.9). In the SecB pathway, the peptides are bound with the molecular chaperone SecB and targeted to the cytoplasmic membrane to be transported across through a dedicated protein-conducting channel ('translocon') called the SecY complex (Figure 3.9a). On the other hand, in the SRP pathway, the *N*-terminal hydrophobic signal peptide is bound by the SRP at the initial stage of translation. With the aid of the membrane-bound receptor FtsY, the SRP-bound translation machinery is targeted to the SecY complex to be exported (Figure 3.9b). An ATPase, SecA, provides the energy needed for translocation with the hydrolysis of ATP. During the translocation process the membrane enzyme, signal peptidase, cleaves the *N*-terminal signal peptide. In Gram-negative bacteria, the outer membrane proteins are excreted exclusively through the SecB pathway, and the cytoplasmic membrane proteins are embedded into the membrane by the SRP pathway. Proteins are excreted in Gram-positive bacteria by a similar mechanism to the SecB pathway, but SecB has not been identified. Many more proteins are involved in the GSP process, including periplasmic chaperones.

3.8.1.2 Twin-arginine translocation (TAT) pathway

In *Escherichia coli*, dimethyl sulfoxide reductase and formate dehydrogenase are cytoplasmic membrane proteins containing cofactors such as molybdenum and iron–sulfur clusters. They are embedded in the membrane in a mechanism called the twin-arginine translocation (TAT) pathway, which is different from GSP and ABC. The best known (or predicted) substrates of the TAT pathway are periplasmic proteins that bind a range of redox cofactors, including molybdopterin, Fe-S, Ni-Fe centres and others. These cofactors can be inserted into the peptide in the cytoplasm only and this requires substantial or complete folding of the mature protein, so it is believed that fully or largely folded structures are transported across the cytoplasmic membrane through this pathway. This system is also used for the translocation of other proteins that do not contain any cofactors, as these might fold too rapidly for the Sec system to handle.

Proteins transported through the TAT pathway contain an *N*-terminal signal sequence. The TAT signal sequence is longer and less hydrophobic than the Sec signal sequence. In all TAT signal sequences a twin-arginine motif is present and this motif is an absolute requirement to route a protein through this pathway.

The TAT apparatus consists of cytoplasmic TatD, and membrane-bound TatA, TatB, Tat C and Tat E. TatA and TatE seem to have overlapping functions, and TatE is dispensable. TatA and TatB have a single transmembrane helix while TatC has six helices (Figure 3.10). ATP is not required at any stage of this pathway, and ΔpH is consumed to provide the energy needed for protein translocation.

The TAT pathway was discovered in plants as a protein translocation mechanism into the thylakoids. A wide variety of prokaryotes have the unique ability to transport folded proteins through tightly sealed membranes. These include tetrachloroethene dehalogenase in *Dehalospirillum multivorans*, hydrogenase in *Desulfovibrio vulgaris*, *Escherichia coli* and *Wolinella succinogenes*, nitrous

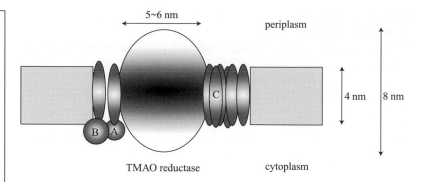

Figure 3.10 Export of trimethylamine *N*-oxide (TMAO) reductase in folded form through the twin-arginine translocation (TAT) pathway.

(*Nat. Rev. Mol. Cell Biol.* **2**:350–356, 2001)

TMAO reductase folds and binds cofactors containing molybdenum, iron, zinc and acid-labile sulfur in the cytoplasm before being transported through the TAT pathway. The diagram shows a model of TAT-dependent protein export. The TatA and TatB subunits are shown as single-spanning with very small (TatA) or slightly larger (TatB) cytoplasmic domains; TatC is shown as containing six transmembrane spans.

oxide reductase in *Pseudomonas stutzeri* and glucose–fructose oxidoreductase in *Zymomonas mobilis*. The *tat* genes were not found in the complete genome sequences of mycoplasmas and certain methanogens.

Some complex proteins are exported through the TAT pathway. A TAT signal sequence in one of the subunits facilitates the transport of the entire complex. The periplasmic nickel-containing hydrogenase in *Escherichia coli* is composed of a small subunit containing a TAT signal sequence and a large subunit devoid of an export signal. This hydrogenase complex is formed in the cytoplasm before being transported.

3.8.1.3 ATP-binding cassette (ABC) pathway

ABC transporters are evolutionarily related and have various functions. This pathway is known not only in bacteria but also in eukaryotes and archaea. As described previously (Section 3.4), ABC facilitates the transport of various nutrients including sugars, amino acids and ions in prokaryotes. Lipopolysaccharide is assembled in the periplasm and exported to the outer membrane (Section 6.9.3) through the outer membrane by the ABC pathway, as is capsular polysaccharide. The ABC pathway is also involved in the extrusion of noxious substances such as antibiotics, extracellular toxins and proteins and the targeting of membrane and surface structures. ABC transporters have been most studied in Gram-negative bacteria, and are discussed in Section 3.8.2.2.

3.8.2 Protein translocation across the outer membrane in Gram-negative bacteria

In Gram-negative bacteria, some proteins have to be transported through the outer membrane after they cross the cytoplasmic membrane. Proteins are excreted through one of the following pathways:

(1) Chaperone/usher pathway
(2) Type I pathway: ATP-binding cassette (ABC) pathway
(3) Type II pathway

(4) Type III pathway
(5) Type IV pathway
(6) Type V pathway: autotransporter and proteins requiring single accessory factors.

Proteins crossing the cytoplasmic membrane by the general secretory pathway (GSP) are translocated by the chaperone/usher, type II, type IV or type V pathways. Proteins transported by the GSP are synthesized with a hydrophobic *N*-terminal signal peptide that is removed at the cytoplasmic membrane: various Sec proteins involved in GSP are found in the cytoplasm and cytoplasmic membrane. Sec proteins are not involved in type I and III pathways. Type I protein translocation machinery is within the family of ATP-binding cassettes (ABC) described previously. Many Gram-negative plant and animal pathogenic bacteria utilize a specialized type III secretion system as a molecular syringe to inject effector proteins directly into host cells. Flagellin (Section 2.3.1) is exported in a similar way to the type III secretion system. The type IV secretion system is used to release proteins through the outer membrane in some pathogenic bacteria after they are exported through the cytoplasmic membrane by GSP.

3.8.2.1   Chaperone/usher pathway

Certain cell surface proteins associated with virulence are exported through the chaperone/usher pathway. They are translocated across the cytoplasmic membrane by the GSP system before binding to a periplasmic chaperone, PapD, which prevents premature folding of the peptides. Periplasmic chaperone–peptide complexes are targeted to an outer membrane protein called the usher, through which the peptide is secreted. Energy is not needed in this process. Pili are formed in this pathway (Figure 3.11).

3.8.2.2   Type I pathway: ATP-binding cassette (ABC) pathway

The ABC pathway is employed to secrete extracellular proteins such as hydrolytic enzymes (proteases and lipases) and toxins in Gram-negative bacteria. The type I pathway is GSP independent, and peptides exported by this pathway do not have a signal peptide. The cytoplasmic membrane protein complex ABC pushes the peptide through a membrane fusion protein (MFP) and an outer membrane protein (OMP) with the hydrolysis of ATP providing energy needed for the translocation. MFP anchors within the cytoplasmic membrane and spans the periplasm (Figure 3.12).

3.8.2.3   Type II pathway

This pathway represents a third branch of the GSP system. The type II pathway is employed for the secretion of a wide variety of proteins in Gram-negative bacteria including pullulanase by *Klebsiella oxytoca*. More than 12 different proteins are involved in this pathway. They are collectively referred to as the secreton. Their location

Figure 3.11 **Model for biogenesis of pili by the chaperone/usher pathway.**

(*Curr. Opin. Cell Biol.* **12**:420–430, 2000)

(a) Pilus subunits cross the cytoplasmic membrane via the GSP system, followed by cleavage of their *N*-terminal signal peptide. The periplasmic chaperone PapD binds to each subunit via a conserved carboxy-terminal subunit motif (white box), allowing proper subunit folding and preventing premature subunit–subunit binding. (b) The PapD–pilin complex is targeted to the outer membrane protein, PapC, the usher protein for assembly into pili and secretion across the outer membrane. *C*-terminal (white box) motifs of the secreted pilin interact with the *N*-terminal (black box) motif of the subunit at the tip of the pilus.

D, chaperone PapD; C, usher PapC; Sec: translocon.

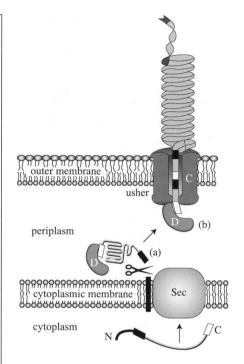

Figure 3.12 **Protein secretion in Gram-negative bacteria through the ABC (Type I) pathway.**

(*Curr. Opin. Cell Biol.* **12**:420–430, 2000)

Proteins exported by the ABC pathway do not have an *N*-terminal signal sequence but have a *C*-terminal secretion sequence. An ABC exporter consists of cytoplasmic membrane ABC, periplasmic membrane fusion protein (MFP) and an outer membrane protein (OMP). ABC hydrolyzes ATP to export proteins through the MFP and OMP.

ABC, ATP-binding cassette; MFP, membrane fusion protein; NTP, nucleotide triphosphate (ATP).

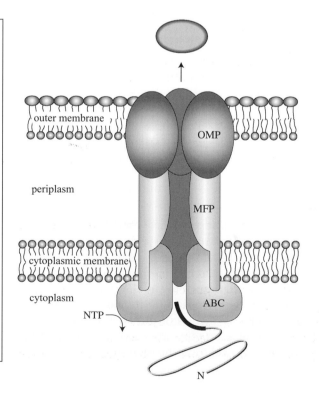

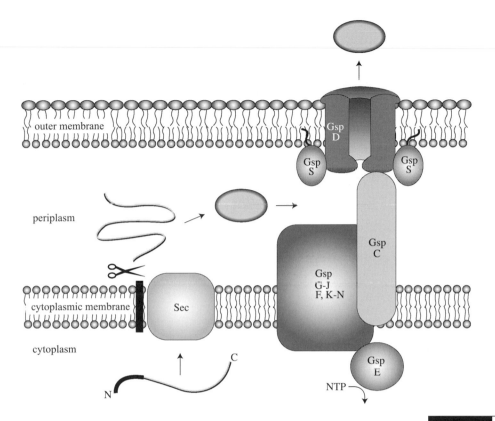

extends from the cytoplasmic membrane to the outer membrane. An outer membrane secreton protein called secretin (GspD) forms highly stable ring-shaped complexes of 12–14 subunits with central channels ranging from 5–10 nm in diameter, large enough to accommodate folded substrates. The cytoplasmic secreton protein (GspE) has a conserved ATP-binding motif, and interacts with cytoplasmic membrane secreton proteins. GspE hydrolyzes ATP to provide energy for the secretion through conformational changes in the cytoplasmic membrane secreton proteins. Peptides exported to the periplasm fold before being secreted through the secretin complex (Figure 3.13). In *Pseudomonas aeruginosa*, the type 4 pilin is transported in a similar mechanism.

3.8.2.4 Type III pathway

A number of animal and plant pathogens inject their anti-host factors directly into the host cytoplasm. This pathway is GSP independent, and requires up to 20 secretion components that assemble into a large structure spanning both bacterial membranes and the host membrane. The channel-forming protein on the outer membrane shares homology with secretin of the type II pathway. Proteins are excreted through a tube-like structure called the needle. ATP hydrolysis provides the energy needed for the excretion (Figure 3.14). Flagella are very similar in structure to the type III excretion

Figure 3.13 The type II pathway for secretion of proteins through the outer membrane in **Gram-negative bacteria.**

(*Curr. Opin. Cell Biol.* 12:420–430, 2000)

Type II substrates cross the CM via the Sec system followed by signal-sequence cleavage and protein folding in the periplasm. The GspD secretin, indicated here as a complex with the GspS lipoprotein, serves as a gated channel for secretion of substrates to the cell surface. GspC may transmit energy from the CM, presumably generated by the cytoplasmic GspE nucleotide-binding protein, to the OM complex.

Gsp, general secretion pathway protein; NTP, nucleotide triphosphate such as ATP.

Figure 3.14 **Protein injection into a host cell through the type III excretion pathway.**

(*Curr. Opin. Cell Biol.* **12**:420–430, 2000)

The figure shows the type III apparatus in species of *Yersinia*. Some type III secretion substrates contain two amino-terminal signal sequences for targeting to the secretion machinery, one encoded by the mRNA and the second serving as a binding site for cytoplasmic Syc chaperones. By homology with flagellar proteins, YscR-U and LcrD may form a central secretion apparatus, energized by YscN. The YscC secretin presumably provides a secretion channel across the OM and the surface-localized YopN protein is thought to serve as a channel gate. By analogy with flagella, type III substrates (Yop proteins) may travel through a central channel in the type III needle. Translocation of Yop proteins into the target eukaryotic cell may take place via a channel formed in the plasma membrane by YopB and YopD.

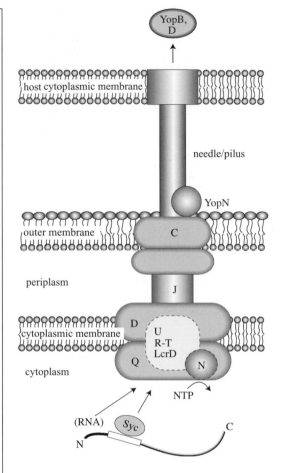

machinery and flagellin is excreted in a similar mechanism to the type III pathway.

Type III secretion is highly regulated and two *N*-terminal secretion signals have been identified for export. The first signal appears to reside in the mRNA and may target the RNA–ribosome complex to the type III machinery for coupled translation and secretion. The second secretion signal serves as the binding site for cytoplasmic chaperones termed Syc proteins and may specifically target protein to the type III machinery for translocation into host cells.

3.8.2.5 Type IV pathway

The type IV pathway has been identified in *Legionella pneumophila* and *Helicobacter pylori* to secrete toxin proteins into the host cells. Peptide secreted through this pathway is exported through the cytoplasmic membrane by the GSP system. The periplasmic intermediate is secreted into the host cell by a similar mechanism as in the type III pathway. However, the outer membrane channel of the type IV pathway is not homologous with that of the type III pathway (Figure 3.15).

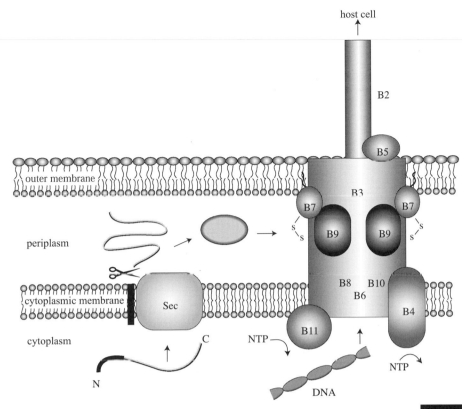

Figure 3.15 Model of type IV secretion for injection of toxin proteins into the eukaryotic host cell.

(*Curr. Opin. Cell Biol.* **12**:420–430, 2000)

This is a protein secretion pathway identified in pathogenic *Legionella pneumophila* and *Helicobacter pylori*: DNA is transported from *Agrobacterium tumefaciens* into plant host cells by a similar mechanism. Protein secretion by the type IV pathway may take place via a periplasmic intermediate, with substrates first transported through the GSP into the periplasm (left). DNA secretion probably takes place from the cytoplasm in a single step without a periplasmic intermediate. The cytoplasmic membrane components contain nucleotide-binding activity and probably energize aspects of the secretion process.

This type IV pathway is similar to the mechanism that facilitates the translocation of DNA from *Agrobacterium tumefaciens* into the plant host cell. DNA is transported directly from the bacterial cytoplasm. Similarity is also found with bacterial conjugation. The cytoplasmic membrane type IV components hydrolyze ATP to provide energy for the secretion.

3.8.2.6 Type V pathway: autotransporter and proteins requiring single accessory factors

Proteases, toxins and cell surface proteins such as those used for adhesion to solid surfaces (adhesins) and for invasion into the host cell (invasins) are excreted through the outer membrane by this pathway. These proteins contain three domains: a *N*-terminal signal sequence for secretion across the cytoplasmic membrane by the GSP system, an internal passenger (functional) domain and a *C*-terminal β-domain. This β-domain forms a porin-like structure at the outer membrane through which the passenger domain passes to the cell surface. The passenger domain is released into the extracellular medium by proteolysis (Figure 3.16). In some cases proteins lacking a β-domain are secreted through a β-domain formed by a separate protein (a single accessory factor). Foreign proteins were released

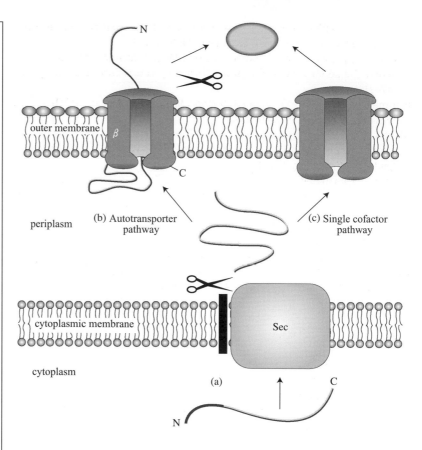

Fig. 3.16 **Models for the autotransporter and single accessory pathways.**

(*Curr. Opin. Cell Biol.* 12:420–430, 2000)

(a) Both pathways are branches of the GSP system, and proteins cross the cytoplasmic membrane via the GSP system, followed by cleavage of their *N*-terminal signal peptide in the periplasm by signal peptidase. (b) For autotransporters, the *C*-terminal or β-domain of the protein inserts into the outer membrane to form a β-barrel secretion channel through which the internal passenger domain passes to the cell surface. The passenger domain is released into the extracellular medium by proteolysis. (c) In pathways requiring a single accessory factor, a separate protein forms the β-barrel outer membrane channel which may be gated.

C, *C*-terminal; N, *N*-terminal; OM, outer membrane; Sec, translocon.

into the environment when a segment of DNA containing genes for the protein, signal peptide and *C*-terminal β-domain was introduced into *Escherichia coli*, showing that the autotransporter is not selective for the passenger proteins.

FURTHER READING

General subjects in transport

Black, P. N. & DiRusso, C. C. (2003). Transmembrane movement of exogenous long-chain fatty acids: proteins, enzymes, and vectorial esterification. *Microbiology and Molecular Biology Reviews* **67**, 454–472.

Busch, W. & Saier, M. H., Jr. (2002). The transporter classification (TC) system, 2002. *Critical Reviews in Biochemistry and Molecular Biology* **37**, 287–337.

Calamita, G. (2000). The *Escherichia coli* aquaporin-Z water channel. *Molecular Microbiology* **37**, 254–262.

Eggeling, L. & Sahm, H. (2003). New ubiquitous translocators: amino acid export by *Corynebacterium glutamicum* and *Escherichia coli*. *Archives of Microbiology* **180**, 155–160.

Hagenbuch, B. & Meier, P. J. (2003). The superfamily of organic anion transporting polypeptides. *Biochimica et Biophysica Acta – Biomembranes* **1609**, 1–18.

Harold, F. M. (2005). Molecules into cells: specifying spatial architecture. *Microbiology and Molecular Biology Reviews* **69**, 544–564.

Hedfalk, K., Tornroth-Horsefield, S., Nyblom, M., Johanson, U., Kjellbom, P. & Neutze, R. (2006). Aquaporin gating. *Current Opinion in Structural Biology* **16**, 447–456.

Hohmann, I., Bill, R. M., Kayingo, I. & Prior, B. A. (2000). Microbial MIP channels. *Trends in Microbiology* **8**, 33–38.

Klebba, P. E. & Newton, S. M. C. (1998). Mechanisms of solute transport through outer membrane porins: burning down the house. *Current Opinion in Microbiology* **1**, 238–247.

Lloubes, R., Cascales, E., Walburger, A., Bouveret, E., Lazdunski, C., Bernadac, A. & Journet, L. (2001). The Tol-Pal proteins of the *Escherichia coli* cell envelope: an energized system required for outer membrane integrity? *Research in Microbiology* **152**, 523–529.

Pallen, M. J., Chaudhuri, R. R. & Henderson, I. R. (2003). Genomic analysis of secretion systems. *Current Opinion in Microbiology* **6**, 519–527.

Paulsen, I. T. (2003). Multidrug efflux pumps and resistance: regulation and evolution. *Current Opinion in Microbiology* **6**, 446–451.

Pugsley, A. P., Francetic, O., Driessen, A. J. & de Lorenzo, V. (2004). Getting out: protein traffic in prokaryotes. *Molecular Microbiology* **52**, 3–11.

Saier, M. H. (2000). Families of transmembrane sugar transport proteins. *Molecular Microbiology* **35**, 699–710.

Saier, M. H. & Reizer, J. (1992). Proposed uniform nomenclature for the proteins and protein domains of the bacterial phosphoenolpyruvate – sugar phosphotransferase system. *Journal of Bacteriology* **174**, 1433–1438.

Sobczak, I. & Lolkema, J. S. (2005). Structural and mechanistic diversity of secondary transporters. *Current Opinion in Microbiology* **8**, 161–167.

Active transport

Kakinuma, Y. (1998). Inorganic cation transport and energy transduction in *Enterococcus hirae* and other streptococci. *Microbiology and Molecular Biology Reviews* **62**, 1021–1045.

Lemieux, M. J., Huang, Y. & Wang, D.-N. (2004). The structural basis of substrate translocation by the *Escherichia coli* glycerol-3-phosphate transporter: a member of the major facilitator superfamily. *Current Opinion in Structural Biology* **14**, 405–412.

Lolkema, J. S., Poolman, B. & Konings, W. N. (1998). Bacterial solute uptake and efflux systems. *Current Opinion in Microbiology* **1**, 248–253.

Reizer, J., Reizer, A. & Saier, M. H. (1994). A functional superfamily of sodium/ solute symporters. *Biochimica et Biophysica Acta – Biomembranes* **1197**, 133–166.

Saier, M. H. (2000). Families of transmembrane transporters selective for amino acids and their derivatives. *Microbiology–UK* **146**, 1775–1795.

Saier, M. H. (2000). Vectorial metabolism and the evolution of transport systems. *Journal of Bacteriology* **182**, 5029–5035.

Sobczak, I. & Lolkema, J. S. (2005). The 2-hydroxycarboxylate transporter family: physiology, structure, and mechanism. *Microbiology and Molecular Biology Reviews* **69**, 665–695.

Vanveen, H. W. (1997). Phosphate transport in prokaryotes: molecules, mediators and mechanisms. *Antonie van Leeuwenhoek* **72**, 299–315.

Group translocation

Barabote, R. D. & Saier, M. H., Jr. (2005). Comparative genomic analyses of the bacterial phosphotransferase system. *Microbiology and Molecular Biology Reviews* **69**, 608–634.

Vadeboncoeur, C. & Pelletier, M. (1997). The phosphoenolpyruvate:sugar phosphotransferase system of oral streptococci and its role in the control of sugar metabolism. *FEMS Microbiology Reviews* **19**, 187–207.

ATP-binding cassette (ABC) pathway

Albers, S. V., Koning, S. M., Konings, W. N. & Driessen, A. J. (2004). Insights into ABC transport in archaea. *Journal of Bioenergetics and Biomembranes* **36**, 5–15.

Cabezon, E. & de la Cruz, F. (2006). TrwB: an F1-ATPase-like molecular motor involved in DNA transport during bacterial conjugation. *Research in Microbiology* **157**, 299–305.

Dassa, E. & Bouige, P. (2001). The ABC of ABCs: a phylogenetic and functional classification of ABC systems in living organisms. *Research in Microbiology* **152**, 211–229.

Davidson. A. L. (2002). Mechanism of coupling of transport to hydrolysis in bacterial ATP-binding cassette transporters. *Journal of Bacteriology* **184**, 1225–1233.

Davidson, A. L. & Chen, J. (2004). ATP-binding cassette transporter in bacteria. *Annual Review of Biochemistry* **73**, 241–268.

Doerrler, W. T. (2006). Lipid trafficking to the outer membrane of Gram-negative bacteria. *Molecular Microbiology* **60**, 542–552.

Elferink, M. G., Albers, S. V., Konings, W. N. & Driessen, A. J. (2001). Sugar transport in *Sulfolobus solfataricus* is mediated by two families of binding protein-dependent ABC transporters. *Molecular Microbiology* **39**, 1494–1503.

Higgins, C. F. (2001). ABC transporters: physiology, structure and mechanism: an overview. *Research in Microbiology* **152**, 205–210.

Horn, C., Jenewein, S., Sohn-Bosser, L., Bremer, E. & Schmitt, L. (2005). Biochemical and structural analysis of the *Bacillus subtilis* ABC transporter OpuA and its isolated subunits. *Journal of Molecular Microbiology and Biotechnology* **10**, 76–91.

Jones, P. M. & George, A. M. (1999). Subunit interactions in ABC transporters: towards a functional architecture. *FEMS Microbiology Letters* **179**, 187–202.

Koning, S. M., Albers, S. V., Konings, W. N. & Driessen, A. J. M. (2002). Sugar transport in hyper-thermophilic archaea. *Research in Microbiology* **153**, 61–67.

Locher, K. P. (2004). Structure and mechanism of ABC transporters. *Current Opinion in Structural Biology* **14**, 426–431.

Narita, S., Matsuyama, S. & Tokuda, H. (2004). Lipoprotein trafficking in *Escherichia coli*. *Archives of Microbiology* **182**, 1–6.

Pohl, A., Devaux, P. F. & Herrmann, A. (2005). Function of prokaryotic and eukaryotic ABC proteins in lipid transport. *Biochimica et Biophysica Acta – Molecular and Cell Biology of Lipids* **1733**, 29–52.

Ranquin, A. & van Gelder, P. (2004). Maltoporin: sugar for physics and biology. *Research in Microbiology* **155**, 611–616.

Silver, R. P., Prior, K., Nsahlai, C. & Wright, L. F. (2001). ABC transporters and the export of capsular polysaccharides from Gram-negative bacteria. *Research in Microbiology* **152**, 357–364.

Tripartite ATP-independent periplasmic (TRAP) transporters

Kelly, D. J. & Thomas, G. H. (2001). The tripartite ATP-independent periplasmic (TRAP) transporters of bacteria and archaea. *FEMS Microbiology Reviews* **25**, 405–424.

Winnen, B., Hvorup, R. N. & Saier, M. H., Jr. (2003). The tripartite tricarboxylate transporter (TTT) family. *Research in Microbiology* **154**, 457–465.

Iron uptake and siderophores

Braun, V. & Braun, M. (2002). Active transport of iron and siderophore antibiotics. *Current Opinion in Microbiology* **5**, 194–201.

Cornelis, P. & Matthijs, S. (2002). Diversity of siderophore-mediated iron uptake systems in fluorescent pseudomonads: not only pyoverdines. *Environmental Microbiology* **4**, 787–798.

Llamas, M. A. & Bitter, W. (2006). Iron gate: the translocation system. *Journal of Bacteriology* **188**, 3172–3174.

Postle, K. & Kadner, R. J. (2003). Touch and go: tying TonB to transport. *Molecular Microbiology* **49**, 869–882.

Schalk, I. J., Yue, W. W. & Buchanan, S. K. (2004). Recognition of iron-free siderophores by TonB-dependent iron transporters. *Molecular Microbiology* **54**, 14–22.

Visca, P., Leoni, L., Wilson, M. J. & Lamont, I. L. (2002). Iron transport and regulation, cell signalling and genomics: lessons from *Escherichia coli* and *Pseudomonas*. *Molecular Microbiology* **45**, 1177–1190.

Wandersman, C. & Delepelaire, P. (2004). Bacterial iron sources: from siderophores to hemophores. *Annual Review of Microbiology* **58**, 611–647.

Protein translocation through general secretion pathway (GSP)

Albers, S. V., Szabo, Z. & Driessen, A. J. M. (2006). Protein secretion in the Archaea: multiple paths towards a unique cell surface. *Nature Reviews Microbiology* **4**, 537–547.

Alder, N. N. & Theg, S. M. (2003). Energy use by biological protein transport pathways. *Trends in Biochemical Sciences* **28**, 442–451.

Buist, G., Ridder, A. N. J. A., Kok, J. & Kuipers, O. P. (2006). Different subcellular locations of secretome components of Gram-positive bacteria. *Microbiology-UK* **152**, 2867–2874.

Clemons, W. M., Jr, Menetret, J.-F., Akey, C. W. & Rapoport, T. A. (2004). Structural insight into the protein translocation channel. *Current Opinion in Structural Biology* **14**, 390–396.

Dalbey, R. E. & Chen, M. (2004). Sec-translocase mediated membrane protein biogenesis. *Biochimica et Biophysica Acta – Molecular Cell Research* **1694**, 37–53.

Desvaux, M., Parham, N. J., Scott-Tucker, A. & Henderson, I. R. (2004). The general secretory pathway: a general misnomer? *Trends in Microbiology* **12**, 306–309.

Ellgaard, L., Molinari, M. & Helenius, A. (1999). Setting the standards: quality control in the secretory pathway. *Science* **286**, 1882–1888.

Hegde, R. S. & Bernstein, H. D. (2006). The surprising complexity of signal sequences. *Trends in Biochemical Sciences* **31**, 563–571.

Holland, I. B. (2004). Translocation of bacterial proteins – an overview. *Biochimica et Biophysica Acta – Molecular Cell Research* **1694**, 5–16.

Lory, S. (1998). Secretion of proteins and assembly of bacterial surface organelles: shared pathways of extracellular protein targeting. *Current Opinion in Microbiology* **1**, 27–35.

Luirink, J. & Sinning, I. (2004). SRP-mediated protein targeting: structure and function revisited. *Biochimica et Biophysica Acta – Molecular Cell Research* **1694**, 17–35.

Manting, E. H. & Driessen, A. J. M. (2000). *Escherichia coli* translocase: the unravelling of a molecular machine. *Molecular Microbiology* **37**, 226–238.

Nakatogawa, H., Murakami, A. & Ito, K. (2004). Control of SecA and SecM translation by protein secretion. *Current Opinion in Microbiology* **7**, 145–150.

Pohlschroeder, M., Gimenez, M. I. & Jarrell, K. F. (2005). Protein transport in Archaea: Sec and twin arginine translocation pathways. *Current Opinion in Microbiology* **8**, 713–719.

Pohlschroeder, M., Hartmann, E., Hand, N. J., Dilks, K. & Haddad, A. (2005). Diversity and evolution of protein translocation. *Annual Review of Microbiology* **59**, 91–111.

Shental-Bechor, D., Fleishman, S. J. & Ben-Tal, N. (2006). Has the code for protein translocation been broken? *Trends in Biochemical Sciences* **31**, 192–196.

Tjalsma, H., Bolhuis, A., Jongbloed, J. D. H., Bron, S. & van Dijl, J. M. (2000). Signal peptide-dependent protein transport in *Bacillus subtilis*: a genome-based survey of the secretome. *Microbiology and Molecular Biology Reviews* **64**, 515–547.

Tjalsma, H., Antelmann, H., Jongbloed, J. D. H., Braun, P. G., Darmon, E., Dorenbos, R., Dubois, J.-Y. F., Westers, H., Zanen, G., Quax, W. J., Kuipers, O. P., Bron, S., Hecker, M. & van Dijl, J. M. (2004). Proteomics of protein secretion by *Bacillus subtilis*: separating the 'secrets' of the secretome. *Microbiology and Molecular Biology Reviews* **68**, 207–233.

van der Sluis, E. O. & Driessen, A. J. M. (2006). Stepwise evolution of the Sec machinery in Proteobacteria. *Trends in Microbiology* **14**, 105–108.

van Wely, K. H. M., Swaving, J., Freudl, R. & Driessen, A. J. M. (2001). Translocation of proteins across the cell envelope of Gram-positive bacteria. *FEMS Microbiology Reviews* **25**, 437–454.

White, S. H. & von Heijne, G. (2005). Transmembrane helices before, during, and after insertion. *Current Opinion in Structural Biology* **15**, 378–386.

Wild, K., Rosendal, K. R. & Sinning, I. (2004). A structural step into the SRP cycle. *Molecular Microbiology* **53**, 357–363.

Protein translocation through ABC pathway

Detmers, F. J. M., Lanfermeijer, F. C. & Poolman, B. (2001). Peptides and ATP binding cassette peptide transporters. *Research in Microbiology* **152**, 245–258.

Omori, K. & Idei, A. (2003). Gram-negative bacterial ATP-binding cassette protein exporter family and diverse secretory proteins. *Journal of Bioscience and Bioengineering* **95**, 1–12.

Twin-arginine translocation (TAT) pathway

Berks, B. C., Sargent, F. & Palmer, T. (2000). The Tat protein export pathway. *Molecular Microbiology* **35**, 260–274.

Berks, B. C., Palmer, T. & Sargent, F. (2003). The Tat protein translocation pathway and its role in microbial physiology. *Advances in Microbial Physiology* **47**, 187–254.

Berks, B. C., Palmer, T. & Sargent, F. (2005). Protein targeting by the bacterial twin-arginine translocation (Tat) pathway. *Current Opinion in Microbiology* **8**, 174–181.

Bronstein, P., Marrichi, M. & DeLisa, M. P. (2004). Dissecting the twin-arginine translocation pathway using genome-wide analysis. *Research in Microbiology* **155**, 803–810.

Halic, M. & Beckmann, R. (2005). The signal recognition particle and its interactions during protein targeting. *Current Opinion in Structural Biology* **15**, 116–125.

Lee, P. A., Tullman-Ercek, D. & Georgiou, G. (2006). The bacterial twin-arginine translocation pathway. *Annual Review of Microbiology* **60**, 373–395.

Meloni, S., Rey, L., Sidler, S., Imperial, J., Ruiz-Argueso, T. & Palacios, J. M. (2003). The twin-arginine translocation (Tat) system is essential for *Rhizobium*-legume symbiosis. *Molecular Microbiology* **48**, 1195–1207.

Muller, M. (2005). Twin-arginine-specific protein export in *Escherichia coli*. *Research in Microbiology* **156**, 131–136.

Palmer, T., Sargent, F. & Berks, B. C. (2005). Export of complex cofactor-containing proteins by the bacterial Tat pathway. *Trends in Microbiology* **13**, 175–180.

Robinson, C. & Bolhuis, A. (2001). Protein targeting by the twin-arginine translocation pathway. *Nature Reviews Molecular Cell Biology* **2**, 350–356.

Robinson, C. & Bolhuis, A. (2004). Tat-dependent protein targeting in prokaryotes and chloroplasts. *Biochimica et Biophysica Acta – Molecular Cell Research* **1694**, 135–147.

Sargent, F., Berks, B. C. & Palmer, T. (2002). Assembly of membrane-bound respiratory complexes by the Tat protein-transport system. *Archives of Microbiology* **178**, 77–84.

Yen, M. R., Tseng, Y. H., Nguyen, E. H., Wu, L. F. & Saier, M. H., Jr. (2002). Sequence and phylogenetic analyses of the twin-arginine targeting (Tat) protein export system. *Archives of Microbiology* **177**, 441–450.

Protein translocation in Gram-negative bacteria

Aizawa, S. (2001). Bacterial flagella and type III secretion systems. *FEMS Microbiology Letters* **202**, 157–164.

Burns, D. L. (2003). Type IV transporters of pathogenic bacteria. *Current Opinion in Microbiology* **6**, 29–34.

Buttner, D. & Bonas, U. (2002). Port of entry – the type III secretion translocon. *Trends in Microbiology* **10**, 186–192.

Cascales, E. & Christie, P. J. (2004). The versatile bacterial type IV secretion systems. *Nature Review Microbiology* **1**, 137–149.

Christie, P. J., Atmakuri, K., Krishnamoorthy, V., Jakubowski, S. & Cascales, E. (2005). Biogenesis, architecture, and function of bacterial type IV secretion systems. *Annual Review of Microbiology* **59**, 451–485.

Cianciotto, N. P. (2005). Type II secretion: a protein secretion system for all seasons. *Trends in Microbiology* **13**, 581–588.

Coombes, B. K. & Finlay, B. B. (2005). Insertion of the bacterial type III translocon: not your average needle stick. *Trends in Microbiology* **13**, 92–95.

Cotter, S. E., Surana, N. K. & Geme, J. W., III (2005). Trimeric autotransporters: a distinct subfamily of autotransporter proteins. *Trends in Microbiology* **13**, 199–205.

de Gier, J. W. & Luirink, J. (2001). Biogenesis of inner membrane proteins in *Escherichia coli*. *Molecular Microbiology* **40**, 314–322.

Desvaux, M., Parham, N. J. & Henderson, I. R. (2004). The autotransporter secretion system. *Research in Microbiology* **155**, 53–60.

Facey, S. J. & Kuhn, A. (2004). Membrane integration of *Escherichia coli* model membrane proteins. *Biochimica et Biophysica Acta – Molecular Cell Research* **1694**, 55–66.

Fekkes, P. & Driessen, A. J. M. (1999). Protein targeting to the bacterial cytoplasmic membrane. *Microbiology and Molecular Biology Reviews* **63**, 161–173.

Filloux, A. (2004). The underlying mechanisms of type II protein secretion. *Biochimica et Biophysica Acta – Molecular Cell Research* **1694**, 163–179.

Frankel, G. (2002). Coiled-coil proteins associated with type III secretion systems: a versatile domain revisited. *Molecular Microbiology* **45**, 905–916.

Ghosh, P. (2004). Process of protein transport by the type III secretion system. *Microbiology and Molecular Biology Reviews* **68**, 771–795.

Girard, V. & Mourez, M. (2006). Adhesion mediated by autotransporters of Gram-negative bacteria: structural and functional features. *Research in Microbiology* **157**, 407–416.

Henderson, I. R., Cappello, R. & Nataro, J. P. (2000). Autotransporter proteins, evolution and redefining protein secretion. *Trends in Microbiology* **8**, 529–532.

Henderson, I. R., Navarro-Garcia, F., Desvaux, M., Fernandez, R. C. & Ala'Aldeen, D. (2004). Type V protein secretion pathway: the autotransporter story. *Microbiology and Molecular Biology Reviews* **68**, 692–744.

Johnson, T. L., Abendroth, J., Hol, W. G. J. & Sandkvist, M. (2006).Type II secretion: from structure to function. *FEMS Microbiology Letters* **255**, 175–186.

Kim, D. S. H., Chao, Y. & Saier, M. H., Jr. (2006). Protein-translocating trimeric autotransporters of Gram-negative bacteria. *Journal of Bacteriology* **188**, 5655–5667.

Lammertyn, E. & Anne, J. (2004). Protein secretion in *Legionella pneumophila* and its relation to virulence. *FEMS Microbiology Letters* **238**, 273–279.

Lawley, T. D., Klimke, W. A., Gubbins, M. J. & Frost, L. S. (2003). F factor conjugation is a true type IV secretion system. *FEMS Microbiology Letters* **224**, 1–15.

Linke, D., Riess, T., Autenrieth, I. B., Lupas, A. & Kempf, V. A. J. (2006). Trimeric autotransporter adhesins: variable structure, common function. *Trends in Microbiology* **14**, 264–270.

Luirink, J., von Heijne, G., Houben, E. & de Gier, J. W. (2006). Biogenesis of inner membrane proteins in *Escherichi coli*. *Annual Review of Microbiology* **59**, 329–355.

Macnab, R. M. (2003). How bacteria assemble flagella. *Annual Review of Microbiology* **57**, 77–100.

Mogensen, J. E. & Otzen, D. E. (2005). Interactions between folding factors and bacterial outer membrane proteins. *Molecular Microbiology* **57**, 326–346.

Page, A. L. & Parsot, C. (2002). Chaperones of the type III secretion pathway: jacks of all trades. *Molecular Microbiology* **46**, 1–11.

Parsot, C., Hamiaux, C. & Page, A. L. (2003). The various and varying roles of specific chaperones in type III secretion systems. *Current Opinion in Microbiology* **6**, 7–14.

Peabody, C. R., Chung, Y. J., Yen, M. R., Vidal-Ingigliardi, D., Pugsley, A. P. & Saier, M. H., Jr. (2003). Type II protein secretion and its relationship to bacterial type IV pili and archaeal flagella. *Microbiology-UK* **149**, 3051–3072.

Plano, G. V., Day, J. B. & Ferracci, F. (2001). Type III export: new uses for an old pathway. *Molecular Microbiology* **40**, 284–293.

Russel, M. (1998). Macromolecular assembly and secretion across the bacterial cell envelope – type II protein secretion system. *Journal of Molecular Biology* **279**, 485–499.

Saier, M. H., Jr. (2004). Evolution of bacterial type III protein secretion systems. *Trends in Microbiology* **12**, 113–115.

Sandkvist, M. (2001). Biology of type II secretion. *Molecular Microbiology* **40**, 271–283.

Thanassi, D. G. & Hultgren, S. J. (2000). Multiple pathways allow protein secretion across the bacterial outer membrane. *Current Opinion in Cell Biology* **12**, 420–430.

Thomas, N. A. & Brett Finlay, B. (2003). Establishing order for type III secretion substrates – a hierarchical process. *Trends in Microbiology* **11**, 398–403.

Tokuda, H. & Matsuyama, S. (2004). Sorting of lipoproteins to the outer membrane in *E. coli*. *Biochimica et Biophysica Acta – Molecular Cell Research* **1693**, 5–13.

Ton-That, H. & Schneewind, O. (2004). Assembly of pili in Gram-positive bacteria. *Trends in Microbiology* **12**, 228–234.

Ulsen, P. & Tommassen, J. (2006). Protein secretion and secreted proteins in pathogenic *Neisseriaceae*. *FEMS Microbiology Reviews* **30**, 292–319.

Yip, C. K. & Strynadka, N. C. J. (2006). New structural insights into the bacterial type III secretion system. *Trends in Biochemical Sciences* **31**, 223–230.

4

Glycolysis

Escherichia coli can grow on a simple medium containing glucose and mineral salts and this bacterium can synthesize all cell constituents using materials provided in this medium. Glucose is metabolized through the Embden–Meyerhof–Parnas (EMP) pathway and hexose monophosphate (HMP) pathway and the metabolic product, pyruvate, is decarboxylated oxidatively to acetyl-CoA to be oxidized through the tricarboxylic acid (TCA) cycle. Twelve intermediates of these pathways are used as carbon skeletons for biosynthesis (Table 4.1). Heterotrophs that utilize organic compounds other than carbohydrates convert their substrates into one or more of these intermediates. For this reason, glucose metabolism through glycolysis and the TCA cycle is called central metabolism.

Eukaryotes metabolize glucose through the EMP pathway to generate ATP, pyruvate and NADH, and the HMP pathway is needed to supply the metabolic intermediates not available from the EMP pathway such as pentose-5-phosphate and erythrose-4-phosphate, and NADPH. Most prokaryotes employ similar mechanisms, but some prokaryotes metabolize glucose through unique pathways known only in prokaryotes, e.g. the Entner–Doudoroff (ED) pathway and phosphoketolase (PK) pathway. Some prokaryotes have genes for the ED pathway in addition to the EMP pathway: genes for these pathways are expressed at the same time in several prokaryotes including a thermophilic bacterium (*Thermotoga maritima*), a thermophilic archaeon (*Thermoproteus tenax*) and a halophilic archaeon (*Halococcus saccharolyticus*). *Escherichia coli* metabolizes glucose via the EMP pathway, but gluconate is oxidized through the ED pathway. Modified EMP and ED pathways are quite common in archaea.

Carbohydrates are phosphorylated before they are metabolized in most cases. It is believed that phosphorylated intermediates are less likely to diffuse away through the cytoplasmic membrane. Some bacteria and archaea also phosphorylate intermediates of glucose metabolism in modified glycolytic pathways.

This chapter describes glucose oxidation to pyruvate and related metabolic pathways. Pyruvate metabolism will be further discussed in Chapters 5, 8 and 9.

Table 4.1. | *Metabolic intermediates used as carbon skeletons for biosynthesis*

Carbon skeleton	From	Precursor for
Glucose-6-phosphate	EMP	polysaccharides
Fructose-6-phosphate	EMP	murein
Ribose-5-phosphate	HMP	nucleic acids
Erythrose-4-phosphate	HMP	amino acids
Triose-phosphate	EMP	lipids
3-phosphoglycerate	EMP	amino acids
Phosphoenolpyruvate	EMP	amino acids
Pyruvate	EMP	amino acids
Acetyl-CoA	Pyruvate	fatty acids
2-ketoglutarate	TCA	amino acids
Succinyl-CoA	TCA	amino acids
Oxaloacetate	TCA	amino acids

4.1 | EMP pathway

Many anaerobic and enteric bacteria transport glucose via group translocation (phosphotransferase system, PTS, Section 3.5) in the form of glucose-6-phosphate. Glucose transported through active transport is phosphorylated by hexokinase:

$$\text{glucose} + \text{ATP} \xrightarrow{\text{hexokinase}} \text{glucose-6-phosphate} + \text{ADP}$$

Hexokinase can phosphorylate other hexoses such as mannose, and requires Mg^{2+} for activity. The enzyme cannot catalyze the reverse reaction.

Glucose-6-phosphate can also be obtained from glycogen:

$$\underset{\text{(glycogen)}}{[\text{glucose}]_n + P_i} \xrightarrow{\text{phosphorylase}} \underset{\text{(glycogen)}}{[\text{glucose}]_{n-1} + \text{glucose-1-phosphate}}$$

$$\text{glucose-1-phosphate} \xrightarrow{\text{phosphoglucomutase}} \text{glucose-6-phosphate}$$

Glucose-6-phosphate is a precursor for the biosynthesis of polysaccharides as well as a substrate of the EMP pathway (Figure 4.1), which is the commonest glycolytic pathway in all kinds of organisms.

4.1.1 Phosphofructokinase (PFK): key enzyme of the EMP pathway

Glucose-6-phosphate is isomerized to fructose-6-phosphate before being phosphorylated to fructose-1,6-diphosphate by the action of phosphofructokinase (PFK). These two reactions require Mg^{2+}.

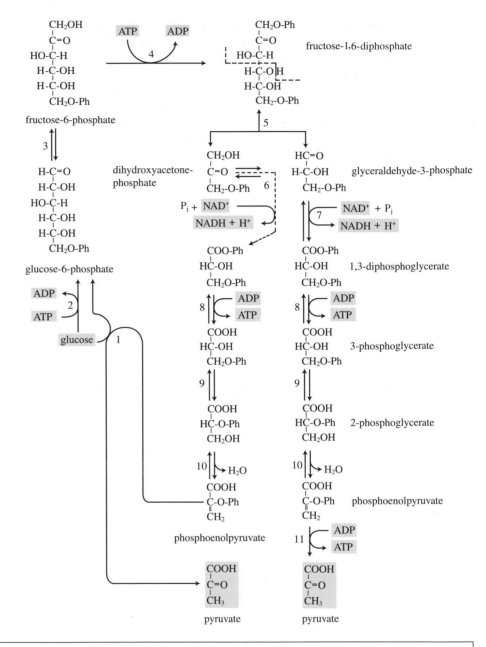

Figure 4.1 Glucose oxidation via the Embden–Meyerhof–Parnas pathway.

Glucose is phosphorylated to glucose-6-phosphate by PEP:glucose phosphotransferase (1) during group translocation (phosphotransferase system, PTS) or by hexokinase (2) after uptake via active transport.

3, glucose-6-phosphate isomerase; 4, phosphofructokinase; 5, fructose diphosphate aldolase; 6, triose-phosphate isomerase; 7, glyceraldehyde-3-phosphate dehydrogenase; 8, 3-phosphoglycerate kinase; 9, phosphoglycerate mutase; 10, enolase; 11, pyruvate kinase.

Glucose-6-phosphate isomerase catalyzes the reverse reaction, but phosphofructokinase does not. The irreversibility of an enzyme is due to thermodynamic reasons, and many enzymes that do not catalyze the reversible reaction are regulated. PFK is the key enzyme of the EMP pathway. If this enzyme is present in a given prokaryote, it can be assumed that this organism catabolizes glucose through the EMP pathway. Fructose-6-phosphate is the precursor of amino sugars and their polymers such as murein (Section 6.8.2).

Fructose-1,6-diphosphate aldolase cleaves fructose-1,6-diphosphate to two molecules of triose-phosphate. This aldolase catalyzes the reverse reaction, and participates in gluconeogenesis (Section 4.2), producing hexose-phosphate when non-carbohydrate substrates are used as carbon sources.

4.1.2   ATP synthesis and production of pyruvate

Triose-phosphate isomerase equilibrates dihydroxyacetone phosphate and glyceraldehyde-3-phosphate produced from fructose-1,6-diphosphate. Under standard conditions the equilibrium shifts to the formation of dihydroxyacetone phosphate ($\Delta G^{0'} = -7.7$ kJ/mol glyceraldehyde-3-phosphate), but the reverse reaction is favoured because glyceraldehyde-3-phosphate is continuously consumed in subsequent reactions. Phospholipids are synthesized from glyceraldehyde-3-phosphate (Section 6.6.2).

Glyceraldehyde-3-phosphate is oxidized to 1,3-diphosphoglycerate by glyceraldehyde-3-phosphate dehydrogenase. This endergonic reaction ($\Delta G^{0'} = +6.3$ kJ/mol glyceraldehyde-3-phosphate) is efficiently pulled by the following exergonic reaction catalyzed by 3-phosphoglycerate kinase ($\Delta G^{0'} = -12.5$ kJ/mol 1,3-phosphoglycerate). This enzyme requires Mg^{2+}, as do most kinases, and ATP generation in this reaction is an example of substrate-level-phosphorylation. 3-phosphoglycerate is the starting material for the synthesis of amino acids, serine, glycine and cysteine (Section 6.4.2).

$$\text{1,3-diphosphoglycerate} + \text{ADP} \xrightleftharpoons{\substack{\text{3-phosphoglycerate} \\ \text{kinase}}} \text{3-phosphoglycerate} + \text{ATP}$$

$$(\Delta G^{0'} = -12.5 \text{ kJ/mol 1,3-diphosphoglycerate})$$

3-phosphoglycerate is converted to 2-phosphoglycerate by phosphoglycerate mutase which requires 2,3-diphosphoglycerate as a coenzyme. 2-phosphoglycerate is dehydrated to phosphoenolpyruvate (PEP) by an enolase in the presence of divalent cations such as Mg^{2+} and Mn^{2+}. PEP is used to generate ATP with the reaction of the last enzyme in the EMP pathway in the presence of Mg^{2+} and K^+. PEP supplies energy in group translocation, and is used to synthesize aromatic amino acids (Section 6.4.4). Glyceraldehyde-3-phosphate is an intermediate in the HMP and ED pathways and the reactions from this triose-phosphate are shared with both these pathways.

Four ATPs are synthesized and two high energy phosphate bonds are consumed in this pathway, resulting in a net gain of two ATPs per glucose oxidized. The NADH reduced in the glycolytic pathway is reoxidized in aerobic (Section 5.8) and anaerobic respiration (Chapter 9), and in fermentation (Chapter 8), reducing various electron acceptors depending on the organism and on their availability.

Figure 4.2 **The Methylglyoxal By pass, a modified EMP pathway under phosphate-limited conditions.**

(Gottschalk, G. 1986, *Bacterial Metabolism*, 2nd edn., Figure 5.15. Springer, New York)

Under phosphate-limited conditions, bacteria such as *Escherichia coli* metabolize dihydroxyacetone phosphate to pyruvate to supply precursors for biosynthesis with a reduced ATP yield.

1, methylglyoxal synthase; 2, glyoxalase; 3, lactate oxidase.

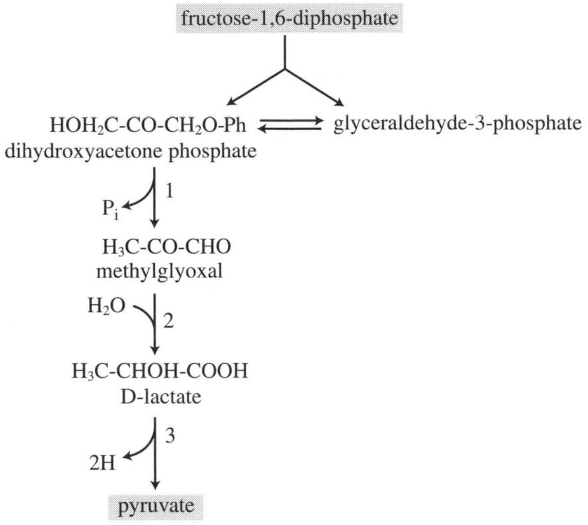

fructose-1,6-diphosphate

$HOH_2C\text{-}CO\text{-}CH_2O\text{-}Ph$ ⇌ glyceraldehyde-3-phosphate
dihydroxyacetone phosphate

P_i ← 1

$H_3C\text{-}CO\text{-}CHO$
methylglyoxal

H_2O ← 2

$H_3C\text{-}CHOH\text{-}COOH$
D-lactate

2H ← 3

pyruvate

4.1.3 Modified EMP pathways

Some prokaryotes, unlike eukaryotes, metabolize glucose through modified EMP pathways depending on the growth conditions.

4.1.3.1 Methylglyoxal bypass

Under phosphate-limited conditions, *Escherichia coli* and *Pseudomonas saccharophila* oxidize dihydroxyacetone phosphate to pyruvate through methylglyoxal (Figure 4.2). This diversion enables acetyl-CoA synthesis through pyruvate when glyceraldehyde-3-phosphate dehydrogenase cannot function due to a low concentration of inorganic phosphate (P_i), one of its substrates, with a reduced ATP yield.

Methylglyoxal is very reactive, destroying nucleic acids and proteins by reacting with guanine and adenine, and with amino acids such as arginine, lysine and cysteine. When this toxic compound accumulates in the cell, various proteins are synthesized, including membrane proteins to excrete it, and glyoxalase to increase the rate of methylglyoxal conversion to lactate.

Methylglyoxal synthase activity is regulated by P_i through feedback inhibition, and is activated by its substrate, dihydroxyacetone phosphate. Because the concentration of dihydroxyacetone phosphate in the cell is around 0.5 mM, lower than the K_m value for methylglyoxal synthase even under phosphate-limited conditions, the methylglyoxal concentration does not reach toxic levels. However, toxic levels of methylglyoxal accumulate with a limited supply of phosphate when other nutrients such as nitrogen are limited. For example, a rumen bacterium, *Prevotella ruminicola*, loses viability when excess carbohydrate is supplied under nitrogen- and phosphate-limited conditions due to methylglyoxal accumulation.

A thermophilic anaerobic bacterium, *Clostridium thermosaccharolyticum*, reduces methylglyoxal to 1,2-propanediol via acetol (CH_3COCH_2OH).

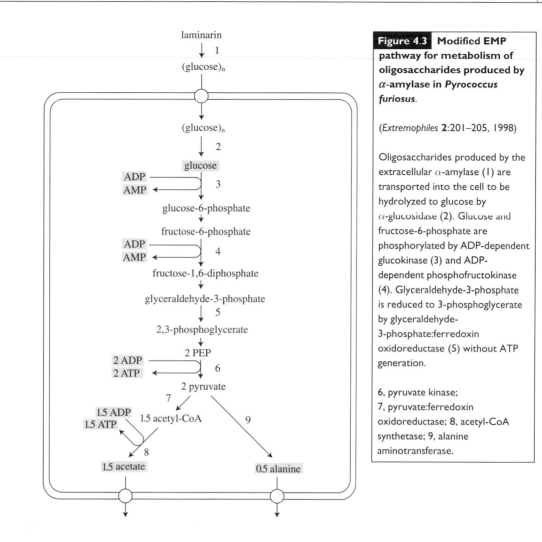

Figure 4.3 Modified EMP pathway for metabolism of oligosaccharides produced by α-amylase in *Pyrococcus furiosus*.

(*Extremophiles* **2**:201–205, 1998)

Oligosaccharides produced by the extracellular α-amylase (1) are transported into the cell to be hydrolyzed to glucose by α-glucosidase (2). Glucose and fructose-6-phosphate are phosphorylated by ADP-dependent glucokinase (3) and ADP-dependent phosphofructokinase (4). Glyceraldehyde-3-phosphate is reduced to 3-phosphoglycerate by glyceraldehyde-3-phosphate:ferredoxin oxidoreductase (5) without ATP generation.

6, pyruvate kinase; 7, pyruvate:ferredoxin oxidoreductase; 8, acetyl-CoA synthetase; 9, alanine aminotransferase.

4.1.3.2 Modified EMP pathways in archaea

Most archaea utilizing carbohydrates employ modified ED pathways (Section 4.4.3.2), but a few metabolize sugars in modified EMP pathways. The halophilic archaeon *Haloarcula vallismortis* transports fructose through an active transport mechanism and a ketohexokinase phosphorylates the free sugar to fructose-1-phosphate, which is transformed to fructose-1,6-diphosphate before being oxidized through the normal EMP pathway (Section 7.2.1).

$$\text{fructose} + \text{ATP} \xrightarrow{\text{ketohexokinase}} \text{fructose-1-phosphate} + \text{ADP}$$

$$\text{fructose-1-phosphate} + \text{ATP} \xrightarrow{\text{1-phosphofructokinase}} \text{fructose-1,6-diphosphate} + \text{ADP}$$

Hyperthermophilic archaea such as *Pyrococcus furiosus*, *Thermococcus celer*, and *Desulfurococcus amylolyticus* employ yet another modified EMP pathway to ferment·sugars (Figure 4.3). *Pyrococcus furiosus* does not use glucose, but ferments starch. This organism transports oligosaccharides produced from starch by α-amylase. Oligosaccharides are hydrolyzed to glucose by α-glucosidase. The differences from glycolysis in this organism are (1) phosphorylation of glucose and fructose-6-phosphate is catalyzed by ADP-dependent kinases, and (2) glyceraldehyde-3-phosphate is reduced to 3-phosphoglycerate by glyceraldehyde-3-phosphate:ferredoxin oxidoreductase without generating ATP. It has been hypothesized that this organism has evolved in high temperature environments to use ADP in the place of ATP, since ADP is more stable than ATP under such conditions. The reason why energy is not conserved during the oxidation of glyceraldehyde-3-phosphate to 1,3-diphosphoglycerate might be due to the use of ferredoxin as the electron acceptor, which has a lower redox potential than NAD^+.

In addition to these ADP-dependent kinases, a pyrophosphate (PP_i)-dependent phosphofructokinase is known in *Thermoproteus tenax*. The PP_i-dependent phosphofructokinase catalyzes the reverse reaction and is not regulated by AMP and PEP unlike the ATP (ADP)-dependent enzymes. This enzyme is probably a component of the gluconeogenesis pathway (Section 4.2) in the thermophile, which has a modified ED pathway as the main glycolytic pathway.

Pyruvate is reduced as an electron acceptor to alanine in *Pyrococcus furiosus* (Section 8.10).

4.1.4 Regulation of the EMP pathway

The EMP pathway serves not only to generate ATP but also to provide precursors for biosynthesis (Table 4.1). The EMP pathway is regulated by the energy status of the cell as well as by the concentration of certain metabolic intermediates. A parameter, the adenylate energy charge (EC), was devised to describe the energy status in a culture (Section 5.6.2). Generally the glycolytic pathway is activated at low EC values, and repressed at high EC values. As stated earlier, the enzymes that do not catalyze the reverse reaction are regulated. In the EMP pathway, these enzymes are phosphofructokinase and pyruvate kinase. In organisms where glucose is transported by active transport, hexokinase is repressed by its product, glucose-6-phosphate.

4.1.4.1 Regulation of phosphofructokinase

The EMP pathway is controlled mainly by regulating phosphofructokinase activity. This enzyme is activated by ADP and GDP, and repressed by PEP in bacteria. On the other hand, AMP activates, and ATP and citrate repress this enzyme in yeast. In general, the enzyme is repressed when the EC value (Section 5.6.2) and the concentration of intermediates used as precursors for biosynthesis are high, and activated when the organism needs ATP and biosynthetic precursors.

Fructose-1,6-diphosphate is dephosphorylated to fructose-6-phosphate and P_i in gluconeogenesis by fructose-1,6-diphosphatase. If fructose-1,6-diphosphatase and phosphofructokinase are both active in a cell, ATP is wasted. This is called a futile cycle and is avoided through an elaborate regulatory mechanism (Section 4.2.4).

4.1.4.2 Regulation of pyruvate kinase

Pyruvate kinase is activated when fructose-1,6-diphosphate is accumulated in the cell. This kind of regulation is termed feedforward activation or precursor activation. The term feedforward activation is used to describe the activation of the last enzyme of a metabolic pathway by the substrate of the first enzyme of the same pathway. Feedback inhibition describes the inhibition of the first enzyme by the final product in an anabolic pathway (Section 12.3.1).

4.1.4.3 Global regulation

In addition to regulation of individual enzymes, glycolysis is regulated as a part of global regulation. CcpA (catabolite control protein A) activates transcription of the genes for glycolytic enzymes in Gram-positive bacteria such as *Bacillus subtilis* (Section 12.1.3) and CsrA (carbon storage regulator A) activates the expression of the glycolytic genes, and represses genes for gluconeogenesis and glycogen synthesis in *Escherichia coli* (Section 12.1.9.3). In *Bacillus subtilis* and related low G + C Gram-positive bacteria, genes for enzymes catalyzing glyceraldehyde-3-phosphate to phosphoenolpyruvate form an operon. This operon is repressed by CggR (central glycolytic gene regulator) under gluconeogenic conditions, and CggR activity is repressed by fructose-1,6-diphosphate (Section 12.1.3.3).

4.2 | Glucose-6-phosphate synthesis: gluconeogenesis

Hexose-6-phosphates are the precursors of polysaccharide synthesis. Microbes growing on carbon sources other than sugars (Chapter 7) need to synthesize glucose-6-phosphate from their substrate. Since phosphofructokinase and pyruvate kinase do not catalyze their reverse reactions, hexose-6-phosphate cannot be synthesized by a reversal of the EMP pathway.

$$\text{fructose-6-phosphate} + \text{ATP} \longrightarrow \text{fructose-1,6-diphosphate} + \text{ADP}$$

$$(\Delta G^{0\prime} = -14.2 \text{ kJ/mol fructose-6-phosphate})$$

$$\text{phosphoenolpyruvate} + \text{ADP} \longrightarrow \text{pyruvate} + \text{ATP}$$

$$(\Delta G^{0\prime} = -23.8 \text{ kJ/mol phosphoenolpyruvate})$$

The above reactions are exergonic, and the reverse reactions are thermodynamically unfavourable. Separate enzymes are therefore used in gluconeogenesis to overcome this problem.

4.2.1 PEP synthesis

Microbes growing on non-carbohydrate compounds produce PEP through pyruvate, oxaloacetate or malate (Chapter 7).

Phosphoenolpyruvate (PEP) synthetase is widespread in microbes:

$$\text{pyruvate} + \text{ATP} \xrightarrow{\text{PEP synthetase}} \text{PEP} + \text{AMP} + \text{P}_i$$
$$(\Delta G^{0'} = -8.4 \, \text{kJ/mol pyruvate})$$

Pyruvate-phosphate dikinase is another enzyme that synthesizes PEP from pyruvate. This enzyme is found in *Acetobacter xylinum* growing on ethanol and in *Propionibacterium shermanii* growing on lactate:

$$\text{pyruvate} + \text{ATP} + \text{P}_i \underset{\text{pyruvate phosphate}}{\overset{\text{dikinase}}{\rightleftharpoons}} \text{PEP} + \text{AMP} + \text{PP}_i$$

Phosphoenolpyruvate carboxykinase produces PEP from a TCA cycle intermediate, oxaloacetate:

$$\text{oxaloacetate} + \text{ATP} \underset{}{\overset{\text{PEP carboxykinase}}{\rightleftharpoons}} \text{PEP} + \text{CO}_2 + \text{ADP}$$

Malate can be converted either through oxaloacetate or directly by the action of malate enzyme:

$$\text{malate} + \text{NAD}^+ + \text{H}^+ \underset{}{\overset{\text{malate enzyme}}{\rightleftharpoons}} \text{pyruvate} + \text{NADH} + \text{CO}_2$$

4.2.2 Fructose diphosphatase

PEP can be converted to fructose-1,6-diphosphate by the EMP pathway enzymes since they catalyze the reverse reactions. However, phosphofructokinase does not catalyze the reverse reaction and fructose-1,6-diphosphate is dephosphorylated to fructose-6-phosphate and P_i by fructose diphosphatase:

$$\text{fructose-1,6-diphosphate} \xrightarrow{\substack{\text{fructose} \\ \text{diphosphatase}}} \text{fructose-6-phosphate} + \text{P}_i$$

Pyrophosphate (PP_i)-dependent phosphofructokinase is found in a hyperthermophilic archaeon *Thermoproteus tenax* and this enzyme catalyzes the reverse reaction. Similar reactions are known in plants and bacteria. This enzyme may be a component of gluconeogenesis:

$$\text{fructose-1,6-diphosphate} + \text{P}_i \underset{}{\overset{\substack{\text{PP}_i\text{-dependent} \\ \text{phosphofructokinase}}}{\rightleftharpoons}} \text{fructose-6-phosphate} + \text{PP}_i$$

4.2.3 Gluconeogenesis in archaea

Methanogenic archaea do not use sugars, and most of the halophilic and hyperthermophilic archaea metabolize sugars through modified ED pathways (Section 4.4.3). Nevertheless, they have a similar gluconeogenic system to bacteria. It is not known what the advantage is to have two different sets of genes for sugar metabolism. Some bacteria have genes for both EMP and ED pathways (Section 4.4.1).

Some methanogens accumulate cyclic 2,3-diphosphoglycerate (cDPG) in the cell. This compound is not known in other organisms and is produced by phosphorylation of 2-phosphoglycerate, an intermediate of gluconeogenesis. Poly-3-phosphoglycerate is synthesized from cDPG as a storage compound.

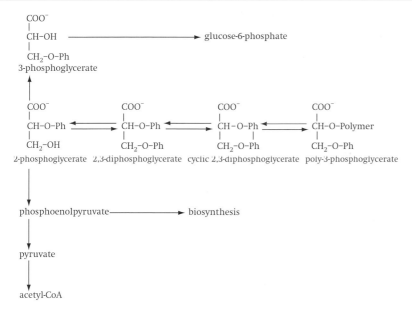

4.2.4 Regulation of gluconeogenesis

Gluconeogenesis is regulated through the control of fructose diphosphatase activity. When the energy status is good, this enzyme is activated by ATP to supply carbon skeletons for growth. AMP represses the enzyme activity when the energy status is too low for growth. A futile cycle is avoided by the opposite controls of glycolysis and gluconeogenesis.

4.3 | Hexose monophosphate (HMP) pathway

When *Escherichia coli* grows on glucose as the sole carbon and energy source, about 72% of the substrate is metabolized through the EMP pathway, and the HMP pathway consumes the remaining 28%. This is because the EMP cannot meet all the requirements for biosynthesis. The HMP pathway provides the biosynthetic metabolism with pentose-5-phosphate, erythrose-4-phosphate and NADPH. This pathway is also called the pentose phosphate pathway. NADPH is used to supply reducing power in biosynthetic processes. $NADP^+$ is reduced only by isocitrate dehydrogenase (Section 5.2) when glucose is metabolized through the EMP pathway and TCA cycle. NADH is seldom used in biosynthetic reactions. Most of the NADPH needed for biosynthesis arises from the HMP pathway.

4.3.1 HMP pathway in three steps

For convenience, the HMP pathway can be discussed in three steps (Figure 4.4). During the initial step of the HMP pathway,

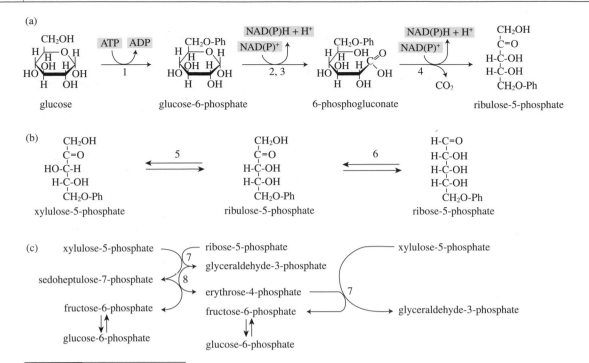

glucose-6-phosphate is oxidized to ribulose-5-phosphate and CO_2, reducing $NADP^+$. Glucose-6-phosphate dehydrogenase, lactonase and 6-phosphogluconate dehydrogenase catalyze these reactions. In the following reactions, ribulose-5-phosphate is converted to ribose-5-phosphate and xylulose-5-phosphate by the action of isomerase and epimerase. Finally, the pentose-5-phosphates are transformed to glucose-6-phosphate and glyceraldehyde-3-phosphate through carbon rearrangement by transaldolase and transketolase. A transaldolase transfers a 3-carbon fragment, and a 2-carbon fragment transfer is catalyzed by a transketolase. HMP can be summarized as:

$$\text{glucose-6-phosphate} + 2NADP^+ \longrightarrow \text{glyceraldehyde-3-phosphate} + 3CO_2 + 2NADPH + 2H^+$$

Ribose-5-phosphate is the precursor for nucleotide synthesis, and aromatic amino acids are produced from erythrose-4-phosphate. NADPH supplies reducing power during biosynthesis (Section 6.4.4).

Some eukaryotic microorganisms metabolize more glucose through the HMP pathway when they use nitrate as their nitrogen source. They use NADPH in assimilatory nitrate reduction (Section 6.2.2).

4.3.2 Additional functions of the HMP pathway

In addition to supplying precursors and reducing power for biosynthesis from glucose, the HMP and related pathways have some other functions. HMP is the major glycolytic metabolism in microbes that (1) utilize pentoses, and (2) do not possess other glycolytic activities. The HMP cycle is also employed for the complete oxidation of sugars in bacteria lacking a functional TCA cycle.

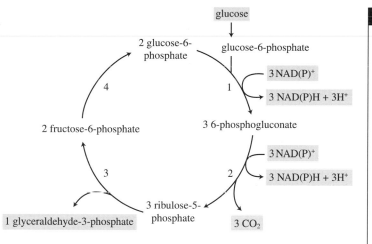

Figure 4.5 **Oxidative HMP cycle.**

Organisms lacking functional EMP or ED pathways, or a functional TCA cycle, oxidize glucose through the oxidative HMP cycle. Glyceraldehyde-3-phosphate is oxidized to pyruvate as in the EMP pathway.

1, glucose-6-phosphate dehydrogenase; 2, 6-phosphogluconate dehydrogenase; 3, carbon rearrangement as in the HMP pathway; 4, glucose-6-phosphate isomerase.

4.3.2.1 Utilization of pentoses

Pentoses are phosphorylated and metabolized to fructose-6-phosphate and glyceraldehyde-3-phosphate through steps 2 and 3 of the HMP pathway. This will be discussed in Chapter 7 (Section 7.2.2).

4.3.2.2 Oxidative HMP cycle

Thiobacillus novellus and *Brucella abortus* oxidize glucose completely although they lack a functional EMP or ED pathway. Glucose is oxidized through the oxidative HMP cycle (Figure 4.5). Glyceraldehyde-3-phosphate is oxidized to pyruvate as in the EMP pathway. The HMP cycle is found in species of *Gluconobacter* which do not have a functional TCA cycle. These bacteria possess the incomplete TCA fork to meet the supply of biosynthetic precursors (Section 5.4.1).

4.3.3 Regulation of the HMP pathway

Lactonase is the only enzyme unable to catalyze the reverse reaction among the enzymes of the HMP pathway (Figure 4.4), but regulation of the pathway is exerted through control of glucose-6-phosphate dehydrogenase and 6-phosphogluconate dehydrogenase. These enzymes are inhibited with the accumulation of NADPH and NADH.

Dehydrogenation of glucose-6-phosphate is a common reaction in the HMP and ED pathways, but is catalyzed by separate enzymes in each. The enzyme involved in the ED pathway uses NAD^+ as the electron acceptor and is inhibited by ATP and PEP. On the other hand, the $NADP^+$-dependent HMP pathway enzyme is inhibited by NAD(P)H but not by ATP.

4.3.4 F_{420}-dependent glucose-6-phosphate dehydrogenase

F_{420} was first identified in methanogens, and was regarded as one of the methanogen-specific coenzymes (Section 9.4.2). This electron carrier is found not only in hyperthermophilic archaea such as species of *Sulfolobus* and *Archaeoglobus*, but also in bacteria. F_{420} was found to be involved in tetracycline biosynthesis in a *Streptomyces* sp., and a

F_{420}-dependent glucose-6-phosphate dehydrogenase has been identified in *Mycobacterium smegmatis*. This enzyme is believed to be a part of the HMP pathway in this bacterium. Since F_{420} is not known in animals, this enzyme is a potential target for antibacterial drugs against medically important species of *Mycobacterium*.

In addition to F_{420}, other cofactors, originally found in methanogens, have been identified in bacteria. Coenzyme M is involved in the oxidation of propylene by *Rhodococcus rhodochrous* and a species of *Xanthobacter* (Figure 7.23, Section 7.7). Tetrahydromethanopterin occurs in a methylotroph, *Methylobacterium extorquens* (Figure 7.32, Section 7.9.2.2).

4.4 | Entner–Doudoroff (ED) pathway

4.4.1 Glycolytic pathways in some Gram-negative bacteria

In addition to the EMP pathway, unique glycolytic pathways are also known in prokaryotes. The ED pathway was first identified in *Pseudomonas saccharophila* by Entner and Doudoroff, and this turned out to be the main glycolytic pathway in prokaryotes that do not possess enzymes of the EMP pathway. In addition to species of *Pseudomonas*, the ED pathway functions as the main glycolytic pathway in other Gram-negative bacteria such as *Zymomonas* and *Azotobacter* species, and gluconate is metabolized through this pathway in some other Gram-negative bacteria including *Escherichia coli*, and in some coryneform bacteria such as species of *Arthrobacter* and *Cellulomonas* (Table 4.2).

4.4.2 Key enzymes of the ED pathway

In the first two reactions of the ED pathway, glucose-6-phosphate is converted to 6-phosphogluconate via phosphogluconolactone, as in the HMP pathway. 6-phosphogluconate is dehydrated to 2-keto-3-deoxy-6-phosphogluconate (KDPG) by 6-phosphogluconate dehydratase. KDPG aldolase splits its substrate into pyruvate and glyceraldehyde-3-phosphate (Figure 4.6). The latter is oxidized to pyruvate as in the EMP pathway. The key enzymes of this pathway are 6-phosphogluconate dehydratase and KDPG aldolase.

Two ATP are generated and one high energy phosphate bond is consumed with a net gain of one ATP per glucose oxidized in this pathway.

4.4.3 Modified ED pathways

Unusually, some prokaryotes oxidize glucose and the intermediates are phosphorylated before being metabolized in a similar manner as in the ED pathway.

4.4.3.1 Extracellular oxidation of glucose by Gram-negative bacteria

Sugars are metabolized after they are phosphorylated, the latter process probably preventing their loss through membrane diffusion. Negatively charged phosphorylated sugars and their intermediates

Table 4.2. *Major glycolytic pathways in prokaryotes*

Organism	EMP	ED
Arthrobacter species	+	+/−[a]
Azotobacter chroococcum	+	−
Ralstonia eutropha (Alcaligenes eutrophus)	−	+
Bacillus subtilis	+	−
Cellulomonas flavigena	+	+/−[a]
Escherichia coli and enteric bacteria	+	+/−[a]
Pseudomonas saccharophila	−	+
Rhizobium japonicum	−	+
Thiobacillus ferrooxidans	−	+
Xanthomonas phaseoli	−	+
Thermotoga maritima	+	+[b]
Thermoproteus tenax	+[c]	+[b,d]
Halococcus saccharolyticus	+[c,e]	+[b,d,f]
Halobacterium saccharovorum	−	+[d]
Clostridium aceticum	−	+[d]
Sulfolobus acidocaldarius	−	+[d]

+, present; −, absent.
[a] When gluconate is used as energy and carbon source.
[b] Enzymes for EMP and ED pathways are expressed simultaneously.
[c] Modified EMP pathway.
[d] Modified ED pathway.
[e] Fructose.
[f] Glucose.

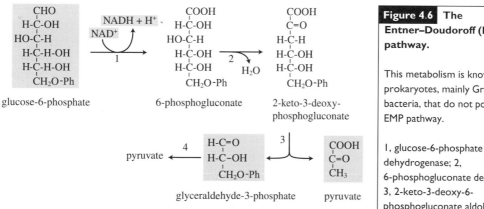

glucose-6-phosphate

6-phosphogluconate

2-keto-3-deoxy-phosphogluconate

glyceraldehyde-3-phosphate   pyruvate

pyruvate

Figure 4.6 The Entner–Doudoroff (ED) pathway.

This metabolism is known only in prokaryotes, mainly Gram-negative bacteria, that do not possess the EMP pathway.

1, glucose-6-phosphate dehydrogenase; 2, 6-phosphogluconate dehydratase; 3, 2-keto-3-deoxy-6-phosphogluconate aldolase; 4, as in the EMP pathway.

are less permeable to the membrane. However, some strains of *Pseudomonas* oxidize glucose extracellularly when the glucose concentration is high. These bacteria possess glucose dehydrogenase and gluconate dehydrogenase on the periplasmic face of the cytoplasmic membrane. When glucose is depleted, gluconate and 2-ketogluconate are transported through specific transporters and

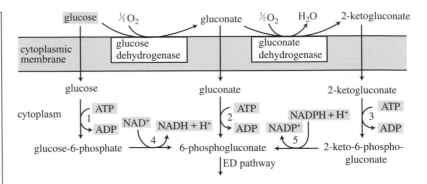

Figure 4.7 **Glucose utilization by some species of *Pseudomonas* in a high glucose environment.**

(Dawes, E. A. 1986, *Microbial Energetics*, Figure 3.5. Blackie & Son, Glasgow)

When the glucose concentration is high, some strains of *Pseudomonas* oxidize glucose at the periplasmic region. Glucose dehydrogenase and gluconate dehydrogenase are cytoplasmic membrane quinoproteins containing pyrroloquinoline quinone (PQQ). The reduced PQQ transfers electrons to cytochrome *c*, and gluconate and 2-ketogluconate are transported through specific transporters to be metabolized in a similar way as in the ED pathway when glucose is depleted.

1, hexokinase; 2, gluconate kinase; 3, 2-ketogluconate kinase; 4, glucose-6-phosphate dehydrogenase; 5, 2-keto-6-phosphogluconate reductase.

phosphorylated, consuming ATP (Figure 4.7). 2-keto-6-phosphogluconate is reduced to 6-phosphogluconate by a NADPH-dependent reductase.

Glucose dehydrogenase and gluconate dehydrogenase in these bacteria are quinoproteins containing pyrroloquinoline quinone (PQQ, methoxatin) as a prosthetic group (Figure 7.31, Section 7.9.2). The reduced form of PQQ transfers electrons to cytochrome *c* of the electron transport chain.

Gluconate and 2-ketogluconate are uncommon in nature, and few microbes use these compounds. The ability to oxidize glucose and to use its products might therefore be advantageous for those organisms capable of doing this.

A group translocation (phosphotransferase system) negative mutant of *Escherichia coli* synthesizes PQQ-containing glucose dehydrogenase and metabolizes glucose in the same way as the *Pseudomonas* species described above.

4.4.3.2 Modified ED pathways in archaea

Hyperthermophilic archaea belonging to the genera *Sulfolobus*, *Thermoplasma*, and *Thermoproteus* metabolize glucose to pyruvate and glyceraldehyde without phosphorylation in a similar way as the ED pathway, and a halophilic archaeon, *Halobacterium saccharovorum*, oxidizes glucose to 2-keto-3-deoxygluconate, which is phosphorylated before metabolism through the ED pathway (Figure 4.8). The archaeal glucose dehydrogenase is a $NAD(P)^+$-dependent enzyme. Some eubacteria, including *Clostridium aceticum* and *Rhodopseudomonas sphaeroides*, metabolize glucose in a similar mechanism to this halophilic archaeon.

4.5 | Phosphoketolase pathways

Lactate is the sole glucose fermentation product in homofermentative lactic acid bacteria (LAB), while the heterofermentative LAB produce acetate and ethanol in addition to lactate from glucose. The former ferment glucose through the EMP pathway and the phosphoketolase (PK) pathway is employed in the latter and

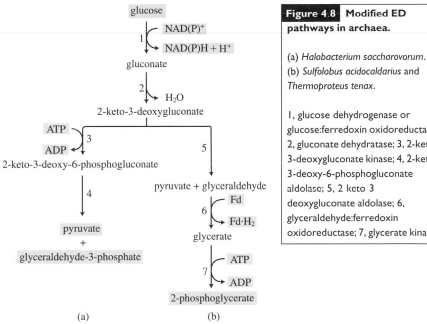

Figure 4.8 Modified ED pathways in archaea.

(a) *Halobacterium saccharovorum*.
(b) *Sulfolobus acidocaldarius* and *Thermoproteus tenax*.

1, glucose dehydrogenase or glucose:ferredoxin oxidoreductase; 2, gluconate dehydratase; 3, 2-keto-3-deoxygluconate kinase; 4, 2-keto-3-deoxy-6-phosphogluconate aldolase; 5, 2 keto 3 deoxygluconate aldolase; 6, glyceraldehyde:ferredoxin oxidoreductase; 7, glycerate kinase.

in bifidus bacteria. A heterofermentative bacterium, *Leuconostoc mesenteroides*, produces lactate and ethanol from glucose through the PK pathway, involving one PK active on xylulose-5-phosphate (Figure 4.9). Lactate and acetate are produced from glucose by *Bifidobacterium bifidum* with two PKs active on fructose-6-phosphate and xylulose-5-phosphate (Figure 4.10). Facultatively homofermentative LAB ferment pentoses and low concentrations of glucose through the PK pathway to produce lactate, acetate and ethanol (Section 8.4).

4.5.1 Glucose fermentation by *Leuconostoc mesenteroides*

Heterofermentative LAB, including, *Leuconostoc mesenteroides*, ferment glucose to lactate, ethanol and carbon dioxide:

$$C_6H_{12}O_6 \longrightarrow CH_3CHOHCOOH + CH_3CH_2OH + CO_2$$

These bacteria oxidize glucose-6-phosphate to ribulose-5-phosphate as in the HMP pathway before being converted to xylulose-5-phosphate by an epimerase. Phosphoketolase splits the pentose-5-phosphate to glyceraldehyde-3-phosphate and acetyl-phosphate. Acetyl-phosphate is reduced to ethanol to regenerate NAD^+ and pyruvate is produced from the triose-phosphate as in the latter part of the EMP pathway.

Since LAB have a restricted electron transport chain they use pyruvate and acetyl-phosphate as electron acceptors in the reactions catalyzed by lactate dehydrogenase, acetaldehyde dehydrogenase and alcohol dehydrogenase to regenerate NAD^+ from NADH (Section 8.4). For this reason *Leuconostoc mesenteroides* synthesizes one more ATP from pentoses than hexoses. In the hexose fermentation,

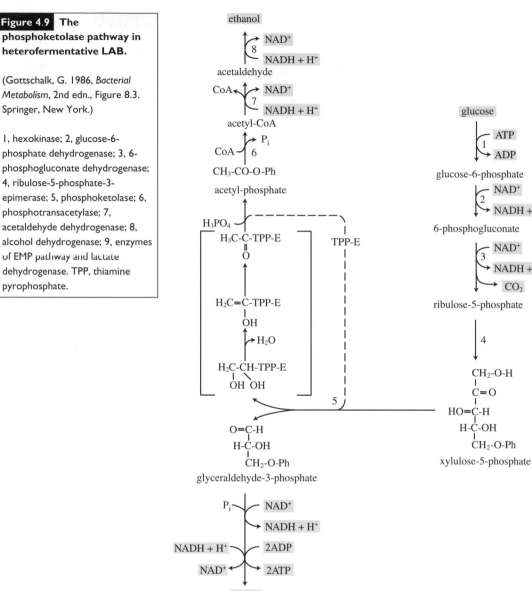

Figure 4.9 The phosphoketolase pathway in heterofermentative LAB.

(Gottschalk, G. 1986, *Bacterial Metabolism*, 2nd edn., Figure 8.3. Springer, New York.)

1, hexokinase; 2, glucose-6-phosphate dehydrogenase; 3, 6-phosphogluconate dehydrogenase; 4, ribulose-5-phosphate-3-epimerase; 5, phosphoketolase; 6, phosphotransacetylase; 7, acetaldehyde dehydrogenase; 8, alcohol dehydrogenase; 9, enzymes of EMP pathway and lactate dehydrogenase. TPP, thiamine pyrophosphate.

acetyl-phosphate is reduced to ethanol to oxidize NADH (reactions 7 and 8 in Figure 4.9), which is reduced by glucose-6-phosphate dehydrogenase (reaction 2 in Figure 4.9) and 6-phosphogluconate dehydrogenase (reaction 3 in Figure 4.9). Acetyl-phosphate is used to synthesize ATP in the reaction catalyzed by acetate kinase in pentose fermentation.

Thiamine pyrophosphate (TPP) is a prosthetic group in phosphoketolase. TPP binds glycoaldehyde, and the complex is dehydrated before being phosphorylated to acetyl-phosphate. PK in *Leuconostoc mesenteroides* is active on xylulose-5-phosphate, but not fructose-6-phosphate. Hexose fermentation results in the net gain of 1 ATP.

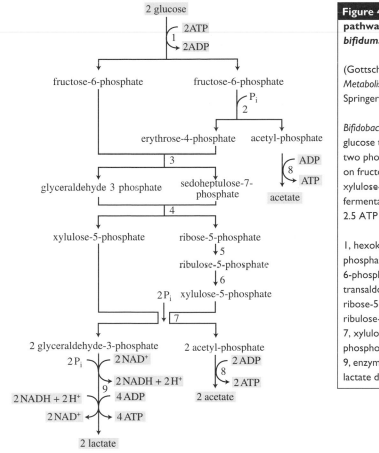

Figure 4.10 The bifidum pathway in *Bifidobacterium bifidum*.

(Gottschalk, G. 1986, *Bacterial Metabolism*, 2nd Edn., Figure 8.4. Springer, New York)

Bifidobacterium bifidum ferments glucose to lactate and acetate with two phosphoketolases each active on fructose-6-phosphate and xylulose-5-phosphate. This fermentation results in a net gain of 2.5 ATP per glucose fermented.

1, hexokinase and glucose-6-phosphate isomerase; 2, fructose-6-phosphate phosphoketolase; 3, transaldolase; 4, transketolase; 5, ribose-5-phosphate isomerase; 6, ribulose-5-phosphate-3-epimerase; 7, xylulose-5-phosphate phosphoketolase; 8, acetate kinase; 9, enzymes of the EMP pathway and lactate dehydrogenase.

4.5.2 Bifidum pathway

Bifidobacterium bifidum has two phosphoketolases, each active on fructose-6-phosphate and xylulose-5-phosphate, and ferments glucose in the bifidum pathway, which is different from that of *Leuconostoc mesenteroides*. As shown in Figure 4.10, this bacterium ferments 2 molecules of glucose to 2 molecules of lactate and 3 molecules of acetate. Fructose-6-phosphate phosphoketolase splits its substrate into erythrose-4-phosphate and acetyl-phosphate. Transketolase and transaldolase rearrange erythrose-4-phosphate and a second molecule of fructose-6-phosphate to 2 molecules of xylulose-5-phosphate as in the HMP pathway. Xylulose-5-phosphate is metabolized to lactate and acetate as in *Leuconostoc mesenteroides*. From the fermentation of 2 molecules of glucose, 7 ATP are synthesized and 2 ATP are consumed with a net gain of 2.5 ATP per glucose fermented.

Fructose-6-phosphate phosphoketolase activity is known only in bifidus bacteria apart from an opportunistic pathogen, *Gardnerella vaginalis*.

4.6 | Use of radiorespirometry to determine glycolytic pathways

In addition to the EMP pathway, prokaryotes metabolize glucose in the ED and PK pathways. The glycolytic pathway that occurs in a given organism can be determined through analysis of key enzyme activities and radioactive tracer experiments. For example, if the cell-free extract of an organism shows activities of 6-phosphogluconate dehydrogenase and 2-keto-3-deoxy-6-phosphogluconate aldolase but not that of phosphofructokinase, the given organism possesses the ED pathway.

A radiorespirometric approach is another method to determine the glycolytic pathway using ^{14}C- glucose labelled in various positions. The results of these experiments show not only what the main glycolytic pathway is but also quantify the proportion of glucose metabolized through the main glycolytic pathway and the HMP pathway. The basis of this method is that certain carbons of glucose are liberated earlier as CO_2 than the others depending on the pathway. For example, 1-position carbon is evolved first when glucose is metabolized through the HMP or PK pathways in the reaction catalyzed by 6-phosphogluconate dehydrogenase. In the EMP and ED pathways, the carboxylic group of pyruvate is converted to CO_2 by the oxidative decarboxylation of pyruvate (Section 5.1). The EMP pathway converts 3- and 4-position carbons to the carboxylic group of pyruvate, while 1- and 4-position carbons become carboxylic groups when glucose is metabolized through the ED pathway (Figure 4.11).

The amount of CO_2 liberated from cultures incubated with ^{14}C-glucose labelled at the 1, 2, or 3 and 4 carbons is determined by measuring the radioactivity collected in the alkaline CO_2-trapping solution. Figure 4.12 shows CO_2 evolution from cultures of *Escherichia coli* grown on glucose labelled at various positions. The culture incubated with glucose labelled at C3 and C4 releases CO_2 at an early stage, showing that this bacterium degrades glucose through the EMP

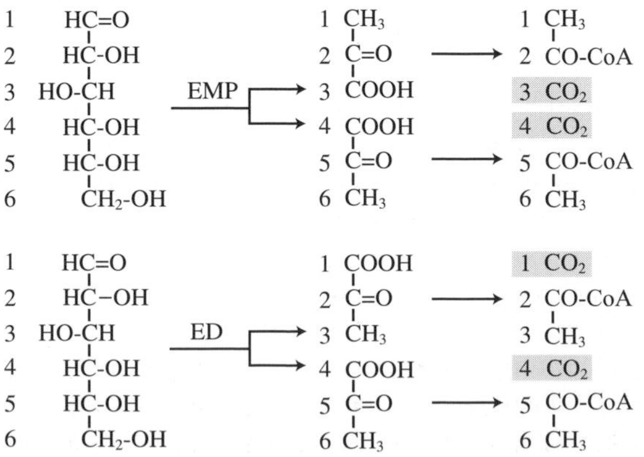

Figure 4.11 The origin of the carboxylic carbon of pyruvate from glucose when metabolized through the EMP and ED pathways.

(Gottschalk, G. 1986, *Bacterial Metabolism*, 2nd edn., Figure 5.10. Springer, New York)

Pyruvate dehydrogenase decarboxylates pyruvate to acetyl-CoA generating CO_2. The carboxylic group of pyruvate originates from the 3- and 4-position carbons of glucose when metabolized through the EMP pathway, and from the 1- and 4-position carbons in the ED pathway.

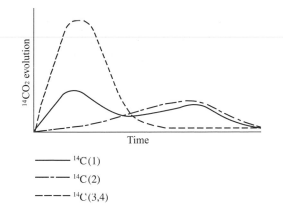

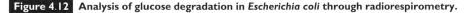

| Figure 4.12 | Analysis of glucose degradation in *Escherichia coli* through radiorespirometry. |

(Gottschalk, G. 1986, *Bacterial Metabolism*, 2nd edn., Figure 3.15. Springer, New York)

Through determination of the radioactivity of CO_2 evolved from cultures grown on $^{14}C(1)$, $^{14}C(2)$, or $^{14}C(3)$ and $^{14}C(4)$-labelled glucose, the glycolytic pathways can be characterized. CO_2 released at an early stage from $^{14}C(1)$ is due to the reaction catalyzed by 6-phosphogluconate dehydrogenase of the HMP pathway. CO_2 originating from pyruvate by the action of pyruvate dehydrogenase of the EMP pathway is from $^{14}C(3)$ and $^{14}C(4)$-labelled glucose. CO_2 from the oxidation of acetyl-CoA through the TCA cycle appears at the later stage from $^{14}C(1)$ and $^{14}C(2)$-labelled glucose.

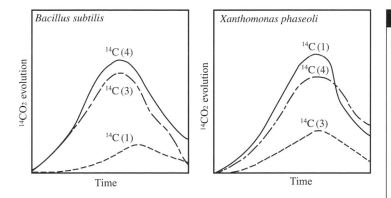

Figure 4.13 **Determination of glycolysis in *Bacillus subtilis* and *Xanthomonas phaseoli* through radiorespirometry.**

(Gottschalk, G. 1986, *Bacterial Metabolism*, 2nd edn., Figure 5.11. Springer, New York)

Bacillus subtilis degrades glucose through the EMP pathway, producing pyruvate with carboxylic groups originating from C3 and C4, which are released as CO_2 by the reaction catalyzed by pyruvate dehydrogenase. In contrast, the pyruvates produced by *Xanthomonas phaseoli* have carboxylic groups originating from C1 and C4.

pathway. On the other hand, CO_2 is released twice from C1-labelled glucose. The first release indicates the degradation of glucose through the HMP pathway, and the second release is from the TCA cycle. Through comparison of CO_2 released from the culture incubated with glucose labelled at C3 and C4 with that of C1-labelled glucose, the amount of glucose metabolized through the EMP and HMP pathways can be calculated as 72% and 28%, respectively (Section 4.3).

Results shown in Figure 4.13 are obtained from radioactivity measurements of CO_2 evolved from cultures of *Bacillus subtilis* and *Xanthomonas phaseoli* grown on $^{14}C(1)$, $^{14}C(3)$ and $^{14}C(4)$-labelled glucose. The former organism possesses the EMP pathway as its main glycolytic pathway: the ED pathway is the main glycolytic mechanism in the latter. *Bacillus subtilis* evolves CO_2 faster from $^{14}C(3)$ and

^{14}C(4)-labelled glucose than from ^{14}C(1)-glucose, while more CO_2 is evolved from ^{14}C(1) and ^{14}C(4)-labelled glucose than ^{14}C(3)-glucose by *Xanthomonas phaseoli*.

FURTHER READING

General

Bruckner, R. & Titgemeyer, F. (2002). Carbon catabolite repression in bacteria: choice of the carbon source and autoregulatory limitation of sugar utilization. *FEMS Microbiology Letters* **209**, 141–148.

Plantinga, T. H., van der Does, C. & Driessen, A. J. M. (2004). Transporter's evolution and carbohydrate metabolic clusters. *Trends in Microbiology* **12**, 4–7.

EMP and modified pathways

Asanuma, N. & Hino, T. (2006). Presence of NADP$^+$-specific glyceraldehyde-3-phosphate dehydrogenase and CcpA-dependent transcription of its gene in the ruminal bacterium *Streptococcus bovis*. *FEMS Microbiology Letters* **257**, 17–23.

Brunner, N. A., Siebers, B. & Hensel, R. (2001). Role of two different glyceraldehyde-3-phosphate dehydrogenases in controlling the reversible Embden-Meyerhof-Parnas pathway in *Thermoproteus tenax*: regulation on protein and transcript level. *Extremophiles* **5**, 101–109.

Hansen, T. & Schoenheit, P. (2001). Sequence, expression, and characterization of the first archaeal ATP-dependent 6-phosphofructokinase, a non-allosteric enzyme related to the phosphofructokinase-B sugar kinase family, from the hyperthermophilic crenarchaeote *Aeropyrum pernix*. *Archives of Microbiology* **177**, 62–69.

Hansen, T. & Schonheit, P. (2004). ADP-dependent 6-phosphofructokinase, an extremely thermophilic, non-allosteric enzyme from the hyperthermophilic, sulfate-reducing archaeon *Archaeoglobus fulgidus* strain 7324. *Extremophiles* **8**, 29–35.

Hansen, T., Reichstein, B., Schmid, R. & Schonheit, P. (2002). The first archaeal ATP-dependent glucokinase, from the hyperthermophilic crenarchaeon *Aeropyrum pernix*, represents a monomeric, extremely thermophilic ROK glucokinase with broad hexose specificity. *Journal of Bacteriology* **184**, 5955–5965.

Imanaka, H., Yamatsu, A., Fukui, T., Atomi, H. & Imanaka, T. (2006). Phosphoenolpyruvate synthase plays an essential role for glycolysis in the modified Embden-Meyerhof pathway in *Thermococcus kodakarensis*. *Molecular Microbiology* **61**, 898–909.

Kengen, S. W. M., Stams, A. J. M. & Devos, W. M. (1996). Sugar metabolism of hyperthermophiles. *FEMS Microbiology Reviews* **18**, 119–137.

Kim, J. W. & Dang, C. V. (2005). Multifaceted roles of glycolytic enzymes. *Trends in Biochemical Sciences* **30**, 142–150.

Labes, A. & Schonheit, P. (2001). Sugar utilization in the hyperthermophilic, sulfate-reducing archaeon *Archaeoglobus fulgidus* strain 7324: starch degradation to acetate and CO_2 via a modified Embden-Meyerhof pathway and acetyl-CoA synthetase (ADP-forming). *Archives of Microbiology* **176**, 329–338.

Labes, A. & Schonheit, P. (2003). ADP-dependent glucokinase from the hyperthermophilic sulfate-reducing archaeon *Archaeoglobus fulgidus* strain 7324. *Archives of Microbiology* **180**, 69–75.

Marsh, J. J. & Lebherz, H. G. (1992). Fructose-bisphosphate aldolases: an evolutionary history. *Trends in Biochemical Sciences* **17**, 110–113.

Nishimasu, H., Fushinobu, S., Shoun, H. & Wakagi, T. (2006). Identification and characterization of an ATP-dependent hexokinase with broad substrate specificity from the hyperthermophilic archaeon *Sulfolobus tokodaii*. *Journal of Bacteriology* **188**, 2014–2019.

Sakuraba, H. & Ohshima, T. (2002). Novel energy metabolism in anaerobic hyperthermophilic archaea: a modified Embden-Meyerhof pathway. *Journal of Bioscience and Bioengineering* **93**, 441–448.

Tanaka, S., Lee, S. O., Hamaoka, K., Kato, J., Takiguchi, N., Nakamura, K., Ohtake, H. & Kuroda, A. (2003). Strictly polyphosphate-dependent glucokinase in a polyphosphate-accumulating bacterium, *Microlunatus phosphovorus*. *Journal of Bacteriology* **185**, 5654–5656.

Tjaden, B., Plagens, A., Dorr, C., Siebers, B. & Hensel, R. (2006). Phosphoenolpyruvate synthetase and pyruvate, phosphate dikinase of *Thermoproteus tenax*: key pieces in the puzzle of archaeal carbohydrate metabolism. *Molecular Microbiology* **60**, 287–298.

Verhees, C. H., Tuininga, J. E., Kengen, S. W. M., Stams, A. J. M., van der Oost, J. & de Vos, W. M. (2001). ADP-dependent phosphofructokinases in mesophilic and thermophilic methanogenic archaea. *Journal of Bacteriology* **183**, 7145–7153.

Villamon, E., Villalba, V., Nogueras, M. M., Tomas, J. M., Gozalbo, D. & Gil, M. L. (2003). Glyceraldehyde-3-phosphate dehydrogenase, a glycolytic enzyme present in the periplasm of *Aeromonas hydrophila*. *Antonie van Leeuwenhoek* **84**, 31–38.

Zakharchuk, L. M., Egorova, M. A., Tsaplina, I. A., Bogdanova, T. I., Krasil'nikova, E. N., Melamud, V. S. & Karavaiko, G. I. (2003). Activity of the enzymes of carbon metabolism in *Sulfobacillus sibiricus* under various conditions of cultivation. *Microbiology-Moscow* **72**, 553–557.

Zheng, P., Sun, J., van den Heuvel, J. & Zeng, A. (2006). Discovery and investigation of a new, second triose phosphate isomerase in *Klebsiella pneumoniae*. *Journal of Biotechnology* **125**, 462–473.

Methylglyoxal bypass

Ferguson, G. P., Totemeyer, S., Maclean, M. J. & Booth, I. R. (1998). Methylglyoxal production in bacteria: suicide or survival. *Archives of Microbiology* **170**, 209–219.

Grant, A. W., Steel, G., Waugh, H. & Ellis, E. M. (2003). A novel aldo-keto reductase from *Escherichia coli* can increase resistance to methylglyoxal toxicity. *FEMS Microbiology Letters* **218**, 93–99.

Ko, J., Kim, I., Yoo, S., Min, B., Kim, K. & Park, C. (2005). Conversion of methylglyoxal to acetol by *Escherichia coli* aldo-keto reductases. *Journal of Bacteriology* **187**, 5782–5789.

Liyanage, H., Kashket, S., Young, M. & Kashket, E. R. (2001). *Clostridium beijerinckii* and *Clostridium difficile* detoxify methylglyoxal by a novel mechanism involving glycerol dehydrogenase. *Applied and Environmental Microbiology* **67**, 2004–2010.

Maiden, M. F. J., Pham, C. & Kashket, S. (2004). Glucose toxicity effect and accumulation of methylglyoxal by the periodontal anaerobe *Bacteroides forsythus*. *Anaerobe* **10**, 27–32.

Weber, J., Kayser, A. & Rinas, U. (2005). Metabolic flux analysis of *Escherichia coli* in glucose-limited continuous culture. II. Dynamic response to famine and feast, activation of the methylglyoxal pathway and oscillatory behaviour. *Microbiology-UK* **151**, 707–716.

Xu, D., Liu, X., Guo, C. & Zhao, J. (2006). Methylglyoxal detoxification by an aldo-keto reductase in the cyanobacterium *Synechococcus* sp. PCC 7002. *Microbiology-UK* **152**, 2013–2021.

Gluconeogenesis

Alves, A. M. C. R., Euverink, G. J. W., Santos, H. & Dijkhuizen, L. (2001). Different physiological roles of ATP- and PP$_i$-dependent phosphofructokinase isoenzymes in the methylotrophic actinomycete *Amycolatopsis methanolica*. *Journal of Bacteriology* **183**, 7231–7240.

Fillinger, S., Boschi-Muller, S., Azza, S., Dervyn, E., Branlant, G. & Aymerich, S. (2000). Two glyceraldehyde-3-phosphate dehydrogenases with opposite physiological roles in a nonphotosynthetic bacterium. *Journal of Biological Chemistry* **275**, 14031–14037.

Grochowski, L. L., Xu, H. & White, R. H. (2005). Ribose-5-phosphate biosynthesis in *Methanocaldococcus jannaschii* occurs in the absence of a pentose-phosphate pathway. *Journal of Bacteriology* **187**, 7382–7389.

Pernestig, A. K., Georgellis, D., Romeo, T., Suzuki, K., Tomenius, H., Normark, S. & Melefors, O. (2003). The *Escherichia coli* BarA-UvrY two-component system is needed for efficient switching between glycolytic and gluconeogenic carbon sources. *Journal of Bacteriology* **185**, 843–853.

Tjaden, B., Plagens, A., Dorr, C., Siebers, B. & Hensel, R. (2006). Phosphoenolpyruvate synthetase and pyruvate, phosphate dikinase of *Thermoproteus tenax*: key pieces in the puzzle of archaeal carbohydrate metabolism. *Molecular Microbiology* **60**, 287–298.

Verhees, C. H., Akerboom, J., Schiltz, E., de Vos, W. M. & van der Oost, J. (2002). Molecular and biochemical characterization of a distinct type of fructose-1,6-bisphosphatase from *Pyrococcus furiosus*. *Journal of Bacteriology* **184**, 3401–3405.

Wolfe, A. J. (2005). The acetate switch. *Microbiology and Molecular Biology Reviews* **69**, 12–50.

HMP pathway

Christensen, J., Christiansen, T., Gombert, A. K., Thykaer, J. & Nielsen, J. (2001). Simple and robust method for estimation of the split between the oxidative pentose phosphate pathway and the Embden-Meyerhof-Parnas pathway in microorganisms. *Biotechnology and Bioengineering* **74**, 517–523.

Gibson, J. L. & Tabita, F. R. (1996). The molecular regulation of the reductive pentose phosphate pathway in proteobacteria and cyanobacteria. *Archives in Microbiology* **166**, 141–150.

Orita, I., Sato, T., Yurimoto, H., Kato, N., Atomi, H., Imanaka, T. & Sakai, Y. (2006). The ribulose monophosphate pathway substitutes for the missing pentose phosphate pathway in the archaeon *Thermococcus kodakaraensis*. *Journal of Bacteriology* **188**, 4698–4704.

Takayama, S., McGarvey, G. J. & Wong, C. H. (1997). Microbial aldolases and transketolases: new biocatalytic approaches to simple and complex sugars. *Annual Review of Microbiology* **51**, 285–310.

ED and modified ED pathways

Conway, T. (1992). The Entner-Doudoroff pathway: history, physiology and molecular biology. *FEMS Microbiology Reviews* **103**, 1–28.

Egan, S. E., Fliege, R., Tong, S. X., Shibata, A., Wolf, R. E. & Conway, T. (1992). Molecular characterization of the Entner-Doudoroff pathway in *Escherichia coli* – sequence analysis and localization of promoters for the *edd-eda* operon. *Journal of Bacteriology* **174**, 4638–4646.

Fuhrer, T., Fischer, E. & Sauer, U. (2005). Experimental identification and quantification of glucose metabolism in seven bacterial species. *Journal of Bacteriology* **187**, 1581–1590.

Garnova, E. S. & Krasil'nikova, E. N. (2003). Carbohydrate metabolism of the saccharolytic alkaliphilic anaerobes *Halonatronum saccharophilum*, *Amphibacillus fermentum*, and *Amphibacillus tropicus*. *Microbiology-Moscow* **72**, 558–563.

Gunnarsson, N., Mortensen, U. H., Sosio, M. & Nielsen, J. (2004). Identification of the Entner-Doudoroff pathway in an antibiotic-producing actinomycete species. *Molecular Microbiology* **52**, 895–902.

Hong, S. H., Park, S. J., Moon, S. Y., Park, J. P. & Lee, S. Y. (2003). *In silico* prediction and validation of the importance of the Entner-Doudoroff pathway in poly(3-hydroxybutyrate) production by metabolically engineered *Escherichia coli*. *Biotechnology and Bioengineering* **83**, 854–863.

Johnsen, U., Selig, M., Xavier, K. B., Santos, H. & Schonheit, P. (2001). Different glycolytic pathways for glucose and fructose in the halophilic archaeon *Halococcus saccharolyticus*. *Archives of Microbiology* **175**, 52–61.

Murray, E. L. & Conway, T. (2005). Multiple regulators control expression of the Entner-Doudoroff aldolase (Eda) of *Escherichia coli*. *Journal of Bacteriology* **187**, 991–1000.

Oren, A. & Mana, L. (2003). Sugar metabolism in the extremely halophilic bacterium *Salinibacter ruber*. *FEMS Microbiology Letters* **223**, 83–87.

Peekhaus, N. & Conway, T. (1998). What's for dinner – Entner-Doudoroff metabolism in *Escherichia coli*. *Journal of Bacteriology* **180**, 3495–3502.

Pruss, B. M., Campbell, J. W., van Dyk, T. K., Zhu, C., Kogan, Y. & Matsumura, P. (2003). FlhD/FlhC is a regulator of anaerobic respiration and the Entner-Doudoroff pathway through induction of the methyl-accepting chemotaxis protein Aer. *Journal of Bacteriology* **185**, 534–543.

Reher, M., Bott, M. & Schonheit, P. (2006). Characterization of glycerate kinase (2-phosphoglycerate forming), a key enzyme of the nonphosphorylative Entner-Doudoroff pathway, from the thermoacidophilic euryarchaeon *Picrophilus torridus*. *FEMS Microbiology Letters* **259**, 113–119.

Siebers, B., Tjaden, B., Michalke, K., Dorr, C., Ahmed, H., Zaparty, M., Gordon, P., Sensen, C. W., Zibat, A., Klenk, H.-P., Schuster, S. C. & Hensel, R. (2004). Reconstruction of the central carbohydrate metabolism of *Thermoproteus tenax* by use of genomic and biochemical data. *Journal of Bacteriology* **186**, 2179–2194.

PK pathways

Posthuma, C. C., Bader, R., Engelmann, R., Postma, P. W., Hengstenberg, W. & Pouwels, P. H. (2002). Expression of the xylulose 5-phosphate phosphoketolase gene, *xpkA*, from *Lactobacillus pentosus* MD363 is induced by sugars that are fermented via the phosphoketolase pathway and is repressed by glucose mediated by CcpA and the mannose phosphoenolpyruvate phosphotransferase system. *Applied and Environmental Microbiology* **68**, 831–837.

Richter, H., Hamann, I. & Unden, G. (2003). Use of the mannitol pathway in fructose fermentation of *Oenococcus oeni* due to limiting redox regeneration capacity of the ethanol pathway. *Archives of Microbiology* **179**, 227–233.

Metabolic analysis

Cocaign-Bousquet, M., Even, S., Lindley, N. D. & Loubiere, P. (2002). Anaerobic sugar catabolism in *Lactococcus lactis*: genetic regulation and enzyme control over pathway flux. *Applied Microbiology and Biotechnology* **60**, 24–32.

de Vos, W. M., Kengen, S. W., Voorhorst, W. G. & van der Oost, J. (1998). Sugar utilization and its control in hyperthermophiles. *Extremophiles* **2**, 201–205.

Klamt, S. & Stelling, J. (2003). Two approaches for metabolic pathway analysis? *Trends in Biotechnology* **21**, 64–69.

Portais, J. C. & Delort, A. M. (2002). Carbohydrate cycling in microorganisms: what can ^{13}C-NMR tell us? *FEMS Microbiology Reviews* **26**, 375–402.

Romano, A. H. & Conway, T. (1996). Evolution of carbohydrate metabolic pathways. *Research in Microbiology* **147**, 448–455.

Schilling, C. H., Schuster, S., Palsson, B. O. & Heinrich, R. (1999). Metabolic pathway analysis: basic concepts and scientific applications in the postgenomic era. *Biotechnology Progress* **15**, 296–303.

Siebers, B. & Schonheit, P. (2005). Unusual pathways and enzymes of central carbohydrate metabolism in archaea. *Current Opinion in Microbiology* **8**, 695–705.

Tricarboxylic acid (TCA) cycle, electron transport and oxidative phosphorylation

Pyruvate produced from glycolysis and other metabolic pathways is metabolized in various ways depending on the organism and growth conditions. Pyruvate is either used as a precursor for biosynthesis, or is oxidized completely to CO_2 under aerobic conditions. This chapter is devoted to the mechanisms of pyruvate oxidation, electron transport and oxidative phosphorylation.

5.1 | Oxidative decarboxylation of pyruvate

Pyruvate is oxidized by the pyruvate dehydrogenase complex to acetyl-CoA and CO_2 reducing NAD^+ under aerobic conditions. The pyruvate dehydrogenase complex consists of 24 molecules of pyruvate dehydrogenase containing thiamine pyrophosphate (TPP), 24 molecules of dihydrolipoate acetyltransferase containing dihydrolipoate and 12 molecules of dihydrolipoate dehydrogenase containing flavin adenine dinucleotide (FAD). In addition, NAD^+ and coenzyme A participate in the reaction (Figure 5.1). This reaction is irreversible, and takes place in the mitochondrion in eukaryotic cells. The reaction can be summarized as:

$$\text{pyruvate} + \text{CoA-SH} + NAD^+ \longrightarrow \text{acetyl-CoA} + \text{NADH} + H^+$$
$$(\Delta G^{0'} = -33.5 \text{ kJ/mol pyruvate})$$

The pyruvate dehydrogenase complex shares its properties with other 2-keto acid dehydrogenase complexes that produce acyl-CoA such as 2-ketoglutarate dehydrogenase (reaction 4 in Figure 5.2) and 2-ketobutyrate dehydrogenase (Figure 7.14, Section 7.5.6), but their activities are controlled differently.

Pyruvate dehydrogenase activity is controlled by its substrate, products and the adenylate energy charge. Pyruvate and AMP activate enzyme activity, and acetyl-CoA, NADH and ATP repress it. Pyruvate and acetyl-CoA are precursors for amino acid (Section 6.4.1) and fatty acid (Section 6.6.1) biosynthesis, respectively.

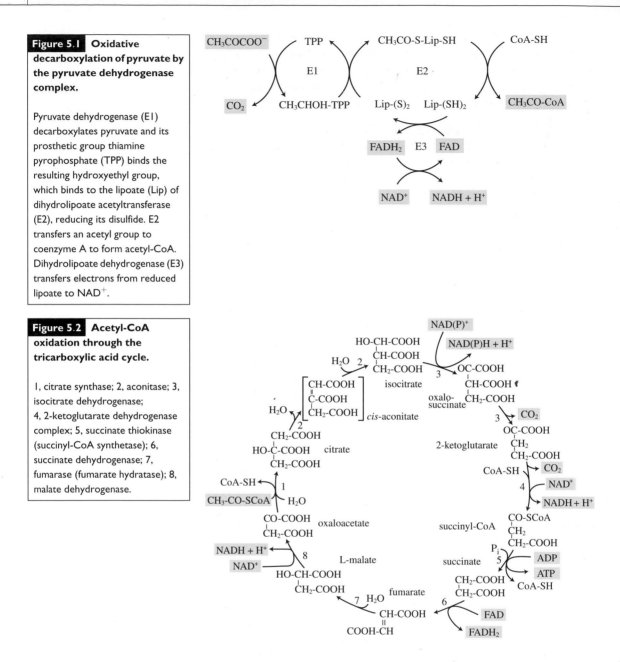

Figure 5.1 Oxidative decarboxylation of pyruvate by the pyruvate dehydrogenase complex.

Pyruvate dehydrogenase (E1) decarboxylates pyruvate and its prosthetic group thiamine pyrophosphate (TPP) binds the resulting hydroxyethyl group, which binds to the lipoate (Lip) of dihydrolipoate acetyltransferase (E2), reducing its disulfide. E2 transfers an acetyl group to coenzyme A to form acetyl-CoA. Dihydrolipoate dehydrogenase (E3) transfers electrons from reduced lipoate to NAD^+.

Figure 5.2 Acetyl-CoA oxidation through the tricarboxylic acid cycle.

1, citrate synthase; 2, aconitase; 3, isocitrate dehydrogenase; 4, 2-ketoglutarate dehydrogenase complex; 5, succinate thiokinase (succinyl-CoA synthetase); 6, succinate dehydrogenase; 7, fumarase (fumarate hydratase); 8, malate dehydrogenase.

5.2 | Tricarboxylic acid (TCA) cycle

This cyclic metabolic pathway was discovered by Krebs and his colleagues in animal tissue. It is referred to as the tricarboxylic acid (TCA) cycle, Krebs cycle or citric acid cycle. Acetyl-CoA produced by the pyruvate dehydrogenase complex is completely oxidized to CO_2, reducing NAD^+, $NADP^+$ and FAD. These reduced electron carriers are oxidized through the processes of electron transport and oxidative phosphorylation to form the proton motive force and to synthesize ATP.

The TCA cycle provides not only reducing equivalents for ATP synthesis but also precursors for biosynthesis. The TCA cycle or related metabolic processes are indispensable, providing biosynthetic precursors for all forms of cells except mycoplasmas which take required materials from their host animal cells.

5.2.1 Citrate synthesis and the TCA cycle

Citrate synthase synthesizes citrate from acetyl-CoA and oxaloacetate (Figure 5.2). This is an exergonic reaction ($\Delta G^{0'} = -32.2$ kJ/mol acetyl-CoA) and irreversible. The reverse reaction is catalyzed by a separate enzyme, ATP:citrate lyase, in the reductive TCA cycle (Section 5.4.2). Citrate is converted to isocitrate catalyzed by aconitase. Isocitrate is oxidized to 2-ketoglutarate by isocitrate dehydrogenase. In most bacteria this enzyme is $NADP^+$ dependent, but two separate enzymes are found in eukaryotes using $NADP^+$ or NAD^+.

The 2-ketoglutarate dehydrogenase complex oxidizes its substrate to succinyl-CoA. As with the pyruvate dehydrogenase complex, this enzyme complex consists of many peptides and cofactors, and catalyzes oxidative decarboxylation producing acyl-CoA. This is another irreversible reaction in the TCA cycle. The reverse reaction is catalyzed by 2-ketoglutarate synthase (2-ketoglutarate:ferredoxin oxidoreductase) in the reductive TCA cycle to fix CO_2 (Section 5.4.2). Some anaerobic fermentative bacteria do not have this enzyme. They supply the precursors for biosynthesis through the incomplete TCA fork (Section 5.4.1). Glutamate and related amino acids are synthesized from 2-ketoglutarate (Section 6.4.3). Succinyl-CoA is the precursor for porphyrin synthesis which provides the chemical nucleus of cytochromes and chlorophyll (Section 6.7). Both carbons of acetyl-CoA are liberated as CO_2 in the two reduction reactions. Like all acyl-CoA derivatives, succinyl-CoA has a high energy linkage. This energy is conserved as ATP through the succinate thiokinase (succinyl-CoA synthetase) reaction producing succinate. This is an example of substrate-level phosphorylation. Guanosine triphosphate (GTP) is synthesized in the mitochondrion by these reactions in eukaryotic cells.

Succinate is oxidized to fumarate by succinate dehydrogenase. Since the redox potential of fumarate/succinate (-0.03 V) is considerably higher than $NAD^+/NADH$ (-0.32 V), $NAD(P)^+$ cannot be reduced in this reaction. The prosthetic group of succinate dehydrogenase, FAD, is reduced. Electrons of the reduced succinate dehydrogenase are transferred to coenzyme Q of the electron transport chain (Section 5.8.2). Fumarate is hydrated to malate by fumarase before being reduced to oxaloacetate by malate dehydrogenase reducing NAD^+. Oxaloacetate is then ready to accept acetyl-CoA for the next round of the cycle. Oxaloacetate is used to synthesize amino acids (Section 6.4.1), and decarboxylated to phosphoenolpyruvate (PEP) in gluconeogenesis (Section 4.2.1). The TCA cycle can be summarized as:

$$CH_3-CO-CoA + 3NAD(P)^+ + FAD + ADP + P_i + 3H_2O \longrightarrow$$
$$2CO_2 + 3NAD(P)H + FADH_2 + ATP + 3H^+ + CoA\text{-}SH$$

The reduced electron carriers channel electrons to the electron transport chain to synthesize ATP through the proton motive force (Section 5.8).

5.2.2 Regulation of the TCA cycle

The TCA cycle is an amphibolic pathway serving anabolic needs by producing ATP as well as catabolic needs by providing precursors for biosynthesis. Consequently, this metabolism is regulated by the energy status of the cell and the availability of biosynthetic precursors. In addition, oxygen regulates the TCA cycle since the reduced electron carriers are recycled, consuming oxygen as the electron acceptor.

Oxygen controls the expression of genes for TCA cycle enzymes. Facultative anaerobes do not synthesize 2-ketoglutarate dehydrogenase under anaerobic conditions without alternative electron acceptors such as nitrate. The activity is lower with nitrate than with oxygen as the electron acceptor. In Gram-negative bacteria, including *Escherichia coli*, the regulatory proteins FNR and Arc regulate the transcription of many genes for aerobic and anaerobic metabolism (Section 12.2.4). A FNR protein with a similar function is also known in Gram-positive *Bacillus subtilis*.

Citrate synthase is regulated to control the TCA cycle. Generally this enzyme is repressed with the accumulation of NADH and ATP or 2-ketoglutarate. This accumulation means that the cell has enough energy and precursors for biosynthesis. Gram-negative bacteria have two different citrate synthase enzymes, one repressed by NADH and the other unaffected. Gram-positive bacteria have only one enzyme and this is not repressed by NADH. Instead, ATP inhibits this enzyme. AMP activates the citrate synthase inhibited by NADH in some bacteria.

5.3 | Replenishment of TCA cycle intermediates

Some intermediates of the TCA cycle serve as precursors for biosynthesis. For efficient operation of this cyclic metabolism, the intermediates used for biosynthesis should be replenished otherwise the concentration of oxaloacetate would be too low to start the TCA cycle. Oxaloacetate is replenished through a process called the anaplerotic sequence (Figure 5.3).

5.3.1 Anaplerotic sequence

Bacteria growing on carbohydrates synthesize oxaloacetate from pyruvate or phosphoenolpyruvate (PEP) as shown in Figure 5.3.

Many organisms, from bacteria to mammals, carboxylate pyruvate to oxaloacetate and this is catalyzed by pyruvate carboxylase consuming ATP:

$$\text{pyruvate} + \text{HCO}_3^- + \text{ATP} \xrightleftharpoons{\text{pyruvate carboxylase}} \text{oxaloacetate} + \text{ADP} + \text{P}_i$$

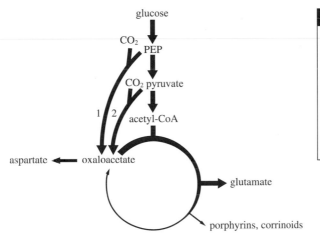

Figure 5.3 **Anaplerotic sequence in bacteria growing on carbohydrates.**

(Gottschalk, G. 1986, *Bacterial Metabolism*, 2nd edn., Figure 3.45. Springer, New York)

1, PEP carboxylase; 2, pyruvate carboxylase.

This enzyme needs biotin. Acetyl-CoA activates this enzyme in many bacteria, as in animals, but some bacteria such as *Pseudomonas aeruginosa* have a pyruvate carboxylase that is not activated by acetyl-CoA.

A PEP carboxylase mutant of *Escherichia coli* is unable to grow in a glucose–mineral salts medium, but can grow when supplemented with TCA cycle intermediates. This bacterium has PEP carboxylase as the anaplerotic sequence, not pyruvate carboxylase. This property is shared by many other bacteria including *Bacillus anthracis*, *Thiobacillus novellus*, *Acetobacter xylinum* and *Azotobacter vinelandii*.

$$\text{PEP} + \text{HCO}_3^- \xrightleftharpoons{\text{PEP carboxylase}} \text{oxaloacetate} + \text{P}_i$$

5.3.2 Glyoxylate cycle

Bacteria growing on carbon sources that are not metabolized through pyruvate or PEP cannot replenish TCA cycle intermediates through the anaplerotic sequence, and need even more oxaloacetate to produce PEP for gluconeogenesis (Section 4.2.1). They therefore need another mechanism, the glyoxylate cycle, for this purpose.

Escherichia coli growing on acetate synthesizes isocitrate lyase and malate synthase for a functional glyoxylate cycle (Figure 5.4). These enzymes convert two molecules of acetyl-CoA to a molecule of malate in conjunction with TCA cycle enzymes. Acetyl-CoA is converted to isocitrate through the TCA cycle, and isocitrate lyase cleaves isocitrate to succinate and glyoxylate:

$$\text{isocitrate} \xrightarrow{\text{isocitrate lyase}} \text{succinate} + \text{glyoxylate}$$

Succinate is oxidized to oxaloacetate through the TCA cycle, and glyoxylate is used in the synthesis of malate by malate synthase with a second molecule of acetyl-CoA:

$$\text{glyoxylate} + \text{acetyl-CoA} \xrightarrow{\text{malate synthase}} \text{malate} + \text{CoA-SH}$$

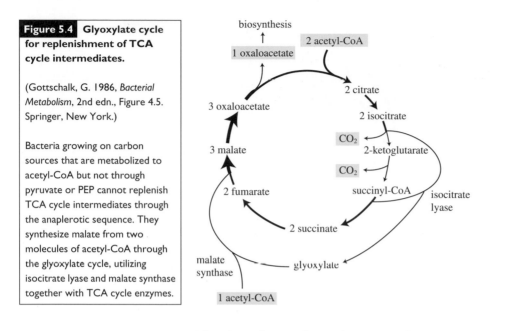

Figure 5.4 **Glyoxylate cycle for replenishment of TCA cycle intermediates.**

(Gottschalk, G. 1986, *Bacterial Metabolism*, 2nd edn., Figure 4.5. Springer, New York.)

Bacteria growing on carbon sources that are metabolized to acetyl-CoA but not through pyruvate or PEP cannot replenish TCA cycle intermediates through the anaplerotic sequence. They synthesize malate from two molecules of acetyl-CoA through the glyoxylate cycle, utilizing isocitrate lyase and malate synthase together with TCA cycle enzymes.

The glyoxylate cycle can be summarized as:

$$2CH_3\text{-CO-CoA} + 3H_2O + FAD \longrightarrow C_4H_6O_5 + FADH_2 + 2CoA\text{-SH}$$

5.3.2.1 Regulation of the glyoxylate cycle

When an organism is growing on carbon sources that are not metabolized through pyruvate or PEP, the TCA cycle supplies energy while the glyoxylate cycle supplies precursors for biosynthesis. Genes for isocitrate lyase and malate synthase are transcribed with the accumulation of acetyl-CoA. The Cra (catabolite repressor/activator) protein is involved in this regulation in *Escherichia coli* (Section 12.1.3.2). The activity of isocitrate lyase is inhibited by PEP, succinate and pyruvate.

Since isocitrate is a branch point of the TCA cycle and the glyoxylate cycle, the activities of isocitrate lyase and isocitrate dehydrogenase which act on this common substrate should be regulated to control the flux. The bacteria solve the problem through differences in affinity for the substrate and by controlling the activity of the enzyme with the higher affinity. The dehydrogenase has a much higher affinity ($K_m = 1-2\,\mu M$) for the substrate than that of the lyase ($K_m = 3\,mM$). When the TCA cycle is needed to generate energy, isocitrate dehydrogenase is activated, but this enzyme is inactivated when precursors for biosynthesis should be synthesized through the glyoxylate cycle. An enzyme with kinase-phosphatase activity controls the activity of isocitrate dehydrogenase (Figure 5.5). The kinase-phosphatase removes phosphate from the inactive phosphorylated isocitrate dehydrogenase to induce flux through the TCA cycle when metabolic intermediates such as isocitrate, PEP, OAA, 2-KG and 3-phosphoglycerate are in high concentration, and when more ATP is needed with the accumulation of AMP and ADP. Under the opposite conditions, the kinase-phosphatase phosphorylates the enzyme protein to inactivate it. When NADPH accumulates, isocitrate is directed toward the glyoxylate cycle. In

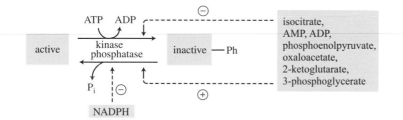

Figure 5.5 | Control of isocitrate dehydrogenase activity by a kinase-phosphatase.

(Dawes, E. A. 1986, *Microbial Energetics*, Figure 3.10. Blackie & Son, Glasgow.)

Since isocitrate dehydrogenase has a much higher affinity for the common substrate than the isocitrate lyase of the glyoxylate cycle, the activity of the former is inactivated when the cell needs precursors for biosynthesis. The enzyme is in an active form when precursors accumulate. Under the opposite conditions where the precursor concentration is high and ATP synthesis is required with the accumulation of AMP and ADP, the enzyme is in an active form to direct the flux to the TCA cycle. A kinase/phosphatase interchanges active and inactive forms of isocitrate dehydrogenase. The phosphorylated enzyme protein is inactive, and the free enzyme is active. When metabolic intermediates and AMP/ADP accumulate, isocitrate dehydrogenase is dephosphorylated for proper functioning of the TCA cycle. The kinase/phosphatase gene forms a single operon with genes for the glyoxylate cycle enzymes, isocitrate lyase and malate synthase.

Escherichia coli, genes for the kinase-phosphatase are located in the same operon with the genes for isocitrate lyase and malate synthase.

The TCA cycle is controlled by citrate synthase activity, and isocitrate dehydrogenase activity is regulated to control the glyoxylate cycle.

5.4 | Incomplete TCA fork and reductive TCA cycle

The TCA cycle is an amphibolic metabolism serving catabolic and anabolic needs. Parts of the TCA cycle, known as the incomplete TCA fork, are maintained in fermentative cells which cannot use NADH to produce ATP through oxidative phosphorylation in order to obtain precursors for biosynthesis. Cyanobacteria that synthesize ATP through photosynthetic electron transport (Section 11.4.3) do so in a similar fashion. In some chemolithotrophs, CO_2 is fixed through the reductive TCA cycle (Section 10.8.2).

5.4.1 Incomplete TCA fork
Some bacteria do not have a functional TCA cycle under certain growth conditions. Enteric bacteria, including *Escherichia coli*, do not synthesize 2-ketoglutarate dehydrogenase under fermentative conditions (Figure 12.5, Section 12.2.4) since NADH cannot be recycled through oxidative phosphorylation under these conditions. Precursors supplied by the TCA cycle are obtained through the incomplete TCA fork consisting of an oxidative fork to produce 2-ketoglutarate and a reductive fork to synthesize succinyl-CoA (Figure 5.6).

Since succinate dehydrogenase cannot reduce fumarate to succinate, fumarate reductase replaces it in the incomplete TCA fork.

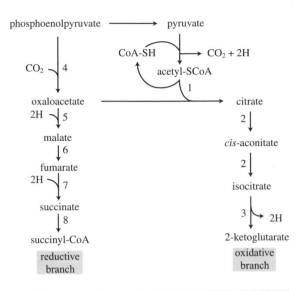

Figure 5.6 Incomplete TCA fork to supply precursors for biosynthesis in organisms that do not synthesize 2-ketoglutarate dehydrogenase.

(Dawes, E. A. 1986, *Microbial Energetics*, Figure 3.8. Blackie & Son, Glasgow)

The TCA cycle is an amphibolic metabolism generating reducing equivalents to synthesize ATP through oxidative phosphorylation as the catabolic function and biosynthetic precursors as the anabolic function. Fermentative anaerobic bacteria are unable to recycle NADH through oxidative phosphorylation under electron acceptor-limited conditions. They do not have 2-ketoglutarate dehydrogenase activity and operate an incomplete TCA fork instead of a functional TCA cycle to supply biosynthetic precursors. Cyanobacteria utilize light energy to synthesize ATP under photosynthetic conditions. They also obtain their precursors through the incomplete TCA fork.

1, pyruvate dehydrogenase and citrate synthase; 2, aconitase; 3, isocitrate dehydrogenase; 4, PEP carboxylase; 5, malate dehydrogenase; 6, fumarase; 7, fumarate reductase; 8, succinyl-CoA synthetase.

Fumarate reductase is one of the anaerobic enzymes expressed under anaerobic conditions in *Escherichia coli* under control by the regulatory protein FNR. The activities of the other TCA cycle enzymes are lower under anaerobic conditions than in aerobic conditions (Section 12.2.4, Table 12.5).

5.4.2 Reductive TCA cycle

The Calvin cycle is employed in most chemolithotrophs to fix CO_2 (Section 10.8.1) but enzymes of the Calvin cycle are not found in some chemolithotrophs, including photosynthetic green sulfur bacteria, a chemolithotrophic bacterium (*Hydrogenobacter thermophilus*) and an archaeon (*Sulfolobus acidocaldarius*). These fix CO_2 through a reversal of the TCA cycle which is referred to as the reductive TCA cycle (Section 10.8.2). As mentioned above, three TCA cycle enzymes cannot catalyze the reverse reactions and separate enzymes substitute for them:

- ATP-citrate lyase: substitutes for citrate synthase
- 2-ketoglutarate: ferredoxin oxidoreductase: substitutes for 2-ketoglutarate dehydrogenase
- fumarate reductase: substitutes for succinate dehydrogenase.

This mechanism of CO_2 fixation is known only in prokaryotes. Other CO_2 fixation pathways known only in prokaryotes are the acetyl-CoA pathway (carbon monoxide dehydrogenase pathway) and the 3-hydroxypropionate pathway. These will be discussed in detail later (Section 10.8).

5.5 | Energy transduction in prokaryotes

Life can be thought of as a process that transforms materials available from the environment into cellular components according to genetic information. Material transformation is also coupled to energy transduction. Energy is needed not only for growth and reproduction but also for the maintenance of viability in processes that include biosynthesis, transport, motility, and many others.

Organisms use energy sources available in their environment. Light and chemical energies are converted into the biological energy for growth and maintenance of viability. Photosynthesis (Chapter 11) is the process where light energy is utilized, and chemical energy is used through fermentation (Chapter 8) and respiration (Section 5.8 and Chapter 9). Organic compounds produced by photosynthesis are used by other organisms in fermentation and respiration. For this reason, photosynthesis is referred to as primary production. Reduced inorganic compounds are also used as energy sources in chemolithotrophs.

Figure 5.7 shows energy transduction processes in biological systems. In these processes, free energy is conserved in the exergonic (free energy producing) reactions and consumed in the endergonic (free energy consuming) reactions. To understand these biological reactions in terms of thermodynamics, the relationship between the biological reactions and the free energy change should be understood.

5.5.1 Free energy
The free energy change in an exergonic reaction is expressed as a negative figure, and a positive figure is used to describe an endergonic

Figure 5.7 **Biological energy transduction processes.**

Photosynthesis: a process converting light energy into biological energy which in turn is consumed to fix CO_2 into organic compounds. Fermentation: a process converting chemical energy into biological energy without the external supply of electron acceptors. Respiration: a process converting chemical energy into biological energy by oxidizing organic and inorganic electron donors coupled with the reduction of externally supplied electron acceptors.

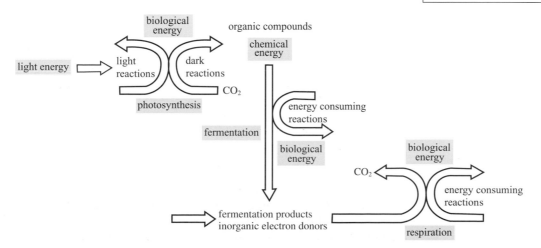

reaction, since free energy leaves the system in an exergonic reaction while the system gains free energy in an endergonic reaction. The free energy change depends on the conditions of a given reaction. The standard condition is defined as the concentration of the reactants and products in one activity unit (when all are in active forms, a 1 M concentration is the same as one activity unit) and at 25 °C for convenience. The free energy change at standard conditions is expressed as ΔG^0, and $\Delta G^{0'}$ and ΔG are used to describe the free energy changes under standard conditions at pH$=7$, and at the given conditions, respectively. Since physiological pH is neutral, $\Delta G^{0'}$ is a frequently used term in biology. $\Delta G^{0'}$ can be calculated in various ways.

5.5.1.1 $\Delta G^{0'}$ from the free energy of formation

The free energy of formation ($\Delta G_f^{0'}$) of common compounds can be found in most chemical data handbooks. $\Delta G^{0'}$ is calculated from $\Delta G_f^{0'}$ using the following equation:

$$\Delta G^{0'} = \Sigma \Delta G_f^{0'} \text{ of products} - \Sigma \Delta G_f^{0'} \text{ of reactants}$$

For example, $\Delta G^{0'}$ is calculated in the reaction of glucose oxidation as:

$$C_6H_{12}O_6 + 6O_2 \longrightarrow 6CO_2 + 6H_2O$$

Where the free energy of formation of each component is:

$$\Delta G_f^{0'} \text{ glucose} = -917.22 \text{ kJ}$$
$$\Delta G_f^{0'} \text{ O}_2 = 0 \text{ kJ}$$
$$\Delta G_f^{0'} \text{ H}_2\text{O} = -237.18 \text{ kJ}$$
$$\Delta G_f^{0'} \text{ CO}_2 = -386.02 \text{ kJ}$$

$$\Delta G^{0'} = [(-237.18 \times 6) + (-386.02 \times 6)] - (-917.22)$$
$$= -2821.98 \text{ kJ/mol glucose}$$

5.5.1.2 $\Delta G^{0'}$ from the equilibrium constant

The equilibrium constant (K'_{eq}) of a reaction, A+B $\rightleftharpoons$ C+D, is expressed as:

$$K'_{eq} = ([C][D])/([A][B])$$

K'_{eq} shows in which direction the reaction takes place at pH$=7.0$ under standard conditions:

When $K'_{eq} > 1.0$, $\Delta G^{0'} < 0$ and the reaction proceeds in the right direction.

When $K'_{eq} = 1.0$, $\Delta G^{0'} = 0$ and the velocity is the same in both directions.

When $K'_{eq} < 1.0$, $\Delta G^{0'} > 0$ and the reaction proceeds in the reverse direction.

The equilibrium constant can be used to calculate $\Delta G^{0'}$ using the following equation:

$$\Delta G^{0\prime} = -2.303\,RT\log K'_{eq}$$

R: gas constant (8.314 J/mol·K)
T: temperature in K (25 °C = 298 K).

5.5.1.3 ΔG from $\Delta G^{0\prime}$

ΔG^0 and $\Delta G^{0\prime}$ express free energy at reactant and product concentrations of one activity unit (=1 M). However, the concentrations inside a cell are much lower than that. ΔG at a given concentration of reactants and products can be calculated from $\Delta G^{0\prime}$ in a reaction of A+B $\rightleftharpoons$ C+D using:

$$\Delta G = \Delta G^{0\prime} + 2.303\,RT\log([C][D]/[A][B])$$

Assuming that the concentration of ATP, ADP and P_i are 2.25, 0.25 and 1.65 mM, respectively, ΔG of ATP hydrolysis to ADP and P_i can be calculated from $\Delta G^{0\prime} = -30.5$ kJ/mol ATP:

$$\Delta G = -30.5 + (8.314 \times 298 \times 2.303)$$
$$\times \log[2.5 \times 10^{-4} \times 1.65 \times 10^{-3}]/[2.25 \times 10^{-3}])$$
$$= -51.8\,\text{kJ/mol ATP}$$

The figures used in the calculation are close to those values found in an actively growing bacterial culture. The free energy change of ATP hydrolysis under physiological conditions is referred to as the phosphorylation potential and is expressed as ΔGp (Section 5.6.3).

5.5.1.4 $\Delta G^{0\prime}$ from ΔG^0

ΔG^0 of a reaction where H^+ is involved is the free energy change at a H^+ concentration of 1M (pH = 0). Biologists are more interested in $\Delta G^{0\prime}$ than ΔG^0. $\Delta G^{0\prime}$ can be calculated from ΔG^0 using:

$$\Delta G^{0\prime} = \Delta G^0 - 2.303\,RT \times 7$$

5.5.2 Free energy of an oxidation/reduction reaction

Energy is generated from an oxidation/reduction reaction. Respiration is a series of oxidation/reduction reactions. The energy from respiratory oxidation/reduction reactions is conserved in biological systems. The amount of energy generated from a reaction is proportional to the difference in the oxidation/reduction potential of the reductant and oxidant.

5.5.2.1 Oxidation/reduction potential

Oxidation is defined as a reaction which loses electron(s) and reduction as a reaction that gains electron(s). Since an electron cannot 'float' in solution, oxidation and reduction reactions are coupled. A given compound can be an oxidant in one reaction and a reductant in another reaction. This property depends on the affinity of the compound for electrons, which is relative to that of other compounds. The affinity for

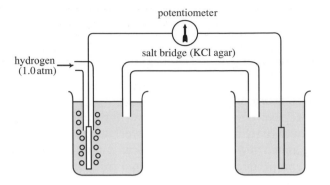

Figure 5.8 **Determination of redox potential.**

A vessel is filled with a solution containing 1M of each of the reduced and oxidized forms of a compound of known redox potential, and the other vessel with 1M of each of reduced and oxidized forms of the test compound. A platinum electrode is placed in each vessel. These two vessels are connected with a KCl salt bridge, and the electrodes with a potentiometer. Due to the differences in the tendency to transfer electrons to the platinum electrode in each vessel, a potential is developed. The potentiometer gives the difference in redox potential of the compounds. A solution of 1M H^+ (oxidized form) gassed with 1 atm H_2 (reduced form) is arbitrarily defined as a reaction that has a redox potential of 0 V.

electrons is referred to as the oxidation/reduction (redox) potential. The higher the affinity for electrons, the higher the redox potential.

Arbitrarily, the redox potential of the half reaction, $\frac{1}{2}H_2 \rightleftharpoons H^+ + e^-$, is defined as 0 V, and the relative values to this reaction are expressed as the redox potential of a given half reaction (Figure 5.8). The redox potential at standard conditions is expressed as E^0 and that at standard conditions, pH 7.0, as $E^{0'}$.

Biologists are often interested in $E^{0'}$, since physiological pH is neutral. $E^{0'}$ can be calculated from E^0 using the following equation:

$$E^{0'} = E^0 - 2.303(RT/nF) \times 7$$

R: gas constant (8.314 J/mol·K)
T: temperature in K (25 °C = 298 K)
n: number of electrons involved in the reaction
F: Faraday constant (96 487 J/V·mol).

$E^{0'}$ of the H^+/H_2 half reaction is calculated as:

$$E^{0'} = E^0 - 2.303 \times (8.314 \times 298/1 \times 96\,487) \times 7 = -0.41V$$

When a temperature of 30 °C is used, the $E^{0'}$ is calculated as 0.42 V. $E^{0'}$ values of some half reactions of biological interest are listed in Table 5.1.

5.5.2.2 Free energy from $\Delta E^{0'}$

Energy is generated from oxidation/reduction reactions, and the amount of energy is directly proportional to the redox potential difference between the reductant and the oxidant ($\Delta E^{0'}$). Free energy from an oxidation/reduction reaction can be calculated using:

$$\Delta G^{0'} = -nF\Delta E^{0'}$$

n: number of electrons involved in the reaction
F: Faraday constant (96 487 J/V·mol)
$\Delta E^{0'}$: oxidation/reduction potential difference between the reductant and oxidant.

The $\Delta G^{0'}$ of succinate oxidation to fumarate with molecular oxygen can be calculated using this equation, as follows:

$$succinate + \frac{1}{2} O_2 \longrightarrow fumarate + H_2O$$

Table 5.1. *Oxidation/reduction potential of compounds of biological interest*

Electron carrier	$E^{0'}$(mV)	Electron donor and acceptor	$E^{0'}$ (mV)
Cytochrome f^a	365	O_2/H_2O	812
Cytochrome a^a	290	Fe^{3+}/Fe^{2+}	771
Cytochrome c^a	254	NO_3^-/NO_2^-	421
Ubiquinone/ubiquinol	113	Crotonyl-CoA/butyryl-CoA	190
Cytochrome b^a	77	Fumarate/succinate	31
Rubredoxin^a	−57	Pyruvate/lactate	−185
FMN/FMNH$_2$	−190	Acetaldehyde/ethanol	−197
Cytochrome $c_3{}^a$	−205	Acetoin/2,3-butanediol	−244
FAD/FADH$_2$	−219	Acetone/isopropanol	−286
Glutathione^a	−230	CO_2/formate	−413
NAD(P)/NAD(P)H	−320	H^+/H_2	−414
Ferredoxin^a	−413	Gluconate/glucose	−440
		CO_2/CO	−540
		Acetate/acetaldehyde	−581

Artificial electron carrier	$E^{0'}$(mV)	Artificial electron carrier	$E^{0'}$(mV)
Toluidine^a	224	Janus green^a	−225
DCPIP^a	217	Neutral red^a	−325
Phenazine methosulfate^a	80	Benzyl viologen^a	−359
Methylene blue^a	11	Methyl viologen^a	−446

^{a} oxidized form/reduced form.
DCPIP, 2,6-dichlorophenolindophenol.

$E^{0'}$ of fumarate/succinate $= +0.03\,\mathrm{V}$
$E^{0'}$ of $1/2\,O_2/H_2O = +0.82\,\mathrm{V}$
$\Delta G^{0'} = -2 \times 96\,487 \times (0.82 - 0.03) = -152.45\,\mathrm{kJ/mol}$ succinate.

5.5.3 Free energy of osmotic pressure

Active transport can concentrate a nutrient in the cell as high as over 100 times the external concentration (Section 3.3). Such a concentration difference produces an osmotic pressure across the semipermeable cytoplasmic membrane. This pressure can be expressed as free energy as in the following equation. The free energy developed by the osmotic pressure is the energy needed to transport one mol of the solute under the given conditions.

$$\Delta G^{0'} = 2.303\,RT \log[S]_i/[S]_o$$

R: gas constant (8.314 J/mol·K)
T: temperature in K (25 °C = 298 K)
$[S]_o$, $[S]_i$: solute concentration outside and inside the cell.

5.5.4 Sum of free energy change in a series of reactions

Excessive heat would be fatal to the cell. Therefore, most biological reactions are catalyzed in multiple steps. The sum of the free energy changes in each step is the same as that obtained in a one-step reaction. If a compound is metabolized through a different series of reactions, the free energy change is constant if the final product(s) are the same. For example, when glucose is metabolized to two pyruvates either through the EMP pathway or the ED pathway, the free energy change is the same.

5.6 | Role of ATP in the biological energy transduction process

Biological reactions are divided into energy-generating catabolism and energy-consuming anabolism. Free energy generated from catabolism is conserved in the form of adenosine triphosphate (ATP) and the proton motive force (Section 5.7) which are consumed in catabolism (Figure 5.9). In this sense, it can be said that ATP and the proton motive force play a central role in biological metabolism linking catabolism and anabolism. ATP is used to supply energy for biosynthesis and transport by the ABC pathway (Section 3.4), while active transport, motility and reverse electron transport processes (Section 10.1) consume the proton motive force. ATP is well suited for this role since the energy needed for its synthesis and released from its hydrolysis is smaller than the energy available from most of the energy-generating catabolic reactions and bigger than most of the energy-consuming anabolic reactions (Section 5.6.1). In addition, ATP is a general intermediate for nucleic acid biosynthesis.

As shown in Figure 5.10, ATP comprises adenosine with three phosphates bound to the 5′-carbon of the ribose residue. The phosphate groups are termed α, β and γ from the group nearest to ribose. ATP is hydrolyzed as shown below with the release of free energy. In most cases, ATP hydrolysis is coupled to energy-consuming reactions. However, pyrophosphate (PP$_i$) is hydrolyzed without energy conservation in most cases by pyrophosphatase to pull the energy-consuming reactions coupled to ATP hydrolysis to AMP and PP$_i$. In some archaea, including *Thermoproteus tenax*, PP$_i$ is used in place of ATP or ADP (Section 4.1.3).

proton motive force: generated by an electron transport chain which acts as a proton pump, using the energy of electrons from an electron carrier to pump protons out across the membrane, separating the charge across the membrane.

Figure 5.9 **Biological energy transduction.**

Free energy generated from catabolism is conserved in the form of ATP and the proton motive force (PMF) which provide the energy needed for anabolism.

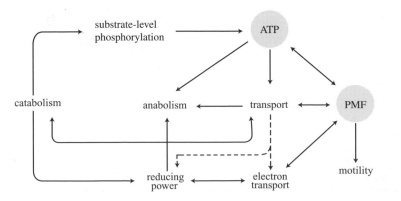

$$\text{ATP} + \text{H}_2\text{O} \longrightarrow \text{AMP} + \text{PP}_i \;(\Delta G^{0'} = -32.2 \,\text{kJ/mol ATP})$$

$$\text{ATP} + \text{H}_2\text{O} \longrightarrow \text{ADP} + \text{P}_i \;(\Delta G^{0'} = -30.5 \,\text{kJ/mol ATP})$$

$$\text{ADP} + \text{H}_2\text{O} \longrightarrow \text{AMP} + \text{P}_i \;(\Delta G^{0'} = -30.5 \,\text{kJ/mol ADP})$$

$$\text{PP}_i + \text{H}_2\text{O} \longrightarrow 2\text{P}_i \;(\Delta G^{0'} = -28.8 \,\text{kJ/mol PP}_i)$$

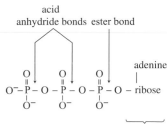

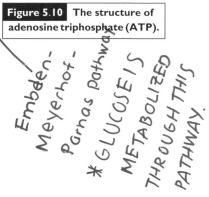

Figure 5.10 The structure of adenosine triphosphate **(ATP)**.

5.6.1 High energy phosphate bonds

It was mentioned previously that free energy is released on the hydrolytic removal of γ-phosphate from ATP and β-phosphate from ADP. For this reason, γ-β and β-α phosphate linkages in ATP are called high energy bonds. There are several other metabolic intermediates with high energy bonds (Table 5.2).

Phosphoenolpyruvate and 1,3-diphosphoglycerate are EMP pathway intermediates and are used to synthesize ATP from ADP in substrate-level phosphorylation (SLP). SLP processes are exergonic reactions because the free energy change ($\Delta G^{0'}$) from the reaction is bigger than that for ATP synthesis. Many ATP consuming anabolic reactions are exergonic. Since both catabolism and anabolism coupled to ATP synthesis and hydrolysis are exergonic reactions, ATP-mediated metabolic reactions are thermodynamically favourable.

The $\Delta G^{0'}$ of ATP hydrolysis is -30.5 kJ/mol. ΔG is much larger than $\Delta G^{0'}$ under physiological conditions, where the ATP concentration is higher than that of ADP in the presence of Mg^{2+}. Mg^{2+} salts of ATP and ADP increase their free energy of hydrolysis.

ATP is synthesized from ADP, and ATP is hydrolyzed to AMP and PP$_i$ in certain energy-consuming reactions. AMP is phosphorylated to ADP by adenylate kinase.

$$\text{ATP} + \text{AMP} \xrightleftharpoons{\text{adenylate kinase}} 2\,\text{ADP} \quad (\Delta G^{0'} = 0\,\text{kJ})$$

Since the $\Delta G^{0'}$ of this reaction is 0 kJ, the direction is determined by the concentration of cellular ATP, ADP and AMP, which reflects the energy status of the cell.

Table 5.2. | *Metabolic intermediates with high energy bonds*

Intermediate	$\Delta G^{0'}$ (kJ/mol)
Phosphoenolpyruvate	−61.9
Carbamoyl phosphate	−51.5
1,3-diphosphoglycerate	−49.4
Acetyl phosphate	−47.3
Creatine phosphate	−43.1
Arginine phosphate	−38.1
Acetyl-CoA	−34.5

5.6.2 Adenylate energy charge

A high ATP concentration means that the energy status is good, and when the energy supply cannot meet demand, the ADP and AMP concentration is high. Arbitrarily, the adenylate energy charge (EC) is a term used to describe the energy status of a cell. EC can be calculated using the following equation:

$$EC = ([ATP] + \tfrac{1}{2}[ADP])/([ATP] + [ADP] + [AMP])$$

The numerator of this equation is the sum of high energy phosphate bonds in the form of ATP, and the denominator the concentration of the total adenylate pool. An EC number of 1 means that the total adenylate is in the ATP form, and 0 in the AMP form.

Since the reaction catalyzed by adenylate kinase is reversible with $\Delta G^{0\prime}$ of 0 kJ/mol, and enzyme activity is high in a cell, the relative concentrations of ATP, ADP and AMP are determined by the EC (Figure 5.11). Many catabolic reactions are repressed by ATP and activated by ADP and/or AMP: anabolic reactions are controlled in a reverse manner (Section 12.4). The reactions are regulated not by the absolute concentration of each adenosine nucleotide, but by their ratio. Thus, overall metabolism is controlled by the EC value (Figure 5.12).

When a bacterium uses acetate as the sole carbon and energy source, acetate should be metabolized through the TCA cycle in catabolism and through the glyoxylate cycle in anabolism to supply carbon compounds for biosynthesis (Section 5.3.2). It has been mentioned that AMP and ADP activate isocitrate dehydrogenase to metabolize the substrate through the TCA cycle (Figure 5.5, Section 5.5). The activity is not controlled by AMP and ADP per se, but by the EC value, activated at EC < 0.8 and inactivated at EC > 0.8. An anabolic enzyme, aspartate kinase (Figure 6.13, Section 6.4.1), is activated at a high EC value and repressed at a low EC value. When a bacterial culture is transferred from a rich medium to a poor medium, AMP is

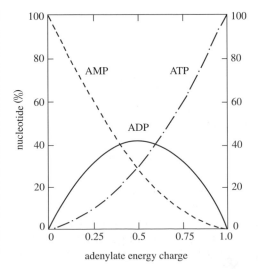

Figure 5.11 Relationship of adenylate energy charge (EC) and the relative concentration of ATP, ADP and AMP.

(Dawes, E. A. 1986, *Microbial Energetics*, Figure 2.2. Blackie & Son, Glasgow)

Since $\Delta G^{0\prime}$ of the reaction catalyzed by adenylate kinase is 0 kJ/mol, the direction of the reaction will be determined by the relative concentrations of ATP, ADP and AMP, which can be expressed by the EC value.

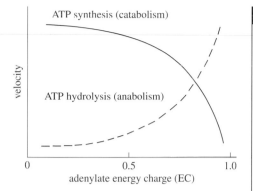

Figure 5.12 Regulation of catabolism and anabolism by the adenylate energy charge (EC).

(Dawes, E. A. 1986, *Microbial Energetics*, Figure 2.3. Blackie & Son, Glasgow)

Adenylate energy charge controls the overall growth of microbes, regulating catabolism which synthesizes ATP, and anabolism which consumes it.

excreted or hydrolyzed to adenosine or adenine to maintain a high EC value at a low rate of ATP synthesis.

A growing bacterial culture maintains an EC value of 0.8–0.95, and the value gradually decreases to around 0.5 and rapidly thereafter when the culture starves. Bacterial cultures with EC values less than 0.5 cannot form colonies. This is not surprising because ATP is essential for viability. Since the EC value is a unitless figure, it does not give any information on the size of the adenylate pool, the concentration of each adenosine nucleotide or the turnover velocity. Since the EC value controls the overall metabolism that is observed during growth, similar EC values are expected in fast-growing and slow-growing cultures, though a fast-growing culture has a bigger adenylate pool and a higher ATP turnover rate than the slow-growing culture.

5.6.3 Phosphorylation potential (ΔGp)

Free energy needed for ATP synthesis or released by its hydrolysis in the cell is referred to as the phosphorylation potential (ΔGp) which is determined by the concentration of ATP, ADP and P_i as in the following equation (Section 5.5.1.3):

$$\Delta Gp = \Delta G^{0'} + 2.303 \log([ADP][P_i]/[ATP])$$

In addition to the concentration of each adenosine nucleotide, ΔGp depends on the concentration of metal ions such as Mg^{2+} which bind the nucleotide. Binding of metal ions increases the ΔGp value. The concentration of metal ions and P_i vary according to the growth conditions. $\Delta G^{0'}$ for ATP hydrolysis is -30.5 kJ/mol, and ΔGp is around -51.8 kJ/mol ATP (Section 5.5.1.3).

5.6.4 Interconversion of ATP and the proton motive force (Δp)

Figure 5.9 shows that ATP and Δp link catabolism and anabolism, and that ATP is converted to Δp, and vice versa. The membrane-bound ATP synthase (ATPase) catalyzes this interconversion (Section 5.8.4). When a microbe grows fermentatively generating ATP through substrate-level phosphorylation, ATP is hydrolyzed to increase the Δp. In contrast, Δp is consumed to synthesize ATP when respiration is the main energy conservation process. ATP and Δp are therefore consumed for different purposes (Figure 5.9).

5.6.5 Substrate-level phosphorylation (SLP)

ATP is synthesized at the cytoplasm as a result of the transfer of phosphate from metabolic intermediates with high energy phosphate bonds to ADP. 1,3-diphosphoglycerate, phosphoenolpyruvate and acyl-phosphate are the metabolic intermediates used to synthesize ATP through SLP. Succinyl-CoA conversion to succinate in the TCA cycle is another example of SLP and is catalyzed by succinyl-CoA synthetase:

$$\text{succinyl-CoA} + P_i + \text{ADP(GDP)} \xrightarrow{\substack{\text{succinyl-CoA} \\ \text{synthetase}}} \text{succinate} + \text{ATP(GTP)} + \text{CoA-SH}$$

5.7 | Proton motive force (Δp)

Chemical and light energies are converted to biological forms of energy through photosynthesis, fermentation and respiration (Figure 5.7). Free energy generated from oxidation–reduction reactions is conserved in the form of ATP through photosynthesis as in respiration. This is referred to as electron transport phosphorylation (ETP) or oxidative phosphorylation. The free energy from ETP is coupled to the expulsion of protons from the cytoplasm to the periplasm and the proton gradient across the cytoplasmic membrane is used to perform useful work including ATP synthesis, active transport, motility and reverse electron transport catalyzed by various membrane proteins (Figure 5.9). Since protons are positively charged, proton expulsion not only builds the proton gradient (ΔpH) but also the inside-negative membrane potential ($\Delta\psi$). ΔpH and $\Delta\psi$ are collectively referred to as the proton motive force (Δp, or $\Delta\tilde{\mu}_{H^+}$).

5.7.1 Proton gradient and membrane potential

With few exceptions, prokaryotic cells maintain an inside alkaline and outside acidic proton gradient, and an inside negative and outside positive membrane potential. This is possible due to the hydrophobic hydrocarbon tails of the membrane phospholipids. Δp consisting of ΔpH and $\Delta\psi$ is expressed in volts as:

$$\Delta p = \Delta\psi + Z\Delta pH$$

$\Delta\psi$:   membrane potential (mV)
ΔpH: pH gradient
Z: 2.303 RT/F
        R: gas constant
        T: temperature in K ($25\,^{\circ}C = 298$ K)
        F: Faraday constant.

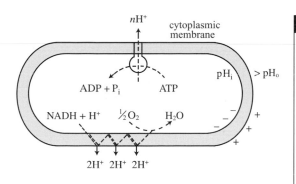

Figure 5.13 **Proton motive force.**

The proton motive force (Δp) is a biological form of energy facilitating ATP synthesis, transport and motility. Δp consists of an inside alkaline and outside acidic proton gradient, and an inside negative and outside positive membrane potential. Electron transport phosphorylation (ETP) during respiration and photosynthesis builds Δp (lower part). Δp is consumed to synthesize ATP. In a fermentative cell, the membrane-bound ATPase hydrolyzes ATP to export H^+ which develops Δp (upper part).

Bacterial cells maintain a Δp of 0.15–0.45 V. ETP generates a Δp that is used to synthesize ATP. In a fermentative cell ATP synthesized through SLP is hydrolyzed to develop the Δp (Figure 5.13). Most of the enzymes in the cytoplasm have an optimum pH of around neutrality while polymers, including DNA, RNA and proteins, are unstable at extremes of pH. For these reasons, the pH of the cytoplasm is maintained around neutrality regardless of the external pH.

5.7.2 Acidophilicity and alkalophilicity
Many microbes grow best at neutral pH, but some have their optimum pH on the extreme acidic or alkaline side. These are referred to as neutrophiles, acidophiles and alkalophiles, respectively. All of them maintain an internal pH near to neutrality. Neutrophiles possess mechanisms to tolerate extreme pH values. When a neutrophilic bacterium, *Salmonella typhimurium*, is acclimated to a mild acidic pH (pH 5.8), over 50 acid shock proteins are synthesized to maintain the internal pH at neutrality. These acid shock proteins render the bacterium able to tolerate a more acidic pH of 3.0. Among the acid shock proteins, amino acid decarboxylases are the best known. At acidic pH in the presence of lysine, the bacterium synthesizes lysine decarboxylase (CadA) and the lysine/cadaverine antiporter (CadB). CadA decarboxylates lysine to cadaverine consuming H^+ to raise the internal pH. CadB exports cadaverine in exchange for lysine.

Many bacteria synthesize a Na^+/H^+ antiporter at alkaline pH values to convert the proton motive force to a sodium motive force to maintain a neutral internal pH as in alkalophiles (Section 5.7.4).

5.7.3 Proton motive force in acidophiles
Thiobacillus ferrooxidans, *Thermoplasma acidophilum* and *Sulfolobus acidocaldarius* are acidophiles with an optimum pH of pH 1–4. These microbes maintain their internal pH around neutrality with a H^+ gradient of 10^3–10^6. They maintain a low or inside positive membrane potential to compensate for the large potential rendered by the H^+ gradient (Figure 5.14). The inside positive membrane potential gives an additional benefit to the organisms by preventing H^+ leakage through the membrane due to the large concentration gradient.

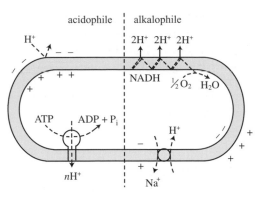

Figure 5.14 Proton motive force in acidophiles and alkalophiles.

Acidophiles maintain a neutral internal pH at an external acidic pH with a large ΔpH. They have a low or inside positive membrane potential to provide an adequate Δp compensating for the large ΔpH. The internal positive membrane potential renders an additional benefit to the bacteria in preventing H^+ leakage due to the large concentration gradient. Alkalophiles grow optimally at alkaline pH with a neutral internal pH. They have a high Na^+/H^+ antiporter activity. H^+ exported by ETP is exchanged with Na^+ to prevent alkalization of the cytoplasm. They maintain a sodium motive force instead of a proton motive force. Alkalophiles and halophiles have Na^+-dependent ETP in addition to the Na^+/H^+ antiporter. This is referred to as a primary sodium pump.

Fatty acids such as acetate inhibit the growth of acidophiles at low concentrations of 5–10 mM. Their optimum pH is lower than the pK_a of the fatty acids. The pK_a of acetate is 4.8. Fatty acids at this pH are mainly in undissociated forms, which are hydrophobic and thus permeable to the membrane phospholipids. The fatty acids diffuse into the cell and dissociate, lowering the internal pH. They function as protonophores (uncouplers, Section 5.8.5). The accumulation of fatty acids in the cell increases the turgor pressure, which may provide another reason for growth inhibition.

Saccharolytic clostridia ferment carbohydrates to fatty acids (Section 8.5) lowering the growth pH to pH 4.5. At these conditions, the cells become very resistant to physical stresses. Such resistance to low pH in the presence of fatty acids and to stress is related to changes in cell surface structure.

During vinegar fermentation, acetic acid bacteria lose their viability when aeration is disrupted. The bacteria pump H^+ out of the cell through ETP with aeration, but the cytoplasm is acidified without H^+ expulsion under oxygen-limited conditions.

5.7.4 Proton motive force and sodium motive force in alkalophiles

Alkalophiles maintain a neutral internal pH at external pH values over 10. Their growth is Na^+ dependent. Na^+ plays an important role in maintaining internal neutral pH in alkalophiles. H^+ exported by ETP is exchanged with Na^+ through the action of the Na^+/H^+ antiporter (Figure 5.14). A Na^+/H^+ antiporter mutant ($nhaC^-$) of alkalophilic *Bacillus firmus* cannot grow at pH 10. In some bacteria, Na^+ is exported instead of H^+ through Na^+-dependent ETP which includes the Na^+-dependent NADH-quinone reductase complex (Section 5.8.2.1). This is referred to as a primary sodium pump.

Alkalophiles maintain a sodium motive force ($\Delta\bar{\mu}_{Na^+}$) instead of a proton motive force, and they have Na^+-dependent ATPase and transporters. Their motility is also Na^+-dependent. Since 1 H^+ is exchanged with nNa^+ ($n > 1.0$), the Na^+/H^+ antiporter increases the membrane potential.

A Na^+/H^+ antiporter is also known in neutrophiles including *Escherichia coli*. Mutants in the antiporter are less tolerant to alkaline pH than wild-type strains. The Na^+/H^+ antiporter renders neutrophiles with some alkaline-resistant characteristics. A sodium motive force is not only found in alkalophiles but also in halophilic archaea. Some decarboxylases are Na^+ dependent (Section 5.8.6).

5.8 | Electron transport (oxidative) phosphorylation

Electron carriers such as $NAD(P)^+$, FAD and PQQ are reduced during glycolysis and the TCA cycle. Electrons from these carriers enter the electron transport chain at different levels. Electron carriers are oxidized, reducing molecular oxygen to water through ETP to conserve free energy as the proton motive force (Δp) (Figure 5.21a).

5.8.1 Chemiosmotic theory

It took many years to elucidate how the free energy generated from ETP is conserved as ATP. Compounds with high energy bonds are not involved in ETP as in substrate-level phosphorylation. ATP is synthesized only with an intact membrane or membrane vesicles, and ATP synthesis is inhibited in the presence of uncouplers or ionophores. From these observations, a chemiosmotic mechanism was proposed. According to this, export of charged particles is coupled to oxidation–reduction reactions to form an electrochemical gradient which is used for ATP synthesis. H^+ are the charged particles exported, and the electrochemical gradient is the proton motive force, consisting of the H^+ gradient (ΔpH) across the membrane and the membrane potential ($\Delta\psi$). The phospholipid membrane is impermeable to H^+ and OH^- and is suitable to maintain the proton gradient. Most of the electron carriers involved in ETP are arranged in the membrane, and the membrane-bound ATP synthase synthesizes ATP from ADP and P_i consuming the proton gradient.

5.8.2 Electron carriers and the electron transport chain

Electron carriers involved in electron transport from NADH to molecular oxygen are localized in the mitochondrial inner membrane in eukaryotic cells and in the cytoplasmic membrane in prokaryotic cells. The mitochondrial electron transport chain is shown in Figure 5.15, and bacterial electron transport systems are shown in Figure 5.19. Bacterial systems are diverse depending on the species and strain as well as on the availability of electron acceptors.

5.8.2.1 Mitochondrial electron transport chain

Eukaryotic electron transport is discussed here as a model to compare with the process in prokaryotes. The mitochondrial electron transport chain consists of complex I, II, III and IV. The overall reaction can be summarized as dehydrogenases (complex I and II) and an oxidase (complex IV) connected by quinone (including complex III).

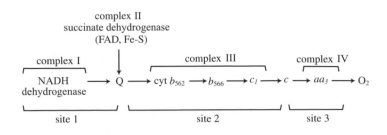

Figure 5.15 The eukaryotic electron transport chain localized in the mitochondrial inner membrane.

(Dawes, E. A. 1986, *Microbial Energetics*, Figure 7.1. Blackie & Son, Glasgow)

The electron transport chain forms four complexes in the mitochondrial inner membrane. These are referred to as complex I, II, III and IV. Complex I has the lowest redox potential and consists of NADH dehydrogenase with FMN as the prosthetic group and [Fe-S] proteins. This complex is called NADH-ubiquinone reductase. Succinate dehydrogenase of the TCA cycle contains FAD as a prosthetic group, and forms complex II (succinate-ubiquinone reductase) with [Fe-S] proteins. Complex I and II transfer a pair of electrons to coenzyme Q from NADH and succinate, respectively. Complex III is ubiquinol-cytochrome c reductase, transferring electrons from reduced coenzyme Q (ubiquinol) to cytochrome c. This complex contains cytochrome b_{562}, cytochrome b_{566}, cytochrome c_1 and [Fe-S] protein. Complex IV is referred to as cytochrome c oxidase. Among the electron transport processes, three steps generate enough free energy to synthesize ATP from ADP and P_i. These are shown in the figure as sites 1, 2 and 3.

NADH dehydrogenase oxidizes NADH, reduced in various catabolic pathways, to NAD^+. This enzyme contains FMN as a prosthetic group and forms with [Fe-S] proteins a complex known as complex I or NADH-ubiquinone reductase. FMN is reduced with the oxidation of NADH and the [Fe-S] proteins mediate electron and proton transfer from $FMNH_2$ to coenzyme Q. This reaction generates enough free energy to synthesize ATP, and the electron transfer from NADH to ubiquinone (coenzyme Q) is referred to as site 1 of ETP. In a mitochondrion and in most bacteria, protons are translocated by this complex, but sodium ions are exported by the complex I of certain bacteria including *Vibrio alginolyticus* (Section 5.7.4).

As a step in the TCA cycle, succinate dehydrogenase oxidizes succinate to fumarate, reducing its prosthetic group, FAD, before electrons are transferred to coenzyme Q. This enzyme forms complex II (or the succinate-ubiquinone reductase complex) of ETP with [Fe-S] proteins, cytochrome b_{558}, and low molecular weight peptide. Other dehydrogenases containing FAD as a prosthetic group reduce coenzyme Q in a similar way. These include glycerol-3-phosphate dehydrogenase and acyl-CoA dehydrogenase.

Electrons from coenzyme Q are transferred to a series of reddish-brown coloured proteins known as cytochromes. Cytochromes involved in mitochondrial electron transport are b_{562}, b_{566}, c_1, c and aa_3 as shown in Figure 5.15. Two separate protein complexes mediate electron transfer from coenzyme Q to molecular oxygen through the cytochromes. These are ubiquinol-cytochrome c reductase (complex III) and cytochrome oxidase (complex IV).

Complex III transfers electrons from coenzyme Q to cytochrome c. This complex consists of [Fe-S] protein, and cytochromes b_{562}, b_{566}

(a)

(b)

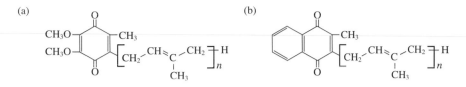

and c_1. At this step, energy is conserved exporting protons (site 2). The cytochrome oxidase complex mediates electron transfer from reduced cytochrome c to molecular oxygen. Energy is also conserved at this step (site 3). This terminal oxidase complex contains cytochrome a and cytochrome a_3.

5.8.2.2 Electron carriers

Electron transport involves various electron carriers including flavoproteins, quinones, [Fe-S] proteins and cytochromes. Flavoproteins are proteins containing riboflavin (vitamin B_2) derivatives as their prosthetic group. They are FMN (flavin mononucleotide) and FAD (flavin adenine dinucleotide). The redox potential of the flavoproteins varies not due to the flavin structure but due to the differences in the protein component.

Two structurally different quinones are involved in the electron transport process, ubiquinone and menaquinone, which serve as coenzyme Q. Quinones are lipid electron carriers, highly hydrophobic and mobile in the semi-solid lipid phase of the membrane. As shown in Figure 5.16, quinones have a side chain of 6, 8 or 10 isoprenoid units. These are named Q_6, Q_8 and Q_{10} according to the number of isoprenoid units. Ubiquinone is found in mitochondria, and bacteria have menaquinone (Figure 5.16). Both forms of quinones are found in Gram-negative facultative anaerobes. The structure of coenzyme Q can be used as one characteristic for bacterial classification. Quinones can carry protons as well as electrons.

[Fe-S] proteins contain [Fe-S] cluster(s), usually [2Fe-2S] or [4Fe-4S]. The non-heme irons are attached to sulfide residues of the cysteines of the protein and acid-labile sulfur (Figure 5.17). The acid-labile sulfur is released as H_2S at an acidic pH. [Fe-S] proteins participating in electron transport can carry protons as well as electrons. There are many different [Fe-S] proteins mediating not only the electron transport process in the membrane, but also various oxidation–reduction reactions in the cytoplasm. The redox potential of different [Fe-S] proteins spans from as low as -410 mV (clostridial ferredoxin, Section 8.5) to $+350$ mV. Many enzymes catalyzing oxidation–reduction reactions are [Fe-S] proteins including hydrogenase, formate dehydrogenase, pyruvate:ferredoxin oxidoreductase and nitrogenase.

Cytochromes are hemoproteins. They are classified according to their prosthetic heme structures (Figure 5.18) and absorb light at 550–650 nm. Cytochrome b_{562} refers to a cytochrome b with the maximum wavelength absorption at 562 nm. Heme is covalently bound to the proteins in cytochrome c, and hemes are non-covalently associated with the protein in other cytochromes. Since cytochromes

Figure 5.16 **Structure of (a) ubiquinone and (b) menaquinone.**

(Gottschalk,G. 1986, *Bacterial Metabolism*, 2nd edn., Figure 2.9. Springer, New York)

$n = 4, 6, 8$ or 10.

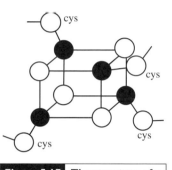

Figure 5.17 **The structure of a [4Fe-4S] cluster.**

(Gottschalk, G. 1986, *Bacterial Metabolism*, 2nd edn., Figure 8.8. Springer, New York)

○, sulfur atom; ●, iron atom.

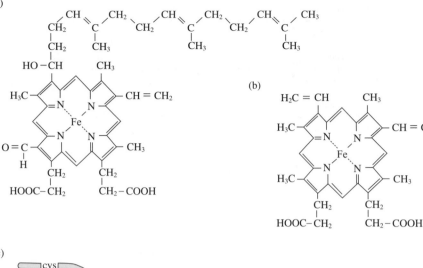

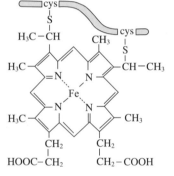

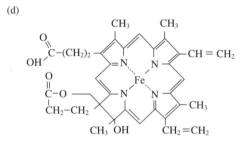

Figure 5.18 Structure of hemes of cytochromes.

(Gottschalk, G. 1986, *Bacterial Metabolism*, 2nd edn., Figure 2.10. Springer, New York)

(a)–(d) show the prosthetic hemes of cytochromes *a*, *b*, *c* and *d* respectively.

carry only one electron, electron transfer from reduced coenzyme Q to cytochrome requires two steps (Section 5.8.3).

5.8.2.3 Diversity of electron transport chains in prokaryotes

As in the mitochondrial electron transport chain, the prokaryotic electron transport chain is organized with dehydrogenase and oxidase connected by quinone. However, electron transport chains in prokaryotes are much more diverse since they use diverse electron donors and electron acceptors including those not used by any eukaryotes. Separate discussion will be made on the diversity of dehydrogenases in Chapter 7 (organic electron donors) and in Chapter 10 (inorganic electron donors, chemolithotrophs). Chapter 9 contains exclusive discussion of the electron transport chains to electron acceptors other than O_2 (anaerobic respiration). In this section, prokaryotic electron transport chains analogous to the mitochondrial system will be described.

The electron carriers involved in prokaryotic electron transport are diverse. As described earlier, menaquinone is used as coenzyme Q in addition to ubiquinone in prokaryotes, and diverse cytochromes

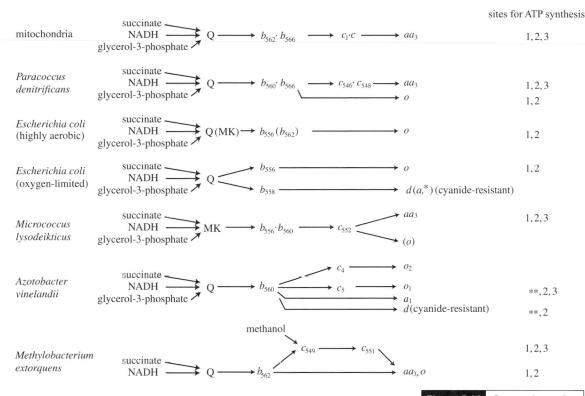

Figure 5.19 Comparison of mitochondrial and bacterial electron transport systems.

(Dawes, E. A. 1986, *Microbial Energetics*, Blackie & Son, Glasgow)

Under nitrogen-fixing conditions with high oxygen tension, *Azotobacter vinelandii* consumes oxygen rapidly without conserving energy at site 1 to protect nitrogenase (Section 6.2.1.4).

participate in prokaryotic electron transport mainly in relation to the terminal oxidase, which leads to branched electron transport chains depending on the availability of O_2.

Some bacteria such as *Paracoccus denitrificans* and *Alcaligenes eutrophus* have very similar electron transport systems to mitochondria. They have cytochrome aa_3 as the terminal oxidase while others use cytochrome d or o in its place. Cytochrome o has a b-type heme, and cytochrome d has a different heme structure (Figure 5.18). They are not only structurally different but also show different responses towards respiratory inhibitors, and form branched electron transport pathways (Figure 5.19). This diversity of electron transport systems is closely related to bacterial growth under a variety of conditions.

The terminal oxidases in bacteria have different affinities for O_2. Under O_2-limited conditions, cytochrome d replaces the normal terminal oxidase, cytochrome aa_3, in *Klebsiella pneumoniae* and *Haemophilus parainfluenzae*. Similarly, *Paracoccus denitrificans* and *Alcaligenes eutrophus* use cytochrome o as their terminal oxidase. The high affinity cytochrome d and o enables the bacteria to use the electron acceptor (O_2) efficiently at low concentrations. In nitrogen-fixing *Azotobacter vinelandii*, cytochrome d functions as the terminal oxidase under nitrogen-fixing conditions with less energy conservation than the normal oxidase (Figure 5.19). Cytochrome d keeps the intracellular O_2 concentration low to protect the O_2-labile nitrogenase (Section 6.2.1).

In branched bacterial electron transport systems, the number of sites for energy conservation is less than that on the mitochondrial system (Section 5.8.4). This might enable a survival strategy under certain conditions but at the expense of reduced energy conservation.

With a few exceptions, the dehydrogenases of chemolithotrophs reduce quinone or cytochrome and this is coupled to the oxidation of their inorganic electron donors. They cannot directly reduce $NAD(P)^+$, which is needed for biosynthetic purposes. They transfer electrons from the reduced quinone or cytochrome to $NAD(P)^+$ in an uphill reaction, consuming the proton motive force, known as reverse electron transport (Section 10.1). Reverse electron transport is possible because complexes I and III can catalyze the reverse reactions. Complex IV cannot catalyze the reverse reaction to use water as the source of electrons. Water is used as the electron source in oxygenic photosynthesis through a different reaction (Section 11.4.3).

5.8.2.4 Inhibitors of electron transport phosphorylation (ETP)

Inhibitors of ETP are grouped into three kinds according to their mechanism of action: electron transport inhibitors, uncouplers and ATPase inhibitors.

Electron transport inhibitors interfere with the enzymes and electron carriers involved in the electron transport system. They inhibit not only ATP synthesis but also oxygen consumption. Rotenone, amytal and piericidin A inhibit NADH dehydrogenase, and 2-n-heptyl-4-hydroquinoline-N-oxide (HQNO), antimycin A, cyanide (CN^-) and azide have their own specific inhibition sites (Figure 5.20).

Uncouplers increase H^+ permeability through the membrane, thus dissipating the proton motive force. The proton motive force becomes too low to be used to synthesize ATP in the presence of uncouplers, but the O_2 consumption rate increases in the presence of the uncouplers. The term uncoupler means that ATP synthesis is not coupled to O_2 consumption. Further discussion occurs later (Section 5.8.5).

ATP synthase inhibitors block the membrane-bound ATPase preventing ATP synthesis even with a high proton motive force. N, N'-dicyclohexylcarbodiimide (DCCD) and oligomycin are well-known ATPase inhibitors. They bind the F_o part of the membrane-bound F_1F_o-ATPase blocking the path for H^+. F_o means the oligomycin-binding component of the F_1F_o-ATPase.

5.8.2.5 Transhydrogenase

Nicotinamide nucleotide transhydrogenase is known in many prokaryotes and catalyzes the following reaction:

$$NADPH + NAD^+ \rightleftharpoons NADP^+ + NADH \quad (\Delta G^{0'} = 0 \text{ kJ/mol NADPH})$$

Though $\Delta G^{0'}$ is 0 kJ/mol, the reaction is exergonic under physiological conditions since the ratio of $NADP^+/NADPH$ is low and $NAD^+/NADH$ is high in the cell. This reaction is coupled to H^+ extrusion increasing the proton motive force. This reaction is referred to as site 0 of ETP.

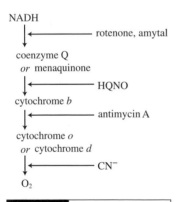

Figure 5.20 Electron transport inhibitors and the sites of their action.

(Gottschalk, G. 1986, *Bacterial Metabolism*, 2nd edn., Figure 2.17. Springer, New York)

HQNO, 2-n-heptyl-4-hydroquinoline-N-oxide.

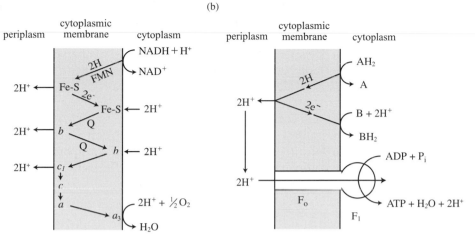

(a)

(b)

Figure 5.21 Formation of proton motive force and ATP synthesis through ETP.

(Dawes, E. A. 1986, *Microbial Energetics*, Figure 7.3. Blackie & Son, Glasgow)

5.8.3 Arrangement of electron carriers in the H^+-translocating membrane

It is not well established how protons are translocated when coupled to electron transport, but there is a consensus that the protons are translocated by exploiting the different properties of the electron carriers (Q-loop and Q-cycle) and through the proton pump. Among the electron carriers, [Fe-S] proteins and cytochromes only carry electrons, while coenzyme Q can carry electrons as well as protons. They are arranged in such a way that H^+ are exported during ETP (Figure 5.21a). The location of these membrane proteins has been studied using various techniques including electron microscopy, immunology and using inhibitors and proteolytic enzymes. These studies have located NADH dehydrogenase to the cytoplasmic side (matrix side in a mitochondrion) of the membrane and cytochrome c to the periplasmic side.

Since the electron carriers are arranged in the mitochondrial inner membrane and prokaryotic cytoplasmic membrane in such a way that electrons move from the inner face to the outer face of the membrane with H^+, and in the reverse direction without H^+, H^+ is exported during the electron transport process (a). The membrane-bound ATPase synthesizes ATP, consuming the proton motive force with H^+ flow to the low H^+ concentration side (b).

5.8.3.1 Q-cycle and Q-loop

To describe H^+ translocation at site 2 catalyzed by ubiquinol-cytochrome c reductase (complex III), a Q-cycle and Q-loop have been proposed. According to the Q-loop mechanism, FMN of NADH dehydrogenase is reduced to $FMNH_2$ by oxidizing NADH on the inside face of the membrane. Electrons from $FMNH_2$ are transferred to a [Fe-S] protein on the outer face of the membrane leaving $2H^+$ outside. This reaction is catalyzed by complex I. The reduced [Fe-S] protein reduces coenzyme Q at the inner face of the membrane consuming $2H^+$. The reduced coenzyme Q moves to the outer face of the membrane to reduce cytochrome c, leaving $2H^+$ outside again through cytochrome b (Figure 5.21a).

The Q-loop mechanism is consistent with experimental results obtained with various bacteria, including *Escherichia coli*, where $2H^+$ are translocated by the ubiquinol-cytochrome c reductase (complex III). However, in a mitochondrion and in some bacteria, $4H^+$ are

Figure 5.22 H$^+$ **translocation through the Q-cycle catalyzed by the ubiquinol-cytochrome reductase complex during ETP.**

(Dawes, E. A. 1986, *Microbial Energetics*, Figure 7.4. Blackie & Son, Glasgow)

In some bacteria including *Escherichia coli*, 2H$^+$ are translocated by the ubiquinol-cytochrome *c* reductase (complex III). In this reaction, known as the Q-loop, coenzyme Q is reduced at the inner face of the membrane and moves to the other side of the membrane, where electrons are transferred to cytochrome leaving protons outside. The Q-cycle has been proposed to explain the translocation of 4H$^+$ in a mitochondrion and some bacteria. In this mechanism the reduced coenzyme Q (QH$_2$) is oxidized to semiquinol (QH) reducing [Fe-S] protein before being fully oxidized to Q, transferring the electron to cytochrome *b* as illustrated in the diagram. The reduced [Fe-S] protein transfers the electron to cytochrome *c*, and the reduced cytochrome *b* returns the electron to Q that takes up 1H$^+$ to be reduced to QH. QH is fully reduced to QH$_2$ taking one electron from [Fe-S] protein and consuming 1H$^+$. During a QH$_2 \rightarrow$ QH$\rightarrow$ Q $\rightarrow$ QH $\rightarrow$ QH$_2$ cycle transferring one electron from Q to cytochrome *c*, 2H$^+$ are translocated.

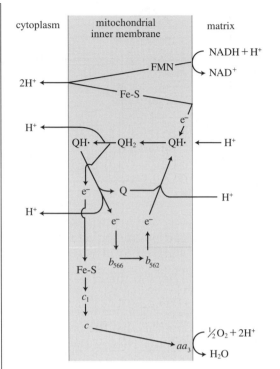

translocated by complex III. To explain how the extra 2H$^+$ are exported, the Q-cycle has been proposed. In this mechanism, reduced coenzyme Q (QH$_2$) is oxidized to semiquinol (QH) reducing [Fe-S] protein before being fully oxidized to Q, transferring the electron to cytochrome *b*. The reduced [Fe-S] protein transfers the electron to cytochrome *c*, and the reduced cytochrome *b* returns the electron to Q that takes up 1H$^+$ to be reduced to QH. QH is fully reduced to QH$_2$ taking one electron from [Fe-S] protein and consuming 1H$^+$ at the inner face of the membrane (Figure 5.22). According to this hypothesis 2H$^+$ are translocated during a QH$_2 \rightarrow$ QH $\rightarrow$ Q $\rightarrow$ QH $\rightarrow$ QH$_2$ cycle, transferring one electron from Q to cytochrome *c*.

5.8.3.2 Proton pump

Complexes I (NADH-ubiquinone reductase) and IV (cytochrome oxidase) translocate H$^+$, but mechanisms similar to the Q-loop or Q-cycle are not known for these complexes. It is believed that these complexes are proton pumps and probably expel H$^+$ by conformational changes in the protein.

5.8.4 ATP synthesis

The membrane-bound ATP synthase (ATPase) synthesizes ATP consuming the proton motive force (Figure 5.21b).

5.8.4.1 ATP synthase

This enzyme is located in the mitochondrial inner membrane or the prokaryotic cytoplasmic membrane. The enzyme consists of a

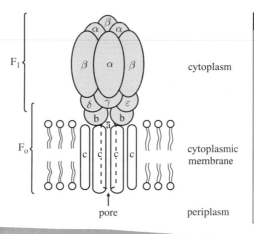

Figure 5.23 Model of the bacterial F_1F_0-ATP synthase.

(Dawes, E. A. 1986, *Microbial Energetics*, Figure 5.3. Blackie & Son, Glasgow)

The F_0 part of ATP synthase consists of highly hydrophobic polypeptides a, b, c, forming a hydrophilic pore for H^+ movement. The catalytic F_1 part consists of five different polypeptides in the ratio of $\alpha_3\beta_3\gamma\delta\varepsilon$.

membrane-embedded F_0 part and a F_1 part protruding into the cytoplasmic side. F_0 is a pore for H^+ to pass, and consists of three different peptides, a, b and c, and ATP synthesis and hydrolysis are catalyzed by the F_1 part consisting of five different peptides ($\alpha_3\beta_3\gamma\delta\varepsilon$) (Figure 5.23).

Oligomycin and dicyclohexylcarbodiimide (DCCD) inhibit ATP synthase by blocking H^+ movement through the pore formed by the F_0 part. F_0 is the oligomycin-binding site. The ATPase of the mitochondrial inner membrane and bacterial cytoplasmic membrane is referred to as F_1F_0-ATPase, F-ATPase or H^+-ATPase. The main function of this enzyme is ATP synthesis while consuming the proton motive force. In many bacteria a similar ATPase synthesizes ATP but consumes the sodium motive force (Section 9.4.3). This is referred to as a Na^+-ATPase.

An ATPase structurally different from the F-ATPase is found at the eukaryotic cytoplasmic membrane and the membrane of organelles other than chloroplasts and mitochondria. The function of this enzyme is development of the proton motive force consuming the ATP synthesized in mitochondria and chloroplasts. This enzyme was first identified in the vacuole of a eukaryotic cell, and named a V-ATPase. V-ATPase has a catalytic V_1 part and membrane-embedded V_0 part. V_1 contains eight different peptides (A, B, C, D, E, F, G, H) and V_0 contains five different peptides (a, d, c, c', c''). A V-type ATPase has been identified in bacteria and archaea including *Clostridium fervidum*, *Enterococcus hirae*, *Thermus thermophilus* and others. This prokaryotic V-ATPase synthesizes ATP while consuming the sodium motive force.

5.8.4.2 H^+/O ratio

Not only is the mechanism of coupling proton translocation to electron transport not well established, but also the number of protons translocated. It is generally accepted that the NADH-ubiquinone reductase complex exports 4 protons, ubiquinol-cytochrome *c* reductase exports 2 or 4 protons, and cytochrome oxidase exports 2 protons with the consumption of 2 electrons (Figure 5.21a). Since 2 electrons reduce 1 oxygen atom, the number is referred to as the H^+/O ratio. A ratio of 10 is accepted for mitochondria, and the ratio is

Figure 5.24 Dissipation of Δp by an uncoupler.

Uncouplers are weak acids or weak bases, hydrophobic not only in undissociated form but also in dissociated form and permeable to the membrane. When an uncoupler is equilibrated at higher internal pH and at low external pH, the concentration of the undissociated form is higher in the external medium and the concentration of the dissociated form is higher in the cell. The undissociated form diffuses into the cell and the dissociated form out of the cell, thus dissipating Δp. ATP cannot be synthesized due to the low Δp, and the cell consumes more O_2 to increase Δp.

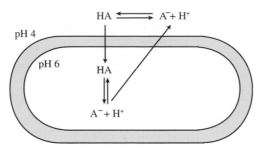

similar or less than 10 in prokaryotes, depending on the electron carriers involved in ETP (Figure 5.15).

5.8.4.3 H^+/ATP stoichiometry

The membrane-bound ATP synthase catalyzes ATP synthesis and hydrolysis consuming and developing the proton motive force (Figure 5.21b). It is not known for certain how many protons move through the enzyme for 1 ATP to be generated (H^+/ATP stoichiometry). H^+/ATP stoichiometry depends on the phosphorylation potential (ΔGp, Section 5.6.3) and the size of the proton motive force (Δp, Section 5.7.1). Various experimental results have shown a H^+/ATP stoichiometry of 2–5 with a stoichiometry of 3 generally being accepted.

This number has an important implication in terms of the minimum free energy conserved in a biological system. Energy is conserved through fermentation (substrate-level phosphorylation, SLP), respiration and photosynthesis (ETP) in biological systems. In SLP a free energy change bigger than ΔGp can be conserved for thermodynamic reasons. Similarly, free energy changes can be conserved during respiration and photosynthesis when the change is bigger than $\Delta Gp/[H^+$/ATP stoichiometry]. In the case of organisms using Na^+ in place of H^+, the minimum amount of energy conserved would be $\Delta Gp/[Na^+$/ATP stoichiometry].

5.8.5 Uncouplers

Many compounds are known to inhibit ATP synthesis with an increase in O_2 consumption during respiration. These are known as uncouplers. They are weak acids or weak bases, and are permeable to the membrane since they are hydrophobic not only in undissociated forms but also in dissociated forms. In a respiring cell, the internal pH is higher than the external pH. Undissociated uncouplers at low external pH diffuse into the cell, where they dissociate at the higher internal pH. Dissociated forms diffuse out of the cell. The net result is H^+ transport into the cell. In this way the uncouplers dissipate Δp (Figure 5.24). In the presence of an uncoupler, ATP cannot be synthesized due to the low Δp, and the cell consumes more O_2 to increase Δp.

Chemicals that increase the permeability of ions are referred to as ionophores. Uncouplers are therefore ionophores for H^+. Alternatively, they can be called protonophores. Figure 5.25 shows some examples of uncouplers.

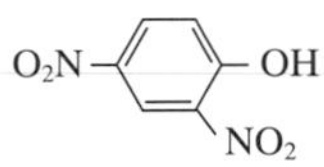

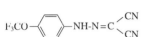

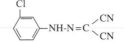

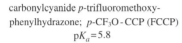

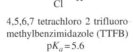

2,4-dinitrophenol
$pK_a = 4.1$

carbonylcyanide-*m*-chlorophenylhydrazone (CCCP)
$pK_a = 6.55$

carbonylcyanide *p*-trifluoromethoxy-
phenylhydrazone; *p*-CF$_3$O-CCP (FCCP)
$pK_a = 5.8$

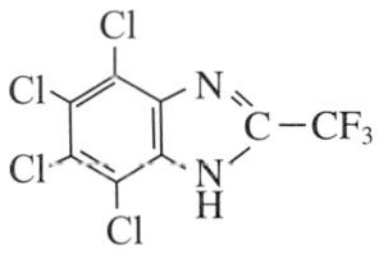

4,5,6,7 tetrachloro 2 trifluoro
methylbenzimidazole (TTFB)
$pK_a = 5.6$

Figure 5.25 Structures of some uncouplers.

5.8.6 Primary H^+ (Na^+) pumps in fermentative metabolism

The free energy changes available from the oxidation–reduction reactions in respiration and photosynthesis are conserved in the form of Δp. In addition to O_2, prokaryotes use various other electron acceptors for their respiration. The term anaerobic respiration is used to describe the energy conservation process using electron acceptors other than O_2. Some primary H^+(Na^+) active transport processes are known in fermentative bacteria where internally supplied electron acceptors are used (Section 8.1).

5.8.6.1 Fumarate reductase

Species of *Propionibacterium* ferment lactate to propionate through the succinate–propionate pathway (Figure 8.16, Section 8.7.1). Fumarate reductase of this pathway reduces fumarate to succinate, extruding H^+. A similar energy conservation process is known in *Vibrio succinogenes*, *Desulfovibrio gigas* and *Clostridium formicoaceticum*. This membrane-bound enzyme uses H_2 and formate as the electron donors and needs menaquinone and cytochrome *b* as coenzymes (Figure 5.26). Strictly speaking, the latter example is anaerobic respiration, since the electron acceptor, fumarate, is externally supplied.

5.8.6.2 Na$^+$-dependent decarboxylase

Klebsiella pneumoniae (*Klebsiella aerogenes*) has a Na$^+$-dependent methyl-malonyl-CoA decarboxylase. This enzyme is membrane-bound, and exports Na$^+$ coupled to the decarboxylation reaction. The free energy change of decarboxylation (-30 kJ/mol methyl-malonyl-CoA) is conserved as a sodium motive force, which is converted to Δp by a Na$^+$/H$^+$ antiporter to synthesize ATP. Figure 5.27 shows a decarboxylation reaction conserving energy as a sodium motive force.

Energy is similarly conserved by glutaconyl-CoA decarboxylase in *Acidamicoccus fermentans* and *Clostridium symbiosum*, and by

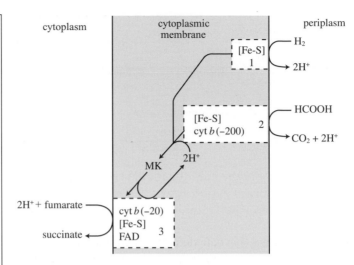

Figure 5.26 ATP synthesis through fumarate reductase.

(Gottschalk, G. 1986, *Bacterial Metabolism*, 2nd edn., Figure 8.22. Springer, New York.)

A membrane-bound electron carrier is reduced oxidizing hydrogen leaving $2H^+$ outside. The electrons are consumed to reduce fumarate consuming $2H^+$ in the cell. Fumarate reductase generates Δp in this mechanism. Species of *Propionibacterium* have a similar energy conservation process during the fermentation of lactate to propionate (Figure 8.16). Numbers in boxes are the redox potentials of the cytochromes: 1, hydrogenase; 2, formate dehydrogenase; 3, fumarate reductase.

succinate decarboxylase in *Veillonella parvula*. Energy conservation is not known in the decarboxylation of oxaloacetate in the citrate fermentation by *Pediococcus halophilus* and *Lactococcus lactis* (Figures 8.4, 8.5, Section 8.4.6).

5.8.6.3 Δp formation through fermentation product/H⁺ symport

A lactic-acid bacterium, *Lactococcus cremoris*, extrudes H^+ by a lactate/H^+ symporter, utilizing the potential energy developed by the lactate gradient due to the high internal lactate concentration in the cell (Figure 5.28). When lactate has accumulated in the environment, H^+ cannot be symported with lactate. In the lactic acid bacterial malolactic fermentation (Section 8.4.6) a malate^{2-}/lactate$^-$ antiporter imports malate in exchange with lactate (Figure 3.6). This reaction generates $\Delta\psi$ since malate^{2-} is imported exporting lactate$^-$. In addition, H^+ is consumed in the malate decarboxylation reaction (Figure 3.6).

5.9 | Other biological energy transduction processes

Figure 5.7 illustrates energy transduction processes in microbes. Chemotrophs convert chemical energy into biological energy, and light energy is used by phototrophs. The biological energy in the form of ATP and Δp is invested in energy-requiring processes including transport (Chapter 3), biosynthesis (Chapter 6) and motility. Not included in the figure is bioluminescence that is found in certain bacteria.

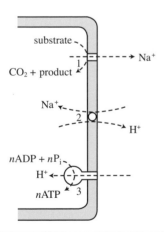

Figure 5.27 Δp generation by Na⁺-dependent decarboxylase.

(Gottschalk, G. 1986, *Bacterial Metabolism*, 2nd edn., Figure 8.17. Springer, New York.)

1, oxaloacetate decarboxylase; 2, Na⁺/H⁺ antiporter; 3, F₁Fₒ-ATPase.

5.9.1 Bacterial bioluminescence

Luminous bacteria convert biological energy into light. Some of these are symbionts and others are free-living (Table 5.3). *Photobacterium*

Table 5.3.	*Luminescent bacteria*		
Habitat	Strain	Symbiosis	
Marine	*Photobacterium fischerii*	fish	
Marine	*Vibrio harveyi*	none	
Fresh water	*Vibrio cholerae* biotype *albensis*	none	
Soil	*Xenorhabdus nematophilus*	nematode	

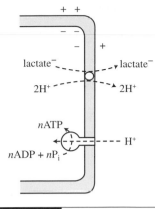

Figure 5.28 Generation of Δp by lactate/ H^+ symport in lactic-acid bacteria.

fischerii lives symbiotically in the light organs of certain fish, while a soil nematode is the habitat of *Xenorhabdus nematophilus*. Some members of the *Vibrio* genus are free-living luminous bacteria in marine and freshwater ecosystems.

Light is emitted when the cell density reaches certain levels. Luminescent bacteria excrete *N*-acyl homoserine lactone as an autoinducer. The components of the bioluminescence reaction are induced only above the threshold concentration of this autoinducer in the environment. This induction is referred to as quorum sensing (Section 12.2.8) and the reaction is strictly O_2 dependent. $FMNH_2$ oxidation provides energy for the luminescence. The bacterial luciferase forms a complex binding $FMNH_2$, O_2 and a long chain aliphatic aldehyde to emit light (Figure 5.29). The bacterial luciferase is a kind of monooxygenase (Section 7.7), and requires aldehyde for light emission. Aldehyde is oxidized to fatty acid by the luciferase, and reduction to aldehyde is coupled to the oxidation of NADH.

5.9.2 Electricity as an energy source in microbes

Chemotrophs use chemical energy as their energy source, and light energy is used by phototrophs. Chemical and light energy are converted into biological energy that is used for various biological functions. These include biosynthesis (chemical energy), transport (potential energy), motility (mechanical energy) and luminescence (light energy). In electric fishes, biological energy is converted into electricity.

There is some evidence which shows that electricity is used as the energy source in denitrification. Strains of *Geobacter* reduce nitrate with an electrode as the electron donor. An enrichment culture can be made using an electrode as the electron donor and nitrate as the electron acceptor. Bacteria can grow in a similar electrochemical device with added nitrate as the electron acceptor.

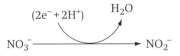

Electrochemically reduced artificial electron carriers such as methyl viologen and neutral red are oxidized by bacteria to produce more reduced products. The artificial electron carriers mediate electron transfer from the electrode to the bacterial cells. *Brevibacterium*

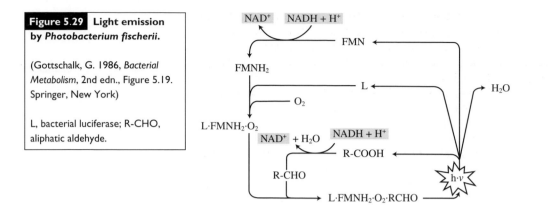

Figure 5.29 **Light emission by *Photobacterium fischerii*.**

(Gottschalk, G. 1986, *Bacterial Metabolism*, 2nd edn., Figure 5.19. Springer, New York)

L, bacterial luciferase; R-CHO, aliphatic aldehyde.

flavum produces more glutamate with neutral red in an electrochemical device. Similarly, anaerobic bacteria including *Clostridium acetobutylicum*, *Desulfovibrio desulfuricans* and *Propionibacterium freudenreichii* incorporate the electrons supplied electrochemically into their normal electron metabolism. Humic substances are known to shuttle electrons between solid surfaces and bacterial cells in natural ecosystems.

FURTHER READING

TCA cycle

Brasen, C. & Schonheit, P. (2004). Unusual ADP-forming acetyl-coenzyme A synthetases from the mesophilic halophilic euryarchaeon *Haloarcula marismortui* and from the hyperthermophilic crenarchaeon *Pyrobaculum aerophilum*. *Archives of Microbiology* **182**, 277–287.

Camacho, M., Rodriguez-Arnedo, A. & Bonete, M. J. (2002). NADP-dependent isocitrate dehydrogenase from the halophilic archaeon *Haloferax volcanii*: cloning, sequence determination and overexpression in *Escherichia coli*. *FEMS Microbiology Letters* **209**, 155–160.

Hu, Y. & Holden, J. F. (2006). Citric acid cycle in the hyperthermophilic archaeon *Pyrobaculum islandicum* grown autotrophically, heterotrophically, and mixotrophically with acetate. *Journal of Bacteriology* **188**, 4350–4355.

Lefebre, M. D., Flannagan, R. S. & Valvano, M. A. (2005). A minor catalase/peroxidase from *Burkholderia cenocepacia* is required for normal aconitase activity. *Microbiology-UK* **151**, 1975–1985.

Makarova, K. S. & Koonin, E. V. (2003). Filling a gap in the central metabolism of archaea: prediction of a novel aconitase by comparative-genomic analysis. *FEMS Microbiology Letters* **227**, 17–23.

Molenaar, D., van der Rest, M. E., Drysch, A. & Yucel, R. (2000). Functions of the membrane-associated and cytoplasmic malate dehydrogenases in the citric acid cycle of *Corynebacterium glutamicum*. *Journal of Bacteriology* **182**, 6884–6891.

Muschko, K., Kienzlen, G., Fiedler, H.-P., Wohlleben, W. & Schwartz, D. (2002). Tricarboxylic acid cycle aconitase activity during the life cycle of *Streptomyces viridochromogenes* Tu494. *Archives of Microbiology* **178**, 499–505.

Nakano, M. M., Zuber, P. & Sonenshein, A. L. (1998). Anaerobic regulation of *Bacillus subtilis* Krebs cycle genes. *Journal of Bacteriology* **180**, 3304–3311.

Serio, A. W., Pechter, K. B. & Sonenshein, A. L. (2006). *Bacillus subtilis* aconitase is required for efficient late-sporulation gene expression. *Journal of Bacteriology* **188**, 6396–6405.

Tang, Y., Guest, J. R., Artymiuk, P. J., Read, R. C. & Green, J. (2004). Post-transcriptional regulation of bacterial motility by aconitase proteins. *Molecular Microbiology* **51**, 1817–1826.

van der Rest, M. E., Frank, C. & Molenaar, D. (2000). Functions of the membrane-associated and cytoplasmic malate dehydrogenases in the citric acid cycle of *Escherichia coli*. *Journal of Bacteriology* **182**, 6892–6899.

Zamboni, N. & Sauer, U. (2003). Knockout of the high-coupling cytochrome aa_3 oxidase reduces TCA cycle fluxes in *Bacillus subtilis*. *FEMS Microbiology Letters* **226**, 121–126.

Anaplerotic sequence

Alber, B. E., Spanheimer, R., Ebenau-Jehle, C. & Fuchs, G. (2006). Study of an alternate glyoxylate cycle for acetate assimilation by *Rhodobacter sphaeroides*. *Molecular Microbiology* **61**, 297–309.

El-Mansi, M., Cozzone, A. J., Shiloach, J. & Eikmanns, B. J. (2006). Control of carbon flux through enzymes of central and intermediary metabolism during growth of *Escherichia coli* on acetate. *Current Opinion in Microbiology* **9**, 173–179.

Ensign, S. A. (2006). Revisiting the glyoxylate cycle: alternate pathways for microbial acetate assimilation. *Molecular Microbiology* **61**, 274–276.

Fukuda, W., Fukui, T., Atomi, H. & Imanaka, T. (2004). First characterization of an archaeal GTP-dependent phosphoenolpyruvate carboxykinase from the hyperthermophilic archaeon *Thermococcus kodakaraensis* KOD1. *Journal of Bacteriology* **186**, 4620–4627.

Gould, T. A., van de Langemheen, H., Munoz-Elias, E. J., McKinney, J. D. & Sacchettini, J. C. (2006). Dual role of isocitrate lyase 1 in the glyoxylate and methylcitrate cycles in *Mycobacterium tuberculosis*. *Molecular Microbiology* **61**, 940–947.

Kim, H. J., Kim, T. H., Kim, Y. & Lee, H. S. (2004). Identification and characterization of *glxR*, a gene involved in regulation of glyoxylate bypass in *Corynebacterium glutamicum*. *Journal of Bacteriology* **186**, 3453–3460.

Koebmann, B. J., Westerhoff, H. V., Snoep, J. L., Nilsson, D. & Jensen, P. R. (2002). The glycolytic flux in *Escherichia coli* is controlled by the demand for ATP. *Journal of Bacteriology* **184**, 3909–3916.

Maharjan, R. P., Yu, P. L., Seeto, S. & Ferenci, T. (2005). The role of isocitrate lyase and the glyoxylate cycle in *Escherichia coli* growing under glucose limitation. *Research in Microbiology* **156**, 178–183.

Netzer, R., Krause, M., Rittmann, D., Peters-Wendisch, P. G., Eggeling, L., Wendisch, V. F. & Sahm, H. (2004). Roles of pyruvate kinase and malic enzyme in *Corynebacterium glutamicum* for growth on carbon sources requiring gluconeogenesis. *Archives of Microbiology* **182**, 354–363.

Sauer, U. & Eikmanns, B. J. (2005). The PEP-pyruvate-oxaloacetate node as the switch point for carbon flux distribution in bacteria. *FEMS Microbiology Reviews* **29**, 765–794.

Wang, Z. X., Bramer, C. O. & Steinbuchel, A. (2003). The glyoxylate bypass of *Ralstonia eutropha*. *FEMS Microbiology Letters* **228**, 63–71.

Incomplete TCA fork and reductive TCA cycle

Atomi, H. (2002). Microbial enzymes involved in carbon dioxide fixation. *Journal of Bioscience and Bioengineering* **94**, 497–505.

Campbell, B. J. & Cary, S. C. (2004). Abundance of reverse tricarboxylic acid cycle genes in free-living microorganisms at deep-sea hydrothermal vents. *Applied and Environmental Microbiology* **70**, 6282–6289.

Yamamoto, M., Arai, H., Ishii, M. & Igarashi, Y. (2006). Role of two 2-oxoglutarate:ferredoxin oxidoreductases in *Hydrogenobacter thermophilus* under aerobic and anaerobic conditions. *FEMS Microbiology Letters* **263**, 189–193.

Energy transduction in prokaryotes

Amend, J. P. & Shock, E. L. (2001). Energetics of overall metabolic reactions of thermophilic and hyperthermophilic Archaea and Bacteria. *FEMS Microbiology Reviews* **25**, 175–243.

Battistuzzi, G., D'Onofrio, M., Borsari, M., Sola, M., Macedo, A. L., Moura, J. J. G. & Rodrigues, P. (2000). Redox thermodynamics of low-potential iron-sulfur proteins. *Journal of Biological Inorganic Chemistry* **5**, 748–760.

Neijssel, O. M. & Demattos, M. J. T. (1994). The energetics of bacterial growth: a reassessment. *Molecular Microbiology* **13**, 179–182.

Schaefer, G., Engelhard, M. & Mueller, V. (1999). Bioenergetics of the Archaea. *Microbiology and Molecular Biology Reviews* **63**, 570–620.

von Stockar, U., Maskow, T., Liu, J., Marison, I. W. & Patino, R. (2006). Thermodynamics of microbial growth and metabolism: an analysis of the current situation. *Journal of Bacteriology* **121**, 517–533.

Yumoto, I. (2002). Bioenergetics of alkaliphilic *Bacillus* spp. *Journal of Bioenergetics and Biomembranes* **93**, 342–353.

Adenosine triphosphate (ATP) and ATPase

Au, K. M., Barabote, R. D., Hu, K. Y. & Saier, M. H. J. (2006). Evolutionary appearance of H^+-translocating pyrophosphatases. *Microbiology-UK* **152**, 1243–1247.

Barriuso-Iglesias, M., Barreiro, C., Flechoso, F. & Martin, J. F. (2006). Transcriptional analysis of the F_oF_1 ATPase operon of *Corynebacterium glutamicum* ATCC 13032 reveals strong induction by alkaline pH. *Microbiology-UK* **152**, 11–21.

Brusilow, W. S. A. (1993). Assembly of the *Escherichia coli* F_1F_o ATPase, a large multimeric membrane-bound enzyme. *Molecular Microbiology* **9**, 419–424.

Capaldi, R. & Aggeler, R. (2002). Mechanism of the F_1F_o-type ATP synthase, a biological rotary motor. *Trends in Biochemical Sciences* **27**, 154–160.

Deckershebestreit, G. & Altendorf, K. (1996). The F_1F_o-type ATP synthases of bacteria: structure and function of the F_o complex. *Annual Review of Microbiology* **50**, 791–824.

Dimroth, P. & Cook, G. M. (2004). Bacterial Na^+- or H^+-coupled ATP synthases operating at low electrochemical potential. *Advances in Microbial Physiology* **49**, 175–218.

Ferguson, S. A., Keis, S. & Cook, G. M. (2006). Biochemical and molecular characterization of a Na^+-translocating F_1F_o-ATPase from the thermoalkaliphilic bacterium *Clostridium paradoxum*. *Journal of Bacteriology* **188**, 5045–5054.

Iida, T., Inatomi, K., Kamagata, Y. & Maruyama, T. (2002). F- and V-type ATPases in the hyperthermophilic bacterium *Thermotoga neapolitana*. *Extremophiles* **6**, 369–375.

Kinosita, K. Jr., Adachi, K., & Itoh, H. (2004). Rotation of F_1-ATPase: how an ATP-driven molecular machine may work. *Annual Review of Biophysics and Biomolecular Structure* **33**, 245–268.

Lapierre, P., Shial, R. & Gogarten, J. P. (2006). Distribution of F- and A/V-type ATPases in *Thermus scotoductus* and other closely related species. *Systematic and Applied Microbiology* **29**, 15–23.

Muller, V., Lemker, T., Lingl, A., Weidner, C., Coskun, U. & Gruber, G. (2005). Bioenergetics of archaea: ATP synthesis under harsh environmental conditions. *Journal of Molecular Microbiology and Biotechnology* **10**, 167–180.

Noda, S., Takezawa, Y., Mizutani, T., Asakura, T., Nishiumi, E., Onoe, K., Wada, M., Tomita, F., Matsushita, K. & Yokota, A. (2006). Alterations of cellular physiology in *Escherichia coli* in response to oxidative phosphorylation impaired by defective F_1-ATPase. *Journal of Bacteriology* **188**, 6869–6876.

Pitryuk, A. V. & Pusheva, M. A. (2001). Different ionic specificities of ATP synthesis in extremely alkaliphilic sulfate-reducing and acetogenic bacteria. *Microbiology-Moscow* **70**, 398–402.

Schafer, I. B., Bailer, S. M., Duser, M. G., Borsch, M., Bernal, R. A., Stock, D. & Gruber, G. (2006). Crystal structure of the archaeal A_1A_o ATP synthase subunit B from *Methanosarcina mazei* Go1: implications of nucleotide-binding differences in the major A_1A_o subunits A and B. *Journal of Molecular Biology* **358**, 725–740.

Proton (sodium) motive force, and acid and alkali tolerance

Arnold, C. N., McElhanon, J., Lee, A., Leonhart, R. & Siegele, D. A. (2001). Global analysis of *Escherichia coli* gene expression during the acetate-induced acid tolerance response. *Journal of Bacteriology* **183**, 2178–2186.

Azcarate-Peril, M. A., Altermann, E., Hoover-Fitzula, R. L., Cano, R. J. & Klaenhammer, T. R. (2004). Identification and inactivation of genetic loci involved with *Lactobacillus acidophilus* acid tolerance. *Applied and Environmental Microbiology* **70**, 5315–5322.

Cotter, P. D. & Hill, C. (2003). Surviving the acid test: responses of Gram-positive bacteria to low pH. *Microbiology and Molecular Biology Reviews* **67**, 429–453.

Dover, N. & Padan, E. (2001). Transcription of *nhaA*, the main Na^+/H^+ antiporter of *Escherichia coli*, is regulated by Na^+ and growth phase. *Journal of Bacteriology* **183**, 644–653.

Flythe, M. D. & Russell, J. B. (2005). The ability of acidic pH, growth inhibitors, and glucose to increase the proton motive force and energy spilling of amino acid-fermenting *Clostridium sporogenes* MD1 cultures. *Archives of Microbiology* **183**, 236–242.

Foster, J. W. (1999). When protons attack: microbial strategies of acid adaptation. *Current Opinion in Microbiology* **2**, 170–174.

Fozoa, E. M., Kajfasza, J. K. & Quivey, R. G., Jr. (2004). Low pH-induced membrane fatty acid alterations in oral bacteria. *FEMS Microbiology Letters* **238**, 291–295.

Herz, K., Vimont, S., Padan, E. & Berche, P. (2003). Roles of NhaA, NhaB, and NhaD Na^+/H^+ antiporters in survival of *Vibrio cholerae* in a saline environment. *Journal of Bacteriology* **185**, 1236–1244.

Hunte, C., Screpanti, E., Venturi, M., Rimon, A., Padan, E. & Michel, H. (2005). Structure of a Na^+/H^+ antiporter and insights into mechanism of action and regulation by pH. *Nature* **435**, 1197–1202.

Kieboom, J. & Abee, T. (2006). Arginine-dependent acid resistance in *Salmonella enterica* serovar *typhimurium*. *Journal of Bacteriology* **188**, 5650–5653.

Kim, J. S., Sung, M. H., Kho, D. H. & Lee, J. K. (2005). Induction of manganese-containing superoxide dismutase is required for acid tolerance in *Vibrio vulnificus*. *Journal of Bacteriology* **187**, 5984–5995.

Kitada, M., Kosono, S. & Kudo, T. (2000). The Na^+/H^+ antiporter of alkaliphilic *Bacillus* sp. *Extremophiles* **4**, 253–258.

Leaphart, A. B., Thompson, D. K., Huang, K., Alm, E., Wan, X. F., Arkin, A., Brown, S. D., Wu, L., Yan, T., Liu, X., Wickham, G. S. & Zhou, J. (2006). Transcriptome profiling of *Shewanella oneidensis* gene expression following exposure to acidic and alkaline pH. *Journal of Bacteriology* **188**, 1633–1642.

Liu, J., Xue, Y., Wang, Q., Wei, Y., Swartz, T. H., Hicks, D. B., Ito, M., Ma, Y. & Krulwich, T. A. (2005). The activity profile of the NhaD-type $Na^+(Li^+)/H^+$ antiporter from the soda lake haloalkaliphile *Alkalimonas amylolytica* is adaptive for the extreme environment. *Journal of Bacteriology* **187**, 7589–7595.

Ma, Z., Richard, H., Tucker, D. L., Conway, T. & Foster, J. W. (2002). Collaborative regulation of *Escherichia coli* glutamate-dependent acid resistance by two AraC-like regulators, GadX and GadW (YhiW). *Journal of Bacteriology* **184**, 7001–7012.

Ma, Z., Gong, S., Richard, H., Tucker, D. L., Conway, T. & Foster, J. W. (2003). GadE (YhiE) activates glutamate decarboxylase-dependent acid resistance in *Escherichia coli* K-12. *Molecular Microbiology* **49**, 1309–1320.

Ma, Z., Richard, H. & Foster, J. W. (2003). pH-dependent modulation of cyclic AMP levels and GadW-dependent repression of RpoS affect synthesis of the GadX regulator and *Escherichia coli* acid resistance. *Journal of Bacteriology* **185**, 6852–6859.

Martin-Galiano, A. J., Overweg, K., Ferrandiz, M. J., Reuter, M., Wells, J. M. & de la Campa, A. G. (2005). Transcriptional analysis of the acid tolerance response in *Streptococcus pneumoniae*. *Microbiology-UK* **151**, 3935–3946.

Miwa, T., Abe, T., Fukuda, S., Ohkawara, S. & Hino, T. (2001). Regulation of H^+-ATPase synthesis in response to reduced pH in ruminal bacteria. *Current Microbiology* **42**, 106–110.

Padan, E., Tzubery, T., Herz, K., Kozachkov, L., Rimon, A. & Galili, L. (2004). NhaA of *Escherichia coli*, as a model of a pH-regulated Na^+/H^+ antiporter. *Biochimica et Biophysica Acta – Bioenergetics* **1658**, 2–13.

Padan, E., Bibi, E., Ito, M. & Krulwich, T. A. (2005). Alkaline pH homeostasis in bacteria: new insights. *Biochimica et Biophysica Acta – Biomembranes* **1717**, 67–88.

Palmer, G. & Reedijk, J. (1992). Nomenclature of electron-transfer proteins. Recommendations 1989. *Journal of Biological Chemistry* **267**, 665–677.

Rhee, J. E., Kim, K. S. & Choi, S. H. (2005). CadC activates pH-dependent expression of the *Vibrio vulnificus cadBA* operon at a distance through direct binding to an upstream region. *Journal of Bacteriology* **187**, 7870–7875.

Rhee, J. E., Jeong, H. G., Lee, J. H. & Choi, S. H. (2006). AphB influences acid tolerance of *Vibrio vulnificus* by activating expression of the positive regulator CadC. *Journal of Bacteriology* **188**, 6490–6497.

Small, P. L. & Waterman, S. R. (1998). Acid stress, anaerobiosis and *gadCB*: lessons from *Lactococcus lactis* and *Escherichia coli*. *Trends in Microbiology* **6**, 214–216.

Electron transport phosphorylation

Bott, M. & Niebisch, A. (2003). The respiratory chain of *Corynebacterium glutamicum*. *Journal of Biotechnology* **104**, 129–153.

Branden, G., Gennis, R.B. & Brzezinski, P. (2006). Transmembrane proton translocation by cytochrome *c* oxidase. *Biochimica et Biophysica Acta – Bioenergetics* **1757**, 1052–1063.

Brandt, U. (2006). Energy converting NADH:quinone oxidoreductase (complex I). *Annual Review of Biochemistry* **75**, 69–92.

Brandt, U. & Trumpower, B. (1994). The protonmotive Q cycle in mitochondria and bacteria. *Critical Reviews in Biochemistry and Molecular Biology* **29**, 165–197.

Brzezinski, P. & Adelroth, P. (2006). Design principles of proton-pumping haem-copper oxidases. *Current Opinion in Structural Biology* **16**, 465–472.

Brzezinski, P. & Larsson, G. (2003). Redox-driven proton pumping by heme-copper oxidases. *Biochimica et Biophysica Acta – Bioenergetics* **1605**, 1–13.

Cosseau, C. & Batut, J. (2004). Genomics of the *ccoNOQP*-encoded *cbb* 3 oxidase complex in bacteria. *Archives of Microbiology* **181**, 89–96.

Crofts, A.R. (2004). The cytochrome bc_1 complex: function in the context of structure. *Annual Review of Physiology* **66**, 689–733.

Crofts, A.R., Lhee, S., Crofts, S.B., Cheng, J. & Rose, S. (2006). Proton pumping in the bc_1 complex: a new gating mechanism that prevents short circuits. *Biochimica et Biophysica Acta – Bioenergetics* **1757**, 1019–1034.

Degier, J.W.L., Lubben, M., Reijnders, W.N.M., Tipker, C.A., Slotboom, D.J., Vanspanning, R.J.M., Stouthamer, A.H. & Vanderoost, J. (1994). The terminal oxidases of *Paracoccus denitrificans*. *Molecular Microbiology* **13**, 183–196.

Faxen, K., Gilderson, G., Adelroth, P. & Brzezinski, P. (2005). A mechanistic principle for proton pumping by cytochrome *c* oxidase. *Nature* **437**, 286–289.

Friedrich, T. & Bottcher, B. (2004). The gross structure of the respiratory complex I: a Lego System. *Biochimica et Biophysica Acta – Bioenergetics* **1608**, 1–9.

Hickman, J.W., Barber, R.D., Skaar, E.P. & Donohue, T.J. (2002). Link between the membrane-bound pyridine nucleotide transhydrogenase and glutathione-dependent processes in *Rhodobacter sphaeroides*. *Journal of Bacteriology* **184**, 400–409.

Hinsley, A.P. & Berks, B.C. (2002). Specificity of respiratory pathways involved in the reduction of sulfur compounds by *Salmonella enterica*. *Microbiology-UK* **148**, 3631–3638.

Hosler, J.P., Ferguson-Miller, S. & Mills, D.A. (2006). Energy transduction: proton transfer through the respiratory complexes. *Annual Review of Biochemistry* **75**, 165–187.

Hunte, C., Palsdottir, H. & Trumpower, B.L. (2003). Protonmotive pathways and mechanisms in the cytochrome bc_1 complex. *FEBS Letters* **545**, 39–46.

Ishikawa, R., Ishido, Y., Tachikawa, A., Kawasaki, H., Matsuzawa, H. & Wakagi, T. (2002). *Aeropyrum pernix* K1, a strictly aerobic and hyperthermophilic archaeon, has two terminal oxidases, cytochrome ba_3 and cytochrome aa_3. *Archives of Microbiology* **179**, 42–49.

Johnson, D.C., Dean, D.R., Smith, A.D. & Johnson, M.K. (2005). Structure, function and formation of iron-sulfur clusters. *Annual Review of Biochemistry* **74**, 247–281.

Lalucque, H. & Silar, P. (2003). NADPH oxidase: an enzyme for multicellularity? *Trends in Microbiology* **11**, 9–12.

Lanyi, J. K. & Pohorille, A. (2001). Proton pumps: mechanism of action and applications. *Trends in Biotechnology* **19**, 140–144.

Link, T. A. (1997). The role of the 'Rieske' iron sulfur protein in the hydroquinone oxidation (Q_P) site of the cytochrome bc_1 complex. The 'proton-gated affinity change' mechanism. *FEBS Letters* **412**, 257–264.

Ludwig, R. A. (2004). Microaerophilic bacteria transduce energy via oxidative metabolic gearing. *Research in Microbiology* **155**, 61–70.

Melo, A., Bandeiras, T. & Teixeira, M. (2004). New insights into Type II NAD(P)H:quinone oxidoreductases. *Microbiology and Molecular Biology Reviews* **68**, 603–616.

Mendz, G. L., Smith, M. A., Finel, M. & Korolik, V. (2000). Characteristics of the aerobic respiratory chains of the microaerophiles *Campylobacter jejuni* and *Helicobacter pylori*. *Archives of Microbiology* **174**, 1–10.

Minohara, S., Sakamoto, J. & Sone, N. (2002). Improved H^+/O ratio and cell yield of *Escherichia coli* with genetically altered terminal quinol oxidases. *Journal of Bioscience and Bioengineering* **93**, 464–469.

Morales, G., Ugidos, A. & Rojo, F. (2006). Inactivation of the *Pseudomonas putida* cytochrome *o* ubiquinol oxidase leads to a significant change in the transcriptome and to increased expression of the CIO and *cbb*3-1 terminal oxidases. *Environmental Microbiology* **8**, 1764–1774.

Mulkidjanian, A. Y. (2005). Ubiquinol oxidation in the cytochrome bc_1 complex: reaction mechanism and prevention of short-circuiting. *Biochimica et Biophysica Acta – Bioenergetics* **1709**, 5–34.

Osyczka, A., Moser, C. C. & Dutton, P. L. (2005). Fixing the Q cycle. *Trends in Biochemical Sciences* **30**, 176–182.

Otten, M. F., Stork, D. R., Reijnders, W. N. M., Westerhoff, H. V. & van Spanning, R. J. M. (2001). Regulation of expression of terminal oxidases in *Paracoccus denitrificans*. *European Journal of Biochemistry* **268**, 2486–2497.

Rich, P. R. (1986). A perspective on Q-cycles. *Journal of Bioenergetics and Biomembranes* **18**, 145–156.

Rich, P. R. (2004). The quinone chemistry of *bc* complexes. *Biochimica et Biophysica Acta – Bioenergetics* **1658**, 165–171.

Richardson, D. J. (2000). Bacterial respiration: a flexible process for a changing environment. *Microbiology-UK* **146**, 551–571.

Schneider, D. & Schmidt, C. L. (2005). Multiple Rieske proteins in prokaryotes: where and why? *Biochimica et Biophysica Acta – Bioenergetics* **1710**, 1–12.

Shimada, H., Shida, Y., Nemoto, N., Oshima, T. & Yamagishi, A. (2001). Quinone profiles of *Thermoplasma acidophilum* HO-62. *Journal of Bacteriology* **183**, 1462–1465.

Snyder, C. H., Merbitz-Zahradnik, T., Link, T. A. & Trumpower, B. L. (1999). Role of the Rieske iron-sulfur protein midpoint potential in the proton-motive Q-cycle mechanism of the cytochrome bc_1 complex. *Journal of Bioenergetics and Biomembranes* **31**, 235–242.

Trumpower, B. L. & Gennis, R. B. (1994). Energy transduction by cytochrome complexes in mitochondrial and bacterial respiration: the enzymology of coupling electron transfer reactions to transmembrane proton translocation. *Annual Review of Biochemistry* **63**, 675–716.

Wikstrom, M. & Verkhovsky, M. I. (2006). Towards the mechanism of proton pumping by the haem-copper oxidases. *Biochimica et Biophysica Acta – Bioenergetics* **1757**, 1047–1051.

Williams, H. D., Zlosnik, J. E. A. & Ryall, B. (2006). Oxygen, cyanide and energy generation in the cystic fibrosis pathogen *Pseudomonas aeruginosa*. *Advances in Microbial Physiology* **52**, 1–71.

Zlosnik, J. E. A., Tavankar, G. R., Bundy, J. G., Mossialos, D., O'Toole, R. & Williams, H. D. (2006). Investigation of the physiological relationship between the cyanide-insensitive oxidase and cyanide production in *Pseudomonas aeruginosa*. *Microbiology-UK* **152**, 1407–1415.

Other prokaryotic energy transduction

Abe, K., Ohnishi, F., Yagi, K., Nakajima, T., Higuchi, T., Sano, M., Machida, M., Sarker, R. I. & Maloney, P. C. (2002). Plasmid-encoded *asp* operon confers a proton motive metabolic cycle catalyzed by an aspartate-alanine exchange reaction. *Journal of Bacteriology* **184**, 2906–2913.

Dimroth, P. & Schink, B. (1998). Energy conservation in the decarboxylation of dicarboxylic acids in fermenting bacteria. *Archives of Microbiology* **170**, 69–77.

Gregory, K. B., Bond, D. R. & Lovley, D. R. (2004). Graphite electrodes as electron donors for anaerobic respiration. *Environmental Microbiology* **6**, 596–604.

Iyer, R., Williams, C. & Miller, C. (2003). Arginine-agmatine antiporter in extreme acid resistance in *Escherichia coli*. *Journal of Bacteriology* **185**, 6556–6561.

Kroger, A., Geisler, V., Lemma, E., Theis, F. & Lenger, R. (1992). Bacterial fumarate respiration. *Archives of Microbiology* **158**, 311–314.

Okane, D. J. & Prasher, D. C. (1992). Evolutionary origins of bacterial bioluminescence. *Molecular Microbiology* **6**, 443–449.

Poolman, B. (1993). Energy transduction in lactic acid bacteria. *FEMS Microbiology Reviews* **12**, 125–147.

Wolken, W. A. M., Lucas, P. M., Lonvaud-Funel, A. & Lolkema, J. S. (2006). The mechanism of the tyrosine transporter TyrP supports a proton motive tyrosine decarboxylation pathway in *Lactobacillus brevis*. *Journal of Bacteriology* **188**, 2198–2206.

6

Biosynthesis and microbial growth

Chapters 4 and 5 describe and explain the anabolic reactions that supply carbon skeletons, reducing equivalent (NADPH) and adenosine 5'-triphosphate (ATP) needed for biosynthesis. This chapter summarizes how the products of such anabolic reactions are used in biosynthesis and growth, ranging from monomer synthesis to the assembly of macromolecules within cells. Chemoheterotrophs, such as *Escherichia coli*, use approximately half of the glucose consumed to synthesize cell materials while the other half is oxidized to carbon dioxide under aerobic conditions.

6.1 | Molecular composition of bacterial cells

The elemental composition of microbial cells was discussed in Chapter 2 in order to help understand what materials the bacteria use as their nutrients. These elements make up a range of molecules with various functions. Cellular molecular composition varies depending on the strain and growth conditions. As an example, Table 6.1 lists the molecular composition of *Escherichia coli* during the logarithmic phase when grown on a glucose–mineral salts medium. The moisture content is over 70%, and protein is most abundant, occupying 55% of the dry cell weight, followed by RNA at about 20%. It is understandable that proteins are abundant since they catalyze cellular reactions. The DNA content is least variable, while the RNA content is higher at a higher growth rate. Not shown in Table 6.1 are storage materials, such as poly-β-hydroxybutyrate and glycogen, which vary profoundly within cells depending on growth conditions, and can comprise up to 70% of the cell dry weight (Section 13.2).

The biosynthetic process, known as anabolism, can be discussed in three steps:

monomer biosynthesis
polymerization of monomers
assembly of polymers into cellular structure.

Table 6.1. | *Molecular composition of an* Escherichia coli *cell*[a]

Component	Content (%)	Average molecular weight	Molecules/cell	Variety
Protein	55.0	40 000	2 360 000	1 050
RNA	20.5			
23S rRNA		1 000 000	18 700	1
16S rRNA		500 000	18 700	1
5S rRNA		39 000	18 700	1
tRNA		25 000	205 000	60
mRNA		1 000 000	1 380	400
DNA	3.1	2.5×10^9	2.13	1
Lipid	9.1	705	22 000 000	4[b]
LPS	3.4	4 346	1 200 000	1[b]
Murein	2.5	?	1	1
Glycogen	2.5	1 000 000	4 360	1
Low molecular weight organics	2.9	?	?	?
Inorganics	1.0	?	?	?

Dry weight of a single cell, 0.28 picogram (pg).
Moisture content of the cell, 0.67 pg.
Total weight of a single cell, 0.95 pg.
[a] Composition of the cells in the log phase. Cells were grown in glucose–mineral salts medium at 37 °C (mass doubling time 40 min).
[b] Phospholipids in four groups regardless of the fatty acid composition.
LPS, lipopolysaccharide.

Figure 6.1 summarizes the catabolism that supplies carbon skeletons for monomer synthesis followed by their polymerization and assembly into cell structure. For anabolism, nitrogen, sulfur, phosphorus, and certain other elements are needed in addition to the carbon skeletons.

6.2 | Assimilation of inorganic nitrogen

Many cell constituents are nitrogenous compounds and include amino acids and nucleic acid bases. Nitrogen exists in various redox states ranging from −5 to +3. Organic nitrogen is used preferentially over inorganic nitrogen by almost all microbes. When organic nitrogen, ammonia or nitrate is not available, some prokaryotes (within the bacteria and archaea) can reduce gaseous nitrogen to ammonia to meet their nitrogen requirements (Table 6.2). This process is known as nitrogen fixation and is not found in eukaryotes.

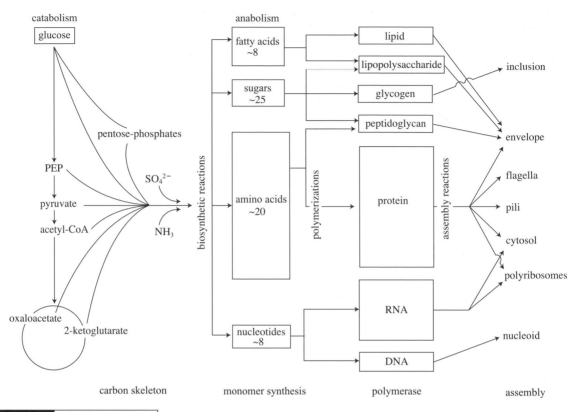

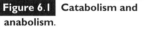

Figure 6.1 Catabolism and anabolism.

Monomers are synthesized from carbon skeletons and inorganic substances such as ammonia, sulfate and phosphate that are obtained from the growth medium or cellular environment. The monomers are polymerized into proteins, nucleic acids, polysaccharides, phospholipids and other macromolecules. Catabolism supplies not only carbon skeletons, but also ATP and NADPH required for anabolic processes. The size of the squares in the diagram represents the relative content in the *Escherichia coli* cell summarized in Table 6.1. The numbers of individual monomers are shown in the squares.

6.2.1 Nitrogen fixation

Nitrogen is incorporated into cell constituents through transamination reactions using glutamate or glutamine as the amino group donor. Glutamate and glutamine are synthesized from ammonia. Gaseous nitrogen (N_2) is very stable as it possesses a triple bond. When reduced or oxidized forms of nitrogen (fixed nitrogen) are not available, some prokaryotes reduce the structurally stable N_2 to ammonia to use as the nitrogen source, investing a large amount of energy in the form of ATP and reduced electron carriers. Ammonia fixed in this way serves as the nitrogen source for all forms of organisms, just as photosynthesis supplies organic materials as the energy source for many organisms. Biological N_2 fixation is estimated as high as 1.3×10^{14} g a year, which is more than twice the amount fixed industrially and naturally by lightning (5×10^{13} g). The fixed nitrogen returns to gaseous nitrogen through nitrification (Section 10.2) and denitrification (Section 9.1), which, together with biological nitrogen fixation, constitute the nitrogen cycle (Figure 6.2).

6.2.1.1 N_2-fixing organisms

A wide variety of prokaryotes have the ability to fix N_2 (Table 6.2). These include certain photosynthetic prokaryotes, anaerobic and aerobic bacteria, and archaea. Some fix N_2 in a symbiotic relationship with plants, while others can do so in the free-living state. Bacteria belonging

Table 6.2.	*Examples of nitrogen-fixing prokaryotes*
Bacteria	
Cyanobacteria	*Anabaena azollae*
	Gloeocapsa spp.
	Mastigocladus laminosus
Photosynthetic bacteria	*Chromatium vinosum*
	Rhodopseudomonas viridis
	Rhodospirillum rubrum
	Heliobacterium chlorum
Strict anaerobes	*Acetobacterium woodii*
	Clostridium pasteurianum
	Desulfovibrio vulgaris
	Desulfotomaculum ruminis
Aerobes and facultative anaerobes	*Azotobacter paspali*
	Azotobacter vinelandii
	Azospirillum lipoferum
	Bacillus polymyxa
	Beijerinkia indica
	Derxia gummosa
	Frankia alni
	Halobacterium halobium
	Klebsiella pneumoniae
	Methylococcus capsulatus
	Methylosinus trichosporium
	Mycobacterium flavum
	Pseudomonas azotogensis
	Rhizobium japonicum
	Thiobacillus ferrooxidans
Archaea	
Methanogens	*Methanosarcina barkeri*
	Methanococcus maripaludis
	Methanobacterium thermoautotrophicum

to the genus *Rhizobium* are well known as symbiotic nitrogen fixers with legumes. Alder trees host nitrogen-fixing *Frankia alni*, an actinomycete-like organism. *Anabaena azollae* is an example of a symbiotic nitrogen-fixing cyanobacterium associated with a fern, *Azolla*.

6.2.1.2 Biochemistry of N₂ fixation

Nitrogen reduction to ammonia can be expressed as:

$$N_2 + 3H_2 + 2H^+ \longrightarrow 2NH_4^+ \quad (\Delta G^{0'} = -39.3 \text{ kJ/mol NH}_4^+)$$

This is an exergonic reaction, but requires a high activation energy due to the stable triple bond in N_2. For this reason, the chemical N_2 fixation process employs a high temperature (300–600 °C) and high pressure (200–800 atm). Nitrogen-fixing microbes produce the

Figure 6.2 **The nitrogen cycle.**

(Sprent, J. I. 1979, *The Biology of Nitrogen-fixing Organisms*, Figure 1.1. McGraw-Hill, Maidenhead)

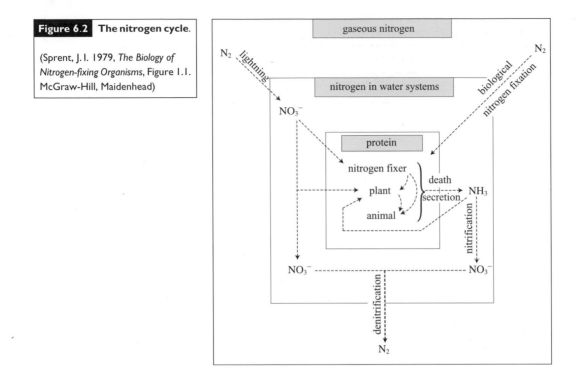

enzyme nitrogenase, which reduces nitrogen under normal physiological conditions.

NITROGENASE

N_2 fixation is catalyzed by nitrogenase. Nitrogenase is a complex protein consisting of azoferredoxin and molybdoferredoxin in a 2:1 ratio. Both of the enzymes are [Fe-S] proteins, and molybdenum is contained in molybdoferredoxin. Azoferredoxin is a homodimer of a protein containing a [4Fe-4S] cluster, and molybdoferredoxin consists of 2 molecules of 2 proteins containing 2 molybdenum and 28 iron and sulfur atoms (Table 6.3). Molybdoferredoxin reduces nitrogen with the reducing equivalents provided by azoferredoxin. Based on their functions, molybdoferredoxin is termed dinitrogenase, and azoferredoxin is termed dinitrogenase reductase. Since the redox potential of azoferredoxin is low ($-0.43\,V$), ferredoxin or flavodoxin is believed to be the electron donor for its reduction. In this process, dissociated azoferredoxin from molybdoferredoxin is reduced, accepting the electrons from low redox potential ferredoxin or flavodoxin followed by ATP binding. Molybdoferredoxin binds N_2 before forming a nitrogenase complex with the ATP-reduced azoferredoxin complex. At this point, electrons are transferred from azoferredoxin to molybdoferredoxin with ATP hydrolysis. Since azoferredoxin is a one electron carrier, the reduction of a nitrogen molecule requires six oxidation–reduction cycles with the hydrolysis of at least 16 ATP molecules (Figure 6.3).

Table 6.3. | *The nitrogenase complex of* Rhizobium *spp.*

Characteristics	Azoferredoxin (dinitrogenase reductase)	Molybdoferredoxin (dinitrogenase)
Molecular weight	70 000	230 000
Subunit	2	4[a]
Iron	8[b]	28
Molybdenum	0	2
Acid-labile sulfide	8	28

[a] Two subunits each of two peptides with molecular weights of 55 000 and 60 000.
[b] One [4Fe-4S] for each subunit.

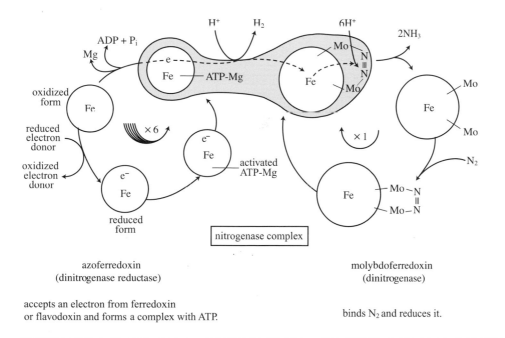

azoferredoxin
(dinitrogenase reductase)

accepts an electron from ferredoxin
or flavodoxin and forms a complex with ATP.

molybdoferredoxin
(dinitrogenase)

binds N_2 and reduces it.

Figure 6.3 **N_2 reduction by the nitrogenase complex.**

(Sprent, J. I. 1979, *The Biology of Nitrogen-fixing Organisms*, Figure 2.1. McGraw-Hill, Maidenhead.)

Azoferredoxin (dinitrogenase reductase) is reduced, coupled with the oxidation of ferredoxin or flavodoxin, and binds ATP. Molybdoferredoxin (dinitrogenase) binds N_2 and forms a nitrogenase complex with the reduced azoferredoxin-ATP complex. Electrons required to reduce N_2 are transferred from azoferredoxin to molybdoferredoxin, and this reaction repeats six times to reduce one molecule of N_2.

Nitrogenase can reduce various other substances in addition to dinitrogen (Table 6.4). The ability to reduce acetylene to ethylene is exploited in a simple nitrogenase assay method. Protons are reduced by the nitrogenase complex to H_2 during normal N_2 fixation.

Table 6.4.	*Substances reduced by the nitrogenase complex*
Substrate	**Product(s)**
N_2	$2NH_4^+$
N_3^-	N_2, NH_4^+
N_2O	N_2
HCN	CH_4, NH_4^+, CH_3NH_2
CH_3CN	C_2H_6, NH_4^+
CH_2CHCN	C_3H_6, NH_4^+, C_3H_8
C_2H_2	C_2H_4
$2H^+$	H_2

Most nitrogen fixers synthesize alternative nitrogenases containing vanadium and iron, or iron only under molybdenum-limited conditions. These have a lower activity than molybdenum-containing nitrogenase.

ELECTRON CARRIERS

Ferredoxin or flavodoxin supply the electrons required for nitrogen reduction. Photosythetic organisms reduce them by light reactions, and obligate anaerobes reduce ferredoxin by pyruvate:ferredoxin oxidoreductase or hydrogenase. Aerobes reduce them through a reverse electron transport mechanism (Section 10.1) using reduced pyridine nucleotides. H_2 produced from the nitrogenase reaction is used to reduce ferredoxin by the action of the hydrogenase.

6.2.1.3 Bioenergetics of N_2 fixation

yield of ATP

The Y_{ATP} of *Klebsiella pneumoniae* was measured to be 4.2 ± 0.2 g/mol ATP under nitrogen-fixing conditions and 10.9 ± 1.5 g/mol ATP with ammonia. These figures show that nitrogenase consumes 29 ATP to reduce one dinitrogen. This figure is much higher than the predicted ATP consumption by the nitrogenase complex. This might be because of the energy consumed in the reverse electron transport process to reduce the low redox potential electron carriers.

6.2.1.4 Molecular oxygen and N_2 fixation

All nitrogenases known to date are inactivated irreversibly by molecular oxygen. O_2 is required to synthesize ATP needed for N_2 fixation through aerobic respiration, and photosystem II (Section 11.4) produces O_2 in N_2-fixing cyanobacteria. To avoid irreversible inactivation of nitrogenase by O_2, N_2-fixing organisms therefore employ various protection mechanisms against O_2.

Species of *Rhizobium* fix N_2 in a symbiotic association with legumes. When they infect the plant root, nodules are formed through transformation of the plant cells, and the bacterial cells become irregularly formed (bacteroidal) instead of rod-shaped. They synthesize ATP through aerobic respiration using energy source(s) and O_2 supplied by the host plants. The O_2 is supplied bound to leghemoglobin which is

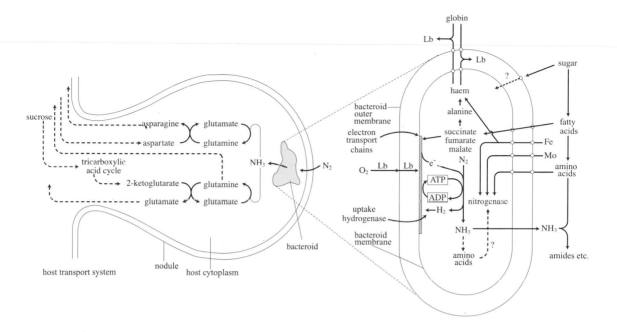

Figure 6.4 **Roles of the host plant and *Rhizobium* in symbiotic N₂ fixation.**

When *Rhizobium* infects the legume plant the host root cells form nodules, and the bacterial cells become irregular bacteroids. The host supplies the carbon and energy source, and O₂ in the leghaemoglobin (Lb) bound form. The bacterium fixes and supplies nitrogen to the host plant.

similar in structure and function to myoglobin in animals. Molecular oxygen is not involved in the respiratory process. The fixed nitrogen is transferred to the host plant (Figure 6.4). Other symbiotic N₂-fixers have a similar relationship with the host plant.

Cyanobacteria obtain the reducing equivalents and ATP for anabolism through oxygenic photosynthesis generating molecular oxygen. Heterocystous cyanobacteria such as *Anabaena* and *Nostoc* spp. transform 5–10% of normal vegetative cells within the filaments into heterocysts which lack oxygenic photosystem II to protect nitrogenase from molecular oxygen produced by photosystem II. Heterocysts fix nitrogen using electrons transported from neighbouring normal cells, and in return the heterocysts supply fixed nitrogen (Figure 6.5).

Unicellular cyanobacteria operate photosystem I only when nitrogenase activity is high, and photosystem II appears with the accumulation of fixed nitrogen within the cell. Under N₂-fixing conditions, O₂ is not generated (Figure 6.6).

Aerobic bacteria protect their nitrogenase through different mechanisms. *Azotobacter vinelandii* keeps the O₂ concentration low using a high affinity terminal oxidase, cytochrome *d*, under nitrogen-fixing conditions instead of the normal cytochrome *o* (Section 5.8.2.3). Some nitrogenases of aerobes such as *Azotobacter chroococcum* and *Derxia gummosa* reversibly change their structure in the presence of O₂ (Figure 6.7).

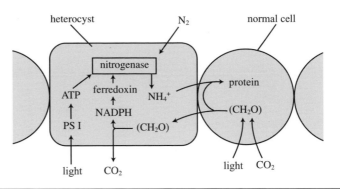

Figure 6.5 | **N_2 fixation in heterocysts of heterocystous cyanobacteria**.

(Gottschalk, G. 1986, *Bacterial Metabolism*, 2nd edn., Figure 10.3. Springer, New York)

Under N_2-fixing conditions, heterocystous cyanobacteria transform 5–10% of the cells in the filament into heterocysts which lack photosystem II to protect the nitrogenase from O_2. Heterocysts fix N_2 using carbon and energy sources obtained from normal vegetative cells, and supply fixed nitrogen to the normal cells.

PSI, photosystem I.

Figure 6.6 **Growth and N_2 fixation in unicellular cyanobacteria.**

(Sprent, J.I. 1979, *The Biology of Nitrogen-fixing Organisms*, Figure 2.4, McGraw-Hill, Maidenhead)

Unicellular cyanobacteria do not operate photosystem II under N_2-fixing conditions to protect nitrogenase from O_2. When fixed nitrogen is accumulated they grow normally with photosystems I and II.

○ – ○, nitrogenase activity; Δ – Δ, O_2 evolution; ● – ●, chlorophyll/phycocyanin ratio (phycocyanin is a constituent of the antenna molecule of photosystem II).

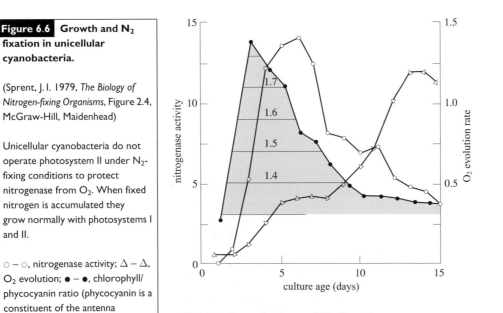

6.2.1.5 Regulation of N_2 fixation

When an organism faces a surplus of fixed nitrogen or is starved of energy sources, nitrogenase activity is not needed. To avoid the waste of energy under these conditions, N_2 fixation is regulated both at the transcriptional level of the genes and by controlling enzyme activity. Nitrogenase activity is inhibited by ammonia and under starvation conditions with a low adenylate energy charge (EC). Inhibition by ammonia is reversible and the response is very quick. This regulation is referred to as the ammonia switch. When ammonia accumulates, an arginine residue of azoferredoxin (dinitrogenase reductase) is

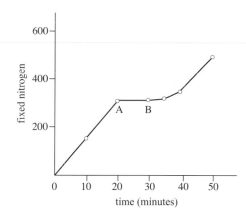

Figure 6.7 Relationship between dissolved O$_2$ concentration and nitrogenase activity in a free-living N$_2$-fixing bacteriums.

(Sprent, J. I. 1979, *The Biology of Nitrogen-fixing Organisms*, Figure 2.5, McGraw-Hill, Maidenhead)

A free-living bacterium, *Azotobacter chroococcum*, fixes N$_2$ under low dissolved O$_2$ (DO) conditions, and nitrogenase activity disappears immediately when the DO concentration increases (A). The enzyme becomes active when the DO concentration decreases (B). At high DO concentrations, the nitrogenase is protected.

bound with ADP-ribose available from NAD$^+$. The nitrogenase complex is inactive with ADP-ribose. Nitrogenase activity is controlled to less than 10% when ATP/ADP is around 1 (EC value of around 0.6).

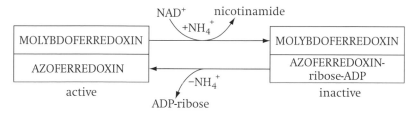

Ammonia inhibits the transcription of genes of the nitrogenase complex and this regulation is elaborate. Genes for the nitrogenase complex (*nif* regulon) consist of 7 operons with 20 genes in *Klebsiella pneumoniae* (Figure 6.8). In addition to these genes, nitrogen control genes, *ntr* (Section 12.2.2), are also involved in their regulation.

Nitrogen control genes consist of *ntrA*, *B*, and *C*. NtrA is a sigma factor of the RNA polymerase (σ^N, σ^{54}), NtrC and NtrB are NR$_I$ and NR$_{II}$, respectively (Figure 12.22, Section 12.2.2). They regulate the transcription and activity of enzymes related to ammonia metabolism. When the ammonia concentration is low, four molecules of UMP bind with NR$_{II}$ which in turn phosphorylates NR$_I$. The phosphorylated NR$_I$ activates NtrA(σ^N) to transcribe *nifA* and *nifL*. NifA is an activator for the transcription of other *nif* genes, NifL is a repressor protein (Figure 6.9).

NtrA, NtrB and NtrC regulate not only the *nif* regulon but also other enzymes related to ammonia metabolism such as glutamine synthetase, and the transport and utilization of arginine, proline and histidine. The regulation of many operons with different functions by a single regulator is referred to as a global control system or multigene system. This is discussed later (Section 12.2).

6.2.2 Nitrate reduction

Many microbes use ammonia and nitrate as their nitrogen source when organic nitrogen is not available. Nitrate is reduced to ammonia before being incorporated into cell materials. These reactions are catalyzed by nitrate reductase and nitrite reductase. They are

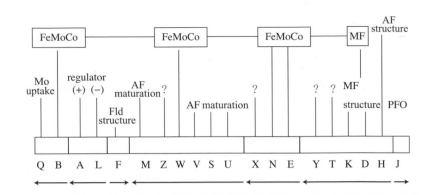

Figure 6.8 Structure of the *nif* regulon in *Klebsiella pneumoniae*.

The *nif* regulon consists of 7 operons with 20 genes. The genes include not only the structural genes for azoferredoxin (AF) and molybdoferredoxin (MF) but also genes for related functions including the synthesis and insertion of cofactor (FeMoCo), and proteins for electron metabolism in N_2 fixation and for regulation. They are pyruvate:flavodoxin oxidoreductase (PFO), flavodoxin(Fld), and *nifA* and *nifL*. The arrows on the bottom indicate the direction of transcription.

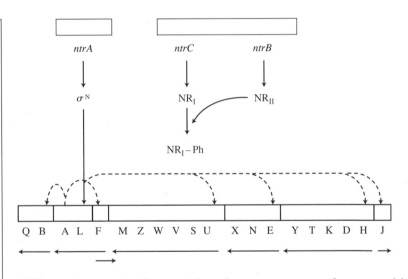

Figure 6.9 Regulation of the *nif* regulon in *Klebsiella pneumoniae*.

The *nif* regulon consists of 20 genes in 7 operons. The nitrogen control gene products, NtrA, NtrB and NtrC, regulate the expression of the *nif* gene: *nifA* and *nifL* participate in the regulation. When the ammonia concentration is low, NtrB (NR_{II}) is uridylylated, and $NR_{II}(UMP)_4$ phosphorylates NR_I. The phosphorylated NR_I binds the enhancer region of *nifA* and *nifL* to induce transcription by NtrA(σ^N). NifA and NifL regulate the expression of the remaining *nif* genes as activator and repressor, respectively. When ammonia accumulates, NR_I–P_i is dephosphorylated, losing the ability to bind the enhancer region.

different from a similar reaction that can occur under anaerobic conditions where nitrate is used as an electron acceptor in an anaerobic respiratory process known as dissimilatory nitrate reduction (Section 9.1). The use of nitrate as a nitrogen source is referred to as assimilatory nitrate reduction.

$$NO_3^- + XH_2 \xrightarrow{\text{nitrate reductase}} NO_2^- + H_2O + X$$

$$NO_2^- + 3NADH + 4H^+ \xrightarrow{\text{nitrite reductase}} NH_3 + 3NAD^+ + 2H_2O$$

NAD(P)H is used as the reducing equivalent by the assimilatory nitrate reductase in many microbes. Reduced cytochrome provides electrons for nitrate reductase in some strains of *Pseudomonas*. Nitrite is further reduced to ammonia by nitrite reductase in a one-step reaction. NADH is the cosubstrate of the enzyme.

Dissimilatory nitrate reduction is strongly inhibited by O_2 (Section 9.1.3), but O_2 does not inhibit assimilatory nitrate reduction, which is inhibited by ammonia (Section 12.2.2). Nitrate is reduced to N_2 by most denitrifying bacteria, but some bacteria, including *Desulfovibrio gigas*, reduce NO_3^- to NH_4^+. Since this reaction is coupled to the formation of a proton motive force and not inhibited by NH_4^+, this reaction is called dissimilatory nitrate reduction (Section 9.1.4).

6.2.3 Ammonia assimilation

Ammonia is assimilated as glutamate by means of two different reactions:

$$\text{2-ketoglutarate} + NH_3 + NADPH + H^+ \xrightarrow{\text{glutamate dehydrogenase}} \text{glutamate} + NADP^+ + H_2O$$

and

$$\text{glutamate} + NH_3 + ATP \xrightarrow{\text{glutamine synthetase}} \text{glutamine} + ADP + P_i$$

$$\text{glutamine} + NADPH + H^+ + \text{2-ketoglutarate} \xrightarrow{\text{glutamate synthase}} 2\text{ glutamate} + NADP^+$$

At a high ammonia concentration, glutamate dehydrogenase can assimilate ammonia without consuming ATP since this enzyme has a low affinity for the substrate ($K_m = 0.1\,M$). ATP is consumed in the assimilation of ammonia at low concentrations by the action of glutamine synthetase which has a high substrate affinity ($K_m = 0.1\,mM$). The amino group of glutamine is transferred to 2-ketoglutarate by glutamate synthase. This enzyme is generally called glutamine: 2-oxoglutarate aminotranferase (GOGAT).

Glutamate and glutamine donate amino groups in various synthetic reactions catalyzed by transaminases (Figure 6.10). In *Escherichia coli*, low specificity transaminase A, B and C synthesize more than ten amino acids. In addition to their role in amino acid synthesis, glutamate and glutamine are also used as -NH_2 donors in the biosynthetic reactions of various other cell constituents including nucleic acid bases, *N*-acetyl-glucosamine and the *N*-acetylmuramic acid of murein. In *Escherichia coli* about 85% of organic nitrogen originates from glutamate and the remaining 15% arises from glutamine.

Since glutamine synthetase consumes ATP to assimilate ammonia at low concentrations, its expression as well as its activity is tightly controlled according to ammonia availability to avoid ATP consumption under ammonia-rich conditions. An enzyme with activity to

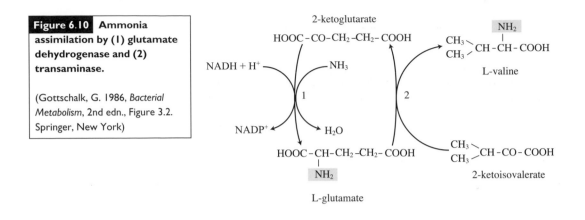

Figure 6.10 Ammonia assimilation by (1) glutamate dehydrogenase and (2) transaminase.

(Gottschalk, G. 1986, *Bacterial Metabolism*, 2nd edn., Figure 3.2. Springer, New York)

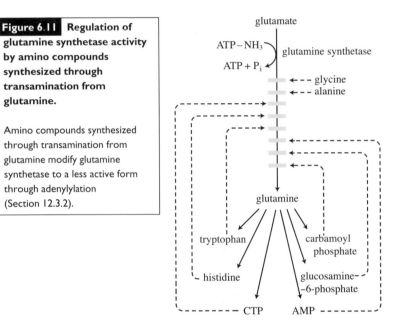

Figure 6.11 Regulation of glutamine synthetase activity by amino compounds synthesized through transamination from glutamine.

Amino compounds synthesized through transamination from glutamine modify glutamine synthetase to a less active form through adenylylation (Section 12.3.2).

uridylylate/deuridylyate PII protein (uridylyltransferase/uridylyl-removing enzyme, UTase/UR, *glnD* product) senses ammonia availability, and this signal is transduced to control the gene expression (Section 12.2.2) and enzyme activity through the adenylylation of the enzyme molecule (Section 12.3.2.2). The enzyme is adenylylated by adenylyltransferase when glutamine is accumulated and the adenylylated enzyme is less active than the native form. Enzyme activity is not completely inhibited under ammonia-rich conditions, because glutamine should be synthesized as an amino group donor for the biosynthesis of nucleotides and some amino acids. Glutamine synthetase activity is also controlled through cumulative feedback inhibition (Section 12.3.1) by various metabolites synthesized from glutamine (Figure 6.11). The regulation involved is very complex and discussed in Chapter 12 (Section 12.2.2).

6.3 | Sulfate assimilation

Sulfur is a constituent of certain amino acids, e.g. methionine and cysteine, as well as various coenzymes, and also plays an important role in the electron transport chain in iron-sulfur proteins. Sulfate is the major inorganic sulfur source in microbes. Sulfate needs to be reduced to sulfide for the biosynthesis of organic sulfur compounds. Sulfate can be used as the terminal electron acceptor in a group of anaerobic bacteria known as sulfate-reducing bacteria (Section 9.3). This process is referred to as dissimilatory sulfate reduction, whereas assimilatory sulfate reduction is used to describe the process that uses sulfate as the sulfur source. Sulfate is activated by adenosine-5'-phosphosulfate (APS) with ATP (Figure 6.12) as in dissimilatory sulfate reduction (Section 9.3). The redox potential ($E^{0'}$) of HSO_4^-/HSO_3^- is $-454\,mV$, which is lower than any known natural electron carriers. Unlike in dissimilatory reduction, APS is further activated to adenosine-3'-phosphate-5'-phosphosulfate (PAPS).

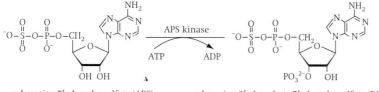

adenosine-5'-phosphosulfate (APS)   adenosine-3'-phosphate-5'-phosphosulfate (PAPS)

Sulfate transported into the cell is converted to adenine-5'-phosphosulfate (APS) in a reaction catalyzed by ATP sulfurylase (sulfate adenyltransferase). PP_i produced in the reaction is hydrolyzed to $2P_i$ by phosphatase to pull the reaction to the synthesis of APS. APS is phosphorylated to adenosine-3'-phosphate-5'-phosphosulfate (PAPS) by APS kinase, consuming ATP. PAPS reductase reduces PAPS to sulfite,

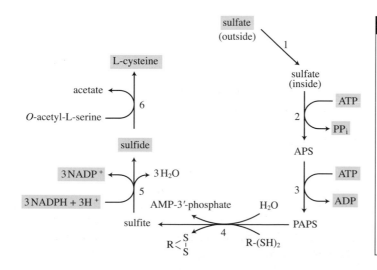

Figure 6.12 **Assimilatory sulfate reduction and cysteine synthesis.**

(Gottschalk, G. 1986, *Bacterial Metabolism*, 2nd edn., Figure 3.4. Springer, New York)

1, active transport; 2, ATP sulfurylase; 3, APS phosphokinase; 4, PAPS reductase; 5, sulfite reductase; 6, *O*-acetyl-L-serine sulfhydrylase. APS, adenosine-5'-phosphosulfate; PAPS, adenosine-3'-phosphate-5'-phosphosulfate; $R(SH)_2$, reduced thioredoxin.

liberating AMP-3'-phosphate, using electrons from reduced thioredoxin. Sulfite reductase further reduces sulfite to sulfide using NADPH as a coenzyme. O-acetyl-L-serine sulfhydrylase incorporates sulfide into cysteine (Figure 6.12). Just as glutamate and glutamine function as -NH₂ donors, cysteine is used as a -SH donor in biosynthesis.

Many bacteria oxidize organic sulfonate, thiols, sulfide or thiophene to sulfate to use as their sulfur source in the absence of sulfate or sulfur-containing amino acids. These properties have been studied as a means to remove sulfur from petroleum and coal.

In *Escherichia coli* a number of proteins have been identified that are produced under sulfate-limited conditions. These are called the sulfate starvation induced (SSI) stimulon (Section 12.2.10).

6.4 | Amino acid biosynthesis

Amino acids are synthesized using carbon skeletons available from central metabolism. These are pyruvate, oxaloacetate, 2-ketoglutarate, 3-phosphoglycerate, phosphoenolpyruvate, erythrose-4-phosphate and ribose-5-phosphate (Table 6.5). Some amino acids are synthesized by different pathways depending on the organism. For convenience, those of *Escherichia coli* are discussed below.

6.4.1 The pyruvate and oxaloacetate families

Pyruvate and oxaloacetate are converted to alanine and aspartate through reactions catalyzed by transaminase. In these reactions, glutamate is used as the -NH₂ donor. Asparagine synthetase synthesizes asparagine from aspartate and ammonia consuming energy in the form of ATP in a similar reaction to that catalyzed by glutamine synthetase.

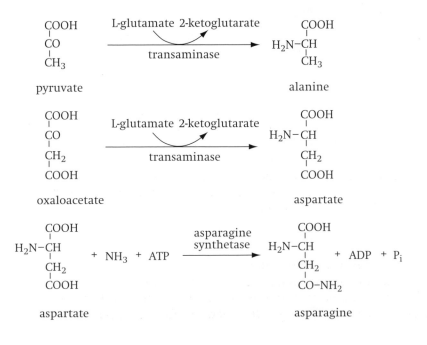

Table 6.5.	*Carbon skeletons used for amino acid biosynthesis*
Precursor	Amino acid
Pyruvate	alanine, valine, leucine
Oxaloacetate	aspartate, asparagine, methionine, lysine, isoleucine, threonine
2-ketoglutarate	glutamate, glutamine, arginine, proline
3-phosphglycerate	serine, glycine, cysteine
PEP and erythrose-4-phosphate	phenylalanine, tyrosine, tryptophan
Ribose-5-phosphate	histidine

Threonine, methionine and lysine are produced from aspartate in addition to asparagine (Figure 6.13). An intermediate of lysine biosynthesis, diaminopimelate, is not found in protein, but is a precursor of murein, and so is ornithine, an intermediate in arginine biosynthesis (Figure 6.17). Most prokaryotes employ the diaminopimelate pathway to synthesize lysine (Figure 6.13), but yeasts and fungi synthesize this amino acid from 2-ketoglutarate through the 2-aminoadipate pathway (Figure 6.14).

Threonine produced from aspartate is deaminated to 2-ketobutyrate, which is used as the precursor for isoleucine biosynthesis. Two molecules of pyruvate are condensed to 2-acetolactate for the biosynthesis of valine and leucine (Figure 6.15). Since pyruvate and 2-ketobutyrate are similar in structure, a series of the same enzymes is used to synthesize isoleucine and valine.

6.4.2 The phosphoglycerate family
The EMP pathway intermediate 3-phosphoglycerate is converted to serine and then further to glycine and cysteine (Figure 6.16).

6.4.3 The 2-ketoglutarate family
Glutamate synthesized from 2-ketoglutarate through the reactions catalyzed by glutamate dehydrogenase or GOGAT is the precursor for the synthesis of proline, arginine and glutamine (Figure 6.17). *N*-acetylornithine deacetylase (reaction 9, Figure 6.17) has not been detected in coryneform bacteria, *Pseudomonas aeruginosa* and the yeast *Saccharomyces cerevisiae*. Instead, the reaction is catalyzed by *N*-acetylglutamate-acetylornithine acetyltransferase, coupling reactions 6 and 9 in Figure 6.17 in these organisms.

6.4.4 Aromatic amino acids
The benzene ring of aromatic amino acids is formed from shikimate, which is produced from the condensation of erythrose-4-phosphate and phosphoenolpyruvate. Shikimate is further metabolized to

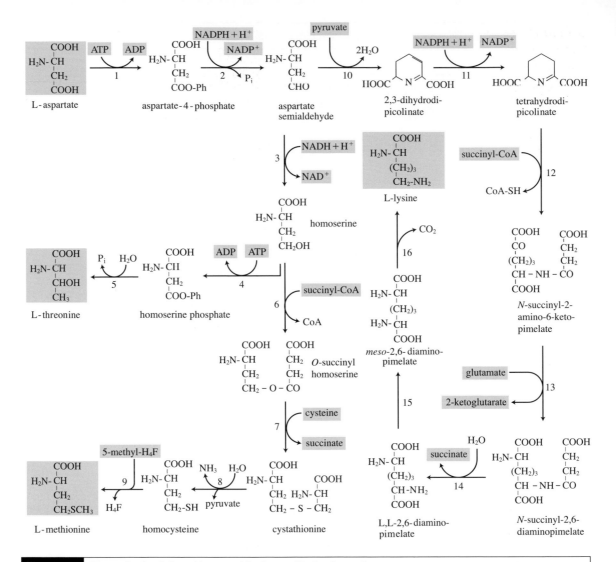

Figure 6.13 Biosynthesis of threonine, methionine and lysine from the common precursor, aspartate.

(Gottschalk, G. 1986, *Bacterial Metabolism*, 2nd edn., Figure 3.6. Springer, New York)

1, aspartate kinase; 2, aspartate semialdehyde dehydrogenase; 3, homoserine dehydrogenase; 4, homoserine kinase; 5, threonine synthase; 6, homoserine a cyltransferase; 7, cystathionine synthase; 8, cystathionine lyase; 9, homocysteine: 5-methyl-tetrahydrofolate methyltransferase; 10, dihydrodipicolinate synthase; 11, dihydrodipicolinate reductase; 12, tetrahydrodipicolinate succinylase; 13, glutamate: succinyl-diaminopimelate a minotransferase; 14, succinyl-diaminopimelate desuccinylase; 15, diaminopimelate epimerase; 16, diaminopimelate decarboxylase. H4F, tetrahydrofolate.

phenylpyruvate and *p*-hydroxyphenylpyruvate before being transaminated to phenylalanine and tyrosine, respectively. Transaminase catalyzes these reactions using glutamate as the -NH$_2$ donor. Tryptophan is synthesized from indole-3-glycerol phosphate catalyzed by tryptophan synthase using serine as the -NH$_2$ donor (Figure 6.18).

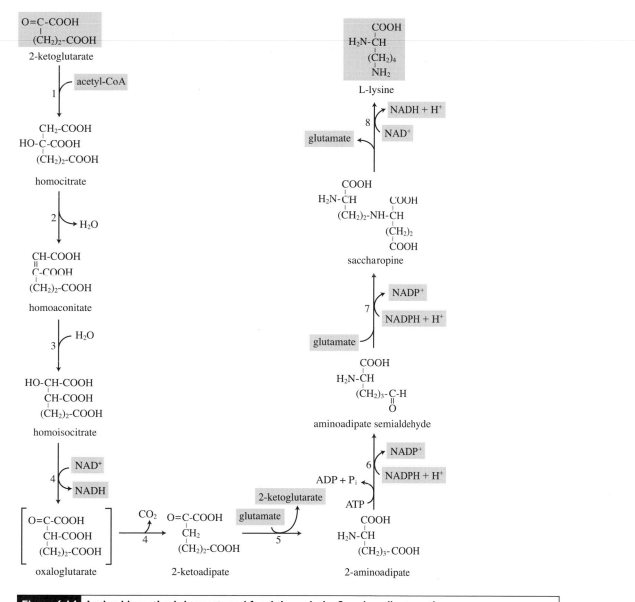

Figure 6.14 **Lysine biosynthesis in yeasts and fungi through the 2-aminoadipate pathway.**

(Gottschalk, G. 1986, *Bacterial Metabolism*, 2nd edn., Figure 3.7. Springer, New York)

1, homocitrate synthase; 2, homocitrate dehydratase; 3, homoaconitate hydratase; 4, homoisocitrate dehydrogenase; 5, aminoadipate aminotransferase; 6, 2-aminoadipate semialdehyde dehydrogenase; 7, saccharopine dehydrogenase (glutamate-forming); 8, saccharopine dehydrogenase (lysine-forming).

Indole-3-glycerol phosphate is formed through the condensation of anthranilate and 5-phospho-D-ribosyl-1-pyrophosphate (PRPP). PRPP is synthesized from ribose-5-phosphate taking pyrophosphate from ATP catalyzed by PRPP synthetase as below:

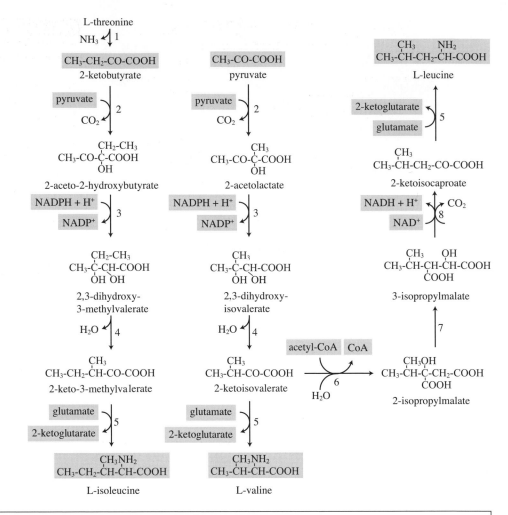

Figure 6.15 Biosynthesis of isoleucine, valine and leucine.

(Gottschalk, G. 1986, *Bacterial Metabolism*, 2nd edn., Figure 3.8. Springer, New York)

Isoleucine and valine are synthesized from 2-ketobutyrate and pyruvate, respectively. Since the precursors are similar in structure, the same enzymes catalyze the reactions.

1, threonine dehydratase; 2, acetohydroxy acid synthase; 3, acetohydroxy acid isomeroreductase; 4, dihydroxy acid dehydratase; 5, transaminase; 6, alpha-isopropylmalate synthase; 7, isopropylmalate isomerase; 8, beta-isopropylmalate dehydrogenase.

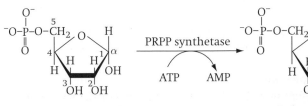

D-ribose-5-phosphate   5-phospho-D-ribosyl-
                              1-pyrophosphate (PRPP)

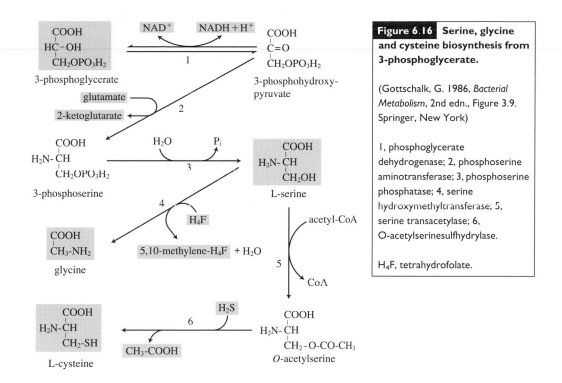

Figure 6.16 Serine, glycine and cysteine biosynthesis from 3-phosphoglycerate.

(Gottschalk, G. 1986, *Bacterial Metabolism*, 2nd edn., Figure 3.9. Springer, New York)

1, phosphoglycerate dehydrogenase; 2, phosphoserine aminotransferase; 3, phosphoserine phosphatase; 4, serine hydroxymethyltransferase; 5, serine transacetylase; 6, O-acetylserinesulfhydrylase.

H_4F, tetrahydrofolate.

PRPP is the precursor for the synthesis of histidine and nucleotides. PRPP synthetase is regulated by a feedback inhibition mechanism by its biosynthetic products.

6.4.5 Histidine biosynthesis
Histidine is produced from PRPP (Figure 6.19).

6.4.6 Regulation of amino acid biosynthesis
Biosynthesis of amino acids is regulated according to their concentration. Enzyme activity is regulated by feedback inhibition while the transcription of genes is regulated by various mechanisms. This subject is discussed later (Section 12.1).

6.5 | Nucleotide biosynthesis

Nucleotides are synthesized either in a *de novo* pathway from the beginning or in a salvage pathway, thereby recycling bases.

6.5.1 Salvage pathway
Normally the half-life of mRNA is short because of continuous synthesis and degradation. The bases arising from nucleic acid turnover are recycled to nucleotide in a salvage pathway. In this reaction, a base

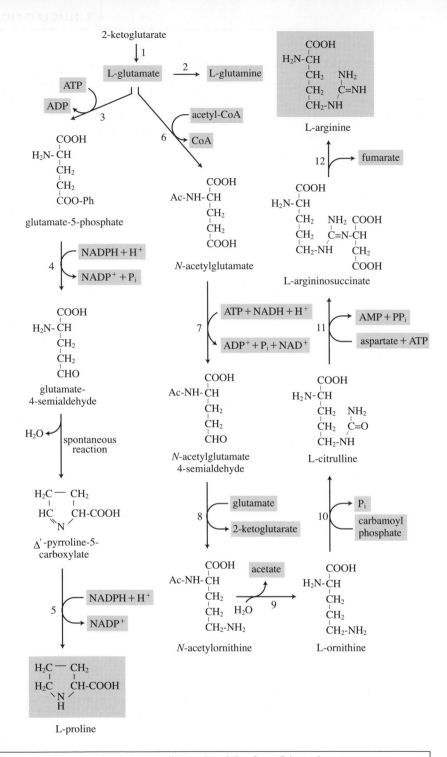

Figure 6.17 Biosynthesis of glutamate, glutamine, proline and arginine from 2-ketoglutarate.

(Gottschalk, G. 1986, *Bacterial Metabolism*, 2nd edn., Figure 3.10. Springer, New York)

1, glutamate dehydrogenase or glutamate synthase; 2, glutamate synthetase; 3, glutamate kinase; 4, glutamate semialdehyde dehydrogenase; 5, Δ^1-pyrroline-5-carboxylate reductase; 6, amino acid acetyltransferase; 7, N-acetylglutamate kinase and N-acetylglutamate semialdehyde dehydrogenase; 8, N-acetylornithine transaminase; 9, N-acetylornithine deacetylase; 10, ornithine transcarbamoylase; 11, argininosuccinate synthetase; 12, argininosuccinate lyase.

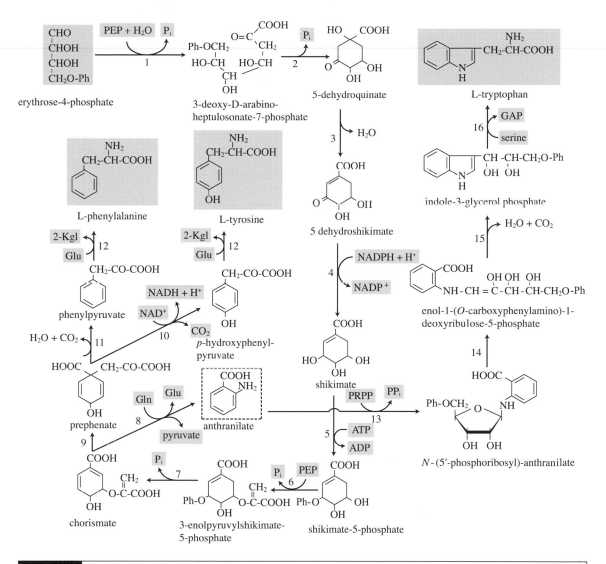

Figure 6.18 Biosynthesis of aromatic amino acids.

(Gottschalk, G. 1986, *Bacterial Metabolism*, 2nd edn., Figure 3.11. Springer, New York)

1, 3-deoxy-D-arabinoheptulosonate synthase; 2, 5-dehydroquinate synthase; 3, 5-dehydroquinate dehydratase; 4, shikimate dehydrogenase; 5, shikimate kinase; 6, enolpyruvylshikimate-5-phosphate synthase; 7, chorismate synthetase; 8, anthranilate synthase; 9, chorismate mutase; 10, prephenate dehydrogenase; 11, prephenate dehydratase; 12, transaminase B; 13, anthranilate phosphoribosyl transferase; 14, phosphoribosyl-anthranilate isomerase; 15, indole-3-glycerol phosphate synthase; 16, tryptophan synthase.

PEP, phosphoenolpyruvate; gln, glutamine; glu, glutamate; 2-kgl, 2-ketoglutarate; PRPP, 5-phosphoribosyl-1-pyrophosphate; GAP, glyceraldehyde-3-phosphate.

replaces the $1'$-PP_i of PRPP. This reaction is catalyzed by the base-specific phosphoribosyltransferase:

$$\text{adenine} + \text{PRPP} \xrightarrow{\quad \text{adenine phosphoribosyltransferase} \quad} \text{AMP} + PP_i$$

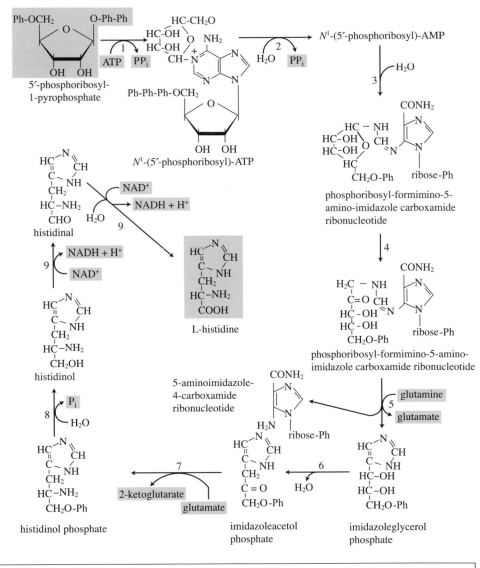

Figure 6.19 Histidine biosynthesis.

(Gottschalk, G. 1986, *Bacterial Metabolism*, 2nd edn., Figure 3.13. Springer, New York)

1, phosphoribosyl; ATP pyrophosphorylase; 2, phosphoribosyl-ATP pyrophosphohydrolase; 3, phosphoribosyl-AMP cyclohydrolase; 4, phosphoribosyl formimino-5-aminoimidazole carboxamide ribonucleotide isomerase; 5, phosphoribosyl-formimino-5-aminoimidazole carboxamide ribonucleotide: glutamine aminotransferase and cyclase; 6, imidazoleglycerol phosphate dehydratase; 7, histidinol phosphate transaminase; 8, histidinol phosphatase; 9, histidinol dehydrogenase (catalyzes both reactions).

6.5.2 Pyrimidine nucleotide biosynthesis through a *de novo* pathway

Nucleotides are also produced through a *de novo* pathway synthesizing new bases. Pyrimidine nucleotide synthesis starts with the synthesis of carbamoyl phosphate from carbonate. ATP provides the energy needed for the reaction and the -NH$_2$ group is from glutamine.

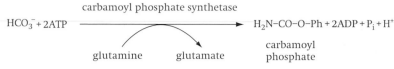

$$HCO_3^- + 2ATP \xrightarrow[\text{glutamine} \quad \text{glutamate}]{\text{carbamoyl phosphate synthetase}} H_2N-CO-O-Ph + 2ADP + P_i + H^+$$

carbamoyl
phosphate

As shown in Figure 6.20, orotate is synthesized as a precursor of pyridine nucleotides before binding to PRPP by the action of phosphoribosyltransferase as in the salvage pathway described previously.

Mononucleotides are phosphorylated in two steps to trinucleotide by nucleotide kinases, consuming ATP, before being used in RNA synthesis. The first enzyme of pyridine nucleotide biosynthesis, aspartate transcarbamoylase, is regulated through feedback inhibition (Section 12.3.1) by cytidine trinucleotide, the final product.

6.5.3 *De novo* synthesis of purine nucleotides

Purine nucleotides are synthesized in a more complicated pathway than pyrimidine nucleotides. Glutamine donates -NH_2 to PRPP before inosine 5′-monophosphate (IMP) is synthesized with the addition of carbons and nitrogens in the forms of glycine, methenyl tetrahydrofolate, glutamine, aspartate and formyl tetrahydrofolate (Figure 6.21). IMP is converted to adenosine 5′-monophosphate (AMP) and guanosine 5′-monophosphate (GMP). The first enzyme of this pathway, PRPP amidotransferase, is regulated through feedback inhibition (Section 12.3.1) by the final products AMP and GMP. These are phosphorylated to dinucleotides in reactions catalyzed by nucleotide kinases that consume ATP. GDP is further phosphorylated to GTP in a similar reaction, and ADP to ATP in the normal ATP synthesis mechanisms either by substrate-level phosphorylation (Section 5.6.5) or by the membrane-bound ATPase (Section 5.8.4).

6.5.4 Synthesis of deoxynucleotides

RNA synthesis requires ATP, guanosine 5′-triphosphate (GTP), cytidine 5′-triphosphate (CTP) and uridine 5′-triphosphate (UTP) as the building blocks, and deoxynucleotide triphosphates are the substrates for DNA polymerase. Ribonucleotides are reduced to deoxynucleotide by ribonucleotide reductase. Ribonucleotide diphosphates are the substrate for the reductase.

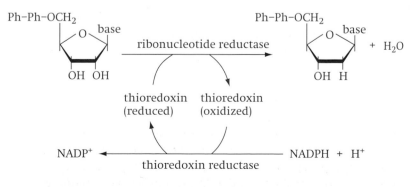

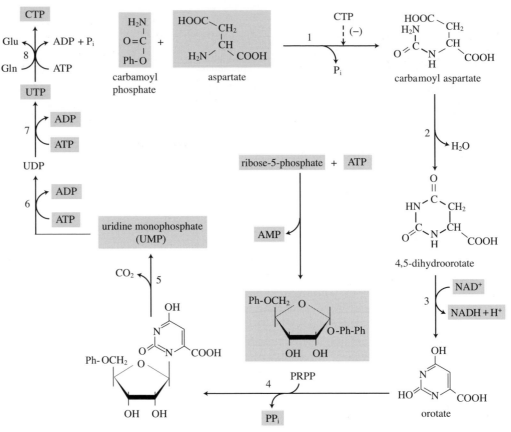

Figure 6.20 *De novo* **biosynthesis of pyridine nucleotides.**

(Gottschalk, G. 1986, *Bacterial Metabolism*, 2nd edn., Figure 3.16. Springer, New York)

1, aspartate transcarbamoylase; 2, dihydroorotase; 3, dihydroorotate dehydrogenase; 4, orotate phosphoribosyl transferase; 5, orotidine-5-phosphate decarboxylase; 6, nucleoside monophosphate kinase; 7, nucleoside diphosphate kinase; 8, CTP synthetase.

Gln, glutamine, Glu, glutamate.

Since ribose is very stable, this reaction requires free radicals. There are three distinct classes of enzyme with different stable free radical amino acids. Type I (Ia and Ib) enzymes require O_2 to generate a tyrosine radical for the reaction. A cofactor, 5′-deoxyadenosylcobalamin (Section 8.7), is the radical in class II enzymes while the class III enzymes are anaerobic and employ a glycyl radical for the reaction (Table 6.6). Eukaryotic cells have class Ia enzymes, and class Ia, Ib and III enzymes are found in *Escherichia coli*. Many aerobic bacteria have class II enzymes, and the anaerobic class III enzymes are functional in anaerobic bacteria and anaerobic methanogenic archaea. Halophilic and thermophilic archaea employ class II enzymes. The anaerobic ribonucleotide reductase (class III) utilizes formate as the electron donor for the reaction, and the other enzyme takes electrons from reduced thioredoxin or glutaredoxin coupling with the oxidation of NADPH. Deoxyuridine diphosphate (dUDP) is dephosphorylated before being methylated to deoxythymidine phosphate (dTMP) in a reaction catalyzed by thymidylate synthase. Deoxyribonucleotides are used in DNA synthesis.

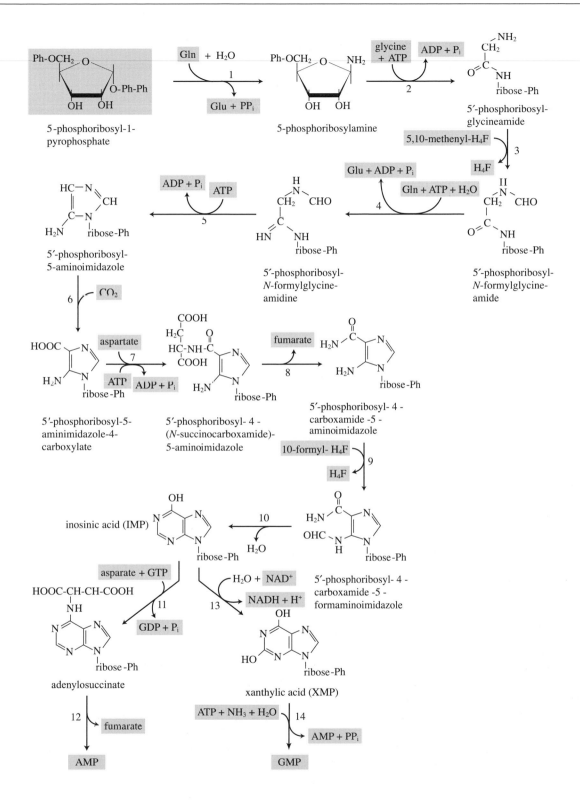

Table 6.6.	Ribonucleotide reductases and their characteristics			
Enzyme	Ia	Ib	II	III
Organism	eukaryotes fac G(-) anaerobic	fac G(-) anaerobic	other bacteria halophilic archaea, thermophilic archaea	methanogens fac G(-) anaerobic
Condition	aerobic	aerobic	aerobic/anaerobic	anaerobic
Radical	tyrosine	tyrosine	coenzyme B_{12}	glycine
Electron donor	thioredoxin I glutaredoxin	thioredoxin I glutaredoxin	thioredoxin I glutaredoxin	formate

fac G(-) anaerobic, facultatively anaerobic Gram-negative bacteria; coenzyme B_{12}, 5′-deoxyadenosylcobalamin.
Source: Trends in Biochemical Sciences, 1997, **22**, 81–85.

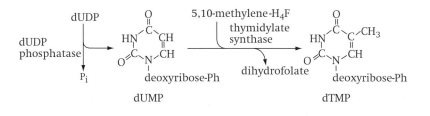

6.6 | Lipid biosynthesis

Phospholipids are essential cellular components as the major part of the membrane. Bacterial phospholipids are based on acylglyceride with an ester link between glycerol and fatty acids as in eukaryotic cells. The archaeal membrane contains phospholipids with an ether linkage between polyalcohol and polyisoprenoid alcohols (Section 2.3.4). Fatty acids and polyisoprenoid alcohols are synthesized from acetyl-CoA.

6.6.1 Fatty acid biosynthesis

Acetyl-CoA is converted to acyl-acyl carrier protein (acyl-ACP) through the action of seven enzymes. In eukaryotes these enzymes

---

Figure 6.21 (cont.) *De novo biosynthesis of purine nucleotides.*

(Gottschalk, G. 1986, *Bacterial Metabolism*, 2nd edn., Figure 3.19. Springer, New York)

1, amidophosphoribosyltransferase; 2, phosphoribosylglycineamide synthetase; 3, phosphoribosylglycineamide formyltransferase; 4, phosphoribosyl-formylglycineamidine synthetase; 5, phosphoribosyl-aminoimidazole synthetase; 6, phosphoribosyl-aminoimidazole carboxylase; 7, phosphoribosylaminoimidazole succinocarboxamide synthetase; 8, adenylosuccinate lyase; 9, phosphoribosylaminoimidazole-carboxamide formyltransferase; 10, IMP cyclohydrolase; 11, adenylosuccinate synthetase; 12, adenylosuccinate lyase; 13, IMP dehydrogenase; 14, GMP synthetase.

H_4F, tetrahydrofolate; Gln, glutamine, Glu, glutamate.

form a complex, but such a complex is not found in prokaryotes. Enzymes directly involved are 3-ketoacyl-ACP synthase, 3-ketoacyl-ACP reductase, 3-hydroxyacyl-ACP dehydratase and enoyl-ACP reductase. Isoenzymes are identified in all of them except 3-ketoacyl-ACP reductase (FabG). These isoenzymes have specific functions. Three isoenzymes of 3-ketoacyl-ACP synthase are I (FabB), II (FabF) and III (FabH). FabA (3-hydroxydecanoyl-ACP dehydratase) and FabZ (3-hydroxyacyl-ACP dehydratase) have a similar function. Enoyl-ACP reductase also has three isozymes: I (FabI), II (FabK) and III (FabL). Fatty acid synthesis is initiated by 3-ketoacyl-ACP synthase III (FabH) catalyzing the formation of acetoacetyl-ACP from malonyl-ACP and acetyl-CoA. The other two 3-ketoacyl-ACP synthases do not react with malonyl-ACP, but catalyze the elongation reaction. Mutants (fabA⁻ and fabB⁻) synthesize saturated fatty acids normally but the synthesis of unsaturated fatty acids is impaired. They are involved in unsaturated fatty acid synthesis.

6.6.1.1 Saturated acyl-ACP

Acetyl-CoA is carboxylated to malonyl-CoA in a reaction catalyzed by acetyl-CoA carboxylase, and malonyl transacylase transfers the malonyl group to ACP. Acetyl-CoA carboxylase is a complex enzyme consisting of one molecule each of biotin carboxylase and biotin carboxyl carrier protein (BCCP), and two molecules of carboxyltransferase.

$$BCCP + HCO_3^- + ATP \xrightarrow{\text{biotin carboxylase}} BCCP\text{-}COO^- + ADP + P_i$$

$$acetyl\text{-}CoA + BCCP\text{-}COO^- \xrightarrow{\text{carboxyltransferase}} malonyl\text{-}CoA + BCCP$$

$$malonyl\text{-}CoA + ACP \xrightarrow{\text{malonyl transacylase}} malonyl\text{-}ACP + CoA\text{-}SH$$

Malonyl-ACP and acetyl-CoA condense to acetoacetyl-ACP, replacing the carboxyl group of malonyl-ACP with an acetyl group catalyzed by 3-ketoacyl-ACP synthase III (FabH). Acetoacetyl-ACP is reduced to 3-hydroxybutyryl-ACP by the action of 3-ketoacyl-ACP reductase (FabG) that uses NADPH as a coenzyme. Crotonyl-ACP is formed from 3-hydroxybutyryl-ACP through a dehydration reaction catalyzed by 3-hydroxyacyl-ACP dehydratase (FabZ). Enoyl-ACP reductase (FabI) reduces crotonyl-ACP to butyryl-ACP using NAD(P)H as a coenzyme. Butyryl-ACP condenses with malonyl-ACP to start the next cycle catalyzed probably by 3-ketoacyl-ACP reductase II (FabF) (Figure 6.22). The initiating 3-ketoacyl-ACP synthase III (FabH) does not catalyze the reverse reaction, and its regulation determines the rate of fatty acid synthesis.

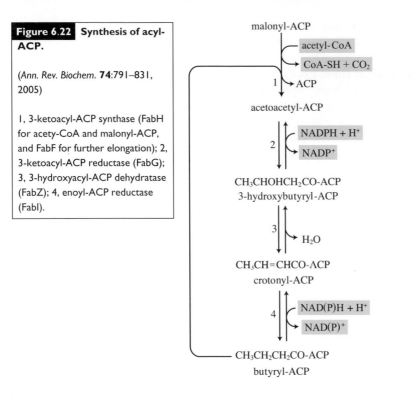

Figure 6.22 **Synthesis of acyl-ACP.**

(*Ann. Rev. Biochem.* **74**:791–831, 2005)

1, 3-ketoacyl-ACP synthase (FabH for acety-CoA and malonyl-ACP, and FabF for further elongation); 2, 3-ketoacyl-ACP reductase (FabG); 3, 3-hydroxyacyl-ACP dehydratase (FabZ); 4, enoyl-ACP reductase (FabI).

6.6.1.2 Branched acyl-ACP

Branched fatty acids are synthesized in two different ways. Branched building blocks such as isobutyryl-ACP or methylmalonyl-ACP result in branched fatty acids from similar reactions as in the straight-chain fatty acid biosynthetic pathway. The other pathway which synthesizes branched fatty acids involves methylation of unsaturated fatty acids as in the formation of cyclopropane fatty acids (Section 6.6.1.4).

6.6.1.3 Unsaturated acyl-ACP

Many unsaturated fatty acid residues are found in biological membranes. They are synthesized by two different mechanisms. These are the aerobic route found both in eukaryotes and prokaryotes and the anaerobic route which occurs in some bacteria. In the anaerobic route the double bond is formed during fatty acid biosynthesis (Figure 6.23). In saturated fatty acid synthesis, 3-hydroxyacyl-ACP dehydratase (FabA) dehydrates 3-hydroxyacyl-ACP to *trans*-2,3-enoyl-ACP that can be reduced by enoyl-ACP reductase (FabI). For unsaturated fatty acid synthesis, *trans*-2,3-enoyl-ACP is isomerized to *cis*-3,4-enoyl-ACP by the bifunctional 3-hydroxyacyl-ACP dehydratase (FabA). Enoyl-ACP reductase (FabI) cannot reduce *cis*-3,4-enoyl-ACP. This *cis*-3,4-enoyl-ACP becomes a substrate for the elongation catalyzed by a separate 3-ketoacyl-ACP synthase (FabF). FabF is expressed constitutively and unstable under physiological temperature. More unsaturated fatty

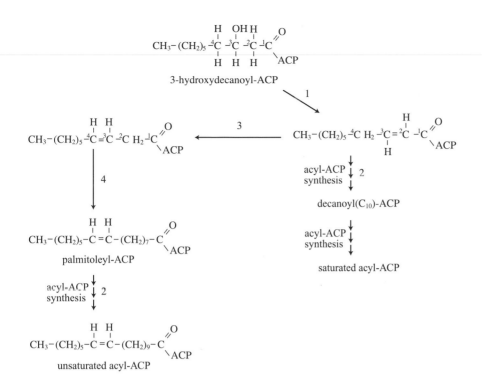

3-hydroxydecanoyl-ACP

Figure 6.23 Unsaturated acyl-ACP synthesis by the anaerobic route.

(*Ann. Rev. Biochem.* 74:791–831, 2005)

The dual function 3-hydroxydecanoyl-ACP dehydratase (FabA) removes a water molecule from 3-hydroxydecanoyl-ACP, producing *trans*-2,3-enoyl-ACP (1). For the synthesis of saturated fatty acid residues the double bond is reduced by 3-ketoacyl-ACP reductase (FabG, 2). On the other hand, the dual function 3-hydroxydecanoyl-ACP dehydratase (FabA) isomerizes *trans*-2,3-enoyl-ACP to *cis*-3,4-enoyl-ACP (3) that cannot be reduced by 3-ketoacyl-ACP reductase (FabG), but is condensed with malonyl-ACP (4) catalyzed by hydroxyacyl-ACP synthase I (FabB). FabA and FabB are the key enzymes of the synthesis of unsaturated fatty acids, and are expressed constitutively. The enzymes are unstable under physiological temperatures, and become stable, producing more unsaturated fatty acid at lower temperatures.

acids are produced at a reduced growth temperature when this enzyme becomes stable (Section 12.2.7).

Saturated fatty acids are oxidized in an aerobic route to produce unsaturated fatty acids. Acyl-ACP oxidase catalyzes this reaction, consuming O_2, to oxidize NADPH to water:

$$\text{palmitoyl-ACP} + \text{NADPH} + \text{H}^+ + \text{O}_2 \xrightarrow{\underset{\text{(fatty acid desaturase)}}{\text{acyl-ACP oxidase}}} \text{palmitoleyl-ACP} + \text{NADP}^+ + 2\text{H}_2\text{O}$$

The fatty acid desaturase is a membrane-bound enzyme in bacteria while eukaryotes have soluble enzymes. The expression of this enzyme is activated when the membrane fluidity becomes low at a low temperature (Section 12.2.7). Some fatty acid desaturases reduce fatty acid residues bound to ACP, and others in the form of phospholipids. The yeast *Saccharomyces cerevisiae* can grow fermentatively on glucose, but cannot grow under strictly anaerobic conditions unless supplemented with unsaturated fatty acids and ergosterol as growth factors. These lipids cannot be synthesized without molecular oxygen in yeast.

6.6.1.4 Cyclopropane fatty acids

Cyclopropane fatty acid residues are found in the bacterial membrane (Section 2.3.4). These are synthesized through the methylation of unsaturated fatty acids catalyzed by cyclopropane fatty acid synthase using *S*-adenosylmethionine as the methyl group donor. These fatty acid residues are called lactobacillic acid, since they were first identified in *Lactobacillus arabinosus*.

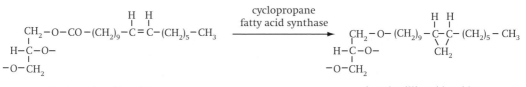

cis-vaccenic acid residue   lactobacillic acid residue

6.6.1.5 Regulation of fatty acid biosynthesis

The fatty acid composition of a membrane varies depending on growth conditions and culture age in a given bacterium. The rate of fatty acid synthesis is determined by the activity of the initiating enzyme, 3-ketoacyl-ACP synthase III (FabH). The length of the fatty acid depends on regulation of the 3-ketoacyl-ACP synthase II (FabF) activity. Bacteria growing at a sub-optimum temperature synthesize more unsaturated fatty acids to maintain membrane fluidity (Section 12.2.7). Unsaturated fatty acid synthesis through the anaerobic route is increased through the increase in stability of the enzymes, 3-hydroxydecanoyl-ACP dehydratase (FabA) and 3-ketoacyl-ACP synthase I (FabB). In the aerobic route, more unsaturated fatty acids are produced through the increased expression of the fatty acid desaturase gene. In *Bacillus subtilis*, a gene for fatty acid desaturase is expressed under cold-shock conditions.

Escherichia coli synthesizes cyclopropane fatty acid (CFA) in the stationary phase. CFA-negative mutants are less resistant to freezing, which suggests that CFA is involved in survival of the bacterium. The CFA synthase gene is recognized by the stationary phase sigma factor of RNA polymerase (σ^s) (Table 12.2, Section 12.1.1).

6.6.2 Phospholipid biosynthesis

The EMP pathway intermediate, dihydroxyacetone phosphate, is reduced to glycerol-3-phosphate oxidizing NADPH:

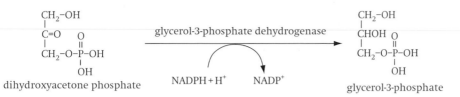

dihydroxyacetone phosphate glycerol-3-phosphate

Glycerol-3-phosphate acyltransferase then synthesizes phosphatidic acid consuming two acyl-ACPs. Phosphatidic acid serves as a precursor for the synthesis of phospholipids and triglycerides.

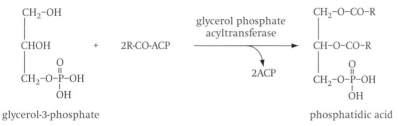

glycerol-3-phosphate phosphatidic acid

The bacterial cytoplasmic membrane contains phosphatidylethanolamine, phosphatidylserine, phosphatidylcholine, phosphatidylinositol, phosphatidylglycerol, and others. These phospholipids are phosphatidic acid derivatives containing alcohol esters linked to the phosphate residue. Phosphatidic acid is activated to CDP-diacylglycerol, consuming cytidine 5′-triphosphate (CTP) to receive alcohols.

$$
\begin{array}{l}
CH_2-CO-R \\
| \\
CH-O-CO-R \quad + \quad CTP \quad \xrightarrow[\text{cytidyltransferase}]{\text{phosphatidate}} \quad CH-O-CO-R \; + \; PP_i \\
| \quad\quad O \\
CH_2O-\overset{\|}{P}-OH \\
\quad\quad OH
\end{array}
$$

phosphatidic acid   CDP-diacylglycerol

Cytidine 5′-monophosphate (CMP) is replaced with alcohols such as serine and inositol in a reaction catalyzed by alcohol-specific phosphatidyl transferase. Phosphatidylserine is decarboxylated to phosphatidylethanolamine before being synthesized to phosphatidylcholine through methylation using S-adenosylmethionine (SAM) as the source of the methyl group. Similar reactions are employed to produce phosphatidylinositol, phosphatidylglycerol and cardiolipin (Figure 6.24).

As in phosphatidylcholine synthesis, C1 units such as methyl groups are transferred in various reactions involving the C1 carriers tetrahydrofolate (H_4F) and SAM. H_4F participates in the reactions which add or remove all forms of C1 units except carbonate, including methyl (-CH_3), methylene (-CH_2-), methenyl (-CH=), formyl (-CHO) and formimino (-CH=NH) as shown in Figure 6.25. SAM functions as a -CH_3 donor. Methanogens have their own C1 carriers such as coenzyme M, tetrahydromethanopterin (H_4MTP) and methanofuran (MF). Some of these are found in other archaea and in some eubacteria (Section 9.4.2).

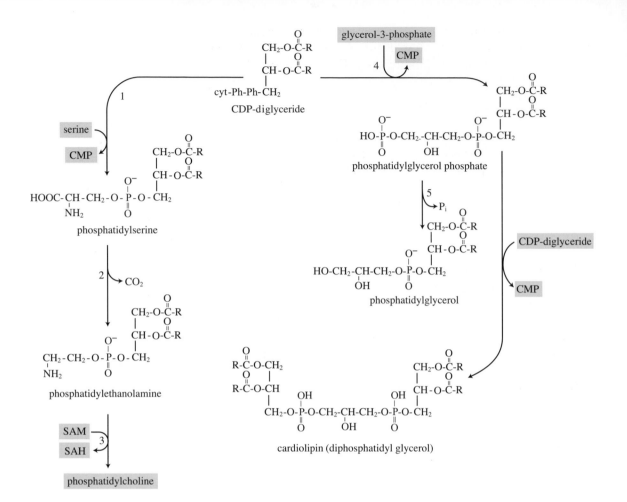

Figure 6.24 **Phospholipid biosynthesis.**

Phosphatidic acid is activated to CDP-diglyceride in a reaction catalyzed by phosphatidate cytidyltransferase before an alcohol replaces CMP.

1, CDP-diacylglyceride:serine O-phosphatidyltransferase; 2, phosphatidylserine decarboxylase; 3, phosphatidylethanolamine methyltransferase; 4, CDP-diacylglyceride:glycerol-3-phosphate 3-phosphatidyltransferase; 5, phosphatidylglycerol phosphatase.

SAM, S-adenosylmethionine; SAH, S-adenosylhomocysteine; CMP, cytidine 5'-monophosphate.

Figure 6.25 **Coenzyme tetrahydrofolate and C1 units carried by tetrahydrofolate.**

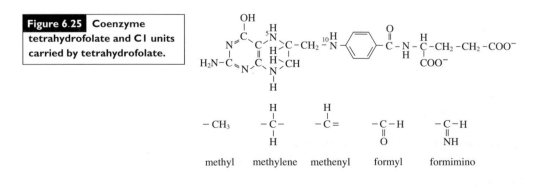

H₄F has a structure consisting of glutamate, *p*-aminobenzoate and folate with reduced pteridine (Figure 6.25). Methionine adenosyltransferase condenses methionine and ATP to produce SAM:

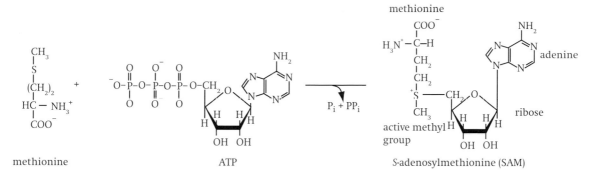

methionine   ATP   *S*-adenosylmethionine (SAM)

6.6.3 Isoprenoid biosynthesis

Various isoprenoid compounds are found in microbial cells, including alcohol residues of archaeal cytoplasmic membrane phospholipids, quinones, and carotenoids. They are synthesized through isopentenyl pyrophosphate (IPP). Two pathways are known to produce IPP. Acetyl-CoA is the starting material in the mevalonate pathway which operates in archaea and eukaryotes. Most bacteria produce IPP from glyceraldehyde-3-phosphate through the mevalonate-independent pathway (Figure 6.26).

6.7 | Heme biosynthesis

Many proteins contain heme, including cytochromes and chlorophylls. 5-aminolevulinate is the common precursor for the synthesis of tetrapyrroles. This precursor is synthesized either from glutamyl-tRNA or from glycine and succinyl-CoA. Aminolevulinate synthase is found only in *Alphaproteobacteria*:

$$\text{succinyl-CoA} + \text{glycine} \xrightarrow{\text{aminolevulinate synthase}} \text{5-aminolevulinate}$$
$$\searrow CO_2$$

In prokaryotes other than *Alphaproteobacteria*, the heme precursor is synthesized from glutamate-tRNA:

glutamyl-tRNA

    ↓ NADPH

    glutamyl-tRNA reductase

    ↓ NADP⁺ + H⁺

glutamate-1-semialdehyde

    glutamate-1-semialdehyde aminotransferase

    ↓

5-aminolevulinate

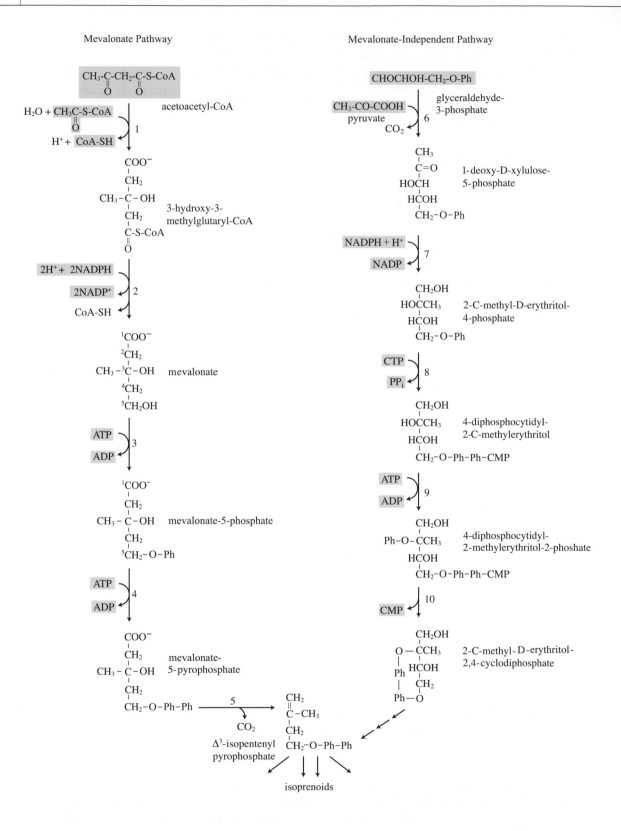

Two 5-aminolevulinate molecules condense to porphobilinogen, which is deaminated to uroporphyrinogen III with the basic structure of tetrapyrrole. Corrinoids and coenzyme F_{430} are synthesized from uroporphyrinogen III. This intermediate is further metabolized to protoporphyrin IX to synthesize chlorophylls and bacteriochlorophylls. Cytochromes and phycobilins are derived from protoheme (Figure 6.27).

6.8 | Synthesis of saccharides and their derivatives

Microbial cells contain various saccharides located in the cell wall, lipopolysaccharide of the outer membrane in Gram-negative bacteria, capsular material and glycogen. These polymers are synthesized from activated monomers derived from glucose-6-phosphate. The latter can be produced not only from sugars but also from non-carbohydrate substrates through gluconeogenesis (Section 4.2).

6.8.1 Hexose phosphate and UDP-sugar
Fructose-6-phosphate is isomerized to mannose-6-phosphate:

$$\text{fructose-6-phosphate} \xrightleftharpoons[\text{isomerase}]{\text{mannose-6-phosphate}} \text{mannose-6-phosphate}$$

Polysaccharides containing galactose are synthesized from UDP-galactose which is converted from glucose-6-phosphate in three steps.

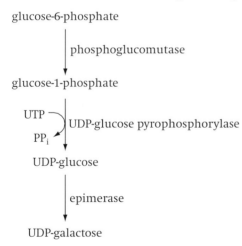

glucose-6-phosphate

│ phosphoglucomutase
▼

glucose-1-phosphate

UTP ─┐
     ╲ UDP-glucose pyrophosphorylase
PP$_i$ ╱
     ▼

UDP-glucose

│ epimerase
▼

UDP-galactose

Figure 6.26 (cont.) **Synthesis of isopentenyl pyrophosphate (IPP), the precursor of isoprenoids through the mevalonate pathway or through the mevalonate-independent pathway.**

(*Mol. Microbiol.* **37**:703–716, 2000)

1, hydroxylmethylglutaryl-CoA synthase; 2, hydroxymethylglutarate reductase; 3, mevalonate kinase; 4, phosphomevalonate kinase; 5, diphosphomevalonate decarboxylase; 6, 1-deoxy-D-xylulose-5-phosphate synthase; 7, 1-deoxy-D-xylulose-5-phosphate reductoisomerase; 8, 4-diphosphocytidyl-2C-methyl-D-erythritol synthase; 9, 4-diphosphocytidyl-2C-methyl-D-erythritol kinase; 10, 2C-methyl-D-erythritol-2,4-cyclodiphosphate synthase.

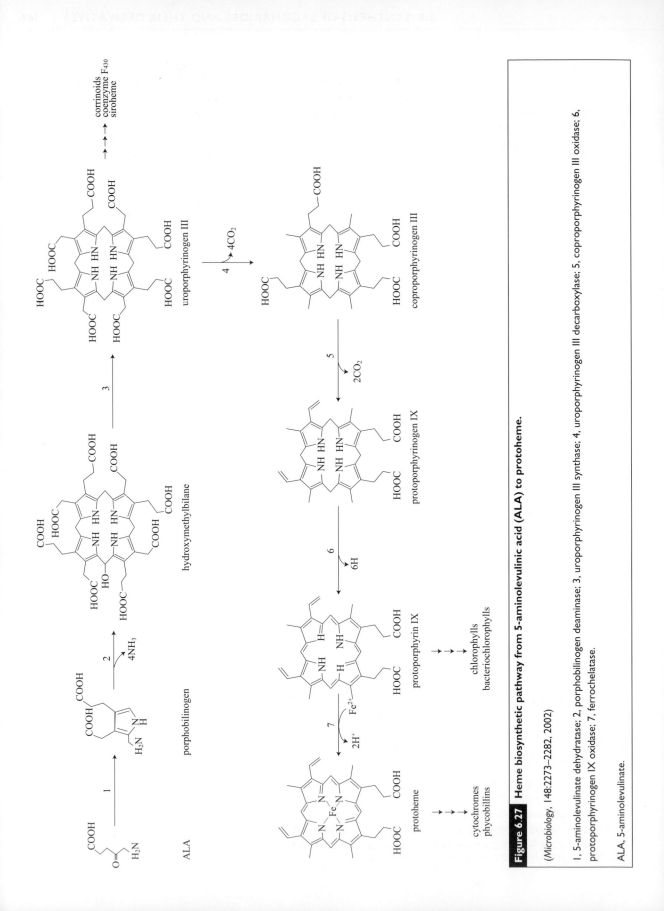

Figure 6.27 Heme biosynthetic pathway from 5-aminolevulinic acid (ALA) to protoheme.

(Microbiology, 148:2273–2282, 2002)

1, 5-aminolevulinate dehydratase; 2, porphobilinogen deaminase; 3, uroporphyrinogen III synthase; 4, uroporphyrinogen III decarboxylase; 5, coproporphyrinogen III oxidase; 6, protoporphyrinogen IX oxidase; 7, ferrochelatase.

ALA, 5-aminolevulinate.

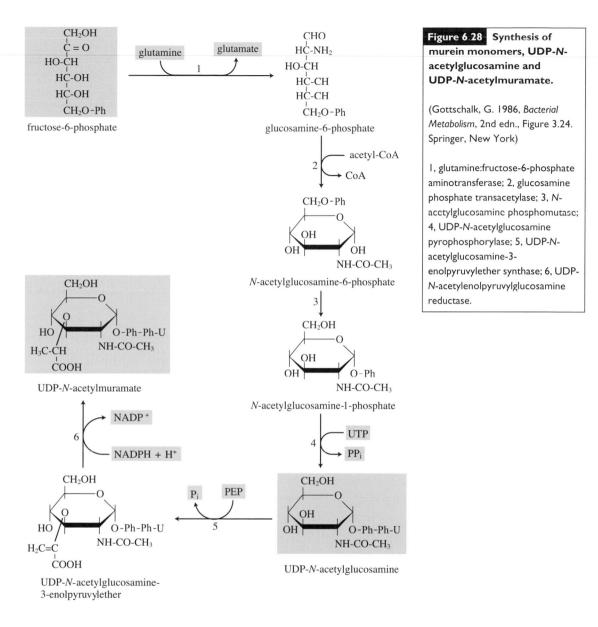

Figure 6.28 Synthesis of murein monomers, UDP-*N*-acetylglucosamine and UDP-*N*-acetylmuramate.

(Gottschalk, G. 1986, *Bacterial Metabolism*, 2nd edn., Figure 3.24. Springer, New York)

1, glutamine:fructose-6-phosphate aminotransferase; 2, glucosamine phosphate transacetylase; 3, *N*-acetylglucosamine phosphomutase; 4, UDP-*N*-acetylglucosamine pyrophosphorylase; 5, UDP-*N*-acetylglucosamine-3-enolpyruvylether synthase; 6, UDP-*N*-acetylenolpyruvylglucosamine reductase.

6.8.2 Monomers of murein

Fructose-6-phosphate is used to synthesize the murein monomers, uridine diphosphate (UDP)-*N*-acetylglucosamine and UDP-*N*-acetylmuramate. The precursor is aminated to glucosamine-6-phosphate using glutamate as the amine group donor before being acetylated to *N*-acetylglucosamine-6-phosphate. The latter is activated to UDP-*N*-acetylglucosamine condensing with UTP. PEP is used to add enolpyruvate to this intermediate before being reduced to UDP-*N*-acetylmuramate consuming NADPH (Figure 6.28).

Amino acids are added to UDP-*N*-acetylmuramate to synthesize UDP-*N*-acetylmuramylpentapeptide. L-alanine, D-glutamate, *meso*-diaminopimelate and D-alanyl-D-alanine form a peptide on the lactyl

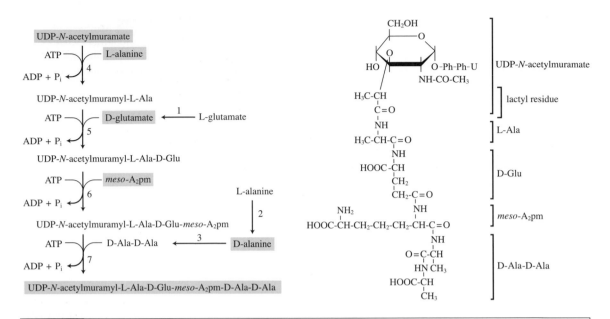

Figure 6.29 Synthesis of UDP-N-acetylmuramylpentapeptide through a non-ribosomal peptide synthesis process adding amino acids to the lactyl group of **UDP-N-acetylmuramate.**

(*Microbiol. Mol. Biol. Rev.* 63:174–229, 1999)

The precursor for the bacterial cell wall murein synthesis, UDP-N-acetylmuramylpentapeptide, is made through a non-ribosomal peptide synthesis process. Amino acids are not activated and mRNA is not required. The amino acid sequence is determined by the enzyme specificity, and ATP is consumed to provide energy needed for the formation of peptide bonds. Different amino acids are found in the second and third positions depending on the bacterial species.

1, glutamate racemase; 2, alanine racemase; 3, D-alanine-D-alanine ligase; 4, UDP-N-acetylmuramate-alanine ligase; 5, UDP-N-acetylmuramyl-alanine-D-glutamate ligase; 6, UDP-N-acetylmuramyl-alanyl-D-glutamate-2,6-diaminopimelate ligase; 7, D-alanyl-D-alanine adding enzyme.

meso-A$_2$pm, 2,6-diaminopimelate.

group of UDP-N-acetylmuramate consuming ATP (Figure 6.29). This reaction is a non-ribosomal peptide synthesis process independent from mRNA and the ribosomes. The amino acid sequence is determined by the enzyme specificity. Murein monomers are synthesized in the cytoplasm.

6.8.3 Monomers of teichoic acid

Ribose-phosphate and glycerol-phosphate are activated to CDP-ribitol and CDP-glycerol, as the precursors of ribitol teichoic acid and glycerol teichoic acid through a similar reaction as in the formation of UDP-sugars, consuming CTP instead of UTP.

6.8.4 Precursor of lipopolysaccharide, O-antigen

Lipopolysaccharide consists of lipid A, core polysaccharide and O-antigen (Figure 2.4, Section 2.3.3.2). O-antigen has a structure based on repeating oligosaccharide. Sugar-nucleotides are added to

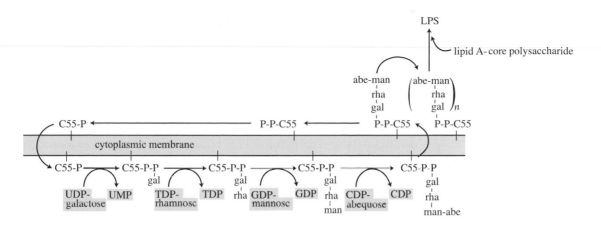

Figure 6.30 Synthesis of O-antigen and lipopolysaccharide in *Salmonella enterica*.

The O-antigen of LPS has a structure of repeating oligosaccharide. The unit oligosaccharide is synthesized onto undecaprenyl phosphate embedded in the cytoplasmic membrane, which receives the sugar moiety from sugar-nucleotides. Undecaprenyl phosphate carries the oligosaccharide across the membrane. LPS is synthesized by the addition of the oligosaccharide to the core polysaccharide that has been transported to the periplasm bound to lipid A. LPS is transported to the outer membrane by the LPS exporter, a member of the ATP-binding cassette (ABC) family.

C55-P, undecaprenyl phosphate.

undecaprenyl (bactoprenol) phosphate embedded in the cytoplasmic membrane (Figure 6.30). Undecaprenyl phosphate transports the oligosaccharide across the membrane (Section 6.9.2.1). O-antigen ligase transfers the oligosaccharide to the core polysaccharide-lipid A in the periplasm to synthesize LPS. Core polysaccharide-lipid A is synthesized in the cytoplasm, and crosses the cytoplasmic membrane. The hydrophobic lipid A is the carrier of the core polysaccharide.

6.9 | Polysaccharide biosynthesis and the assembly of cell surface structures

Polysaccharides in bacterial cells include glycogen, a storage material, and structural polymers such as murein and teichoic acid in the cell wall, and LPS in the outer membrane. The precursors are synthesized in the cytoplasm and murein and LPS are synthesized after the precursors are transported across the cytoplasmic membrane. For this reason, the synthesis and assembly of the cell wall and the outer membrane are closely related to the transport of their precursors.

6.9.1 Glycogen synthesis
UDP-glucose is the precursor for glycogen synthesis in eukaryotes. Glucose-1-phosphate is activated to ADP-glucose in prokaryotes before being polymerized to glycogen by glycogen synthase and glycosyl (4→6) transferase.

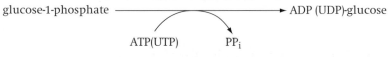

ADP (UDP)-glucose pyrophosphorylase

glucose-1-phosphate ⟶ ADP (UDP)-glucose

ATP(UTP)   PP$_i$

Glycogen synthase transfers the glucose moiety of ADP (UDP)-glucose to the non-reducing end of the existing glycogen to form α-1,4-glucoside (Figure 6.31). When the α-1,4 chain reaches a certain length, glycosyl (4→6) transferase catalyzes a transglycosylation reaction, transferring the α-1,4 chain to form an α-1,6 linkage (Figure 6.32).

Figure 6.31 **Formation of the α-1,4 linkage in glycogen by glycogen synthase.**

(Gottschalk, G. 1986, *Bacterial Metabolism*, 2nd edn., Figure 3.28. Springer, New York)

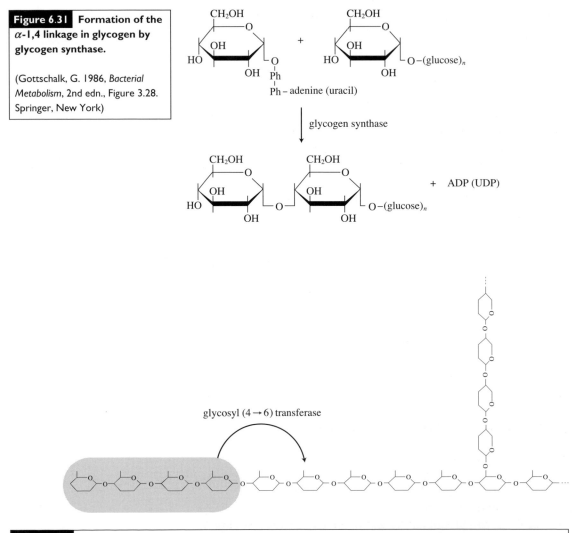

Figure 6.32 **Formation of the α-1,6 side chain in glycogen through the transglycosylation reaction catalyzed by glycosyl(4→6) transferase.**

(Gottschalk, G. 1986, *Bacterial Metabolism*, 2nd edn., Figure 3.29. Springer, New York)

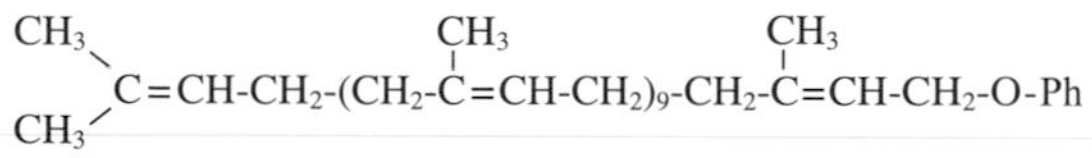

$$CH_3 \diagdown C=CH-CH_2-(CH_2-\overset{CH_3}{\underset{|}{C}}=CH-CH_2)_9-CH_2-\overset{CH_3}{\underset{|}{C}}=CH-CH_2-O-Ph$$
$$CH_3 \diagup$$

Figure 6.33 The structure of undecaprenyl (bactoprenyl) phosphate.

6.9.2 Murein synthesis and cell wall assembly

6.9.2.1 Transport of cell wall precursor components through the membrane

The cell wall consists of murein and teichoic acid in Gram-positive bacteria, and of murein in Gram-negative bacteria. Various proteins are also associated with the cell wall, especially in Gram-positive bacteria. Monomers of the polymeric compounds are synthesized in the cytoplasm before being transported to the periplasm to be polymerized. They are hydrophilic in nature, and cannot diffuse through the cytoplasmic membrane. To overcome this, the hydrophobic membrane compound, undecaprenyl phosphate (Figure 6.33), transports them through the membrane.

6.9.2.2 Murein synthesis

Phospho-N-acetylmuramylpentapeptide (phospho-MurNAc-pentapeptide) is transferred to the undecaprenyl phosphate embedded in the membrane from UDP-MurNAc-pentapeptide (lipid I) separating uridine 5′-monophosphate (UMP). N-acetylglucosamine (GlcNAc) transferase forms undecaprenyl-GlcNAc-MurNAc-pentapeptide pyrophosphate (lipid II) separating UDP from UDP-GlcNAc. In Gram-positive bacteria, lipid II is further modified by the addition of amino acids to the third amino acid position, which is lysine in the pentapeptide. Different amino acids are added depending on the species. Glycyl-tRNA is consumed to add five glycyl units in *Staphylococcus aureus*. Undecaprenyl-GlcNAc-MurNAc-pentapeptide-(gly)$_5$ pyrophosphate crosses the membrane to transfer GlcNAc-N-acetylmuramylpentapeptide-(gly)$_5$ to the existing murein, liberating undecaprenyl pyrophosphate through the action of transglycosylase, and a transpeptidase cross-links the neighbouring chains. A phosphatase converts undecaprenyl pyrophosphate to undecaprenyl phosphate, which starts another round of the same series of reactions (Figure 6.34).

Inhibitors of murein synthesis, ristocetin and vancomycin, inhibit transglycosylase, and bacitracin interferes with the dephosphorylation of undecaprenyl pyrophosphate. β-lactam antibiotics inhibit transpeptidation and carboxypeptidation reactions.

6.9.2.3 Teichoic acid synthesis

Teichoic acid is synthesized in a similar way as murein in Gram-positive bacteria (Figure 6.35). GlcNAc is transferred from UDP-GlcNAc to undecaprenyl phosphate before taking glycerol-phosphate from CDP-glyceride to form glycerol-P-N-acetylglucosamine-P-P-undecaprenyl. Glycerol-P-N-acetylglucosamine-P is transferred to the existing teichoic acid separating undecaprenyl phosphate. Teichoic acid synthesis is not inhibited by bacitracin (which inhibits dephosphorylation of undecaprenyl pyrophosphate in murein synthesis).

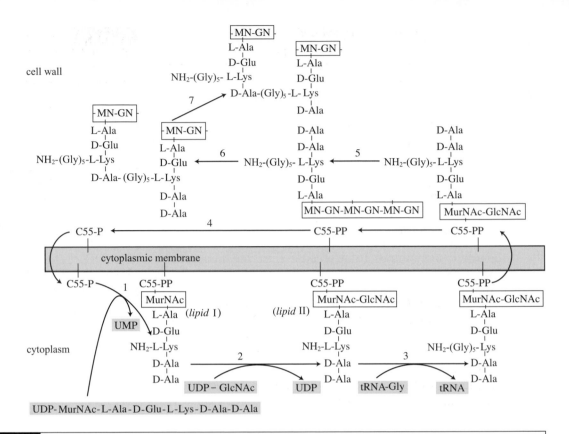

Figure 6.34 Murein synthesis in a **Gram-positive bacterium,** *Staphylococcus aureus.*

(*Microbiol. Mol. Biol. Rev.* **63**:174–229, 1999)

Murein precursors, *N*-acetylglucosamine-UDP (GlcNAc) and UDP-*N*-acetylmuramate (MurNAc)-pentapeptide are synthesized in the cytoplasm (Figures 6.28 and 6.29). UMP is separated from UDP-MurNAc-pentapeptide, transferring phospho-MurNAc-pentapeptide to undecaprenyl phosphate to form MurNAc-pentapeptide undecaprenyl pyrophosphate, which is known as lipid I (1). Subsequently, GlcNAc is transferred to MurNAc-pentapeptide undecaprenyl pyrophosphate from UDP-GlcNAc, separating UDP to form undecaprenyl-GlcNAc-MurNAc-pentapeptide pyrophosphate, also known as lipid II (2). Five glycyl groups bind the third amino acid, lysine, in the pentapeptide, consuming glycyl-tRNAs before being translocated to the outer leaflet of the membrane (3). GlcNAc-MurNAc-pentapeptide-(gly)$_5$ forms a β-1,4-glucoside with the existing murein by a transglycosylation reaction (5), and a transpeptidase cross-link (6). Inhibitors of murein synthesis, ristocetin and vancomycin, inhibit transglycosylase, and bacitracin interferes with the dephosphorylation of undecaprenyl pyrophosphate (4). β-lactam antibiotics inhibit (6) transpeptidation and (7) carboxypeptidation reactions.

1, phospho-*N*-acetylmuramoyl pentapeptide transferase; 2, *N*-acetylglucosamine transferase; 3, glycyl transferase; 4, undecaprenyl diphosphatase; 5, transglycosylase; 6, transpeptidase; 7, carboxypeptidase.

C55–P, undecaprenylphosphate.

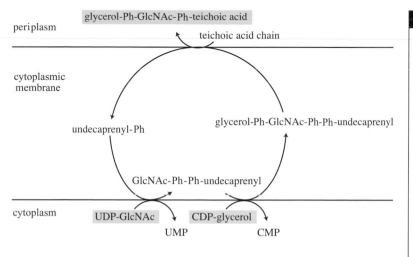

periplasm

glycerol-Ph-GlcNAc-Ph-teichoic acid

teichoic acid chain

cytoplasmic membrane

undecaprenyl-Ph

glycerol-Ph-GlcNAc-Ph-Ph-undecaprenyl

GlcNAc-Ph-Ph-undecaprenyl

cytoplasm

UDP-GlcNAc   CDP-glycerol

UMP   CMP

Figure 6.35 **Synthesis of glycerol teichoic acid.**

Undecaprenyl-phosphate translocates the teichoic acid precursors across the cytoplasmic membrane as in murein synthesis. UDP-N-acetylglucosamine and CDP-glyceride react with undecaprenyl-phosphate at the cytoplasmic side of the membrane to form glycerol-P-N-acctylglucosaminc-P-P-undecaprenyl, which crosses the membrane before glycerol-P-N-acetylglucosamine-P is transferred to the existing teichoic acid separating undecaprenyl-phosphate.

6.9.2.4 Cell wall proteins in Gram-positive bacteria

A range of proteins are anchored to the cell wall of Gram-positive bacteria, including enzymes and virulence-related proteins. Most of these are covalently bonded to the murein. They are transported through the membrane before being bonded to the murein by an enzyme called sortase. Sortase cleaves the substrates at a conserved LPXTG motif near the C-terminal and covalently links them to a penta-glycine crossbridge in murein (Figure 6.36).

6.9.3 Outer membrane assembly

The asymmetrical Gram-negative bacterial outer membrane consists of outer layer LPS, and inner layer phospholipids and proteins. They are synthesized in the cytoplasm and are translocated to their position through the cytoplasmic membrane and the murein before being assembled.

6.9.3.1 Protein translocation

Proteins present in the OM are either lipoproteins or integral proteins. The former are anchored to the OM with a N-terminal lipid tail stabilizing the OM structure by linking to the murein layer. Integral proteins are referred to as outer membrane proteins (OMP) and have a membrane-spanning region (Table 2.4). Precursor polypeptides of the lipoproteins are translocated through the cytoplasmic membrane either through GSP or ABC pathways (Section 3.8) before being folded in the periplasmic region and targeted to their destination in a process similar to the chaperone/usher pathway (Section 3.8.2.1). OMP are translocated and targeted to the OM through GSP and chaperone/usher pathways.

6.9.3.2 Lipopolysaccharide (LPS) translocation

LPS, the outer layer of the outer membrane, is a complex molecule consisting of lipid A, core polysaccharide and O-antigen. O-antigen

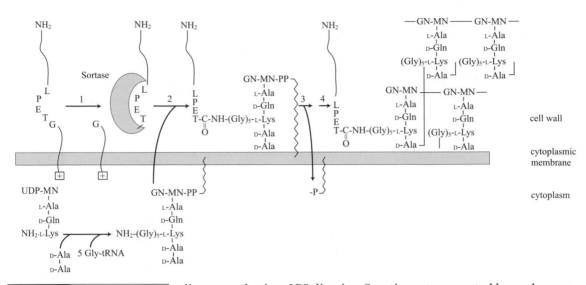

Figure 6.36 Cell wall protein anchoring in Gram-positive bacteria.

(*Trends Microbiol.* 8:148–151, 2000)

The process can be divided into four distinct steps. (1) The full length precursor is exported from the cytoplasm via an amino-terminal leader peptide. (2) The protein is prevented from release into the extracellular milieu by the charged tail and hydrophobic domain. (3) The protein is cleaved by sortase between the threonyl and glycyl residues of the LPXTG motif with the formation of a thioester between the conserved cysteine of sortase and the threonine carboxyl. (4) The newly liberated carboxy terminus of threonine is transferred via an amide bond exchange to an amino group found at the end of the wall murein.

ligase synthesizes LPS, ligating O-antigen, transported by undecaprenyl phosphate through the cytoplasmic membrane, to the lipid A-core polysaccharide complex in the periplasmic region. LPS is targeted to the outer layer of the OM by the LPS exporter, a member of the ATP-binding cassette (ABC) family.

6.9.3.3 Phospholipid translocation
Phospholipids of the inner layer of the outer membrane are transported from the cytoplasmic membrane to their location by a protein called flippase (Section 6.9.4).

6.9.4 Cytoplasmic membrane (CM) assembly
The CM consists of phospholipids and proteins. Protein constitutes up to 65% of the CM with various functions including solute transport and electron transport for energy conservation. CM proteins are embedded in the membrane through the SRP (signal recognition particle) pathway, one of two general secretion pathways (GSP, Section 3.8.1). There is some evidence which suggests that a membrane protein, YidC in *Escherichia coli*, facilitates GSP-independent membrane integration, folding and assembly of energy-transducing membrane protein complexes. Phospholipids form the inner layer of the CM after they are synthesized, and are flipped to the outer layer by the action of flippase.

6.10 | Deoxyribonucleic acid (DNA) replication

DNA carries genetic information in its nucleotide sequence and its synthesis is complex, involving a variety of enzymes referred to as the DNA replicase system or replisome.

6.10.1 DNA replication
DNA forms a double stranded helix with hydrogen bonding between adenine-thymine and guanine-cytosine pairs (Figure 6.37). Most

(a)

(b)

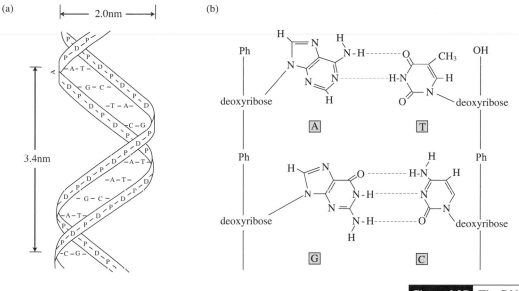

Figure 6.37 The DNA double helix.

(a) The double helix. (b) Specific pairing between adenine (A) and thymine (T), and cytosine (C) and guanine (G) through hydrogen bonding.

D, deoxyribose; P and Ph, phosphate.

prokaryotic DNA occurs in a circular form, although linear chromosomal DNA is found in some *Streptomyces* species. The chromosomal DNA forms a complex with various proteins which participate in replication, transcription and their control. The chromosome is attached to the cytoplasmic membrane through the replication origin (*oriC*), and the DNA is in a supercoiled form. A number of proteins, which are involved in replication and in chromosome segregation during cell division, are associated with the replication origin.

Figure 6.38 shows DNA replication in prokaryotes. Over 30 different proteins are involved in DNA replication, which forms a complex replisome, located roughly at the centre of the cell. The chromosome moves through the replisome to be replicated. DNA topoisomerase converts supercoiled DNA to a relaxed form before the helicase separates the double strand to single strands, thereby breaking the hydrogen bonds to start replication. The DNA binding protein (DBP or helix destabilizing protein, HDP) maintains the single stranded DNA molecules while DNA polymerase replicates through complementary base pairing.

6.10.1.1 RNA primer

DNA polymerase requires a template for base pairing and a primer to add the nucleotide. Single strand DNA is used as the template. To provide primer, RNA polymerase (primase) synthesizes oligoribonucleotides of 10–30 bases by means of complementary base pairing with the template. This is referred to as the RNA primer. DNA polymerase adds deoxynucleotide to the primer (consuming deoxynucleoside triphosphate) based on the base pairing with the template in the direction of 5′–3′.

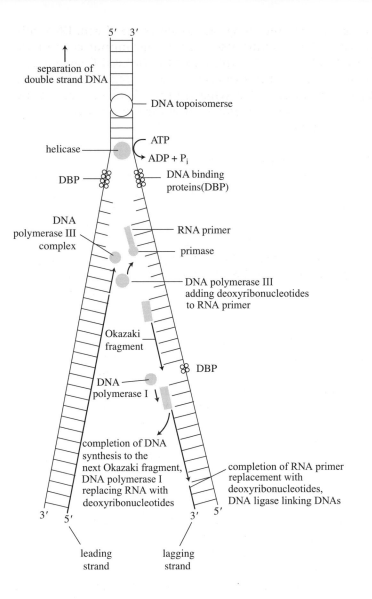

Figure 6.38 **DNA replication.**

Topoisomerase (gyrase) relaxes the supercoiled DNA, and helicase separates the double strand to single strands. Since DNA is synthesized in the direction of 5'–3', synthesis is continuous in one leading strand while the other strand is synthesized in segments (Okazaki fragments) which are ligated later.

6.10.1.2 Okazaki fragment

Helicase separates double stranded DNA into single strands, the leading strand in the direction of 3'–5' and the other, the lagging strand, in the direction of 5'–3'. DNA polymerase adds new nucleotide only to the 3'-OH of the existing DNA molecules. For this reason DNA synthesis takes place continuously on the leading strand while discontinuous DNA synthesis occurs on the lagging strand to a length of 1000–2000 nucleotides. This is referred to as the Okazaki fragment.

6.10.1.3 DNA polymerase

Escherichia coli has three separate DNA polymerases, I, II and III. All of them have DNA polymerase activity in the direction of 5'–3' and exonuclease activity removing nucleotide at the 3′ end. DNA polymerase I and III have exonuclease activity removing nucleotide at the 5′ end. DNA polymerase III synthesizes the Okazaki fragment onto the RNA primer before the RNA primer is removed and replaced with DNA by the action of DNA polymerase I with exonuclease activity. DNA ligase links Okazaki fragments after the primer RNA is completely replaced with DNA.

6.10.2 Spontaneous mutation

Prokaryotes grow much faster than higher organisms and they replicate at a speed of up to 15 000 bases per minute. Errors in DNA replication result in mutations. Mutation rates in prokaryotes are low, being around 10^{-10} errors per base inserted even at this high replication speed. This high accuracy in the replication process is possible because of the exact base pairing and the 'proof reading' by DNA polymerase III, with the exonuclease activity replacing misinserted nucleotides. Failure in these processes will result in spontaneous mutation.

6.10.3 Post-replicational modification

Bacterial DNA contains 5-methyl cytosine and 6-methyl adenine. These are the products of the methylation reactions catalyzed by DNA methyltransferase after DNA is synthesized. This process is referred to as post-replicational modification. The DNA is methylated to protect it from restriction enzymes which degrade foreign DNA invading the cell.

Several hundred restriction enzymes are known in prokaryotes. They recognize specific sequences in the DNA to hydrolyze and to methylate. They are grouped into types I, II and III according their properties. Types I and III not only hydrolyze the DNA sequence they recognize but also methylate adenine and cytosine within the recognized sequence. Type II restriction enzymes have hydrolyzing activity but not methylating activity. For this reason type I and III enzymes are referred to as restriction-modification enzymes. S-adenosylmethionine serves as the methyl group donor for the DNA modification reaction. In addition to the function of restriction-modification these enzymes may participate in other processes such as DNA replication, cell cycle and regulation of gene expression. Topoisomerase converts the modified DNA into a supercoiled form.

6.10.4 Chromosome segregation

DNA replication proceeds in both directions from the replication origin (*oriC*) to the terminator sequence located 180° opposite the origin. When the two replication forks meet at the terminator region, the daughter chromosomes are separated by an enzyme called topoisomerase IV before being segregated to the poles of the cell away from the septum separating the daughter cell by cell division. Several different proteins are involved in the segregation process which is not fully understood. The proteins involved in chromosome segregation include SpoOJ in *Bacillus subtilis* and the ParB protein in *Caulobacter crescentus*. They form a complex with *oriC*.

6.11 | Transcription

Transcription is the process of RNA synthesis from the DNA template. Major RNA types include mRNA, rRNA and tRNA with specific functions in expression of the genetic information. They are all synthesized in a similar way before being modified to their own specific forms through post-transcriptional processing.

6.11.1 RNA synthesis

RNA is synthesized through transcription, a process as complex as replication. RNA polymerase (DNA-dependent RNA polymerase) catalyzes the formation of phosphodiester bonds between ribonucleotides consuming GTP, CTP, ATP or UTP depending on the base pairing within DNA. The synthesis proceeds in the $5'$–$3'$ direction as in replication.

Certain regions in DNA have a strong affinity for RNA polymerase and are referred to as promoters. RNA polymerase recognizes and binds the promoter region of DNA to start RNA synthesis. RNA polymerase consists of five subunits in the ratio of $\alpha_2\beta\beta'\sigma$ in bacteria. The σ-factor of the enzyme recognizes the promoter region of DNA, and the enzyme complex binds to it. At this point, the σ-factor is separated from the complex and the core enzyme $\alpha_2\beta\beta'$ moves along the DNA synthesizing RNA according to a base-pairing mechanism (Figure 6.39). Multiple σ-factors are known in bacteria which recognize different promoter regions. Proteins known as activators and repressors control the activity of some promoters (Section 12.1).

Transcription is terminated either by an intrinsic input mechanism involving a specific DNA region known as a termination site or by an extrinsic input mechanism exerted by a protein, ρ-factor.

6.11.2 Post-transcriptional processing

The transcripts of tRNA and rRNA are longer than the functional RNAs and the latter contain methylated nucleotides. The transcripts are methylated before being cut by endonuclease to their functional size (Figure 6.40). Unusual bases are found in tRNA. The nucleotides in tRNA are further modified in a mechanism called post-transcriptional processing.

It should be noted that unlike eukaryotic transcripts, most prokaryotic transcripts of mRNA do not contain an intron and translation starts while transcription is in progress. mRNA editing is not known in prokaryotes.

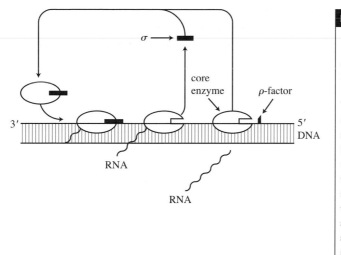

Figure 6.39 **RNA synthesis.**

(Gottschalk, G. 1986, *Bacterial Metabolism*, 2nd edn., Figure 3.40. Springer, New York)

The RNA polymerase complex consists of a core enzyme ($\alpha_2\beta\beta'$) and σ-factor. The enzyme complex binds DNA after the σ-factor recognizes the promoter region on the DNA. When RNA synthesis starts, σ factor is separated from the core enzyme which moves along the DNA synthesizing RNA according to the base-pairing mechanism. Many proteins are involved in the initiation, elongation and termination of the transcription process. An example is ρ-factor which terminates the transcription.

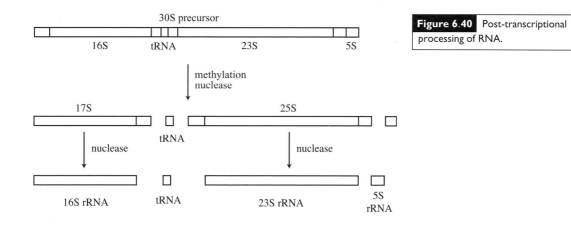

Figure 6.40 Post-transcriptional processing of RNA.

6.12 | Translation

Translation is the process in which the genetic information passed on to mRNA is used to make proteins. As the genetic information carried by DNA is passed to RNA during transcription, the information in mRNA is translated into protein through an amino acid sequence encoded by the sequence of bases in the mRNA (Table 6.7). Amino acids are activated before being polymerized into peptide.

Table 6.7.	*Genetic code of mRNA*						
UUU	Phe	UCU	Ser	UAU	Tyr	UGU	Cys
UUC	Phe	UCC	Ser	UAC	Tyr	UGC	Cys
UUA	Leu	UCA	Ser	UAA	Term[a]	UGA	Term[a]
UUG	Leu	UCG	Ser	UAG	Term[a]	UGG	Trp
CUU	Leu	CCU	Pro	CAU	His	CGU	Arg
CUC	Leu	CCC	Pro	CAC	His	CGC	Arg
CUA	Leu	CCA	Pro	CAA	Gln	CGA	Arg
CUG	Leu	CCG	Pro	CAG	Gln	CGG	Arg
AUU	Ile	ACU	Thr	AAU	Asn	AGU	Ser
AUC	Ile	ACC	Thr	AAC	Asn	AGC	Ser
AUA	Ile	ACA	Thr	AAA	Lys	AGA	Arg
AUG	Met[b]	ACG	Thr	AAG	Lys	AGG	Arg
GUU	Val	GCU	Ala	GAU	Asp	GGU	Gly
GUC	Val	GCC	Ala	GAC	Asp	GGC	Gly
GUA	Val	GCA	Ala	GAA	Glu	GGA	Gly
GUG	Val	GCG	Ala	GAG	Glu	GGG	Gly

[a] UAA, UAG and UGA are non-coding codons where peptide synthesis stops. They are referred to as termination or nonsense codons.

[b] AUG encodes *N*-formylmethionine at the beginning of mRNA in bacteria.

6.12.1 Amino acid activation

Amino acids are activated to aminoacyl-tRNA consuming ATP:

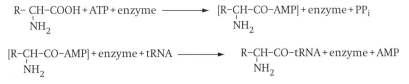

enzyme : aminoacyl-tRNA synthetase

The aminoacyl-tRNA synthetase catalyzing this reaction recognizes not only amino acids but also tRNA. More than one tRNA is needed for each amino acid since most of the amino acids are coded by multiple codons, except methionine and tryptophan (Table 6.7), and a tRNA is needed for each codon base pairing with its specific anticodon (Figure 6.41).

6.12.2 Synthesis of peptide: initiation, elongation and termination

The coding region in mRNA starts with AUG. Peptide synthesis starts with methionine in eukaryotes and with *N*-formylmethionine in bacteria. AUG (TAC on DNA) is referred to as the initiation codon. Though peptide synthesis is a continuous process, translation can be described for convenience as initiation, elongation and termination steps.

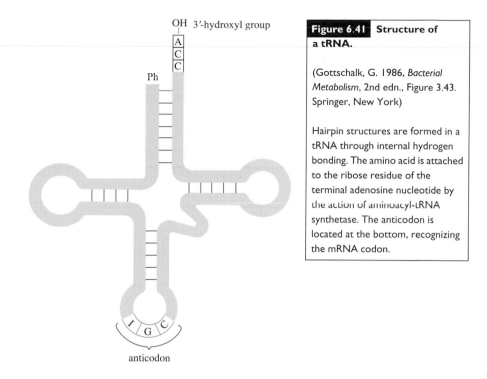

OH 3'-hydroxyl group

A
C
C

Ph

anticodon

I G C

Figure 6.41 Structure of a tRNA.

(Gottschalk, G. 1986, *Bacterial Metabolism*, 2nd edn., Figure 3.43. Springer, New York)

Hairpin structures are formed in a tRNA through internal hydrogen bonding. The amino acid is attached to the ribose residue of the terminal adenosine nucleotide by the action of aminoacyl-tRNA synthetase. The anticodon is located at the bottom, recognizing the mRNA codon.

6.12.2.1 Ribosomes

Ribosomes are the site of peptide synthesis. The prokaryotic ribosome has a size of 70S consisting of a 50S large subunit and 30S small subunit. The large subunit contains 5S and 23S rRNA and about 35 proteins, while the small 30S subunit consists of 16S rRNA and about 21 proteins. The number of ribosomal proteins differs depending on the species. The eukaryotic ribosome has a size of 80S, consisting of 60S and 40S subunits. Mitochondria and chloroplasts have ribosomes similar to prokaryotes.

6.12.2.2 Initiation and elongation

The 30S subunit recognizes the ribosome binding site known as the Shine–Dalgarno sequence to bind the initiation codon on the mRNA. *N*-formylmethionyl-tRNA binds to the peptidyl site (P) occupying the initiation codon before the 50S subunit binds the 30S subunit–mRNA complex to form the initiation complex. Various proteins, including the initiation factors IF-1, IF-2 and IF-3, participate during this initiation step, and energy is provided through the hydrolysis of GTP.

The initiation complex takes up the aminoacyl-tRNA corresponding to the second codon of the mRNA to its aminoacyl site (A site) which occupies the second codon. A peptide bond is formed between the carboxyl group of the *N*-formylmethionyl residue and amino group of the aminoacyl residue occupying the A site. When the peptide bond is formed, the ribosome moves along the mRNA to position the dipeptide at the P site. These steps are repeated during the elongation process where proteins, including the elongation factors Tu, Ts and G, play important roles and energy is supplied as GTP (Figure 6.42).

6.12.2.3 Termination

When the ribosome reaches the termination codon, releasing factors, R1, R2 and S, separate the peptide from the ribosome. Many ribosomes bind a single mRNA during the translation process and the term polysome is used to describe such an mRNA complex with multiple ribosomes.

When translation stalls or stops without a termination codon, e.g. because the message stops prematurely (truncated mRNA) or rare codons are present, the ribosome and truncated peptide cannot be separated from mRNA. In this case, the ribosome should be rescued and the truncated peptide removed and destroyed. A small RNA called SsrA (small stable RNA A) or tmRNA is involved in the translation quality control process. The name tmRNA reflects its properties as both a tRNA and an mRNA. It can be charged with alanine as a tRNA, and has a coding region for 10 amino acids. The alanine-charged tmRNA can enter the empty A site of the ribosome, the charged alanine on its tRNA portion is transferred to the stalled polypeptide chain, and translation resumes, but from the 10-amino-acid ORF internal to the tmRNA. Translation ends at the nonsense codon within the tmRNA reading frame, releasing a polypeptide with an 11-amino-acid C-terminal tag (the SsrA tag). This sequence is sufficient to direct the polypeptide to one of a number of energy-dependent proteases, usually ClpXP, rapidly clearing the cell of this presumably abnormal polypeptide.

It should be noted that there are many small RNAs in all forms of cells having various regulatory functions (Section 12.1.9.4).

6.12.3 Post-translational modification and protein folding

The peptides mature to functional proteins with unique roles through various processes including chemical modification, folding to form secondary and tertiary structures, and complexation with cofactors.

The initiation residue at the N-terminal and terminal residue at the C-terminal are removed from the peptides released from the ribosome. At this stage, unusual amino acids are formed through chemical modification. Extracellular and membrane proteins are exported losing the signal peptide (Section 3.8).

During protein folding, small molecular weight chaperones, e.g. trigger factor, bind the nascent peptide to prevent misfolding during the translation process. Small proteins can be spontaneously folded into native forms. Large proteins are folded into native forms through a process aided by a chaperone or chaperonin (Figure 6.43), both of which require ATP for their function.

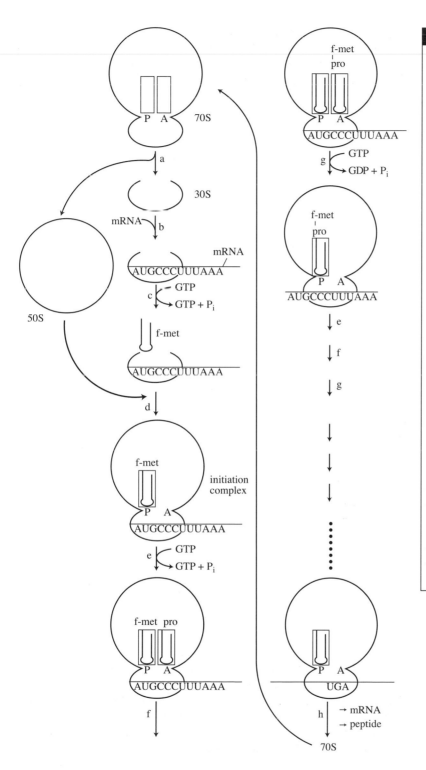

Figure 6.42 **Peptide synthesis – translation.**

A ribosome splits into its subunits, 30S and 50S (a), and the 30S subunit binds the initiation codon on mRNA after recognizing the ribosome-binding site known as the Shine–Dalgarno sequence (b). *N*-formylmethionyl-tRNA binds to the peptidyl site (P) occupying the initiation codon (c). At this stage the 50S subunit forms the initiation complex (d). During this initiation step, initiation factors (IF-1, IF-2, IF-3) are involved and GTP provides energy. The aminoacyl-tRNA corresponding to the second codon of the mRNA binds the aminoacyl site (A site) of the initiation complex which occupies the second codon (e) before a peptide bond is formed between the carboxyl group of the *N*-formylmethionyl residue and amino group of the aminoacyl residue occupying the A site (f). When the peptide bond is formed, the ribosome moves along the mRNA to position the dipeptide to the P site (g). During the repeat of this elongation process, elongation factors (Tu, Ts, G) are needed and energy is supplied in the form of GTP. When the ribosome reaches the termination codon, releasing factors (R1, R2, S) separate the peptide from the ribosome (h).

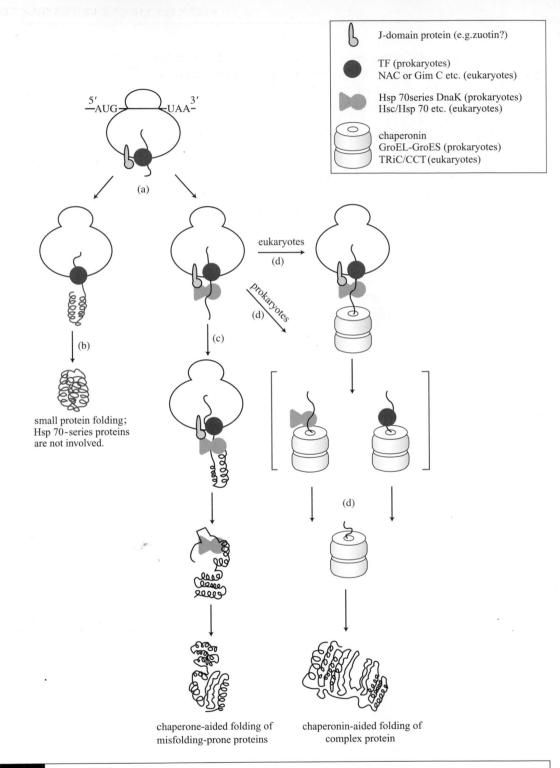

Figure 6.43 Protein folding.

(*Curr. Opin. Struct. Biol.* **10**:26–33, 2000)

During peptide synthesis on a ribosome, misfolding is prevented by small molecular weight chaperones such as trigger factor (TF) (a), while small proteins can fold into the native forms spontaneously (b). For large proteins, folding is aided by a chaperone such as DnaK (c) or by a chaperonin such as GroEL (d). ATP is consumed to supply the energy needed for the function of chaperone and chaperonin.

Thiol:disulfide oxidoreductase forms disulfide bonds oxidizing specific cysteine residues within the peptide. Disulfide bonds of extracellular proteins are formed by periplasmic enzymes.

6.13 | Assembly of cellular structure

For convenience, the assembly of the outer membrane, cell wall and cytoplasmic membrane is dealt with here in relation to biosynthesis of their constituents discussed above. Other distinct cellular structures include flagella, capsules, ribosomes and the nucleoid.

6.13.1 Flagella

A flagellum consists of a basal structure, hook and main filament, all of which are composed of protein. The basal structure includes the basal body, flagellar motor, switch and the exporter apparatus (Figure 2.1, Section 2.3.1). The basal body has a rod structure traversing the periplasmic space and surrounded with three ring structures. These are the cytoplasmic membrane MS ring, the periplasmic P ring and the outer membrane L ring (Section 2.3.1). The flagellar motor can be subdivided into the stator and the rotor. The stator is attached to the murein layer to surround the basal body, and the rotor is attached to the MS ring. A flagellum extends from the cytoplasm to the cell surface and has to be exported and embedded in position. Proteins constituting the flagellum are not found in the cytoplasm, suggesting that they are exported as soon as they are synthesized. Cytoplasmic flagellar proteins are embedded through the signal recognition particle pathway (SRP, Figure 3.9, Section 3.8.1.1). The proteins located beyond the cytoplasmic membrane are translocated through the type III flagellar export pathway (Figure 3.14). The filament is formed spontaneously (Figure 6.44).

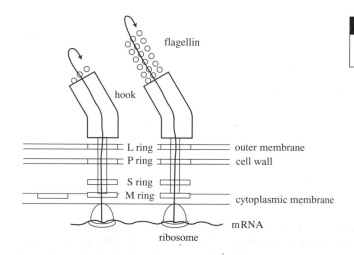

Figure 6.44 Flagellar filament formation in Gram-negative bacteria.

6.13.2 Capsules and slime

These are synthesized and exported in a similar way as the LPS of the outer membrane of Gram-negative bacteria.

6.13.3 Nucleoid assembly

The prokaryotic nucleoid consists of DNA and a large number of proteins with a variety of functions. The proteins bind DNA while the latter is synthesized to form the nucleoid.

6.13.4 Ribosome assembly

As stated previously, the prokaryotic ribosome has a size of 70S, consisting of a 50S large subunit and 30S small subunit. The large subunit contains 5S and 23S rRNA and about 35 proteins, while the small 30S subunit consists of 16S rRNA and about 21 proteins. A mixture of rRNAs and proteins forms functional ribosomes under suitable conditions. This spontaneous assembly is possible since the ribosomal proteins recognize the specific site of the rRNA which complexes with the proteins. Since each subunit of the ribosome consists of many different proteins, a proper assembly is possible with binding of the proteins in the right sequence. For example, the ribosomal proteins of the 30S subunit are divided into three classes: primary, secondary, and tertiary binding proteins depending on their binding sequence. The primary binding proteins bind directly to 16S rRNA in the absence of other ribosomal proteins. Secondary binding proteins require the prior binding of one of the primary binding proteins. Tertiary binding proteins require the prior binding of one or more secondary binding proteins.

6.14 | Growth

We have already discussed the reactions that supply ATP, NADPH and carbon skeletons, and their utilization in biosynthetic processes. When a bacterium is inoculated into a liquid medium, exponential growth is observed after a lag phase. When the substrate is used up or metabolic waste accumulates, the culture enters the stationary phase. The growth rate of microbes is influenced by various environmental factors including nutrient concentration, temperature, pH, etc. Catabolic and anabolic pathways are coordinated in such a way that each cellular component increases in the same proportion for balanced growth at any given condition. Balanced growth is possible due to the elaborate metabolic regulation that will be discussed later (Chapter 12).

In this section, discussion will be made of how microbial growth is related to energy transduction. Microbes consume energy not only for biosynthesis but also for their maintenance during growth.

Energy expenditure for biosynthesis:
 Transport
 Nitrogen and sulfur transformations

Monomer synthesis
Polymerization
Assembly of cell structures

Energy expenditure for maintenance:
pH and osmotic homeostasis
Turnover
Motility

6.14.1 Cell division

Prokaryotes propagate asexually. During propagation, a prokaryotic cell divides at the appropriate time and at the correct location in the cell, and each of the progeny receives a complete complement of genes and cell constituents.

6.14.1.1 Binary fission

Prokaryotes rely on binary fission with a few exceptions. During growth, the cell size increases. In rod-shaped bacteria, the cell elongates. When it reaches about twice its starting length, the cell divides in the middle by binary fission. Concurrent with growth, the genetic material of the cell replicates and segregates into the progeny cells with the cell constituents in a controlled manner. Master cell-cycle regulators (CtrA, cell-cycle transcriptional regulator) determine the timing of replication initiation with respect to the cell-division cycle, the maintenance of nucleoid position and other events during cell division of *Caulobacter crescentus*. After replication, each chromosome segregates to the polar position with the aid of the chromosome partition protein (MukB in *Escherichia coli*), the structural maintenance of chromosome complex (SMC), the rod-shape determining protein (TreB), and other DNA-binding proteins.

For cell division to occur, the division apparatus assembles at the site of future cytoplasmic cleavage. The filamentous temperature sensitive protein Z (FtsZ), a structural homologue of the eukaryotic cytoskeletal element tubulin, assembles into a ring-like structure at the centre of the cell. A temperature-sensitive mutant of *ftsZ* cannot divide at high temperature to form filaments. Other components of the division machinery assemble at the FtsZ ring. These components redirect cell wall growth, and prevent damage to the DNA while the cell envelope invaginates. These include FtsA, I, K, L, N, Q, W proteins, ZipA (Z interacting protein A), MinC (septum site-determining protein C), MinD (septum site-determining protein D), MinE (cell division topological specificity factor E) and SulA (cell division inhibitor A). Finally, the cell divides to form two approximately equivalent offspring. Although many of the genes involved in cell division have been identified, the mechanisms of action of these gene products are still under investigation. FtsZ is highly conserved among prokaryotes propagating through binary fission and is one of the first proteins to assemble at the future cell division site. In some prokaryotes, some genes for binary fission including *ftsZ* are not found. Such organisms multiply by different mechanisms such as multiple intracellular offspring, multiple fission and by budding. In some bacteria, these alternative reproductive strategies are essential for propagation, whereas in others they are used conditionally.

6.14.1.2 Multiple intracellular offspring

Some Gram-positive bacteria (known as firmicutes) form endospores under unsuitable conditions for growth and this has been best described in *Bacillus subtilis*. The sporulation process is initiated through a modified form of binary fission. Instead of dividing at the mid-cell point, a sporulating cell divides near one pole. Asymmetric cell division produces a small forespore and a large parent cell. During this division, approximately one third of one of the chromosomes is trapped in the forespore. The DNA translocase SpoIIIE, which is located in the division septum, then pumps the rest of the chromosome into the forespore. The other copy of the chromosome is retained in the parent cell. Although the spore is formed at only one pole, *Bacillus subtilis* prepares for division at both poles. FtsZ rings assemble at both poles and ring-shaped invaginations can be observed near both cell poles early in sporulation. It has been suggested, however, that construction of the cell-division apparatus at both poles allows the cell a 'second chance' if the first asymmetric division fails to capture a chromosome. Some *Bacillus subtilis* mutants have a 'disporic' phenotype. In these cells, sequential bipolar cell division occurs and the two copies of the chromosome are partitioned into the polar forespores, leaving the parental cell devoid of genetic material, which results in the arrest of sporulation. Therefore, it seems that the mechanisms for bipolar division are present in endospore formers. Once asymmetric division and chromosome translocation are complete, the forespore is engulfed by the parent cell and the internalized forespore matures into a spore. Details of sporulation are discussed later (Section 13.3.1).

Some related firmicutes produce multiple endospores within a parental cell (also known as the mother cell) while binary fission is the normal mode of growth. These include *Anaerobacter polyendosporus*, an anaerobic bacterium isolated from rice paddy soil. Under proper laboratory conditions, this bacterium can produce up to seven endospores. On the other hand, *Metabacterium polyspora*, a yet-to-be-cultured bacterium found in the gastrointestinal tract of the guinea pig, forms multiple (up to nine) endospores in its normal life cycle. After the spores germinate, some vegetative cells propagate through binary fission but the majority sporulate in *Metabacterium polyspora*. The process of endospore formation in *Metabacterium polyspora* differs from that in *Bacillus subtilis* and many other endospore-forming bacteria. The asymmetric cell division of *Metabacterium polyspora* normally takes place at both cell poles. DNA is partitioned into both polar compartments, but some DNA is also retained in the parental cell. After engulfment, the forespores can undergo division to produce multiple forespores that grow and mature into multiple endospores. Unlike sporulating *Bacillus subtilis*, which contains two copies of its genome, *Metabacterium polyspora* must contain three or more copies of its genome. Morphologically similar symbionts have been found in various other rodent species although no *Metabacterium*-like symbiont has been maintained in culture.

Another group of Gram-positive bacteria that use a modified pathway to form endospores are the segmented filamentous bacteria (SFB). SFB have been found in the intestinal tracts of various animals, but the SFB of rodents are by far the best characterized although they cannot be maintained under laboratory conditions. SFB develop as a multicellular filament that is anchored to the gastrointestinal tract surface. Once it is firmly attached, the SFB grows and divides, eventually forming a filament that can be up to 1 mm long, although most filaments are roughly 100 μm in length. The cell-division programme is initiated at the unattached end of the filament and is sequentially triggered in cells that are closer to the holdfast. The smaller cell is then engulfed by the larger parental cell and, after engulfment, the internalized cell divides. At this stage, the intracellular offspring have one of two fates: either differentiation, or maturation to form a spore. Differentiation produces holdfast protrusions. These active offspring are released into the lumen of the intestine after the mother cell lyses and they attach to the intestinal surface to establish new filaments within the host. Maturation results in two intracellular offspring cells that are encased in a common spore coat, which forms an endospore. The endospore provides an effective dispersal mechanism for the SFB. In fact, exposure to airborne endospores alone can establish an SFB population in a host. These alternative forms of offspring (either active or dormant) allow the SFB to maintain local populations and to survive inhospitable environments before colonizing amenable hosts.

Yet another mode of intracellular multiple offspring propagation is known in a yet-to-be-cultured firmicute, *Epulopiscium fishelsoni* (Section 2.3.5), that colonizes the intestinal tracts of certain species of surgeonfish. This cigar-shaped bacterium can reach more than 0.6 mm in length, the largest bacterium identified so far. These bacteria produce multiple intracellular offspring that resemble the large endospores that are formed by *Metabacterium polyspora*. However, *Epulopiscium fishelsoni* produces active (rather than dormant) offspring. Usually two offspring are produced, although up to 12 offspring per parent cell have been observed in certain strains.

6.14.1.3 Multiple offspring by multiple fission
Three lineages of prokaryotes, the cyanobacteria (Section 11.1.1), the proteobacteria including aerobic anoxygenic phototrophic bacteria (Section 11.1.3) and the actinobacteria, propagate through multiple rounds of fission of an enlarged multinucleoid spherical cell, or synchronous division at many sites along the length of a multinucleoid filament which leads to the formation of multiple vegetative

offspring or spores. A multiple-fission reproductive phase is often induced by depletion of a nutrient, and might be used to aid offspring dispersal.

Among cyanobacteria, *Pleurocapsa* and *Stanieria* genera propagate through multiple fission to produce offspring known as baeocytes (Greek, 'small cells') with limited binary fission. The life cycle begins with a baeocyte, which is a spherical cell 1–2 μm in diameter. The baeocyte produces an extracellular matrix that is known as the F-layer and accretion of this layer continues throughout the life cycle. This matrix, probably a polysaccharide, aids attachment of the baeocyte to a solid surface. During vegetative growth, the attached cell enlarges up to 30 μm in diameter. As the cell grows, its genomic DNA replicates and the nucleoids segregate in the cytoplasm. In subsequent reproductive stages, a rapid succession of cytoplasmic fissions leads to multiple baeocyte formation. The number of baeocytes produced (4 to more than 1000) depends on the volume of the reproductive-phase parent cell. Multiple fission is not accompanied by a notable increase in total cytoplasmic volume, and each round of division produces sequentially smaller offspring cells, which distinguishes this process from binary fission. Eventually the extracellular matrix tears open, releasing the baeocytes. It is not known what triggers the reproductive cycle in these organisms. Species of the genus *Synechocystis* use a combination of asymmetric cell division and multiple fission for reproduction.

Multiple spores in the aerial mycelium of streptomycetes are well documented. A spore germinates to give rise to a vegetative cell whose tip growth produces branched filaments that only septate occasionally, forming long cells that contain multiple nucleoids. In response to nutrient depletion, some streptomycetes alter their pattern of growth to produce aerial mycelia and dispersible spores. A complex extracellular signalling cascade, which might provide checkpoints to ensure the coordination of secondary metabolism in a developing colony, controls the transition from vegetative growth to reproduction. In some of the aerial filaments, synchronous cell division occurs at regular intervals to produce uninucleoid cells which develop into spores. Mechanisms which act on Z-ring assembly in other bacteria (such as nucleoid occlusion and the Min system) do not seem to be as conserved in the streptomycetes.

Bdellovibrio spp. are tiny predatory *Deltaproteobacteria* that invade the periplasm of the prey bacterium and systematically consume it. Attack-phase *Bdellovibrio* cells are highly motile as they search for susceptible prey. Once host-contact is made, *Bdellovibrio* cells penetrate the outer membrane of a prey cell. Concealed within the host, *Bdellovibrio* lyses the cell wall, which reduces the prey to a spheroplast, and then the *Bdellovibrio* assimilates organic compounds from the prey cytoplasm. *Bdellovibrio* grows in the prey periplasm, but this growth is not accompanied by cell division – instead, the cell elongates. Once the host cytoplasm is consumed, the *Bdellovibrio* reproduces: the filament divides through multiple fission, and cells differentiate into motile attack-phase cells.

6.14.1.4 Budding

Budding is known in different groups of bacteria, including the firmicutes, the cyanobacteria, the prosthecate proteobacteria, the anoxygenic phototrophic bacteria (Section 11.1.3) and the *Planctomycetes*. Although mechanisms that control bud formation in these and other bacteria have been proposed, the molecular mechanisms that regulate bud formation are not fully characterized.

The budding prosthecate *Alphaproteobacteria* include *Caulobacter crescentus*, *Hyphomonas*, *Pedomicrobium* and *Ancalomicrobium* spp. They are morphologically diverse. Developmental progression in *Caulobacter crescentus*, DNA replication, cell division and polar organelle development, is well defined. Key global response regulators, such as CtrA (cell cycle transcriptional regulator) and GcrA (cell cycle regulator), which govern progression, have been identified. It is possible that the regulatory mechanisms that have been discovered in *Caulobacter crescentus* are conserved in other prosthecate *Alphaproteobacteria* and could be involved in bud-site determination and development.

6.14.2 Growth yield

Generally microbial cell number or cell mass (G) is proportional to the amount of substrate consumed (C), which can be expressed as:

$$G = KC$$

where K is a constant depending on the substrate and expressed as 'g cell yield/mol substrate consumed'. When *Escherichia coli* is cultivated on a glucose–mineral salts medium, 0.5 g cell mass is generated from 1 g glucose consumed. In this case, K is 90 g cell yield/mol glucose. Y is used instead of K to define the molar growth yield. $Y_{glucose}$ is used to define the molar growth yield on glucose.

When the exact number of ATP molecules generated from metabolism of the substrate is known, $Y_{glucose}$ can be converted to gram cell yield/ATP generated (Y_{ATP}). Homolactic acid fermentation generates 2 mol ATP from 1 mol glucose fermented, and Y_{ATP} is half of $Y_{glucose}$. Y_{ATP} varies from 4.7 in *Zymomonas anaerobia* to 20.9 in *Lactobacillus casei* (Table 6.8). In branched pathway fermentation (Section 8.1), the growth conditions determine the fermentation route and therefore differences in ATP generation. *Clostridium butyricum* ferments glucose to acetate and butyrate. Under low partial pressures of hydrogen, this bacterium produces more acetate generating one ATP per acetate produced.

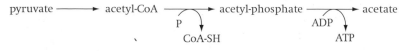

Table 6.8. *Growth yield of fermentative microorganisms*

Microorganism	Substrate	$Y_{substrate}$	ATP/mol substrate	Y_{ATP}
Streptococcus faecalis	glucose	21.8	2.0	10.9
	gluconate	18.7	1.8	10.4
	2-ketogluconate	19.5	2.3	8.5
	ribose	21.0	1.67	12.6
	arginine	10.2	1.0	10.2
	pyruvate	10.4	1.0	10.4
Streptococcus agalactiae	glucose	20.8	2.25	9.3
	pyruvate	7.5	0.72	10.4
Streptococcus pyogenes	glucose	25.5	2.6	9.8
Streptococcus lactis	glucose	19.5	2.0	9.8
Lactobacillus plantarum	glucose	20.4	2.0	10.2
	galactose	32.5	2.97	10.9
Lactobacillus casei	glucose	42.9	2.05	20.9
	mannitol	40.5	2.22	18.2
	citrate	18.2	0.96	19.0
Bifidobacterium bifidum	glucose	37.4	2.85	13.1
	lactose	52.8	5.08	10.4
	galactose	27.8	2.80	9.9
	mannitol	27.8	2.35	11.8
Saccharomyces cerevisiae	glucose	20.4	2.0	10.2
Saccharomyces rosea	glucose	23.2	2.0	11.6
Zymomonas mobilis	glucose	8.5	1.0	8.5
Zymomonas anaerobia	glucose	5.9	1.0	5.9
Sarcina ventriculi	glucose	30.5	2.62	11.7
Aerobacter aerogenes	glucose	30.6	3.0	10.2
	fructose	35.1	3.0	11.7
	mannitol	27.0	2.5	10.8
	gluconate	27.5	2.5	11.0
Aerobacter cloacae	glucose	27.1	2.27	11.9
Escherichia coli	glucose	33.6	3.0	11.2
Ruminococcus flavefaciens	glucose	29.1	2.75	10.6
Actinomyces israeli	glucose	24.7	2.0	12.3
Clostridium tetanomorphum	glutamate	6.8	0.62	10.9
Clostridium aminobutyricum	4-aminobutyrate	7.6	0.5	15.2
	4-hydroxybutyrate	8.9	0.5	17.8
Clostridium glycolicum	ethyleneglycol	7.7	0.5	15.4
Clostridium kluyveri	crotonate	4.8	0.5	9.6
	ethanol + acetate	–	–	9.2
Clostridium pasteurianum	sucrose	73.1	6.64	11.0
Clostridium thermoaceticum	glucose	50.0	3.0	16.6

Butyrate is the main product under high partial pressures of hydrogen, generating less ATP than from acetate production.

$$2\text{pyruvate} \longrightarrow \text{butyryl-CoA} \longrightarrow \text{butyryl-phosphate} \longrightarrow \text{butyrate}$$

6.14.3   Theoretical maximum Y_{ATP}

As discussed earlier, energy in the forms of ATP (GTP) and NAD(P)H are invested to synthesize monomers and to polymerize them into cell materials. It is possible to calculate how much energy is needed to synthesize 1 gram cell material through summing up the investment in monomer synthesis (Table 6.9) and polymerization (Table 6.10). From this figure, the theoretical maximum Y_{ATP} (Y_{ATP}^{max}) can be calculated.

Table 6.11 summarizes Y_{ATP}^{max} under various nutritional conditions ranging from chemolithotrophic to heterotrophic metabolism in a complex medium with added glucose. As expected, the lowest Y_{ATP}^{max} of 6.5 g/mol ATP is calculated for chemolithotrophic growth fixing carbon dioxide, while the Y_{ATP}^{max} is 31.9 g/mol ATP in heterotrophic metabolism in a complex medium. This difference shows that a large amount of energy is consumed for the synthesis of monomers and their polymerization. The values for heterotrophic metabolism on glucose, pyruvate, malate and acetate are 28.8 g, 13.5 g, 15.4 g and 10.0 g cell/mol ATP, respectively. This difference is due to energy consumption by gluconeogenesis.

The Y_{ATP}^{max} values in Table 6.11 only account for direct energy requirements for growth. The actual Y_{ATP} is much lower than these figures since there is energy expenditure for motility, pH and salt homeostasis, and other non-growth-related activities. This is known as the maintenance energy. Since less energy is consumed to synthesize storage materials such as glycogen and polyhydroxyalkanoate (Section 13.2), the growth yield is higher when the organism grows under conditions suitable for their biosynthesis.

6.14.4   Growth yield using different electron acceptors and maintenance energy

Growth yield is dependent on metabolism as shown above. For example $Y_{glucose}$ values for *Proteus mirabilis* are 58.1, 30.1 and 14.0 g dry wt/mol glucose under aerobic, denitrifying and fermentative conditions, respectively. This difference is partly due to the number of ATP molecules generated. Since prokaryotic electron transport chains are diverse with different H^+/O ratios and H^+/ATP stoichiometry varying depending on the energy status of the cell (Section 5.8.4), it is difficult to calculate ATP yield in respiratory metabolism. For this reason Y_{ATP} is calculated from fermentation.

Growth yield in respiratory metabolism can be estimated using the following equation: $1/Y = m(1/\mu) + (1/Y^{max})$, where Y is the growth yield, m is the maintenance energy, μ is the growth rate and Y^{max} is the maximum growth yield. The inverse values of the growth yield

Table 6.9. *ATP and precursors required to produce the monomers needed for the formation of one gram of Escherichia coli cells*

Cell constituent	Content (µmol/g dry wt)	Precursor	Requirement for ATP and precursors (µmol/µmol) ATP	NADH	NADPH	C1	NH$_3$	S
Amino acids								
alanine	488	1 pyr	0	0	1	0	1	0
arginine	281	1 2-kg	7	−1	4	0	4	0
asparagine	229	1 oaa	3	0	1	0	2	0
aspartate	229	1 oaa	0	0	1	0	1	0
cysteine	87	1 pga	4	−1	5	0	1	1
glutamate	250	1 2-kg	0	0	1	0	1	0
glutamine	250	1 2-kg	1	0	1	0	2	0
glycine	282	1 pga	0	−1	1	−1	1	0
histidine	90	1 penP	6	−3	1	1	3	0
isoleucine	276	1 oaa, 1 pyr	2	0	5	0	1	0
leucine	428	2 pyr, 1 acCoA	0	−1	2	0	1	0
lysine	326	1 oaa, 1 pyr	2	0	4	0	2	0
methionine	146	1 oaa	7	0	8	1	1	1
phenylalanine	176	1 eryP, 2 pep	1	0	2	0	1	0
proline	210	1 2-kg	1	0	3	0	1	0
serine	205	1 pga	0	−1	1	0	1	0
threonine	241	1 oaa	2	0	3	0	1	0
tryptophan	54	1 penP, 1 eryP, 1 pep	5	−2	3	0	2	0
tyrosine	131	1 eryP, 2 pep	1	−1	2	0	1	0
valine	402	1 pyr	0	0	2	0	1	0
Ribonucleotides								
ATP	165	1 penP, 1 pga	11	−3	1	1	5	0
GTP	203	1 penP, 1 pga	13	−3	0	1	5	0
CTP	126	1 penP, 1 oaa	9	0	1	0	3	0
UTP	136	1 penP, 1 oaa	7	0	1	0	2	0
Deoxyribonucleotides								
dATP	24.7	1 penP, 1 pga	11	−3	2	1	5	0
dGTP	25.4	1 penP, 1 pga	13	−3	1	1	5	0
dCTP	25.4	1 penP, 1 oaa	9	0	2	0	3	0
dTTP	24.7	1 penP, 1 oaa	10.5	0	3	1	2	0
Lipids								
glycerol phosphate	129	1 triosP	0	0	1	0	0	0
serine	129	1 pga	0	−1	1	0	0	0
average fatty acid	258	8.2 acCoA	7.2	0	14	0	0	0
Lipopolysaccharide								
UDP-glucose	15.7	1 gluP	1	0	0	0	0	0
CDP-ethanolamine	23.5	1 pga	3	−1	1	1	1	0

	Content		Requirement for ATP and precursors (µmol/µmol)					
Cell constituent	(µmol/g dry wt)	Precursor	ATP	NADH	NADPH	CI	NH$_3$	S
fatty acid	47	7 acCoA	6	0	11.5	0	0	0
CMP-KDO	23.5	I penP, I pep	2	0	0	0	0	0
CDP-heptose	23.5	1.5 gluP	1	0	−4	0	0	0
UDP-glucosamine	15.7	I fruP	2	0	0	1	1	0
Murein								
UDP-GlcNAc	27.6	I fruP, I acCoA	3	0	0	0	1	0
UDP-MurNAc	27.6	I fruP, I pep, I acCoA	4	0	1	0	1	0
alanine	55.2	I pyr	0	0	1	0	1	0
diaminopimelate	27.6	I oaa, I pyr	2	0	3	0	2	0
glutamate	27.6	I 2-kg	0	0	1	0	1	0
Polyamine								
ornithine equivalent	59.3	I 2-kg	2	0	3	0	2	0

acCoA, acetyl-CoA; eryP, erythrose-4-phosphate; fruP, fructose-6-phosphate; gluP, glucose-6-phosphate; 2-kg, 2-ketoglutarate; oaa, oxaloacetate; penP, pentose-5-phosphate; pep, phospho-enolpyruvate; pga, 3-phosphoglycerate; pyr, pyruvate; triosP, triose phosphate; KDO, 2-keto – 3-deoxyoctonate; UDP-GlcNAc, UDP-N-acetylgucosamine; UDP-MurNAc, UDP-N-acetylmuramic acid.

Table 6.10. *ATP requirement for polymer synthesis by* Escherichia coli *using glucose–mineral salts medium*

Polymer	ATP required for the synthesis of polymer to make up one gram dry wt cell material (mmol/g dry wt)
Polysaccharide	2.1
Protein	
Glucose to amino acids	1.4
Polymerization (translation)	19.1
Lipid	0.1
RNA	
Glucose to ribonucleotides	3.5
Polymerization (transcription)	0.9
DNA	
Glucose to deoxyribonucleotides	0.9
Polymerization	0.2
Transport	5.2
RNA turnover	1.4
Overall	34.8

Table 6.11. *ATP requirement for growth and Y_{ATP} under different nutritional conditions*

Polymer	Cellular content	ATP requirement (mmol/g cell) in medium						
	(g/100 g dry wt)	A	B	C	D	E	F	G
Polysaccharide	16.6	2.06	2.06	7.18	7.18	5.10	9.20	19.50
Protein	52.4	19.14	20.50	19.14	33.94	28.50	42.70	90.70
Lipid	9.4	0.14	0.14	2.70	2.70	2.50	5.00	17.20
RNA	15.7	2.40	4.37	4.62	7.13	7.00	10.10	17.84
DNA	3.2	0.57	1.05	0.99	1.59	1.30	1.90	3.36
mRNA turnover		1.39	1.39	1.39	1.39	1.39	1.39	1.39
Subtotal		25.70	29.51	36.02	53.93	45.79	70.29	149.99
Transport		5.74	5.20	11.55	20.00	20.00	30.60	5.20
Total		31.44	34.71	47.57	73.93	65.79	100.80	155.19
Y_{ATP} (g cell/mol ATP)		31.80	28.81	21.02	13.53	15.20	9.92	6.44

Cellular content is based on *Escherichia coli*.
Medium:
A, glucose + amino acids − nucleic acid bases + mineral salts.
B, glucose + mineral salts.
C, pyruvate + amino acids + nucleic acid bases + mineral salts.
D, pyruvate + mineral salts.
E, malate + mineral salts.
F, acetate + mineral salts.
G, carbon dioxide + mineral salts (chemolithotrophic metabolism).

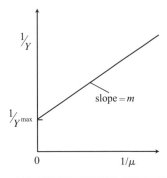

Figure 6.45 Relationship between growth rate, growth yield and maintenance energy.

Growth yield is determined in a chemostat at various dilution (growth) rates (μ). The inverse values are plotted as above to obtain the maintenance energy (m) and maximum growth yield (Y^{max}) at the given conditions.

obtained from a chemostat operated at different growth rates is plotted against the inverse of the growth rate as in Figure 6.45. From this exercise the maintenance energy can be determined. As shown above, the theoretical maximum Y_{ATP} is more than the actual Y_{ATP}. Assuming that the total energy available from metabolism is the sum of the energy for growth and the maintenance energy, it can be seen that more energy is used for the purpose of maintenance.

6.14.5 Maintenance energy

As shown in Figure 6.45, more maintenance energy is required at a lower growth rate. The maintenance energy required is determined by various growth conditions that influence the growth rate, e.g. pH and salt homeostasis and half-life of cellular polymers, among others. Organisms producing fatty acids require more maintenance energy, since fatty acids behave as uncouplers. When under substrate excess conditions, less cell yield is expected with incomplete substrate oxidation and uncoupled oxidation of the substrate. This fact should be considered in the determination of maintenance energy.

FURTHER READING

Nitrogen fixation

Bauer, C. E., Elsen, S. & Bird, T. H. (1999). Mechanisms for redox control of gene expression. *Annual Review of Microbiology* **53**, 495–523.

Dixon, R. & Kahn, D. (2004). Genetic regulation of biological nitrogen fixation. *Nature Reviews Microbiology* **2**, 621–631.

Dubbs, J. M. & Tabita, F. R. (2004). Regulators of nonsulfur purple phototrophic bacteria and the interactive control of CO_2 assimilation, nitrogen fixation, hydrogen metabolism and energy generation. *FEMS Microbiology Reviews* **28**, 353–376.

Elsen, S., Swem, L. R., Swem, D. L. & Bauer, C. E. (2004). RegB/RegA, a highly conserved redox-responding global two-component regulatory system. *Microbiology and Molecular Biology Reviews* **68**, 263–279.

Gonzalez, J. E. & Marketon, M. M. (2003). Quorum sensing in nitrogen-fixing rhizobia. *Microbiology and Molecular Biology Reviews* **67**, 574–592.

Prell, J. & Poole, P. (2006). Metabolic changes of rhizobia in legume nodules. *Trends in Microbiology* **14**, 161–168.

Richardson, D. J. & Watmough, N. J. (1999). Inorganic nitrogen metabolism in bacteria. *Current Opinion in Chemical Biology* **3**, 207–219.

Zehr, J. P., Jenkins, B. D., Short, S. M. & Steward, G. F. (2003). Nitrogenase gene diversity and microbial community structure: a cross-system comparison. *Environmental Microbiology* **5**, 539–554.

Zhang, C. C., Laurent, S., Sakr, S., Peng & Bedu, S. (2006). Heterocyst differentiation and pattern formation in cyanobacteria: a chorus of signals. *Molecular Microbiology* **59**, 367–375.

Assimilation of inorganic nitrogen and sulfur

Arcondeguy, T., Jack, R. & Merrick, M. (2001). P-II signal transduction proteins, pivotal players in microbial nitrogen control. *Microbiology and Molecular Biology Reviews* **65**, 80–105.

Beinert, H. (2000). A tribute to sulfur. *European Journal of Biochemistry* **267**, 5657–5664.

Burkovski, A. (2003). Ammonium assimilation and nitrogen control in *Corynebacterium glutamicum* and its relatives: an example for new regulatory mechanisms in actinomycetes. *FEMS Microbiology Reviews* **27**, 617–628.

Burkovski, A. (2003). I do it my way: regulation of ammonium uptake and ammonium assimilation in *Corynebacterium glutamicum*. *Archives of Microbiology* **179**, 83–88.

Cook, A. M. & Denger, K. (2002). Dissimilation of the C_2 sulfonates. *Archives of Microbiology* **179**, 1–6.

Friedrich, C. G., Rother, D., Bardischewsky, F., Quentmeier, A. & Fischer, J. (2001). Oxidation of reduced inorganic sulfur compounds by bacteria: emergence of a common mechanism? *Applied and Environmental Microbiology* **67**, 2873–2882.

Lin, J. T. & Stewart, V. (1998). Nitrate assimilation by bacteria. *Advances in Microbial Physiology* **39**, 1–30.

Reitzer, L. (2003). Nitrogen assimilation and global regulation in *Escherichia coli*. *Annual Review of Microbiology* **57**, 155–176.

Ye, R. W. & Thomas, S. M. (2001). Microbial nitrogen cycles: physiology, genomics and applications. *Current Opinion in Microbiology* **4**, 307–312.

Monomer synthesis – amino acids and nucleotides

Albertini, A. M. & Galizzi, A. (1999). The sequence of the *trp* operon of *Bacillus subtilis* 168 (trpC2) revisited. *Microbiology-UK* **145**, 3319–3320.

Alifano, P., Fani, R., Lio, P., Lazcano, A., Bazzicalupo, M., Carlomagno, M. S. & Bruni, C. B. (1996). Histidine biosynthetic pathway and genes: structure, regulation, and evolution. *Microbiological Reviews* **60**, 44–69.

Buckel, W. & Golding, B. T. (2006). Radical enzymes in anaerobes. *Annual Review of Microbiology* **60**, 27–49.

He, X. & Liu, H. W. (2002). Mechanisms of enzymatic C–O bond cleavages in deoxyhexose biosynthesis. *Current Opinion in Chemical Biology* **6**, 590–597.

Jordan, A. & Reichard, P. (1998). Ribonucleotide reductase. *Annual Review of Biochemistry* **67**, 71–98.

Lendzian, F. (2005). Structure and interactions of amino acid radicals in class I ribonucleotide reductase studied by ENDOR and high-field EPR spectroscopy. *Biochimica et Biophysica Acta – Bioenergetics* **1707**, 67–90.

Lu, C. D. (2006). Pathways and regulation of bacterial arginine metabolism and perspectives for obtaining arginine overproducing strains. *Applied Microbiology and Biotechnology* **70**, 261–272.

Nordlund, P. & Reichard, P. (2006). Ribonucleotide reductase. *Annual Review of Biochemistry* **75**, 681–706.

Stubbe, J. (2000). Ribonucleotide reductases: the link between an RNA and a DNA world? *Current Opinion in Structural Biology* **10**, 731–736.

Torrents, E., Jordan, A., Karlsson, M. & Gibert, I. (2000). Occurrence of multiple ribonucleotide reductase classes in gamma-proteobacteria species. *Current Microbiology* **41**, 346–351.

Monomer synthesis – lipids

Behrouzian, B. & Buist, P. H. (2002). Fatty acid desaturation: variations on an oxidative theme. *Current Opinion in Chemical Biology* **6**, 577–582.

Cronan, J. E. (2006). A bacterium that has three pathways to regulate membrane lipid fluidity. *Molecular Microbiology* **60**, 256–259.

Kuzuyama, T. (2002). Mevalonate and nonmevalonate pathways for the biosynthesis of isoprene units. *Bioscience, Biochemistry and Biotechnology* **66**, 1619–1627.

Mansilla, M. C. & de Mendoza, D. (2005). The *Bacillus subtilis* desaturase: a model to understand phospholipid modification and temperature sensing. *Archives of Microbiology* **183**, 229–235.

Meganathan, R. (2001). Ubiquinone biosynthesis in microorganisms. *FEMS Microbiology Letters* **203**, 131–139.

Schujman, G. E. & de Mendoza, D. (2005). Transcriptional control of membrane lipid synthesis in bacteria. *Current Opinion in Microbiology* **8**, 149–153.

Monomer synthesis – others

Boucher, Y. & Doolittle, W. F. (2000). The role of lateral gene transfer in the evolution of isoprenoid biosynthesis pathways. *Molecular Microbiology* **37**, 703–716.

Fontecave, M., Atta, M. & Mulliez, E. (2004). *S*-adenosylmethionine: nothing goes to waste. *Trends in Biochemical Sciences* **29**, 243–249.

Frankenberg, N., Moser, J. & Jahn, D. (2003). Bacterial heme biosynthesis and its biotechnological application. *Applied Microbiology and Biotechnology* **63**, 115–127.

O'Brian, M. R. & Thony-Meyer, L. (2002). Biochemistry, regulation and genomics of haem biosynthesis in prokaryotes. *Advances in Microbial Physiology* **46**, 257–318.

Panek, H. & O'Brian, M. R. (2002). A whole genome view of prokaryotic haem biosynthesis. *Microbiology-UK* **148**, 2273–2282.

Roessner, C. A. & Scott, A. I. (2006). Fine-tuning our knowledge of the anaerobic route to cobalamin (Vitamin B$_{12}$). *Journal of Bacteriology* **188**, 7331–7334.

Umeno, D., Tobias, A. V. & Arnold, F. H. (2005). Diversifying carotenoid biosynthetic pathways by directed evolution. *Microbiology and Molecular Biology Reviews* **69**, 51–78.

Polysaccharides and other polymers

Moffitt, M. C. & Neilan, B. A. (2000). The expansion of mechanistic and organismic diversity associated with non-ribosomal peptides. *FEMS Microbiology Letters* **191**, 159–167.

Rocchetta, H. L., Burrows, L. L. & Lam, J. S. (1999). Genetics of O-antigen biosynthesis in *Pseudomonas aeruginosa*. *Microbiology and Molecular Biology Reviews* **63**, 523–553.

Stachelhaus, T. & Marahiel, M. A. (1995). Modular structure of genes encoding multifunctional peptide synthetases required for non-ribosomal peptide synthesis. *FEMS Microbiology Letters* **125**, 3–14.

Stewart, G. C. (2005). Taking shape: control of bacterial cell wall biosynthesis. *Molecular Microbiology* **57**, 1177–1181.

Replication

Bachellerie, J. P. & Cavaille, J. (1997). Guiding ribose methylation of rRNA. *Trends in Biochemical Sciences* **22**, 257–261.

Bravo, A., Serrano-Heras, G. & Salas, M. (2005). Compartmentalization of prokaryotic DNA replication. *FEMS Microbiology Reviews* **29**, 25–47.

Denamur, E. & Matic, I. (2006). Evolution of mutation rates in bacteria. *Molecular Microbiology* **60**, 820–827.

Kaguni, J. M. (2006). DnaA: controlling the initiation of bacterial DNA replication and more. *Annual Review of Microbiology* **60**, 351–371.

Katayama, T. (2001). Feedback controls restrain the initiation of *Escherichia coli* chromosomal replication. *Molecular Microbiology* **41**, 9–17.

Kelman, L. M. & Kelman, Z. (2004). Multiple origins of replication in archaea. *Trends in Microbiology* **12**, 399–401.

MacNeill, S. A. (2001). Understanding the enzymology of archaeal DNA replication: progress in form and function. *Molecular Microbiology* **40**, 520–529.

Paulsson, J. & Chattoraj, D. K. (2006). Origin inactivation in bacterial DNA replication control. *Molecular Microbiology* **61**, 9–15.

Sandler, S. J. & Marians, K. J. (2000). Role of PriA in replication fork reactivation in *Escherichia coli*. *Journal of Bacteriology* **182**, 9–13.

Transcription and post-transcription modification

Bell, S. D. & Jackson, S. P. (1998). Transcription and translation in Archaea: a mosaic of eukaryal and bacterial features. *Trends in Microbiology* **6**, 222–228.

Boeneman, K. & Crooke, E. (2005). Chromosomal replication and the cell membrane. *Current Opinion in Microbiology* **8**, 143–148.

Borukhov, S. & Severinov, K. (2002). Role of the RNA polymerase sigma subunit in transcription initiation. *Research in Microbiology* **153**, 557–562.

Borukhov, S., Lee, J. & Laptenko, O. (2005). Bacterial transcription elongation factors: new insights into molecular mechanism of action. *Molecular Microbiology* **55**, 1315–1324.

Brennicke, A., Marchfelder, A. & Binder, S. (1999). RNA editing. *FEMS Microbiology Reviews* **23**, 297–316.

Hickey, A. J., de Macario, E. C. & Macario, A. J. L. (2002). Transcription in the Archaea: basal factors, regulation, and stress-gene expression. *Critical Reviews in Biochemistry and Molecular Biology* **37**, 537–599.

Stuart, K. & Panigrahi, A. K. (2002). RNA editing: complexity and complications. *Molecular Microbiology* **45**, 591–596.

Translation and protein folding

Baker, D. & Lim, W. (2002). From folding towards function. *Current Opinion in Structural Biology* **12**, 11–13.

Barras, F., Loiseau, L. & Py, B. (2005). How *Escherichia coli* and *Saccharomyces cerevisiae* build Fe/S proteins. *Advances in Microbial Physiology* **50**, 41–101.

Bock, A., King, P. W., Blokesch, M. & Posewitz, M. C. (2006). Maturation of hydrogenases. *Advances in Microbial Physiology* **51**, 1–225.

Boni, I. V. (2006). Diverse molecular mechanisms of translation initiation in prokaryotes. *Molecular Biology* **40**, 587–596.

Booth, P. J. & Curnow, P. (2006). Membrane proteins shape up: understanding *in vitro* folding. *Current Opinion in Structural Biology* **16**, 480–488.

Bowie, J. U. (2005). Solving the membrane protein folding problem. *Nature* **438**, 581–589.

Bukau, B., Deuerling, E., Pfund, C. & Craig, E. A. (2000). Getting newly synthesized proteins into shape. *Cell* **101**, 119–122.

Bulaj, G. (2005). Formation of disulfide bonds in proteins and peptides. *Biotechnology Advances* **23**, 87–92.

Casalot, L. & Rousset, M. (2001). Maturation of the [NiFe] hydrogenases. *Trends in Microbiology* **9**, 228–237.

Cianciotto, N. P., Cornelis, P. & Baysse, C. (2005). Impact of the bacterial type I cytochrome *c* maturation system on different biological processes. *Molecular Microbiology* **56**, 1408–1415.

Cobucci-Ponzano, B., Rossi, M. & Moracci, M. (2005). Recoding in Archaea. *Molecular Microbiology* **55**, 339–348.

Collet, J. F. & Bardwell, J. C. A. (2002). Oxidative protein folding in bacteria. *Molecular Microbiology* **44**, 1–8.

Craig, E. A., Eisenman, H. C. & Hundley, H. A. (2003). Ribosome-tethered molecular chaperones: the first line of defense against protein misfolding? *Current Opinion in Microbiology* **6**, 157–162.

Daggett, V. & Fersht, A. R. (2003). Is there a unifying mechanism for protein folding? *Trends in Biochemical Sciences* **28**, 18–25.

Das, G. & Varshney, U. (2006). Peptidyl-tRNA hydrolase and its critical role in protein biosynthesis. *Microbiology-UK* **152**, 2191–2195.

Deuerling, E. & Bukau, B. (2004). Chaperone-assisted folding of newly synthesized proteins in the cytosol. *Critical Reviews in Biochemistry and Molecular Biology* **39**, 261–277.

Driessen, A. J. M., Fekkes, P. & van der Wolk, J. P. W. (1998). The Sec system. *Current Opinion in Microbiology* **1**, 216–222.

Ellis, R. J. (1994). Protein folding: chaperoning nascent proteins. *Nature* **370**, 96–97.

Feldman, D. E. & Frydman, J. (2000). Protein folding *in vivo*: the importance of molecular chaperones. *Current Opinion in Structural Biology* **10**, 26–33.

Ferguson, N. & Fersht, A. R. (2003). Early events in protein folding. *Current Opinion in Structural Biology* **13**, 75–81.

Finking, R. & Marahiel, M. A. (2004). Biosynthesis of nonribosomal peptides. *Annual Review of Microbiology* **58**, 453–488.

Frazzon, J. & Dean, D. R. (2003). Formation of iron-sulfur clusters in bacteria: an emerging field in bioinorganic chemistry. *Current Opinion in Chemical Biology* **7**, 166 173.

Frydman, J. (2001). Folding of newly translated proteins *in vivo*: the role of molecular chaperones. *Annual Review of Biochemistry* **70**, 603–647.

Ganoza, M. C., Kiel, Mi. C. & Aoki, H. (2002). Evolutionary conservation of reactions in translation. *Microbiology and Molecular Biology Reviews* **66**, 460–485.

Gogarten, J. P., Senejani, A. G., Zhaxybayeva, O., Olendzenski, L. & Hilario, E. (2002). Inteins: structure, function, and evolution. *Annual Review of Microbiology* **56**, 263–287.

Gottesman, M. E. & Hendrickson, W. A. (2000). Protein folding and unfolding by *Escherichia coli* chaperones and chaperonins. *Current Opinion in Microbiology* **3**, 197–202.

Gruebele, M. (2002). Protein folding: the free energy surface. *Current Opinion in Structural Biology* **12**, 161–168.

Gunasekaran, K., Eyles, S. J., Hagler, A. T. & Gierasch, L. M. (2001). Keeping it in the family: folding studies of related proteins. *Current Opinion in Structural Biology* **11**, 83–93.

Hirokawa, G., Demeshkina, N., Iwakura, N., Kaji, H. & Kaji, A. (2006). The ribosome-recycling step: consensus or controversy? *Trends in Biochemical Sciences* **31**, 143–149.

Ibba, M. & Soll, D. (2000). Aminoacyl-tRNA synthesis. *Annual Review of Biochemistry* **69**, 617–650.

Ito, K. (2005). Ribosome-based protein folding systems are structurally divergent but functionally universal across biological kingdoms. *Molecular Microbiology* **57**, 313–317.

Johnson, D. C., Dean, D. R., Smith, A. D. & Johnson, M. K. (2005). Structure, function and formation of iron-sulfur clusters. *Annual Review of Biochemistry* **74**, 247–281.

Kranz, R., Lill, R., Goldman, B., Bonnard, G. & Merchant, S. (1998). Molecular mechanisms of cytochrome *c* biogenesis: three distinct systems. *Molecular Microbiology* **29**, 383–396.

Ladenstein, R. & Ren, B. (2006). Protein disulfides and protein disulfide oxidoreductases in hyperthermophiles. *FEBS Journal* **273**, 4170–4185.

Laursen, B. S., Sorensen, H. P., Mortensen, K. K. & Sperling-Petersen, H. U. (2005). Initiation of protein synthesis in bacteria. *Microbiology and Molecular Biology Review* **69**, 101–123.

Lin, Z. & Rye, H. S. (2006). GroEL-mediated protein folding: making the impossible, possible. *Critical Reviews in Biochemistry and Molecular Biology* **41**, 211–239.

Lund, P. A. (2001). Microbial molecular chaperones. *Advances in Microbial Physiology* **44**, 93–140.

Nakamoto, H. & Bardwell, J. C. A. (2004). Catalysis of disulfide bond formation and isomerization in the *Escherichia coli* periplasm. *Biochimica et Biophysica Acta – Molecular Cell Research* **1694**, 111–119.

O'Donoghue, P. & Luthey-Schulten, Z. (2003). On the evolution of structure in aminoacyl-tRNA synthetases. *Microbiology and Molecular Biology Reviews* **67**, 550–573.

Onuchic, J. N. & Wolynes, P. G. (2004). Theory of protein folding. *Current Opinion in Structural Biology* **14**, 70–75.

Saibil, H. (2000). Molecular chaperones: containers and surfaces for folding, stabilising or unfolding proteins. *Current Opinion in Structural Biology* **10**, 251–258.

Saibil, H. R. & Ranson, N. A. (2002). The chaperonin folding machine. *Trends in Biochemical Sciences* **27**, 627–632.

Tenson, T. & Mankin, A. (2006). Antibiotics and the ribosome. *Molecular Microbiology* **59**, 1664–1677.

Thony-Meyer, L. (1997). Biogenesis of respiratory cytochromes in bacteria. *Microbiology and Molecular Biology Reviews* **61**, 337–376.

Thony-Meyer, L. (2000). Haem-polypeptide interactions during cytochrome *c* maturation. *Biochimica et Biophysica Acta – Bioenergetics* **1459**, 316–324.

Travaglini-Allocatelli, C., Gianni, S. & Brunori, M. (2004). A common folding mechanism in the cytochrome *c* family. *Trends in Biochemical Sciences* **29**, 535–541.

Turkarslan, S., Sanders, C. & Daldal, F. (2006). Extracytoplasmic prosthetic group ligation to apoproteins: maturation of *c*-type cytochromes. *Molecular Microbiology* **60**, 537–541.

Wickner, S., Maurizi, M. R. & Gottesman, S. (1999). Posttranslational quality control: folding, refolding, and degrading proteins. *Science* **286**, 1888–1893.

Williamson, J. R. (2003). After the ribosome structures: how are the subunits assembled? *RNA* **9**, 165–167.

Woesten, M. M. S. M. (1998). Eubacterial sigma-factors. *FEMS Microbiology Reviews* **22**, 127–150.

Zhang, X., Chaney, M., Wigneshweraraj, S. R., Schumacher, J., Bordes, P., Cannon, W. & Buck, M. (2002). Mechanochemical ATPases and transcriptional activation. *Molecular Microbiology* **45**, 895–903.

Assembly of cellular structures and growth

Angert, E. R. (2005). Alternatives to binary fission in bacteria. *Nature Reviews Microbiology* **3**, 214–224.

Bernander, R. (1998). Archaea and the cell cycle. *Molecular Microbiology* **29**, 955–961.

Bernstein, H. D. (2000). The biogenesis and assembly of bacterial membrane proteins. *Current Opinion in Microbiology* **3**, 203–209.

Bhavsar, A. P. & Brown, E. D. (2006). Cell wall assembly in *Bacillus subtilis*: how spirals and spaces challenge paradigms. *Molecular Microbiology* **60**, 1077–1090.

Bignell, C. & Thomas, C. M. (2001). The bacterial ParA-ParB partitioning proteins. *Journal of Biotechnology* **91**, 1–34.

Bos, M. P. & Tommassen, J. (2004). Biogenesis of the Gram-negative bacterial outer membrane. *Current Opinion in Microbiology* **7**, 610–616.

Bouche, J. P. & Pichoff, S. (1998). On the birth and fate of bacterial division sites. *Molecular Microbiology* **29**, 19–26.

Bulthuis, B. A., Koningstein, G. M., Stouthamer, A. H. & Vanverseveld, H. W. (1993). The relation of proton motive force, adenylate energy charge and phosphorylation potential to the specific growth rate and efficiency of energy transduction in *Bacillus licheniformis* under aerobic growth conditions. *Antonie van Leeuwenhoek* **63**, 1–16.

Button, D. K. (1993). Nutrient-limited microbial growth kinetics: overview and recent advances. *Antonie van Leeuwenhoek* **63**, 225–235.

Cabeen, M. T. & Jacobs-Wagner, C. (2005). Bacterial cell shape. *Nature Reviews Microbiology* **3**, 601–610.

Carballido-Lopez, R. (2006). Orchestrating bacterial cell morphogenesis. *Molecular Microbiology* **60**, 815–819.

Cooper, S. (2001). Size, volume, length and the control of the bacterial division cycle. *Microbiology-UK* **147**, 2629–2630.

de Gier, J. W. & Luirink, J. (2001). Biogenesis of inner membrane proteins in *Escherichia coli*. *Molecular Microbiology* **40**, 314–322.

Dewar, S. J. & Dorazi, R. (2000). Control of division gene expression in *Escherichia coli*. *FEMS Microbiology Letters* **187**, 1–7.

Doerrler, W. T. (2006). Lipid trafficking to the outer membrane of Gram-negative bacteria. *Molecular Microbiology* **60**, 542–552.

Donachie, W. D. (2001). Co-ordinate regulation of the *Escherichia coli* cell cycle or the cloud of unknowing. *Molecular Microbiology* **40**, 779–785.

Donachie, W. D. & Blakely, G. W. (2003). Coupling the initiation of chromosome replication to cell size in *Escherichia coli*. *Current Opinion in Microbiology* **6**, 146–150.

Dramsi, S., Trieu-Cuot, P. & Bierne, H. (2005). Sorting sortases: a nomenclature proposal for the various sortases of Gram-positive bacteria. *Research in Microbiology* **156**, 289–297.

Duong, F., Eichler, J., Price, A., Leonard, M. R. & Wickner, W. (1997). Biogenesis of the Gram-negative bacterial envelope. *Cell* **91**, 567–573.

Facey, S. J. & Kuhn, A. (2004). Membrane integration of *Escherichia coli* model membrane proteins. *Biochimica et Biophysica Acta – Molecular Cell Research* **1694**, 55–66.

Fatica, A. & Tollervey, D. (2002). Making ribosomes. *Current Opinion in Cell Biology* **14**, 313–318.

Ferenci, T. (1999). 'Growth of bacterial cultures' 50 years on: towards an uncertainty principle instead of constants in bacterial growth kinetics. *Research in Microbiology* **150**, 431–438.

Fernandez, L. A. & Berenguer, J. (2000). Secretion and assembly of regular surface structures in Gram-negative bacteria. *FEMS Microbiology Reviews* **24**, 21–44.

Ghosh, S. K., Hajra, S., Paek, A. & Jayaram, M. (2006). Mechanisms for chromosome and plasmid segregation. *Annual Review of Biochemistry* **75**, 211–241.

Gordon, G. S. & Wright, A. (2000). DNA segregation in bacteria. *Annual Review of Microbiology* **54**, 681–708.

Hayes, F. & Barilla, D. (2006). Assembling the bacterial segrosome. *Trends in Biochemical Sciences* **31**, 247–250.

Hayes, F. & Barilla, D. (2006). The bacterial segrosome: a dynamic nucleoprotein machine for DNA trafficking and segregation. *Nature Reviews Microbiology* **4**, 133–143.

Henson, M.A. (2003). Dynamic modeling of microbial cell populations. *Current Opinion in Biotechnology* **14**, 460–467.

Hiraga, S. (1992). Chromosome and plasmid partition in *Escherichia coli*. *Annual Review of Biochemistry* **61**, 283–306.

Holms, H. (2001). Flux analysis: a basic tool of microbial physiology. *Advances in Microbial Physiology* **45**, 271–340.

Holtje, J.V. (1995). From growth to autolysis: the murein hydrolases in *Escherichia coli*. *Archives of Microbiology* **164**, 243–254.

Janakiraman, A. & Goldberg, M.B. (2004). Recent advances on the development of bacterial poles. *Trends in Microbiology* **12**, 518–525.

Jannasch, H.W. & Egli, T. (1993). Microbial growth kinetics: a historical perspective. *Antonie van Leeuwenhoek* **63**, 213–224.

Joseleau-Petit, D., Vinella, D. & D'Ari, R. (1999). Metabolic alarms and cell division in *Escherichia coli*. *Journal of Bacteriology* **181**, 9–14.

Koch, A.L. (2000). The bacterium's way for safe enlargement and division. *Applied and Environmental Microbiology* **66**, 3657–3663.

Lesterlin, C., Barre, F.X. & Cornet, F. (2004). Genetic recombination and the cell cycle: what we have learned from chromosome dimers. *Molecular Microbiology* **54**, 1151–1160.

Lewis, P.J. (2001). Bacterial chromosome segregation. *Microbiology-UK* **147**, 519–526.

Lobry, J.R. & Louarn, J.M. (2003). Polarisation of prokaryotic chromosomes. *Current Opinion in Microbiology* **6**, 101–108.

Lundgren, M. & Bernander, R. (2005). Archaeal cell cycle progress. *Current Opinion in Microbiology* **8**, 662–668.

Lutkenhaus, J. (1998). The regulation of bacterial cell division: a time and place for it. *Current Opinion in Microbiology* **1**, 210–215.

Lybarger, S.R. & Maddock, J.R. (2001). Polarity in action: asymmetric protein localization in bacteria. *Journal of Bacteriology* **183**, 3261–3267.

Macnab, R.M. (2003). How bacteria assemble flagella. *Annual Review of Microbiology* **57**, 77–100.

Margolin, W. (2000). Themes and variations in prokaryotic cell division. *FEMS Microbiology Reviews* **24**, 531–548.

Marraffini, L.A., DeDent, A.C. & Schneewind, O. (2006). Sortases and the art of anchoring proteins to the envelopes of Gram-positive bacteria. *Microbiology and Molecular Biology Reviews* **70**, 192–221.

Mazmanian, S.K., Hung, I.T. & Schneewind, O. (2001). Sortase-catalysed anchoring of surface proteins to the cell wall of *Staphylococcus aureus*. *Molecular Microbiology* **40**, 1049–1057.

Mileykovskaya, E. & Dowhan, W. (2005). Role of membrane lipids in bacterial division-site selection. *Current Opinion in Microbiology* **8**, 135–142.

Mol, O. & Oudega, B. (1996). Molecular and structural aspects of fimbriae biosynthesis and assembly in *Escherichia coli*. *FEMS Microbiology Reviews* **19**, 25–52.

Navarre, W.W. & Schneewind, O. (1999). Surface proteins of Gram-positive bacteria and mechanisms of their targeting to the cell wall envelope. *Microbiology and Molecular Biology Reviews* **63**, 174–229.

Page, M.D., Sambongi, Y. & Ferguson, S.J. (1998). Contrasting routes of *c*-type cytochrome assembly in mitochondria, chloroplast and bacteria. *Trends in Biochemical Sciences* **23**, 103–108.

Pallen, M. J., Lam, A. C., Antonio, M. & Dunbar, K. (2001). An embarrassment of sortases: a richness of substrates? *Trends in Microbiology* **9**, 97–101.

Paterson, G. K. & Mitchell, T. J. (2004). The biology of Gram-positive sortase enzymes. *Trends in Microbiology* **12**, 89–95.

Pogliano, K., Pogliano, J. & Becker, E. (2003). Chromosome segregation in Eubacteria. *Current Opinion in Microbiology* **6**, 586–593.

Prozorov, A. A. (2005). The bacterial cell cycle: DNA replication, nucleoid segregation, and cell division. *Microbiology-Moscow* **74**, 375–387.

Romberg, L. & Levin, P. A. (2003). Assembly dynamics of the bacterial cell division protein FTSZ: poised at the edge of stability. *Annual Review of Microbiology* **57**, 125–154.

Rothfield, L. (2003). New insights into the developmental history of the bacterial cell division site. *Journal of Bacteriology* **185**, 1125–1127.

Rothfield, A., Taghbalout, L. & Shih, Y. L. (2005). Spatial control of bacterial division-site placement. *Nature Reviews Microbiology* **3**, 959–968.

Ruiz, N., Kahne, D. & Silhavy, T. J. (2006). Advances in understanding bacterial outer-membrane biogenesis. *Nature Reviews Microbiology* **4**, 57–66.

Sauer, F. G., Barnhart, M., Choudhury, D., Knight, S. D., Waksman, G. & Hultgren, S. J. (2000). Chaperone-assisted pilus assembly and bacterial attachment. *Current Opinion in Structural Biology* **10**, 548–556.

Scheffers, D. J. & Pinho, M. G. (2005). Bacterial cell wall synthesis: new insights from localization studies. *Microbiology and Molecular Biology Reviews* **69**, 585–607.

Sciochetti, S. A. & Piggot, P. J. (2000). A tale of two genomes: resolution of dimeric chromosomes in *Escherichia coli* and *Bacillus subtilis*. *Research in Microbiology* **151**, 503–511.

Scott, J. R. & Barnett, T. C. (2006). Surface proteins of Gram-positive bacteria and how they get there. *Annual Review of Microbiology* **60**, 397–423.

Sherratt, D., Lau, I. & Barre, F. (2001). Chromosome segregation. *Current Opinion in Microbiology* **4**, 653–659.

Silver, R. P., Prior, K., Nsahlai, C. & Wright, L. F. (2001). ABC transporters and the export of capsular polysaccharides from Gram-negative bacteria. *Research in Microbiology* **152**, 357–364.

Smith, C. A. (2006). Structure, function and dynamics in the Mur family of bacterial cell wall ligases. *Journal of Molecular Biology* **362**, 640–655.

Smith, T. J., Blackman, S. A. & Foster, S. J. (2000). Autolysins of *Bacillus subtilis*: multiple enzymes with multiple functions. *Microbiology-UK* **146**, 249–262.

Ton-That, H. & Schneewind, O. (2004). Assembly of pili in Gram-positive bacteria. *Trends in Microbiology* **12**, 228–234.

Viollier, P. H. & Shapiro, L. (2004). Spatial complexity of mechanisms controlling a bacterial cell cycle. *Current Opinion in Microbiology* **7**, 572–578.

Vollmer, W. & Holtje, J. (2001). Morphogenesis of *Escherichia coli*. *Current Opinion in Microbiology* **4**, 625–633.

von Stockar, U., Maskow, T., Liu, J., Marison, I. W. & Patino, R. (2006). Thermodynamics of microbial growth and metabolism: an analysis of the current situation. *Journal of Biotechnology* **121**, 517–533.

White, S. H. & von Heijne, G. (2005). Transmembrane helices before, during, and after insertion. *Current Opinion in Structural Biology* **15**, 378–386.

Whitfield, C. (2006). Biosynthesis and assembly of capsular polysaccharides in *Escherichia coli*. *Annual Review of Biochemistry* **75**, 39–68.

Wirtz, K. W. A. (2006). Phospholipid transfer proteins in perspective. *FEBS Letters* **580**, 5436–5441.

Yonekura, K., Maki-Yonekura, S. & Namba, K. (2002). Growth mechanism of the bacterial flagellar filament. *Research in Microbiology* **153**, 191–197.

Heterotrophic metabolism on substrates other than glucose

It has been described previously how glucose and mineral salts can support the growth of certain heterotrophs. In this case, the organisms obtain ATP, NADPH and carbon skeletons for biosynthesis through central metabolism. Almost all natural organic compounds can be utilized through microbial metabolism. In this chapter, the bacterial metabolism of organic compounds other than glucose is discussed. Since central metabolism is reversible in one way or another, it can be assumed that an organism can use a compound if that compound is converted to intermediates of central metabolism. Some bacteria can use an extensive variety of organic compounds as sole carbon and energy sources, while some organisms can only use limited numbers of organic compounds; for example, *Bacillus fastidiosus* can use only urate.

7.1 | Hydrolysis of polymers

Plant and animal cells consist mainly of polymers. They include polysaccharides, such as starch and cellulose, as well as proteins, nucleic acids, and many others. Such polymers cannot be easily transported into microbial cells but are first hydrolyzed to monomers or oligomers by extracellular enzymes before being transported into the cell.

7.1.1 Starch hydrolysis

Starch is a glucose polymer consisting of amylose and amylopectin. The former has a straight chain structure with α-1,4-glucoside bonds, while the latter has side chains with α-1,6-glucoside bonds. Starch is the commonest storage material in plants, and many prokaryotes produce amylase to utilize it as their energy and carbon source.

Amylases are classified according to their mode of action. α-amylase is an endoglucanase that randomly hydrolyzes α-1,4-glucoside bonds to produce a mixture of dextrin, maltose and glucose, but does not hydrolyze α-1,6-bonds. Many bacteria, including species of *Bacillus*, *Pseudomonas* and *Clostridium*, produce this enzyme. In fact, *Bacillus stearothermophilus* is used to produce α-amylase industrially.

β-amylase is an exoglucanase and removes maltose units from the non-reducing end of amylose. Another exoglucanase, glucoamylase, hydrolyzes glucose units from the non-reducing end of amylose. Many bacteria, including *Bacillus* and *Pseudomonas* spp., produce this enzyme. Glucoamylase is produced industrially using fungi.

Amylases hydrolyze the α-1,4-glucoside bond of amylose, but cannot hydrolyze the α-1,6-glucoside bond. The α-1,6-glucoside bond of amylopectin is hydrolyzed by pullulanase and isoamylase. Pullulanase hydrolyzes pullulan, an α-glucan produced by the fungus *Aureobasidium pullulans*, and isoamylase attacks the α-1,6-glucoside bonds of amylopectin and glycogen. These are referred to as debranching enzymes. *Aerobacter aerogenes*, *Bacillus cereus*, and several other bacteria produce pullulanase; *Bacillus amyloliquefaciens* can produce isoamylase. Some hyperthermophilic bacteria and archaea including *Thermoanaerobacter ethanolicus*, *Pyrococcus furiosus*, and *Thermococcus litoralis* produce an enzyme that can hydrolyze not only α-1,4 but also α-1,6-glucoside bonds. This enzyme is called an amylopullulanase or neopullulanase and is extremely thermostable.

7.1.2 Cellulose hydrolysis

Cellulose is one of the most abundant organic compounds in nature and is a glucose homopolymer like starch. Although starch is easily hydrolyzed, the structural material cellulose is more resistant to hydrolysis. Cellulose is a straight chain of β-1,4-linked glucose units without any side chains unlike starch. Due to the absence of side chains in cellulose, extensive hydrogen bonding between the cellulose molecules forms a crystalline structure. At least three different enzymes are required for the complete hydrolysis of crystalline cellulose. They are endo-β-glucanase, exo-β-glucanase (β-glucan cellobiohydrolase) and cellobiase (β-glucosidase). Some bacteria such as *Cellulomonas flavigena* and *Clostridium thermocellum* produce all three classes of enzymes. These form a cellulosome complex with other proteins. These bacteria produce multiple endo-β-glucanases and exo-β-glucanases but it is not known if they are produced by separate genes or if some of them are partially degraded forms of the enzymes. Some bacteria can only use amorphous cellulose. They lack β-glucan cellobiohydrolase, an enzyme responsible for the digestion of crystalline cellulose.

Cellulases hydrolyze β-1,4-glucoside bonds, which are hidden within the crystalline structure. The enzymes are too big for direct contact with the hydrolytic sites and various hypotheses have been proposed to explain how the crystalline cellulose is hydrolyzed by the enzyme. According to one hypothesis, endo-β-1,4-glucanase hydrolyzes the amorphous region of the substrate, generating large numbers of non-reducing ends for β-glucan cellobiohydrolase, a member of the exo-β-1,4-glucanase enzyme class, to remove cellobiose units (Figure 7.1). Through the concerted action of these enzymes, crystalline cellulose is completely hydrolyzed. Cellobiose is either hydrolyzed to glucose by cellobiase or transported into the cell.

Another hypothesis has also been proposed to explain digestion of crystalline cellulose. A C_1 enzyme decrystallizes the native cellulose to

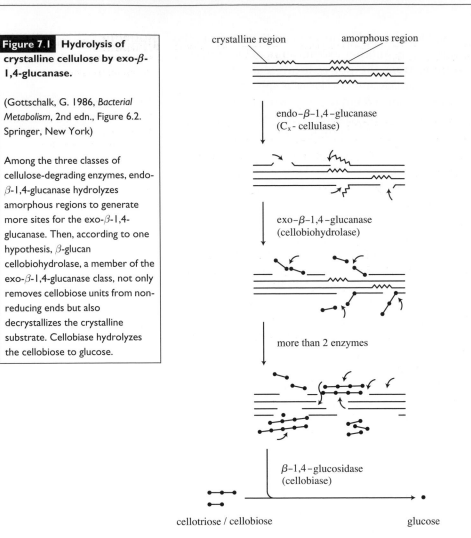

Figure 7.1 **Hydrolysis of crystalline cellulose by exo-β-1,4-glucanase.**

(Gottschalk, G. 1986, *Bacterial Metabolism*, 2nd edn., Figure 6.2. Springer, New York)

Among the three classes of cellulose-degrading enzymes, endo-β-1,4-glucanase hydrolyzes amorphous regions to generate more sites for the exo-β-1,4-glucanase. Then, according to one hypothesis, β-glucan cellobiohydrolase, a member of the exo-β-1,4-glucanase class, not only removes cellobiose units from non-reducing ends but also decrystallizes the crystalline substrate. Cellobiase hydrolyzes the cellobiose to glucose.

the amorphous form with little or no hydrolytic activity before the hydrolytic enzymes attack the β-glucose bonds (Figure 7.2). X-ray crystallography has shown the existence of this enzyme. An enzyme with strong cellulose-binding activity and low hydrolytic ability has been found in bacteria such as *Cellulomonas flavigena*, *Clostridium thermocellum* and *Clostridium cellulovorans*, and in fungi such as *Trichoderma reesei*.

7.1.3 Other polysaccharide hydrolases

In addition to starch and cellulose, other natural polysaccharides include hemicellulose, pectin, and chitin. Hemicellulose is a heteropolysaccharide consisting of various pentoses and hexoses, and their derivatives, linked with β-glycoside bonds. Many microorganisms, including cellulolytic species, can use this heteropolysaccharide as a carbon and energy source with the extracellular enzymes collectively called hemicellulase. Xylose is the most abundant monosaccharide in most hemicelluloses, and hemicellulase is sometimes referred to as xylanase.

$$\text{crystalline cellulose} \xrightarrow[\substack{\text{formation of unstable} \\ \text{glucopyranose ring}}]{C_1} \text{reactive cellulose} \xrightarrow[\text{hydrolysis}]{C_x} \text{soluble sugar}$$

Figure 7.2 Initiation of cellulose degradation by a decrystallizing enzyme.

An alternative hypothesis for the decrystallization of cellulose by an enzyme possessing decrystallizing activity but little or no hydrolytic activity. The decrystallized cellulose is hydrolyzed by hydrolytic enzymes.

Pectin is a methyl ester of α-1,4-polygalacturonate. Pectin esterase hydrolyzes the ester bond to produce methanol, and endo- and exo-type pectinases degrade α-1,4-polygalacturonate to galacturonate. *Bacillus polymyxa*, *Erwinia carotovora* and several other bacteria can produce these enzymes. Industrially, these enzymes can be used to clarify fruit juices such as apple.

Chitin has the structure of poly-β-1,4-N-acetylglucosamine and is the major constituent of fungal cell walls and the exoskeletons of insects and crustaceans. This polysaccharide is the second most abundant in the biosphere after cellulose. Chitin has a crystalline structure and degradation requires more than one enzyme. Chitinase is produced by many soil bacteria including *Chromobacterium violaceum*, *Serratia marcescens*, *Serratia plymuthica*, *Serratia liquefaciens*, *Aeromonas hydrophila*, *Enterobacter agglomerans*, *Pseudomonas aeruginosa*, *Pseudomonas chitinovorans*, *Bacillus circulans*, *Streptomyces lividans*, and *Streptomyces griseus*. The enzyme hydrolyzes β-1,4 bonds to produce N-acetylglucosamine. Chitin deacetylase removes acetate from chitin to chitosan, which is hydrolyzed to glucosamine by chitinase. Chitin deacetylase is found in fungi such as *Absidia cierulea*, *Colletotrichum lindemuthanum* and *Mucor rouxii*. Chitinase producers can inhibit the growth of plant pathogenic fungi and can be used as biocontrol agents.

Various other polysaccharides are found in nature including those constituting the saccharide portion of proteins and lipids. These are also hydrolyzed by extracellular enzymes which are produced by many microorganisms. Polysaccharide-hydrolyzing enzymes are classified according to their amino acid sequence.

7.1.4 Disaccharide phosphorylases

Some disaccharides are imported into the cell and utilized through non-hydrolytic enzymes. They phosphorylate a monosaccharide of the disaccharide using inorganic phosphate liberating the other monosaccharide in the free form. These enzymes are referred to as phosphorylases. Since hexokinase is not involved in metabolism of the phosphorylated sugar, one less ATP is consumed in the metabolism of disaccharides by phosphorylases than by hydrolases.

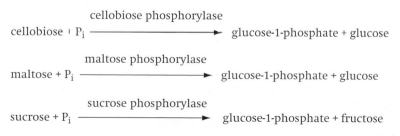

$$\text{cellobiose} + P_i \xrightarrow{\text{cellobiose phosphorylase}} \text{glucose-1-phosphate} + \text{glucose}$$

$$\text{maltose} + P_i \xrightarrow{\text{maltose phosphorylase}} \text{glucose-1-phosphate} + \text{glucose}$$

$$\text{sucrose} + P_i \xrightarrow{\text{sucrose phosphorylase}} \text{glucose-1-phosphate} + \text{fructose}$$

Sucrose phosphorylase is found in *Pseudomonas saccharophila*, and cellobiose is phosphorylated in *Cellulomonas flavigena*.

7.1.5 Hydrolysis of proteins, nucleic acids and lipids

All cells have intracellular polymer-hydrolyzing enzymes for turnover of cellular polymers. These enzymes are not involved in the use of polymers available in the environment as their energy and carbon sources. As mentioned previously, microorganisms that utilize extracellular polymers produce extracellular enzymes.

Extracellular proteases are classified into acidic, neutral and alkaline enzymes according to their optimum pH. Subtilisin produced by *Bacillus licheniformis* is an alkaline protease with an optimum pH of 8–11. Neutral proteases are produced by *Bacillus megaterium* and *Pseudomonas aeruginosa*. Most alkaline proteases are non-specific and hydrolyze any peptide bonds. Many acidic proteases are specific and hydrolyze the peptide bonds of specific amino acids.

DNA and RNA can serve as carbon and energy sources in bacteria that produce extracellular DNase and RNase, respectively. *Staphylococcus aureus* produces extracellular DNase, and RNase is known in *Bacillus subtilis*, *Streptomyces* spp. and other bacteria. Lipids are utilized by microorganisms after hydrolysis by extracellular lipase to fatty acids and glycerol.

7.2 | Utilization of sugars

Extracellular polysaccharide hydrolysis generates various hexoses, pentoses and their derivatives. These substances support the growth of many bacteria and archaea.

7.2.1 Hexose utilization

Figure 7.3 shows how common hexoses are converted to central metabolic intermediates. Fructose is phosphorylated either to fructose-1-phosphate during group translocation or to fructose-6-phosphate by hexokinase after active transport into the cell. Fructose-1-phosphate is further phosphorylated to fructose-1,6-diphosphate by 1-phosphofructokinase.

Mannose is actively transported into the cell before hexokinase converts it to mannose-6-phosphate. Phosphomannoisomerase isomerizes mannose-6-phosphate to fructose-6-phosphate. Galactokinase transfers phosphate from ATP to galactose to form galactose-1-phosphate, and glucose:galactose-1-phosphate uridylyltransferase transfers UMP from UDP-glucose to galactose-1-phosphate to form UDP-galactose and glucose-1-phosphate. UDP-galactose is converted to UDP-glucose through the action of UDP-glucose epimerase. Phosphoglucomutase converts glucose-1-phosphate to glucose-6-phosphate. Glycogen is metabolized through glucose-1-phosphate.

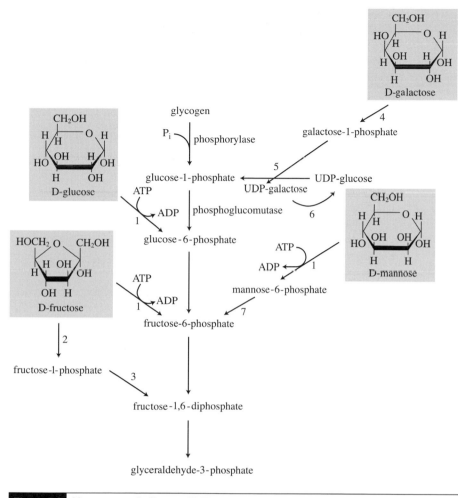

Figure 7.3 Hexose metabolism to EMP pathway intermediates.

1, hexokinase; 2, group translocation system, or ketohexokinase; 3, 1-phosphofructokinase; 4, galactokinase; 5, glucose:galactose-1-phosphate uridylyltransferase; 6, uridine diphosphate glucose epimerase; 7, mannose-6-phosphate isomerase.

Maltose is transported into the cell through the ABC system, and sucrose and cellobiose through active transport. The disaccharides are split either by specific phosphorylases (Section 7.1.4) or by hydrolases, α-glucosidase, invertase or β-glucosidase (cellobiase). Lactose is actively transported before being hydrolyzed to glucose and galactose by β-galactosidase.

7.2.2 Pentose utilization

Xylose, arabinose and ribose are common pentoses in nature. Ribose is phosphorylated to ribose-5-phosphate. Isomerases convert arabinose and xylose to ribulose and xylulose, respectively (Figure 7.4). These ketoses are phosphorylated to be metabolized through the HMP pathway (Section 4.3).

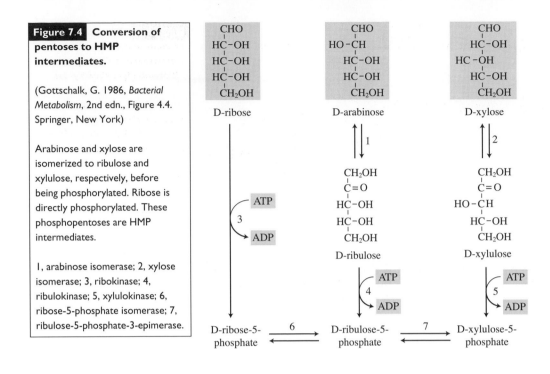

Figure 7.4 Conversion of pentoses to HMP intermediates.

(Gottschalk, G. 1986, *Bacterial Metabolism*, 2nd edn., Figure 4.4. Springer, New York)

Arabinose and xylose are isomerized to ribulose and xylulose, respectively, before being phosphorylated. Ribose is directly phosphorylated. These phosphopentoses are HMP intermediates.

1, arabinose isomerase; 2, xylose isomerase; 3, ribokinase; 4, ribulokinase; 5, xylulokinase; 6, ribose-5-phosphate isomerase; 7, ribulose-5-phosphate-3-epimerase.

7.3 | Organic acid utilization

7.3.1 Fatty acid utilization

Acyl-CoA synthetase forms acyl-CoA from fatty acids and coenzyme-A before acyl-CoA is converted to acetyl-CoA through β-oxidation. The glyoxylate cycle (Section 5.3.2) is employed with the TCA cycle to convert acetyl-CoA into the carbon skeletons needed for biosynthesis (Figure 7.5).

β-oxidation splits a 2-carbon unit in the form of acetyl-CoA from acyl-CoA and this is catalyzed by five enzymes including acyl-CoA synthetase (Figure 7.6). Fatty acids with an even number of carbons result solely in acetyl-CoA while propionyl-CoA remains after β-oxidation of fatty acids with an odd number of carbons.

Propionyl-CoA is a metabolic intermediate of a number of compounds including the amino acids, L-valine and L-isoleucine. The acrylate pathway (Figure 7.7) and the methylmalonyl-CoA pathway (Figure 7.8) were identified as the oxidative metabolism of propionyl-CoA in earlier work. More recent studies have shown that propionyl-CoA is metabolized to pyruvate through the methylcitrate cycle (Figure 7.9) in many bacteria including *Escherichia coli*, *Salmonella typhimurium* and *Ralstonia eutropha*, and in fungi such as *Saccharomyces cerevisiae* and *Aspergillus nidulans*. Propionate can inhibit fungal growth on glucose but can also serve as a carbon and energy source.

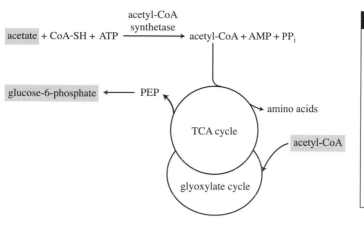

Figure 7.5 **Conversion of acetate to central metabolic intermediates through the TCA and glyoxylate cycles.**

Acetyl-CoA is an intermediate in the metabolism of many compounds. This activated form of acetate is metabolized through the TCA cycle to provide energy and through the glyoxylate cycle to supply carbon skeletons for biosynthesis.

Malonate, a succinate dehydrogenase inhibitor, is activated to malonyl-CoA before being decarboxylated to acetyl-CoA in *Pseudomonas fluorescens* and *Acinetobacter calcoaceticus*:

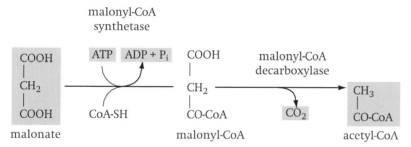

An acyl carrier protein (ACP) replaces coenzyme A in malonate metabolism by some bacteria such as *Klebsiella pneumoniae* and the anaerobe *Malonomonas rubra*:

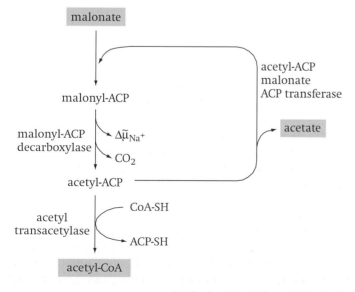

(*Molecular Microbiology*, 1997, **25**, 3–10)

Figure 7.6 Palmitate degradation to acetyl-CoA through β-oxidation.

(Gottschalk, G. 1986, *Bacterial Metabolism*, 2nd edn., Figure 6.5. Springer, New York)

1, acyl-CoA synthetase; 2, fatty acyl-CoA dehydrogenase; 3, 3-hydroxyacyl-CoA hydrolase; 4, 3-hydroxyacylCoA dehydrogenase; 5, acetyl-CoA acetyltransferase.

7.3.2 Organic acids more oxidized than acetate

Acetate can serve as the sole carbon source through the TCA and glyoxylate cycles. C2 compounds more oxidized than acetate cannot be metabolized in the same way. These compounds include glycolate, glyoxylate and oxalate. Glycolate is generated from the dephosphorylation of phosphoglycolate that is produced during photorespiration (Section 10.8.1.2). Purine degradation results in glyoxylate. *Escherichia coli* and *Pseudomonas* spp. use these substances through the dicarboxylic acid cycle to generate energy, and through the glycerate pathway to supply carbon skeletons for biosynthesis (Figure 7.10). The dicarboxylic acid cycle is similar to the glyoxylate cycle described in Chapter 5 (Section 5.3.2). Since phosphoenolpyruvate, an intermediate of the dicarboxylic acid cycle, is used for biosynthesis, it is replenished through the glycerate pathway.

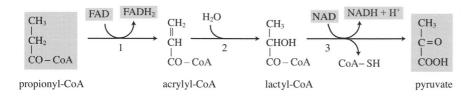

propionyl-CoA   acrylyl-CoA   lactyl-CoA   pyruvate

Figure 7.7   **Acrylate pathway oxidizing propionyl-CoA to pyruvate in *Pseudomonas aeruginosa*.**

(Gottschalk, G. 1986, *Bacterial Metabolism*, 2nd edn., p. 150. Springer, New York)

Pseudomonas aeruginosa oxidizes valine to propionyl-coenzyme A (CoA), as it occurs in animal tissues, followed by the oxidation of propionyl-CoA to acrylyl-CoA, lactyl-CoA, and pyruvate. 1, acyl-CoA dehydrogenase; 2, lactyl-CoA dehydratase; 3, lactyl-CoA dehydrogenase.

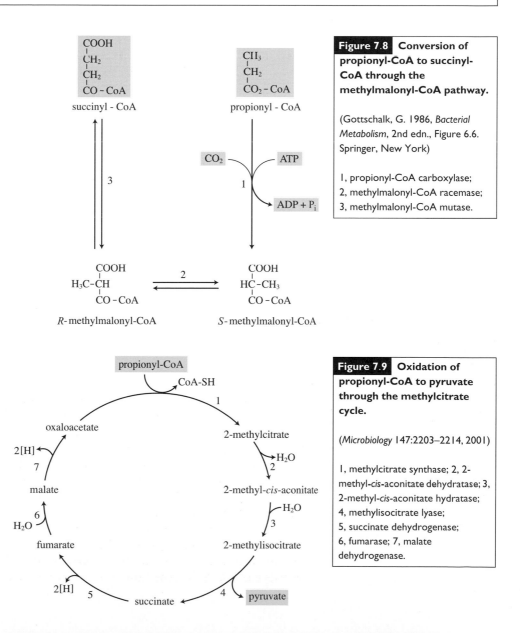

Figure 7.8   **Conversion of propionyl-CoA to succinyl-CoA through the methylmalonyl-CoA pathway.**

(Gottschalk, G. 1986, *Bacterial Metabolism*, 2nd edn., Figure 6.6. Springer, New York)

1, propionyl-CoA carboxylase; 2, methylmalonyl-CoA racemase; 3, methylmalonyl-CoA mutase.

Figure 7.9   **Oxidation of propionyl-CoA to pyruvate through the methylcitrate cycle.**

(*Microbiology* 147:2203–2214, 2001)

1, methylcitrate synthase; 2, 2-methyl-*cis*-aconitate dehydratase; 3, 2-methyl-*cis*-aconitate hydratase; 4, methylisocitrate lyase; 5, succinate dehydrogenase; 6, fumarase; 7, malate dehydrogenase.

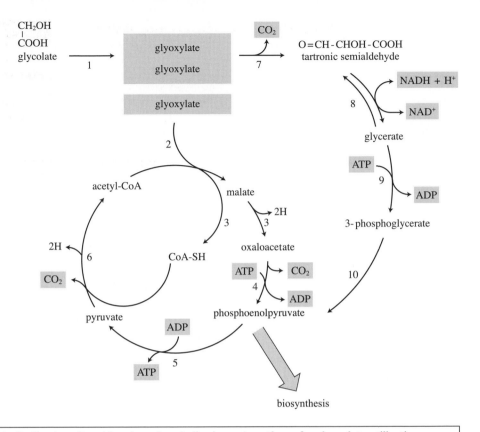

The catabolic dicarboxylic acid cycle and anabolic glycerate pathway for glyoxylate utilization.

Certain bacteria including *Escherichia coli* and species of the genus *Pseudomonas* oxidize glyoxylate through the dicarboxylic acid cycle and convert the substrate to phosphoenolpyruvate through the glycerate pathway.

1, glycolate dehydrogenase; 2, malate synthase; 3, malate dehydrogenase; 4, PEP carboxykinase; 5, pyruvate kinase; 6, pyruvate dehydrogenase; 7, glyoxylate carboligase; 8, tartronate semialdehyde reductase; 9, glycerate kinase; 10, phosphoglycerate mutase and enolase.

Paracoccus denitrificans converts glyoxylate through the 3-hydroxyaspartate pathway to oxaloacetate, a TCA cycle intermediate (Figure 7.11). In the 3-hydroxyaspartate pathway, glyoxylate is converted to glycine before condensing with a second glyoxylate molecule to yield *erythro*-3-hydroxyaspartate. *Erythro*-3-hydroxyaspartate dehydratase deaminates this to oxaloacetate. As in the deamination of amino acids with a hydroxyl group (Section 7.5.3), this enzyme is referred to as a dehydratase, although it deaminates the substrate.

Pseudomonas oxalaticus uses oxalate as sole carbon and energy source. This most oxidized dicarboxylic acid is activated to oxalyl-CoA before a part of it is oxidized to CO_2 as an energy source and the remaining part is reduced to glyoxylate for biosynthesis (Figure 7.12). *Bacillus oxalophilus* and *Methylobacterium extorquens* also use oxalate as their sole carbon and energy source.

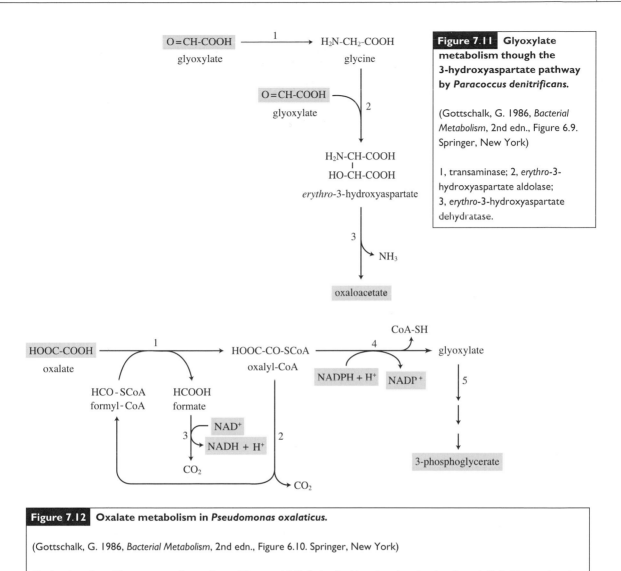

Figure 7.11 Glyoxylate metabolism though the 3-hydroxyaspartate pathway by *Paracoccus denitrificans*.

(Gottschalk, G. 1986, *Bacterial Metabolism*, 2nd edn., Figure 6.9. Springer, New York)

1, transaminase; 2, *erythro*-3-hydroxyaspartate aldolase; 3, *erythro*-3-hydroxyaspartate dehydratase.

Figure 7.12 Oxalate metabolism in *Pseudomonas oxalaticus*.

(Gottschalk, G. 1986, *Bacterial Metabolism*, 2nd edn., Figure 6.10. Springer, New York)

Oxalate is activated by coenzyme A transferase (1) to oxalyl-CoA that is either decarboxylated to formyl-CoA (2) or reduced to glyoxylate (4). Glyoxylate is used for biosynthesis through the glycerate pathway (5) and formyl-CoA is oxidized via formate (3).

7.4 | Utilization of alcohols and ketones

Anaerobic fermentative microorganisms produce alcohols such as ethanol, butanol and others. Alcohol dehydrogenase and aldehyde dehydrogenase oxidize primary alcohols to fatty acids:

$$R\text{-}CH_2\text{-}CH_2OH + NAD^+ \xrightarrow{\text{alcohol dehydrogenase}} R\text{-}CH_2\text{-}CHO + NADH + H^+$$

$$R\text{-}CH_2\text{-}CHO + NAD^+ + H_2O \xrightarrow{\text{aldehyde dehydrogenase}} R\text{-}CH_2\text{-}COOH + NADH + H^+$$

Secondary alcohol dehydrogenase oxidizes a secondary alcohol to a ketone before conversion to a fatty acid and primary alcohol by the action of a monooxygenase and esterase as described later (Section 7.7, Figure 7.22).

Propanediol is a fermentation product of glycerol, and some species of *Bacillus* and facultative anaerobic bacteria produce butane-diol and acetoin from glucose (Section 8.6). Saccharolytic clostridia ferment carbohydrates to yield various fermentation products including acetone and isopropanol (Section 8.5.2).

Salmonella typhymurium oxidizes propanediol to propionyl-CoA, which is metabolized through the methylcitrate cycle and other pathways (Section 7.3.1):

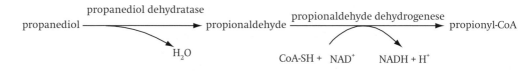

Butanediol is oxidized to acetaldehyde and acetyl-CoA by species of *Bacillus*, *Ralstonia eutropha* (*Alcaligenes eutrophus*) and *Pelobacter carbinolicus* through the action of diol dehydrogenase and the acetoin dehydrogenase complex. The acetoin dehydrogenase enzyme complex is a keto acid dehydrogenase like the pyruvate dehydrogenase complex (Section 5.1).

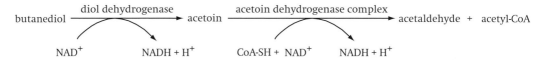

A *Xanthobacter* sp. can oxidize isopropanol to acetone before carboxylating it to acetoacetate. Acetone is carboxylated in *Rhodobacter capsulatus*, *Rhodomicrobium vannielii*, and *Thiosphaera pantotropha*.

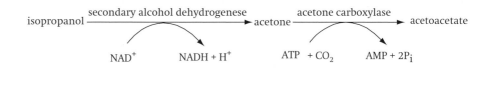

7.5 | Amino acid utilization

Nutrient broth is a common medium for cultivation of many bacteria in the laboratory and contains peptone and beef extract. These substances largely consist of amino acids and peptides. Organisms growing on such a nutrient medium transport and metabolize amino acids and peptides through the central metabolic pathways. Amino acids are used for protein synthesis and are deaminated to the corresponding 2-keto acids. The 2-keto acids are oxidized to acyl-CoA by 2-keto acid dehydrogenases for use as carbon and energy sources. They are deaminated through different mechanisms depending on their nature.

7.5.1 Oxidative deamination

Amino acids are deaminated either by amino acid oxidase reducing its prosthetic flavin or by amino acid dehydrogenase reducing $NAD(P)^+$. Amino acid oxidases have a low specificity for the substrate and a single enzyme can oxidize up to ten different amino acids. Since bacterial cell walls contain D-amino acids, bacteria have L-amino acid as well as D-amino acid oxidase.

$$R-CHNH_2-COOH + H_2O \rightarrow R-CO-COOH + NH_3 + 2e^- + 2H^+$$

Amino acid dehydrogenase oxidizes L-alanine or L-glutamate to pyruvate and 2-ketoglutarate, respectively. Since transaminases convert pyruvate and 2-ketoglutarate to alanine and glutamate, all amino acids can be deaminated by the combination of transaminase and amino acid dehydrogenase.

$$\text{alanine} + NAD^+ + H_2O \xrightleftharpoons{\text{alanine dehydrogenase}} \text{pyruvate} + NADH + NH_4^+$$

$$\text{glutamate} + NAD^+ + H_2O \xrightleftharpoons{\text{glutamate dehydrogenase}} \text{2-ketoglutarate} + NADH + NH_4^+$$

7.5.2 Transamination

Transamination is an enzymic reaction that transfers the $-NH_2$ group from amino acids to 2-keto acids. As shown above, alanine and glutamate dehydrogenases deaminate their substrate. When coupling transamination and dehydrogenation of alanine or glutamate, an amino acid is oxidized to the corresponding 2-keto acid reducing $NAD(P)^+$.

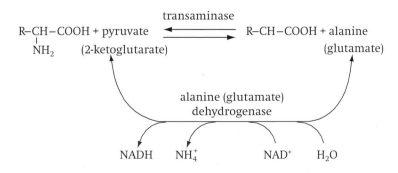

7.5.3 Amino acid dehydratase

Amino acid dehydratases deaminate serine and threonine, removing the hydroxyl group (-OH) at the same time:

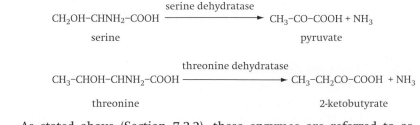

As stated above (Section 7.3.2), these enzymes are referred to as dehydratases although they deaminate their substrates. The reactions take place in two steps, the first step being a dehydration reaction.

$$\text{R-CHOH-CHNH}_2\text{-COOH} \xrightarrow{-\text{H}_2\text{O}} \text{R-CH=CONH}_2\text{-COOH} \xrightarrow{+\text{H}_2\text{O},\ -\text{NH}_3} \text{CH}_2\text{-CO-COOH}$$

Isoleucine biosynthesis starts with the deamination of threonine to 2-ketobutyrate that is catalyzed by threonine dehydratase (Figure 6.15, Section 6.4.1). *Escherichia coli* has separate threonine dehydratases for isoleucine synthesis and the use of threonine as a carbon and energy source. As expected, they are regulated differently.

Aspartate and histidine are deaminated in similar reactions to those catalyzed by dehydratase. Unlike dehydrogenases and oxidases, aspartase and histidase form double bonds between two and three carbons. Water does not take part in these reactions, nor electron carriers.

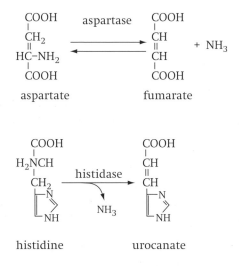

7.5.4 Deamination of cysteine and methionine

Transmethylase removes the methyl group from methionine to yield homocysteine. Desulfhydrase removes amino and sulfide groups simultaneously from cysteine and homocysteine to produce pyruvate and 2-ketobutyrate, respectively. Desulfhydrase is known in many aerobic and facultative anaerobic bacteria including *Escherichia coli*, *Proteus vulgaris* and *Bacillus subtilis*.

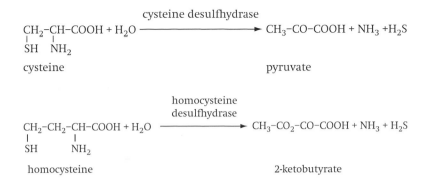

7.5.5 Deamination products of amino acids

Deamination of amino acids yields various organic acids:

glycine → glyoxylate
alanine → pyruvate
cysteine → pyruvate
aspartate → oxaloacetate, fumarate
asparagine → oxaloacetate, fumarate
glutamate → 2-ketoglutarate
glutamine → 2-ketoglutarate
threonine → 2-ketobutyrate
methionine → 2-ketobutyrate
serine → pyruvate
histidine → urocanate
valine → 2-ketoisovalerate
leucine → 2-ketoisocaproate
isoleucine → 2-keto-3-methylvalerate.

Pyruvate, oxaloacetate, fumarate and 2-ketoglutarate are intermediates of central metabolism and can be used both for anabolic and catabolic purposes. Glyoxylate is metabolized through the dicarboxylic acid cycle–glycerate pathway (Section 7.3.2, Figure 7.10) or through the 3-hydroxyaspartate pathway (Section 7.3.2, Figure 7.11) depending on the organism. Urocanate from histidine deamination is metabolized through glutamate, as shown in Figure 7.13. A dehydrogenase complex oxidizes 2-ketobutyrate to propionyl-CoA that is metabolized through the methylcitrate cycle and other routes (Section 7.3.1).

Amino acids with a side chain are oxidized to 2-ketoisovalerate, 2-ketoisocaproate and 2-keto-3-methylvalerate. They are further oxidized to the corresponding acyl-CoA by the 2-keto acid dehydrogenase complex before acyl-CoA dehydrogenase forms a double bond between two and three carbons. The same enzyme catalyzes each of these reactions. Unsaturated fatty acids with side chains are metabolized by a separate enzyme to propionyl-CoA, acetyl-CoA and acetoacetate (Figure 7.14). Acetoacetate yields two acetyl-CoA through acetoacetyl-CoA.

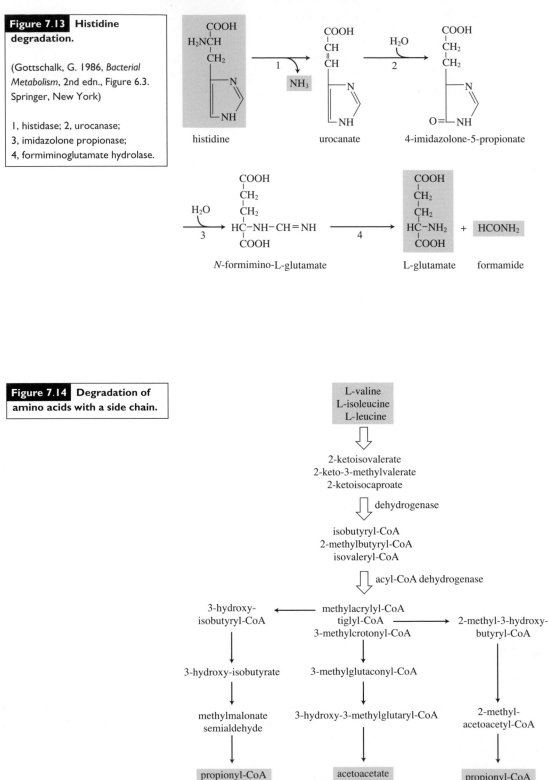

Figure 7.13 **Histidine degradation.**

(Gottschalk, G. 1986, *Bacterial Metabolism*, 2nd edn., Figure 6.3. Springer, New York)

1, histidase; 2, urocanase; 3, imidazolone propionase; 4, formiminoglutamate hydrolase.

histidine   urocanate   4-imidazolone-5-propionate

N-formimino-L-glutamate   L-glutamate   formamide

Figure 7.14 **Degradation of amino acids with a side chain.**

L-valine
L-isoleucine
L-leucine

2-ketoisovalerate
2-keto-3-methylvalerate
2-ketoisocaproate

dehydrogenase

isobutyryl-CoA
2-methylbutyryl-CoA
isovaleryl-CoA

acyl-CoA dehydrogenase

3-hydroxy-isobutyryl-CoA ← methylacrylyl-CoA / tiglyl-CoA / 3-methylcrotonyl-CoA → 2-methyl-3-hydroxy-butyryl-CoA

3-hydroxy-isobutyrate   3-methylglutaconyl-CoA

methylmalonate semialdehyde   3-hydroxy-3-methylglutaryl-CoA   2-methyl-acetoacetyl-CoA

propionyl-CoA   acetoacetate + acetyl-CoA   propionyl-CoA + acetyl-CoA

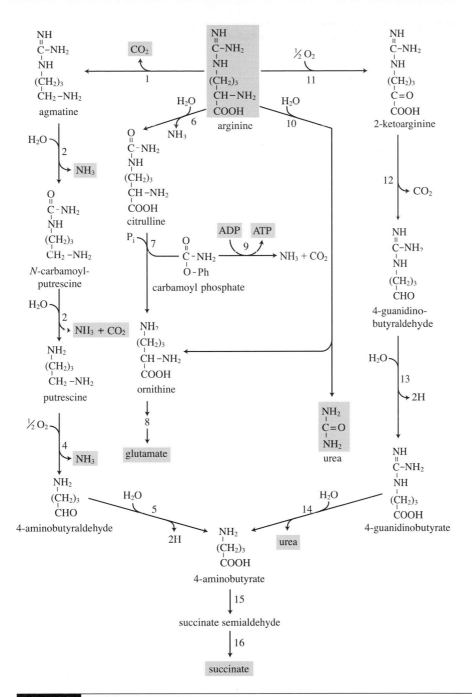

Figure 7.15 Arginine degradation.

(Gottschalk, G. 1986, *Bacterial Metabolism*, 2nd edn., Figure 6.4. Springer, New York)

Arginine is metabolized through various pathways depending upon the organism. Enteric bacteria, including *Escherichia coli*, start the reaction with arginine decarboxylase (1) and arginine oxidase (11) catalyzes the first reaction in *Pseudomonas putida*. Anaerobic bacteria such as *Clostridium perfringens* and lactic-acid bacteria deiminate arginine (6) and *Bacillus subtilis* removes urea from arginine by the action of arginase (10).

1, arginine decarboxylase; 2, agmatine deiminase; 3, *N*-carbamoylputrescine hydrolase; 4, putrescine oxidase; 5, aminobutyraldehyde dehydrogenase; 6, arginine deiminase; 7, ornithine carbamoyltransferase; 8, reverse reaction of arginine biosynthesis (enzymes 6, 7, 8, 9 in Figure 6.17); 9, carbamate kinase; 10, arginase; 11, arginine oxidase; 12, 2-ketoarginine decarboxylase; 13, 4-guanidinobutyraldehyde oxidoreductase; 14, guanidinobutyrase; 15, transaminase; 16, succinate semialdehyde dehydrogenase.

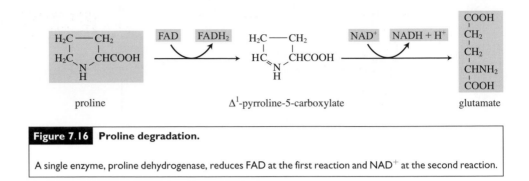

proline Δ^1-pyrroline-5-carboxylate   glutamate

Figure 7.16   **Proline degradation.**

A single enzyme, proline dehydrogenase, reduces FAD at the first reaction and NAD^+ at the second reaction.

7.5.6 Other amino acids

As shown in Figure 7.15, arginine is metabolized to glutamate or succinate through a number of different pathways depending on the organism.

Enteric bacteria convert proline to glutamate through reactions catalyzed by proline dehydrogenase that has proline oxidase as well as Δ^1-pyrroline-5-carboxylate (P5C) dehydrogenase activity. FAD is reduced from the first reaction of proline oxidation, and the same enzyme oxidizes P5C to glutamate reducing NAD^+ (Figure 7.16). This enzyme binds to the cytoplasmic membrane when proline is available as the electron donor, but in the absence of proline, the *put* (proline dehydrogenase and proline permease) operon is the binding site of the enzyme on the chromosome which inhibits transcription of the operon (Figure 12.17, Section 12.1.8).

Lysine is converted to 2-keto-6-aminocaproate through a transamination reaction before degradation to acetyl-CoA via 2-aminoadipate, glutaryl-CoA and acetoacetyl-CoA in a series of enzymic reactions (Figure 7.17).

Tryptophan degradation is the most complex among amino acids (Figure 7.18). Tryptophan-2,3-dioxygenase initiates the reactions, opening the pyrrole ring to yield formylkynurenine, which is degraded to alanine and anthranilate via kynurenine. Anthranilate degradation is discussed in Sections 7.8.1 and 7.8.2 together with phenylalanine, tyrosine and aromatic hydrocarbons.

7.6 | Degradation of nucleic acid bases

Nucleases hydrolyze RNA and DNA to ribonucleoside monophosphate and deoxyribonucleoside monophosphate. They are further hydrolyzed to bases and ribose (deoxyribose) by nucleotidase and nucleosidase.

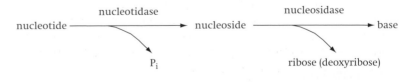

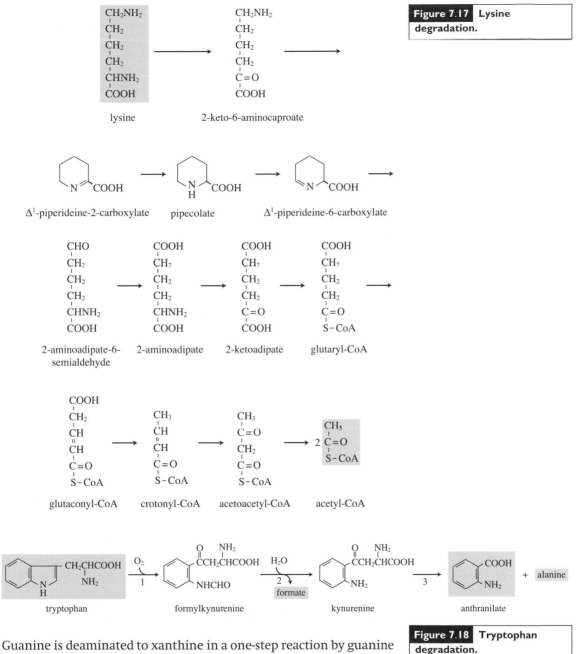

Figure 7.17 Lysine degradation.

Figure 7.18 Tryptophan degradation.

1, tryptophan-2,3-dioxygenase;
2, formylkynureninase;
3, kynureninase.

Guanine is deaminated to xanthine in a one-step reaction by guanine deaminase, while adenine requires two reactions to be oxidized to xanthine via hypoxanthine catalyzed by adenine deaminase and xanthine dehydrogenase (or xanthine oxidase). Hypoxanthine and xanthine are oxidized either by xanthine dehydrogenase or by xanthine oxidase depending on the organism. Xanthine dehydrogenase or oxidase not only oxidizes hypoxanthine to xanthine but also the next reaction to urate. Xanthine dehydrogenase uses $NAD(P)^+$ or ferredoxin as its electron acceptor, while xanthine oxidase reduces molecular oxygen generating superoxide.

Figure 7.19 **Degradation of purine bases.**

1, adenine deaminase; 2, xanthine dehydrogenase or xanthine oxidase; 3, guanine deaminase; 4, uricase; 5, allantoin racemase; 6, $S(+)$-allantoinase; 7, allantoate amidohydrolase; 8, allantoicase; 9, S-ureidoglycolase; 10, urease.

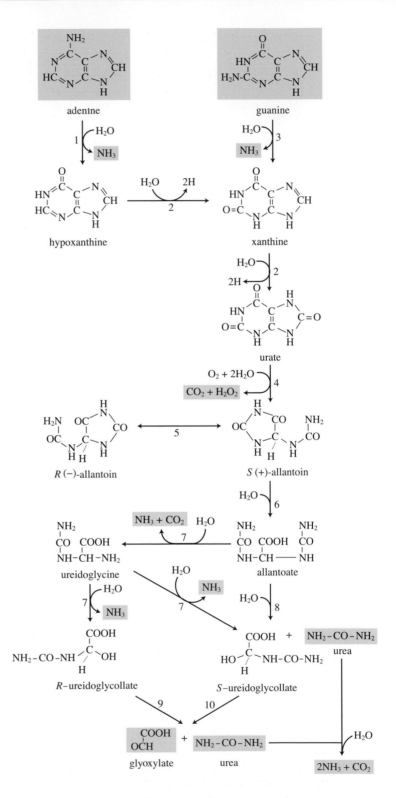

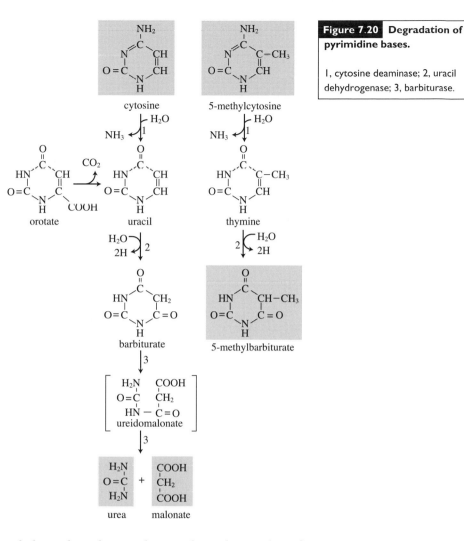

Figure 7.20 Degradation of pyrimidine bases.

1, cytosine deaminase; 2, uracil dehydrogenase; 3, barbiturase.

Urate is degraded to glyoxylate and urea through a series of reactions (Figure 7.19). Glyoxylate is metabolized through the dicarboxylic acid cycle–glycerate pathway (Figure 7.10) or the hydroxyaspartate pathway (Figure 7.11) as described earlier (Section 7.3.2). Some organisms use part of the pathway shown in Figure 7.19 to use purine bases as their carbon or nitrogen sources. Pyrimidine bases are degraded as shown in Figure 7.20.

7.7 | Oxidation of aliphatic hydrocarbons

Many prokaryotic and eukaryotic microbes can use aliphatic hydrocarbons, especially those that are liquid at ambient temperature (Table 7.1). Methane is metabolized through a specialized pathway and is described in a separate section (Section 7.9).

Hydrocarbons are water insoluble and differ from water-soluble substrates in terms of transport. Since microbes cannot thrive in a

Table 7.1.	Some examples of hydrocarbon-utilizing microbes

Bacteria
  Acinetobacter calcoaceticus
  Arthrobacter paraffineus
  Arthrobacter simplex
  Corynebacterium glutamicum
  Mycobacterium smegmatis
  Nocardia petroleophila
  Pseudomonas aeruginosa
  Pseudomonas fluorescens
Fungi
  Candida lipolytica
  Torulopsis colliculosa
  Cephalosporium roseum
  Hormoconis (Cladosporium) resinae

Figure 7.21 **Oxidation of aliphatic hydrocarbons by *Pseudomonas oleovorans*.**

(Gottschalk, G. 1986, *Bacterial Metabolism*, 2nd edn., Figure 6.11. Springer, New York)

1, rubredoxin:NADH oxidoreductase; 2, *n*-alkane monooxygenase; 3, alcohol dehydrogenase; 4, aldehyde dehydrogenase.

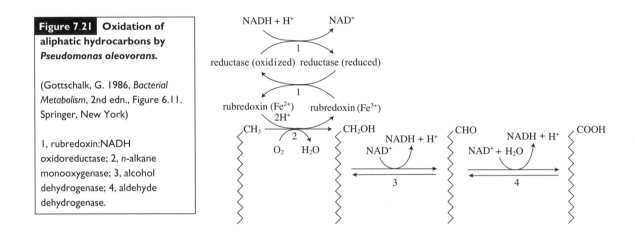

pure oil phase, microbes use hydrocarbons as their carbon and energy source at the water–oil interface. Hydrophobic glycolipids are found on the cell surfaces of bacteria and fungi that use hydrocarbons. This glycolipid solubilizes the hydrocarbon before it is transported into the cell. Many bacteria, including *Acinetobacter calcoaceticus*, produce a surfactant to improve hydrocarbon transport as an emulsion.

Hydrocarbon monooxygenase oxidizes a hydrocarbon to a primary alcohol at the cytoplasmic membrane (Figure 7.21). This enzyme can oxidize two different substrates using one molecule of oxygen. This kind of enzyme is referred to as a monooxygenase, mixed function oxidase or hydroxylase.

$$\text{substrate-H} + AH_2 + O_2 \xrightarrow{\text{monooxygenase}} \text{substrate-OH} + A + H_2O$$

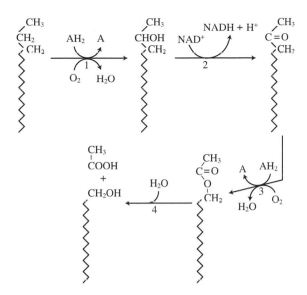

Figure 7.22 Oxidation of aliphatic hydrocarbons by *Nocardia petroleophila.*

(Gottschalk, G. 1986, *Bacterial Metabolism*, 2nd edn., Figure 6.12. Springer, New York)

1, monooxygenase; 2, secondary alcohol dehydrogenase; 3, monooxygenase; 4, acetylesterase.

The alcohol produced from the oxidation of the hydrocarbon is further oxidized to a fatty acid in reactions catalyzed by alcohol dehydrogenase and aldehyde dehydrogenase (Section 7.4).

The monooxygenase of *Nocardia petroleophilia* oxidizes the second carbon of the hydrocarbon to produce a secondary alcohol that is further oxidized to a ketone by a secondary alcohol dehydrogenase. A second monooxygenase and acetylesterase converts the ketone to a primary alcohol and acetate (Figure 7.22).

Rhodococcus rhodochrous metabolizes propylene to acetoacetate (Figure 7.23). Intermediates of this 3-carbon compound are metabolized bound to coenzyme M (2-mercaptoethanesulfonic acid), which is known in methanogenic archaea as a C1 carrier (Section 9.4.2).

7.8 | Oxidation of aromatic compounds

The complex aromatic polymer lignin comprises about 25% of land-based biomass on Earth, and coal and petroleum contain a variety of aromatic compounds. These substances are oxidized mainly under aerobic conditions due to their high structural stability. Aliphatic hydrocarbons are easily oxidized, but the aromatic portion of petroleum is persistent in natural ecosystems. Aromatic hydrocarbon degradation is best known in *Pseudomonas* spp., e.g. *Pseudomonas acidovorans* and *Pseudomonas putida*.

7.8.1 Oxidation of aromatic amino acids

Phenylalanine monooxygenase oxidizes phenylalanine to tyrosine that is deaminated to *p*-hydroxyphenylpyruvate. Dioxygenases are involved in the following reactions (Figure 7.24). Enzymes that incorporate both atoms of molecular oxygen into one substrate

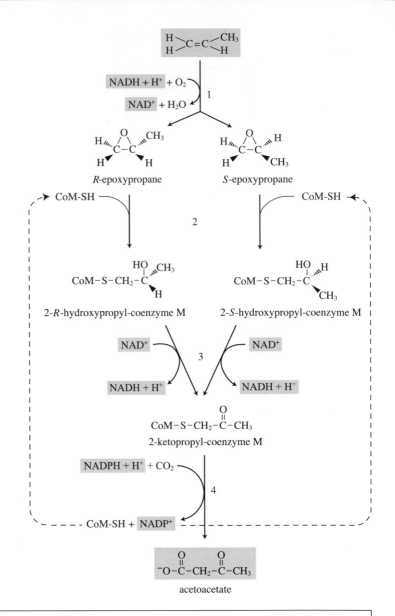

Figure 7.23 **Propylene oxidation by *Rhodococcus rhodochrous*.**

(*J. Bacteriol.* **182**:2629–2634, 2000)

This is a rare example of bacterial metabolism involving coenzyme M, a common coenzyme in methanogenic archaea. Other archaeal coenzymes found in bacteria are F_{420} in tetracycline-producing *Streptomyces* species and *Mycobacterium smegmatis* (Section 4.3.4) and tetrahydromethanopterin in methylotrophs (Section 7.9.2).

1, alkane monooxygenase; 2, epoxyalkane:coenzyme M transferase; 3, hydroxypropyl-CoM dehydrogenase; 4, NADPH:2-ketopropyl-CoM dehydrogenase.

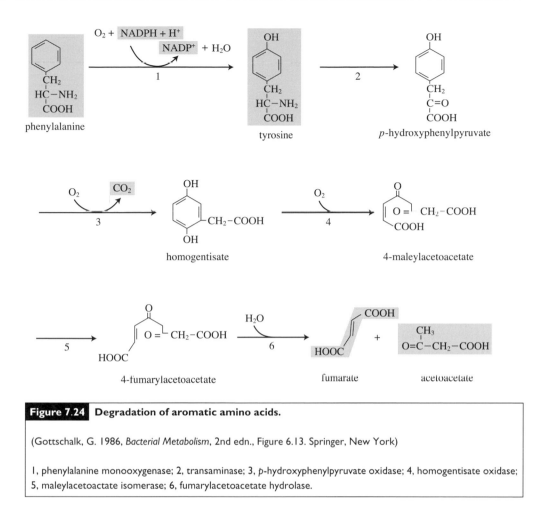

Figure 7.24 **Degradation of aromatic amino acids.**

(Gottschalk, G. 1986, *Bacterial Metabolism*, 2nd edn., Figure 6.13. Springer, New York)

1, phenylalanine monooxygenase; 2, transaminase; 3, *p*-hydroxyphenylpyruvate oxidase; 4, homogentisate oxidase; 5, maleylacetoactate isomerase; 6, fumarylacetoacetate hydrolase.

are referred to as dioxygenases or oxidases. Homogentisate oxidase opens up the benzene ring, finally producing fumarate and acetoacetate.

7.8.2 *Ortho* and *meta* cleavage, and the gentisate pathway

The metabolism of aromatic compounds can be divided into two steps. In the first step monooxygenases incorporate hydroxyl groups into the benzene ring. Through this step, aromatic compounds are converted to one of three intermediates: protocatechuate, catechol and gentisate. Aromatics with a hydroxyl group are mainly converted to protocatechuate (Figure 7.25), and catechol is derived from aromatic hydrocarbons, aromatic compounds with amino groups and lignin monomers (Figure 7.26). Some bacteria generate gentisate from naphthalene, 3-hydroxybenzoate, phenol derivatives, 3,6-dichloro-2-methoxybenzoate, and other substances. The benzene rings of these intermediates are opened up by dioxygenases.

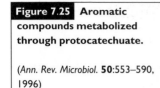

Figure 7.25 Aromatic compounds metabolized through protocatechuate.

(*Ann. Rev. Microbiol.* **50**:553–590, 1996)

Hydroxylated aromatic hydrocarbons are generally metabolized through protocatechuate.

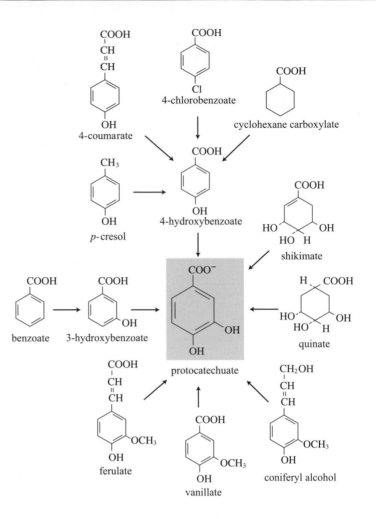

Ring fission is catalyzed by dioxygenases and termed *ortho* cleavage (Figure 7.27) when it occurs between the hydroxyl groups and *meta* cleavage (Figure 7.28) when it occurs adjacent to one of the hydroxyls. Another ring cleavage pathway, the gentisate pathway, is responsible for the oxidation of aromatic compounds with hydroxyl groups at *para* positions (Figure 7.29). After the ring fission reactions, the products are metabolized to succinate, acetyl-CoA, pyruvate and acetaldehyde. Bacteria with the *ortho* cleavage pathway have the genes in their chromosome while the genes of *meta* cleavage are found in plasmids. These plasmids are referred to as degradative plasmids to differentiate them from those with other functions such as antibiotic resistance.

Gentisate and its derivatives are degraded to pyruvate and fumarate through the gentisate pathway initiated by gentisate dioxygenase (Figure 7.29). Depending on the organisms, maleylpyruvate is either isomerized to fumarylpyruvate or degraded directly.

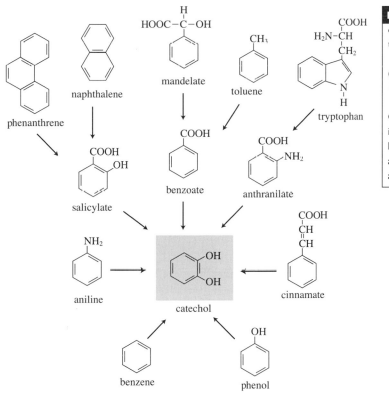

Figure 7.26 Aromatic compounds metabolized through catechol.

(*Ann. Rev. Microbiol.* **50**:553–590, 1996)

Catechol is generally the metabolic intermediate of aromatic hydrocarbons, lignin monomers and aromatic compounds with amine groups.

7.8.3 Oxygenase and aromatic compound oxidation

Since hydrocarbon degradation is initiated by oxygenase in aerobic organisms, it was believed that molecular oxygen is essential for the degradation of aromatic compounds. However, many studies have shown that aromatic compounds can be oxidized under anaerobic conditions (Section 9.9).

7.9 | Utilization of methane and methanol

Methane, methanol and methylamines are naturally available in large quantities. Some bacteria and yeasts are known to use them as their sole carbon and energy sources. These organisms are referred to as methylotrophs and metabolize the C1 compounds through pathways not known in multicarbon compound metabolism. The term 'methylotroph' is used to refer to all C1-utilizing organisms in a broad sense, and also used to describe C1-utilizing organisms that cannot use methane in a narrow sense.

7.9.1 Methanotrophy and methylotrophy

Methyltrophs are divided into methanotrophs and methylotrophs according to their ability to use methane. Methanotrophs use C1

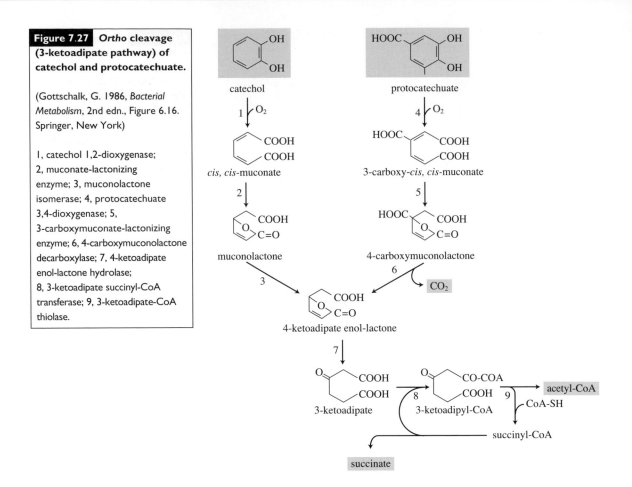

Figure 7.27 *Ortho* cleavage (3-ketoadipate pathway) of catechol and protocatechuate.

(Gottschalk, G. 1986, *Bacterial Metabolism*, 2nd edn., Figure 6.16. Springer, New York)

1, catechol 1,2-dioxygenase; 2, muconate-lactonizing enzyme; 3, muconolactone isomerase; 4, protocatechuate 3,4-dioxygenase; 5, 3-carboxymuconate-lactonizing enzyme; 6, 4-carboxymuconolactone decarboxylase; 7, 4-ketoadipate enol-lactone hydrolase; 8, 3-ketoadipate succinyl-CoA transferase; 9, 3-ketoadipate-CoA thiolase.

compounds but do not use multicarbon compounds. These are termed obligate methylotrophs. Methylotrophs do not use methane. Based on their carbon assimilation metabolism, methylotrophs are divided into heterotrophic methylotrophs and autotrophic methylotrophs. Autotrophic methylotrophs assimilate carbon dioxide through the Calvin cycle while heterotrophic methylotrophs assimilate formaldehyde through the ribulose monophosphate pathway or the serine–isocitrate lyase pathway (Table 7.2). Methanotrophs are not known in eukaryotes, and methylotrophic yeasts assimilate formaldehyde through the xylulose monophosphate pathway.

7.9.2 Methanotrophy

7.9.2.1 Characteristics of methanotrophs

Methanotrophs use C1 compounds as their carbon and energy sources, and are unable to use multicarbon compounds. For these reasons, they are referred to as obligate methylotrophs. In addition to their use of C1 compounds, they have some other characteristics. All of them are Gram-negative and have extensive intracellular

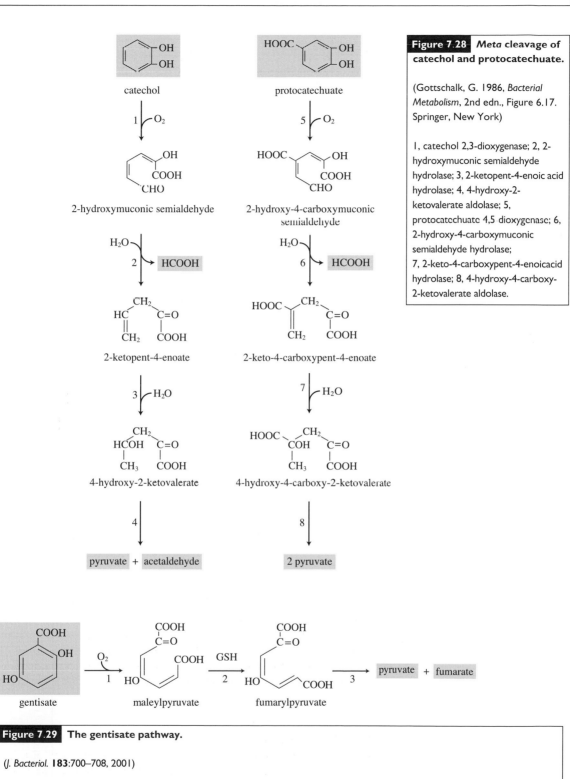

Figure 7.28 *Meta cleavage of catechol and protocatechuate.*

(Gottschalk, G. 1986, *Bacterial Metabolism*, 2nd edn., Figure 6.17. Springer, New York)

1, catechol 2,3-dioxygenase; 2, 2-hydroxymuconic semialdehyde hydrolase; 3, 2-ketopent-4-enoic acid hydrolase; 4, 4-hydroxy-2-ketovalerate aldolase; 5, protocatechuate 4,5 dioxygenase; 6, 2-hydroxy-4-carboxymuconic semialdehyde hydrolase; 7, 2-keto-4-carboxypent-4-enoicacid hydrolase; 8, 4-hydroxy-4-carboxy-2-ketovalerate aldolase.

Figure 7.29 **The gentisate pathway.**

(*J. Bacteriol.* **183**:700–708, 2001)

Naphthalene and other aromatic hydrocarbons are degraded through gentisate. Maleylpyruvate is converted to pyruvate and fumarate either directly or through fumarylpyruvate. 1, gentisate dioxygenase; 2, maleylpyruvate isomerase; 3, fumarylpyruvate hydrolase.

Table 7.2. | *Classification of methylotrophs according to their characteristics*

Physiological characteristics	Carbon assimilation pathway	Organisms	C1 compounds assimilated
Heterotrophic methylotrophy	serine–isocitrate lyase pathway	*Methylobacterium extorquens* AM1 *Pseudomonas* MA *Methylbacterium organophilum*	methanol, MMA, formate MMA methane, methanol
	ribulose monophosphate pathway	*Arthrobacter* P1 *Bacillus* PM6	MMA, DMA, TMA MMA, DMA, TMA, TMO, tetramethylammonium
Autotrophic methylotrophy	Calvin cycle	Group 1. Phototrophic *Rhodopseudomonas* spp.	methanol, CO, formate

		Group 2. Chemoautotrophic
		  Thiobacillus A2
		  Paracoccus denitrificans
		  Pseudomonas carboxydovorans
			methanol, MMA, formate
			methanol, MMA, formate, CO
			CO

		Group 3. *Pseudomonas oxalaticus*	formate
Obligate methyltrophy	serine–isocitrate lyase pathway	*Methylobacterium* spp. *Methylocystis* spp. *Methylomonas methanooxidans* *Methylosinus trichosporium*	methane methane methane, methanol methane, methanol
	ribulose monophosphate pathway	*Methylomonas methanica* *Methylophilus methylotrophus* *Methanococcus capsulatus*	methane, methanol methanol, MMA, DMA TMA

MMA, monomethylamine; DMA, dimethylamine; TMA, trimethylamine; TMO, trimethyl-N-oxide; CO, carbon monoxide.

Methanotrophs	Morphology	Flagella	Resting cell	Intracellular membrane structure	Carbon assimilation pathway	G + C content (%)
Methylomonas	rod	polar	cyst-like body	I	RMP pathway	50–54
Methylobacter	rod	polar	thick-walled cyst	II	RMP pathway	50–54
Methylococcus	coccus	none	cyst-like body	I	RMP pathway	62
Methylosinus	rod, vibrioid	polar tuft	exospore	II	SIL pathway	62–66
Methylocystis	vibrioid	none	PHB-rich cyst	II	SIL pathway	?
Methylobacterium	rod	none	none	II	SIL pathway	58–66

Table 7.3. | *Characteristics of methanotrophs*

I, multilayer membrane structure throughout the cell; II, double-layer membrane structure under the cell surface; RMP pathway, ribulose monophosphate pathway; SIL pathway, serine–isocitrate lyase pathway; PHB, polyhydroxybutyrate.

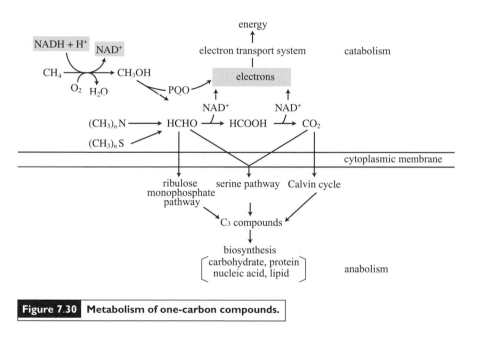

Figure 7.30 Metabolism of one-carbon compounds.

membrane structures as do the nitrifiers. This membrane structure is used in the classification of methanotrophs (Table 7.3). Spore and cyst forms of resting cells are known in all the obligate methylotrophs except species of *Methylobacterium*.

7.9.2.2 Dissimilation of methane by methanotrophs
Methane monooxygenase oxidizes methane to methanol using NADH as the cosubstrate (Figure 7.30). Electrons from methanol oxidation are channelled to the electron transport system for ATP

(a)

(b)

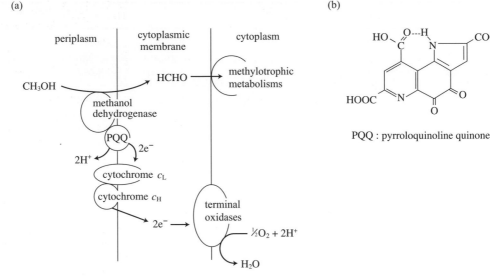

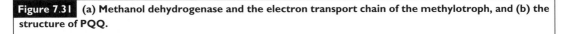

Figure 7.31 (a) Methanol dehydrogenase and the electron transport chain of the methylotroph, and (b) the structure of PQQ.

Cytochrome c_L: cytochrome c of low potential, cytochrome c_H: cytochrome c of high potential.

synthesis. Formaldehyde or carbon dioxide are assimilated for biosynthesis of cell materials.

Methanol dehydrogenase reduces pyrroloquinoline quinone (PQQ, Figure 7.31) coupled to the oxidation of methanol to formaldehyde. In addition to methanol dehydrogenase, PQQ with a redox potential of $+0.12\,V$ serves as the coenzyme of glucose dehydrogenase in *Pseudomonas putida* (Section 4.4.3) and alcohol dehydrogenase in acetic acid bacteria (Section 7.10.1).

Formaldehyde is oxidized to carbon dioxide either in free or in bound forms. Formaldehyde dehydrogenase and formate dehydrogenase oxidize formaldehyde in the free form to CO_2 via formate in Gram-positive methylotrophs such as *Amycolatopsis methanolica* and in the autotrophic methylotroph *Paracoccus denitrificans*.

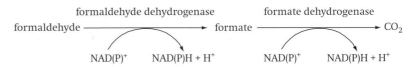

On the other hand, formaldehyde is bound to tetrahydrofolate (H_4F) or to tetrahydromethanopterin (H_4MTP) before being oxidized to CO_2 in many Gram-negative bacteria that convert formaldehyde to cell materials through the ribulose monophosphate or serine–isocitrate lyase pathway (Figure 7.32). H_4MTP is regarded as a methanogen-specific

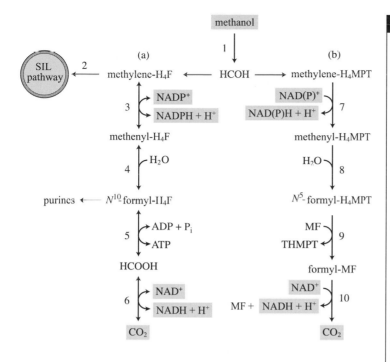

Figure 7.32 **Methanol oxidation by *Methylobacterium extorquens*.**

(*J. Bacteriol.* **180**:5351–5356, 1998)

This methylotrophic Gram-negative bacterium uses tetrahydrofolate (H_4F) and tetrahydromethanopterin (H_4MP) as C1 carriers, and has enzymes active on C1 compounds carried by H_4F and H_4MP. H_4MP (b) is the main C1 carrier in this bacterium while other methylotrophs use H_4F (a) as their main C1 carrier. 1, methanol dehydrogenase; 2, serine H_4F hydroxymethyltransferase; 3, methylene H_4F dehydrogenase; 4, methenyl H_4F cyclohydrolase; 5, formyl H_4F synthetase; 6, formate dehydrogenase; 7, methylene H_4MTP dehydrogenase; 8, methenyl H_4MTP cyclohydrolase; 9, formyl methanofuran H_4MTP formyltransferase; 10, formyl methanofuran (formyl-MF) dehydrogenase.

coenzyme (Section 9.4.2). *Methylobacterium extorquens* has both sets of enzymes that catalyze oxidation of C1 compounds bound either to H_4F or H_4MTP. Since the enzyme activities are higher for C1 compounds bound to H_4MTP, it is believed that formaldehyde is oxidized after being bound to H_4MTP in this bacterium (Figure 7.32b). H_4F is used in methylotrophs that employ the serine–isocitrate lyase pathway to convert formaldehyde to cell materials (Figure 7.32a).

Methylophilus methylotrophus does not possess enzymes that oxidize formaldehyde in the free form or bound to C1 carriers. This bacterium oxidizes formaldehyde to CO_2 through the ribulose monophosphate cycle (Figure 7.33).

7.9.3 Carbon assimilation by methylotrophs

Obligate methylotrophs cannot use multicarbon compounds as their carbon source. Obligate methylotrophs and heterotrophic methylotrophs employ either the ribulose monophosphate (RMP) pathway to assimilate formaldehyde or the serine–isocitrate lyase (SIL) pathway to assimilate formaldehyde and CO_2. Autotrophic methylotrophs fix CO_2 through the Calvin cycle. Methylotrophic yeasts growing on methanol employ yet another novel pathway, the xylulose monophosphate (XMP) pathway.

7.9.3.1 Ribulose monophosphate (RMP) pathway

The RMP pathway that assimilates formaldehyde as triose phosphate is a collection of four different pathways all sharing the first two reactions. RMP accepts formaldehyde to form hexulose-6-phosphate that is isomerized to fructose-6-phosphate (Figure 7.34).

Figure 7.33 Oxidation of formaldehyde through the ribulose monophosphate cycle in *Methylophilus methylotrophus.*

1, hexulose phosphate synthase; 2, hexulose-6-phosphate isomerase; 3, fructose-6-phosphate isomerase; 4, glucose-6-phosphate dehydrogenase; 5, 6-phosphogluconate dehydrogenase.

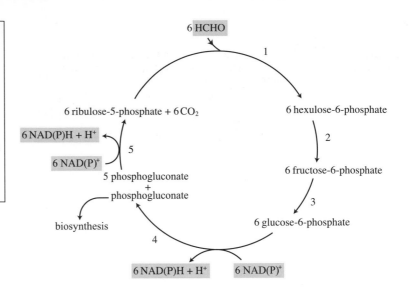

Fructose-6-phosphate is cleaved through two alternative routes, one involving fructose-1,6-diphosphate aldolase (RMP-EMP variant, Figure 7.34a) and the other 2-keto-3-deoxy-6-phosphogluconate aldolase (RMP-ED variant, Figure 7.34b). The resulting triose phosphate and pyruvate are used in assimilatory metabolism. The remaining triose phosphate is used to regenerate ribulose-5-phosphate through carbon rearrangement with two molecules of fructose-6-phosphate. As shown in Figure 7.35, the carbon rearrangement takes place in two different ways depending on the organism: one involves transaldolase and transketolase (TA variant, Figure 7.35a) while the other involves fructose-1,6-diphosphate aldolase (FDA variant, Figure 7.35b).

Each variant of RMP pathway can be summarized as:

EMP–TA variant:	$3HCHO + ATP \rightarrow$ glyceraldehyde-3-phosphate
EMP–FDA variant:	$3HCHO + 2ATP \rightarrow$ dihydroxyacetone-phosphate
ED–TA variant:	$3HCHO \rightarrow$ pyruvate $+ NAD(P)H$
ED–FDA variant:	$3HCHO + ATP \rightarrow$ pyruvate $+ NAD(P)H$

7.9.3.2 Serine–isocitrate lyase (SIL) pathway

Formaldehyde is the source of all the carbons of triose phosphate synthesized through the RMP pathway, but *Methylosinus trichosporium* uses formaldehyde and CO_2 to synthesize 2-phosphoglycerate via acetyl-CoA. This metabolic pathway is referred to as the serine–isocitrate lyase (SIL) pathway (Figure 7.36). SIL can be considered in two parts: (1) acetyl-CoA synthesis from formaldehyde and CO_2 through 2-phosphoglycerate and (2) acetyl-CoA conversion to serine via glyoxylate (Figure 7.37).

Formaldehyde forms methylene-H_4F with tetrahydrofolate (H_4F) before condensing with glycine to serine. An aminotransferase converts

(a)

(b)

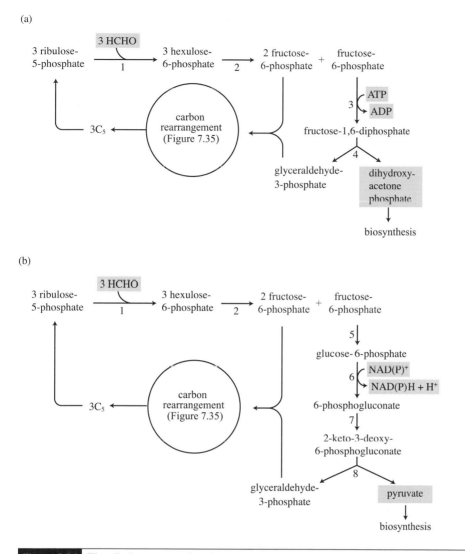

Figure 7.34 **The ribulose monophosphate pathway.**

The carbon rearrangement reactions in the circles are shown in Figure 7.35. (a) RMP – Modified EMP pathway. (b) RMP – Modified ED pathway. 1, hexulose phosphate synthase; 2, hexulose phosphate isomerase; 3, phosphofructokinase; 4, fructose-1,6-diphosphate aldolase; 5, hexose phosphate isomerase; 6, glucose-6-phosphate dehydrogenase; 7, 6-phosphogluconate dehydratase; 8, 2-keto-3-deoxy-6-phosphogluconate aldolase.

serine to hydroxypyruvate coupling the amination of glyoxylate to glycine. Hydroxypyruvate is reduced to PEP to be carboxylated to oxaloacetate catalyzed by PEP carboxylase. Oxaloacetate is reduced and activated to malyl-CoA to be cleaved to glyoxylate and acetyl-CoA. Acetyl-CoA is oxidized to glyoxylate in the next round of reactions (Figure 7.37).

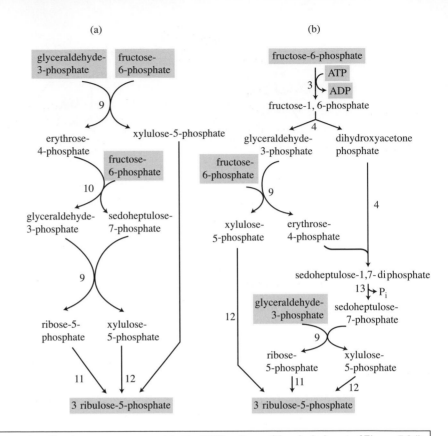

(a)   (b)

Figure 7.35 Two alternative routes of carbon rearrangement in the RMP pathway (the circled part of Figure 7.34).

(a) TA variant involving transketolase and transaldolase. (b) FDA variant involving transketolase and fructose-1,6-diphosphatase aldolase. 3, 4, as in Figure 7.34; 9, transketolase; 10, transaldolase; 11, ribose-5-phosphate isomerase; 12, ribulose-5-phosphate-3-epimerase.

Each step of the SIL pathway can be summarized as:

$$HCHO + CO_2 + 2ATP + 2NAD(P)H + 2H^+ + CoASH \longrightarrow$$
$$CH_3CO\text{-}CoA + 2ADP + 2P_i + 2NAD(P)^+$$

$$CH_3CO\text{-}CoA + NAD^+ + FAD + 2H_2O \longrightarrow$$
$$CHO-COOH + NADH + H^+ + FADH_2$$

$$CHO-COOH + HCHO + NAD(P)H + H^+ + ATP \longrightarrow$$
$$3\text{-phosphoglycerate} + ADP + NAD(P)^+$$

sum: $$2HCHO + CO_2 + 3ATP + 2NAD(P)H + 2H^+ + FAD \longrightarrow$$
$$3\text{-phosphoglycerate} + 2NAD(P)^+ + FADH_2 + 3ADP + 2P_i + H_2O$$

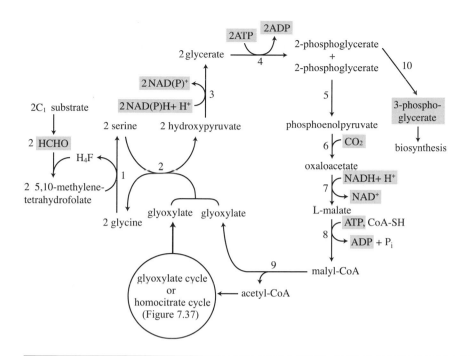

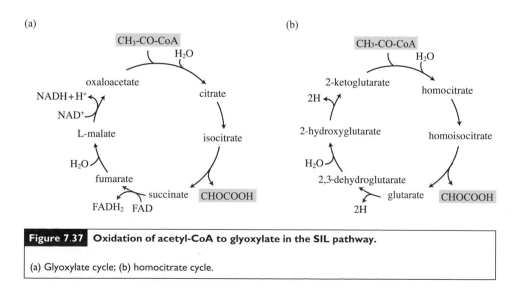

Figure 7.36 **The serine-isocitrate lyase (SIL) pathway.**

The SIL pathway is divided into two steps for convenience. These are (1) acetyl-CoA synthesis from formaldehyde and CO_2 through 2-phosphoglycerate and (2) acetyl-CoA conversion to serine via glyoxylate. The circle shows acetyl-CoA oxidation to glyoxylate as detailed in Figure 7.37.

1, serine hydroxymethyltransferase; 2, serine:glyoxylate aminotransferase; 3, hydroxypyruvate reductase; 4, glycerate kinase; 5, PEP hydratase; 6, PEP carboxylase; 7, malate dehydrogenase; 8, malyl-CoA synthetase; 9, malyl-CoA lyase; 10, phosphoglycerate mutase.

H_4F, tetrahydrofolate.

Figure 7.37 **Oxidation of acetyl-CoA to glyoxylate in the SIL pathway.**

(a) Glyoxylate cycle; (b) homocitrate cycle.

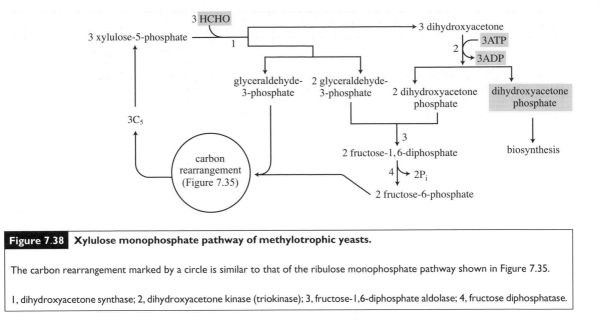

Figure 7.38 **Xylulose monophosphate pathway of methylotrophic yeasts.**

The carbon rearrangement marked by a circle is similar to that of the ribulose monophosphate pathway shown in Figure 7.35.

1, dihydroxyacetone synthase; 2, dihydroxyacetone kinase (triokinase); 3, fructose-1,6-diphosphate aldolase; 4, fructose diphosphatase.

7.9.3.3 Xylulose monophosphate (XMP) pathway

Methylotrophic yeasts assimilate methanol through the xylulose monophosphate (XMP) pathway (Figure 7.38) that is different from the bacterial C1 metabolic pathways. Dihydroxyacetone phosphate is synthesized from formaldehyde in this pathway for biosynthetic purposes.

Dihydroxyacetone synthase condenses formaldehyde with xylulose-5-phosphate to produce glyceraldehyde-3-phosphate and dihydroxyacetone. Dihydroxyacetone synthase is a kind of transketolase that uses formaldehyde as its substrate. Since dihydroxyacetone is an important intermediate, this metabolism is alternatively referred to as the dihydroxyacetone pathway. Dihydroxyacetone is phosphorylated to dihydroxyacetone phosphate by triokinase. One third of the triose phosphate is used for biosynthesis while the remaining molecules condense with equivalent molecules of glyceraldehyde-3-phosphate to produce fructose-6-phosphate. Carbon rearrangement is similar to that of the RMP pathway (Figure 7.35), and generates three molecules of xylulose-5-phosphate as the formaldehyde acceptor in the next round of the pathway from two molecules of fructose-6-phosphate, and a molecule of glyceraldehyde-3-phosphate. The XMP pathway can be summarized as:

$$3\text{HCHO} + 3\text{ATP} \longrightarrow \text{dihydroxyacetone phosphate} + 3\text{ADP} + 2\text{P}_i$$

Methane utilizers are not known in eukaryotes and methanol utilization is restricted to a few yeast strains including species of *Candida*, *Hansenula* and *Torulopsis*. Methanol oxidation is catalyzed by methanol oxidase producing hydrogen peroxide.

Table 7.4. | *Energy efficiency in C1 assimilation*

Pathway	Variant	Substrate	Electron carriers and ATP		
			NAD(P)H	FADH$_2$	ATP
Calvin cycle		3CO$_2$	−5	0	−7
RMP pathway	EMP–TA	3HCHO	+1	0	+1
	EMP–FDA	3HCHO	+1	0	0
	ED–TA	3HCHO	+1	0	0
	ED–FDA	3HCHO	+1	0	−3
SIL pathway		2HCHO + CO$_2$	−2	+1	−2
XMP pathway		3HCHO	+1	0	−1

+, generated; −, consumed.

$$CH_3OH + O_2 \xrightarrow{\text{methanol oxidase}} HCHO + H_2O_2$$

Methanol oxidase and catalase are located in a specific organelle in such yeasts known as the peroxisome.

7.9.4 Energy efficiency in C1 metabolism

Microorganisms assimilate C1 compounds through the RMP, SIL and XMP pathways, and the Calvin cycle. However, energy requirements as ATP and reduced electron carriers differ between the pathways. Since the starting materials and the products are different, calculations can be made normalizing pyruvate as the final product (Table 7.4).

It is seen that the EMP–TA variant of the RMP pathway is the most efficient while the least efficient is the Calvin cycle.

7.10 | Incomplete oxidation

During heterotrophic metabolism under aerobic conditions, part of the substrate is converted into cell materials and the remainder is oxidized to carbon dioxide, providing energy for growth. Certain microbes excrete metabolic intermediates in large quantities due to a lack of enzymes for complete oxidation or because enzymes are repressed under the given conditions. Examples are acetate production by acetic acid bacteria and acetoin and butanediol production by members of the *Bacillus* genus.

7.10.1 Acetic acid bacteria

Acetic acid bacteria are aerobic bacteria that produce acetate from ethanol. When ethanol is completely consumed, species of *Acetobacter* utilize acetate while acetate is not consumed by species belonging to the genus *Gluconobacter*.

As discussed earlier (Section 4.3.2.2) *Gluconobacter* spp. metabolize sugar through the oxidative HMP cycle, and the resulting glyceraldehyde-3-phosphate is oxidized to acetate via pyruvate and acetyl-CoA. Since succinate dehydrogenase is absent in this genus, acetyl-CoA cannot be oxidized through the TCA cycle. An incomplete TCA pathway (Section 5.4.1) operates to supply precursors for biosynthesis where PEP is carboxylated to oxaloacetate.

All members of acetic acid bacteria including *Gluconobacter oxydans* and *Acetobacter aceti* oxidize ethanol to acetate.

$$CH_3-CH_2OH + PQQ \xrightarrow{\text{alcohol dehydrogenase}} CH_3-CHO + PQQH_2$$

ethanol   acetaldehyde

$$CH_3-CHO + PQQ \xrightarrow{\text{aldehyde dehydrogenase}} CH_3-COOH + PQQH_2$$

As shown in the reactions, these enzymes are quinoproteins containing pyrroloquinoline quinone (PQQ). Electrons from reduced PQQ are transferred to coenzyme Q of the bacterial electron transport chain. The mid-point redox potential of PQQ is $+0.12\,V$, which is much higher than that of $NAD^+/NADH$ which is $-0.34\,V$ (Section 7.9.2.2).

Since *Gluconobacter oxydans* is unable to synthesize succinate dehydrogenase (a TCA cycle enzyme), this bacterium cannot oxidize acetate to CO_2. On the other hand, TCA cycle enzymes are repressed in *Acetobacter aceti* and acetate is accumulated when the ethanol concentration is high. When ethanol is exhausted, genes for the TCA cycle enzymes are expressed in *Acetobacter aceti* in order to utilize the acetate. This bacterium requires amino acids to grow on ethanol and acetate, probably due to a limited ability to generate carbon compounds for biosynthesis (Figure 7.39).

Acetic acid bacteria oxidize various alcohols and ketones. This property is exploited for oxidation of D-sorbitol to L-sorbose in the ascorbic acid production process.

7.10.2 Acetoin and butanediol

The expression of TCA cycle enzymes is repressed in *Bacillus* species, including *Bacillus polymyxa*, *Bacillus subtilis* and others, growing on carbohydrates. They do not oxidize the substrate completely during vegetative cell growth, accumulating acetoin and 2,3-butanediol (Figure 7.40). This metabolism is similar to the fermentation in facultative anaerobic enteric bacteria (Section 8.6.2). The enteric bacteria ferment carbohydrates to these compounds under anaerobic conditions, while the *Bacillus* spp. do this under aerobic conditions.

When carbohydrate is exhausted, *Bacillus* species sporulate using acetoin and 2,3-butanediol as energy sources through the TCA cycle

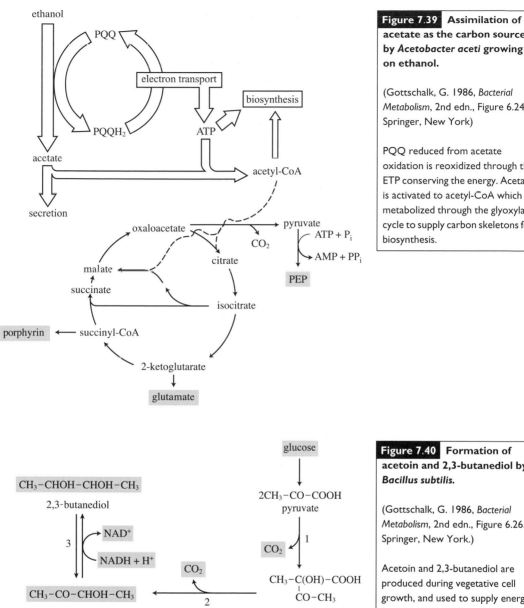

Figure 7.39 Assimilation of acetate as the carbon source by *Acetobacter aceti* growing on ethanol.

(Gottschalk, G. 1986, *Bacterial Metabolism*, 2nd edn., Figure 6.24. Springer, New York)

PQQ reduced from acetate oxidation is reoxidized through the ETP conserving the energy. Acetate is activated to acetyl-CoA which is metabolized through the glyoxylate cycle to supply carbon skeletons for biosynthesis.

Figure 7.40 Formation of acetoin and 2,3-butanediol by *Bacillus subtilis.*

(Gottschalk, G. 1986, *Bacterial Metabolism*, 2nd edn., Figure 6.26. Springer, New York.)

Acetoin and 2,3-butanediol are produced during vegetative cell growth, and used to supply energy in the sporulation process.

1, 2-acetolactate synthase;
2, 2-acetolactate decarboxylase;
3, 2,3-butanediol dehydrogenase.

(Section 7.4). 2,3-butanediol is a compound of industrial interest as a raw material with two functional groups for polymer synthesis.

7.10.3 Other products of aerobic metabolism

Coryneform bacteria such as *Corynebacterium glutamicum* and *Brevibacterium flavum* excrete glutamate in large quantities under aerobic nutrient-rich conditions. The enzymes of the TCA cycle have low activities in this case. This is exploited industrially to produce monosodium glutamate.

FURTHER READING

Depolymerization of polymers

Bae, H. J., Turcotte, G., Chamberland, H., Karita, S. & Vezina, L. P. (2003). A comparative study between an endoglucanase IV and its fused protein complex Cel5-CBM6. *FEMS Microbiology Letters* **227**, 175–181.

Ballschmiter, M., Armbrecht, M., Ivanova, K., Antranikian, G. & Liebl, W. (2005). AmyA, an α-amylase with β-cyclodextrin-forming activity, and AmyB from the thermoalkaliphilic organism *Anaerobranca gottschalkii*: two α-amylases adapted to their different cellular localizations. *Applied and Environmental Microbiology* **71**, 3709–3715.

Bauer, M. W., Driskill, L. E. & Kelly, R. M. (1998). Glycosyl hydrolase from thermophilic microorganisms. *Current Opinion in Biotechnology* **9**, 141–145.

Bertoldo, C. & Antranikian, G. (2002). Starch-hydrolyzing enzymes from thermophilic archaea and bacteria. *Current Opinion in Chemical Biology* **6**, 151–160.

Bertoldo, C., Armbrecht, M., Becker, F., Schafer, T., Antranikian, G. & Liebl, W. (2004). Cloning, sequencing, and characterization of a heat- and alkali-stable type I pullulanase from *Anaerobranca gottschalkii*. *Applied and Environmental Microbiology* **70**, 3407–3416.

Bhat, S. & Owen, E. (2001). Isolation and characterisation of a major cellobiohydrolase (S-8) and a major endoglucanase (S-11) subunit from the cellulosome of *Clostridium thermocellum*. *Anaerobe* **7**, 171–179.

Collins, T., Gerday, C. & Feller, G. (2005). Xylanases, xylanase families and extremophilic xylanases. *FEMS Microbiology Reviews* **29**, 3–23.

Dahiya, N., Tewari, R. & Hoondal, G. S. (2006). Biotechnological aspects of chitinolytic enzymes: a review. *Applied Microbiology and Biotechnology* **71**, 773–782.

Desvaux, M. (2005). The cellulosome of *Clostridium cellulolyticum*. *Enzyme and Microbial Technology* **37**, 373–385.

Felse, P. A. & Panda, T. (1999). Regulation and cloning of microbial chitinase genes. *Applied and Environmental Microbiology* **51**, 141–151.

Folders, J., Algra, J., Roelofs, M. S., van Loon, L. C., Tommassen, J. & Bitter, W. (2001). Characterization of *Pseudomonas aeruginosa* chitinase, a gradually secreted protein. *Journal of Bacteriology* **183**, 7044–7052.

Gao, J., Bauer, M. W., Shockley, K. R., Pysz, M. A. & Kelly, R. M. (2003). Growth of hyperthermophilic archaeon *Pyrococcus furiosus* on chitin involves two family 18 chitinases. *Applied and Environmental Microbiology* **69**, 3119–3128.

Jayani, R. S., Saxena, S. & Gupta, R. (2005). Microbial pectinolytic enzymes: a review. *Process Biochemistry* **40**, 2931–2944.

Kang, S., Vieille, C. & Zeikus, J. G. (2005). Identification of *Pyrococcus furiosus* amylopullulanase catalytic residues. *Applied Microbiology and Biotechnology* **66**, 408–413.

Khalikova, E., Susi, P. & Korpela, T. (2005). Microbial dextran-hydrolyzing enzymes: fundamentals and applications. *Microbiology and Molecular Biology Reviews* **69**, 306–325.

Kong, H., Shimosaka, M., Ando, Y., Nishiyama, K., Fujii, T. & Miyashita, K. (2001). Species-specific distribution of a modular family 19 chitinase gene in *Burkholderia gladioli*. *FEMS Microbiology Ecology* **37**, 135–141.

Lee, H.-S., Shockley, K. R., Schut, G. J., Conners, S. B., Montero, C. I., Johnson, M. R., Chou, C.-J., Bridger, S. L., Wigner, N., Brehm, S. D., Jenney, F. E.,

Comfort, D. A. & Kelly, R. M. (2006). Transcriptional and biochemical analysis of starch metabolism in the hyperthermophilic archaeon *Pyrococcus furiosus*. *Journal of Bacteriology* **188**, 2115–2125.

Lin, F. P. & Leu, K. L. (2002). Cloning, expression, and characterization of thermostable region of amylopullulanase gene from *Thermoanaerobacter ethanolicus* 39E. *Applied Biochemistry and Biotechnology* **97**, 33–44.

Lynd, L. R., Weimer, P. J., van Zyl, W. H. & Pretorius, I. S. (2002). Microbial cellulose utilization: fundamentals and biotechnology. *Microbiology and Molecular Biology Reviews* **66**, 506–577.

Murashima, K., Kosugi, A. & Doi, R. H. (2002). Determination of subunit composition of *Clostridium cellulovorans* cellulosomes that degrade plant cell walls. *Applied and Environmental Microbiology* **68**, 1610–1615.

Murashima, K., Kosugi, A. & Doi, R. H. (2002). Synergistic effects on crystalline cellulose degradation between cellulosomal cellulases from *Clostridium cellulovorans*. *Journal of Bacteriology* **184**, 5088–5095.

Schrempf, H. (2001). Recognition and degradation of chitin by streptomycetes. *Antonie Van Leeuwenhoek* **79**, 285–289.

Vasella, A., Davies, G. J. & Bohm, M. (2002). Glycosidase mechanisms. *Current Opinion in Chemical Biology* **6**, 619–629.

Watanabe, K. (2004). Collagenolytic proteases from bacteria. *Applied Microbiology and Biotechnology* **63**, 520–526.

Zhang, Y. H. P. & Lynd, L. R. (2004). Kinetics and relative importance of phosphorolytic and hydrolytic cleavage of cellodextrins and cellobiose in cell extracts of *Clostridium thermocellum*. *Applied and Environmental Microbiology* **70**, 1563–1569.

Zverlov, V. V., Velikodvorskaya, G. A. & Schwarz, W. H. (2002). A newly described cellulosomal cellobiohydrolase, CelO, from *Clostridium thermocellum*: investigation of the exo-mode of hydrolysis, and binding capacity to crystalline cellulose. *Microbiology-UK* **148**, 247–255.

Carbohydrate and related compound utilization

Bausch, C., Ramsey, M. & Conway, T. (2004). Transcriptional organization and regulation of the L-idonic acid pathway (GntII system) in *Escherichia coli*. *Journal of Bacteriology* **186**, 1388–1397.

Organic acid utilization

Abo-Amer, A. E., Munn, J., Jackson, K., Aktas, M., Golby, P., Kelly, D. J. & Andrews, S. C. (2004). DNA interaction and phosphotransfer of the C4-dicarboxylate-responsive DcuS-DcuR two-component regulatory system from *Escherichia coli*. *Journal of Bacteriology* **186**, 1879–1889.

Aoshima, M., Ishii, M. & Igarashi, Y. (2004). A novel enzyme, citryl-CoA lyase, catalysing the second step of the citrate cleavage reaction in *Hydrogenobacter thermophilus* TK-6. *Molecular Microbiology* **52**, 763–770.

Bramer, C. O. & Steinbuchel, A. (2001). The methylcitric acid pathway in *Ralstonia eutropha*: new genes identified involved in propionate metabolism. *Microbiology-UK* **147**, 2203–2214.

Brasen, C. & Schonheit, P. (2001). Mechanisms of acetate formation and acetate activation in halophilic archaea. *Archives of Microbiology* **175**, 360–368.

Brasen, C. & Schonheit, P. (2004). Unusual ADP-forming acetyl-coenzyme A synthetases from the mesophilic halophilic euryarchaeon *Haloarcula marismortui* and from the hyperthermophilic crenarchaeon *Pyrobaculum aerophilum*. *Archives of Microbiology* **182**, 277–287.

Braun, V., Mahren, S. & Ogierman, M. (2003). Regulation of the FecI-type ECF sigma factor by transmembrane signalling. *Current Opinion in Microbiology* **6**, 173–180.

Claes, W. A., Puhler, A. & Kalinowski, J. (2002). Identification of two *prpDBC* gene clusters in *Corynebacterium glutamicum* and their involvement in propionate degradation via the 2-methylcitrate cycle. *Journal of Bacteriology* **184**, 2728–2739.

Dimroth, P. & Hilbi, H. (1997). Enzymic and genetic basis for bacterial growth on malonate. *Molecular Microbiology* **25**, 3–10.

Ensign, S. A. (2006). Revisiting the glyoxylate cycle: alternate pathways for microbial acetate assimilation. *Molecular Microbiology* **61**, 274–276.

Gerstmeir, R., Wendisch, V. F., Schnicke, S., Ruan, H., Farwick, M., Reinscheid, D. & Eikmanns, B. J. (2003). Acetate metabolism and its regulation in *Corynebacterium glutamicum*. *Journal of Biotechnology* **104**, 99–122.

Ingram-Smith, C., Martin, S. R. & Smith, K. S. (2006). Acetate kinase: not just a bacterial enzyme. *Trends in Microbiology* **14**, 249–253.

Kretzschmar, U., Ruckert, A., Jeoung, J. H. & Gorisch, H. (2002). Malate: quinone oxidoreductase is essential for growth on ethanol or acetate in *Pseudomonas aeruginosa*. *Microbiology-UK* **148**, 3839–3847.

Lau, W. W. Y. & Armbrust, E. V. (2006). Detection of glycolate oxidase gene *glcD* diversity among cultured and environmental marine bacteria. *Environmental Microbiology* **8**, 1688–1702.

Lewis, J. A. & Escalante-Semerena, J. C. (2006). The FAD-dependent tricarballylate dehydrogenase (TcuA) enzyme of *Salmonella enterica* converts tricarballylate into cis-aconitate. *Journal of Bacteriology* **188**, 5479–5486.

Meister, S., Saum, M., Alber, B. E. & Fuchs, G. (2005). L-malyl-coenzyme A/β-methylmalyl-coenzyme A lyase is involved in acetate assimilation of the isocitrate lyase-negative bacterium *Rhodobacter capsulatus*. *Journal of Bacteriology* **187**, 1415–1425.

Munoz-Elias, E. J., Upton, A. M., Cherian, J. & McKinney, J. D. (2006). Role of the methylcitrate cycle in *Mycobacterium tuberculosis* metabolism, intracellular growth, and virulence. *Molecular Microbiology* **60**, 1109–1122.

Palacios, S. & Escalante-Semerena, J. C. (2004). 2-methylcitrate-dependent activation of the propionate catabolic operon (prpBCDE) of *Salmonella enterica* by the PrpR protein. *Microbiology-UK* **150**, 3877–3887.

Palacios, S., Starai, V. J. & Escalante-Semerena, J. C. (2003). Propionyl coenzyme A is a common intermediate in the 1,2-propanediol and propionate catabolic pathways needed for expression of the *prpBCDE* operon during growth of *Salmonella enterica* on 1,2-propanediol. *Journal of Bacteriology* **185**, 2802–2810.

Sahin, N. (2003). Oxalotrophic bacteria. *Research in Microbiology* **154**, 399–407.

Wolfe, A. J. (2005). The acetate switch. *Microbiology and Molecular Biology Reviews* **69**, 12–50.

Alcohol utilization

Ali, N. O., Bignon, J., Rapoport, G. & Debarbouille, M. (2001). Regulation of the acetoin catabolic pathway is controlled by sigma L in *Bacillus subtilis*. *Journal of Bacteriology* **183**, 2497–2504.

Anthony, C. & Ghosh, M. (1997). The structure and function of PQQ-containing quinoproteins. *Current Science* **72**, 716–727.

Chinnawirotpsan, P., Matsushita, K., Toyama, H., Adachi, O., Limtong, S. & Theeragool, G. (2003). Purification and characterization of two NAD-dependent alcohol dehydrogenases (ADHs) induced in the quinoprotein ADH-deficient mutant of *Acetobacter pasteurianus* SKU1108. *Bioscience, Biotechnology, and Biochemistry* **67**, 958–965.

de Faveri, D., Torre, P., Molinari, F. & Converti, A. (2003). Carbon material balances and bioenergetics of 2,3-butanediol bio-oxidation by *Acetobacter hansenii*. *Enzyme and Microbial Technology* **33**, 708–719.

Havemann, G. D. & Bobik, T. A. (2003). Protein content of polyhedral organelles involved in coenzyme B12-dependent degradation of 1,2-propanediol in *Salmonella enterica* serovar *typhimurium* LT2. *Journal of Bacteriology* **185**, 5086–5095.

Havemann, G. D., Sampson, E. M. & Bobik, T. A. (2002). PduA is a shell protein of polyhedral organelles involved in coenzyme B_{12}-dependent degradation of 1,2-propanediol in *Salmonella enterica* serovar *typhimurium* LT2. *Journal of Bacteriology* **184**, 1253–1261.

Hosaka, T., Ohtsuki, T., Mimura, A. & Ohkuma, M. (2001). Characterization of the NADH-linked acetylacetoin reductase/2,3-butanediol dehydrogenase gene from *Bacillus cereus* YUF-4. *Journal of Bioscience and Bioengineering* **91**, 539–544.

Kotani, T., Yamamoto, T., Yurimoto, H., Sakai, Y. & Kato, N. (2003). Propane monooxygenase and NAD^+-dependent secondary alcohol dehydrogenase in propane metabolism by *Gordonia* sp. strain TY-5. *Journal of Bacteriology* **185**, 7120–7128.

Matsushita, K., Toyama, H., Yamada, M. & Adachi, O. (2002). Quinoproteins: structure, function, and biotechnological applications. *Applied Microbiology and Biotechnology* **58**, 13–22.

Peng, X., Taki, H., Komukai, S., Sekine, M., Kanoh, K., Kasai, H., Choi, S. K., Omata, S., Tanikawa, S., Harayama, S. & Misawa, N. (2006). Characterization of four *Rhodococcus* alcohol dehydrogenase genes responsible for the oxidation of aromatic alcohols. *Applied Microbiology and Biotechnology* **71**, 824–832.

Sher, J., Elevi, R., Mana, L. & Oren, A. (2004). Glycerol metabolism in the extremely halophilic bacterium *Salinibacter ruber*. *FEMS Microbiology Letters* **232**, 211–215.

Tachibana, S., Kuba, N., Kawai, F., Duine, J. A. & Yasuda, M. (2003). Involvement of a quinoprotein (PQQ-containing) alcohol dehydrogenase in the degradation of polypropylene glycols by the bacterium *Stenotrophomonas maltophilia*. *FEMS Microbiology Letters* **218**, 345–349.

Amino acid and nucleic acid base utilization

Colabroy, K. L. & Begley, T. P. (2005). Tryptophan catabolism: identification and characterization of a new degradative pathway. *Journal of Bacteriology* **187**, 7866–7869.

Gruber, K. & Kratky, C. (2002). Coenzyme B_{12} dependent glutamate mutase. *Current Opinion in Chemical Biology* **6**, 598–603.

Hoschle, B., Gnau, V. & Jendrossek, D. (2005). Methylcrotonyl-CoA and geranyl-CoA carboxylases are involved in leucine/isovalerate utilization (Liu) and acyclic terpene utilization (Atu), and are encoded by liuB/liuD and atuC/atuF, in *Pseudomonas aeruginosa*. *Microbiology-UK* **151**, 3649–3656.

Lan, J. & Newman, E. B. (2003). A requirement for anaerobically induced redox functions during aerobic growth of *Escherichia coli* with serine, glycine and leucine as carbon source. *Research in Microbiology* **154**, 191–197.

Hydrocarbon utilization

Basu, A., Apte, S. K. & Phale, P. S. (2006). Preferential utilization of aromatic compounds over glucose by *Pseudomonas putida* CSV86. *Applied and Environmental Microbiology* **72**, 2226–2230.

Borzenkov, I., Milekhina, E., Gotoeva, M., Rozanova, E. & Belyaev, S. (2006). The properties of hydrocarbon-oxidizing bacteria isolated from the oil-fields of Tatarstan, western Siberia, and Vietnam. *Microbiology-Moscow* **75**, 66–72.

Corvini, P., Schaeffer, A. & Schlosser, D. (2006). Microbial degradation of nonylphenol and other alkylphenols? Our evolving view. *Applied Microbiology and Biotechnology* **72**, 223–243.

Danko, A. S., Saski, C. A., Tomkins, J. P. & Freedman, D. L. (2006). Involvement of coenzyme M during aerobic biodegradation of vinyl chloride and ethene by *Pseudomonas putida* strain AJ and *Ochrobactrum* sp. strain TD. *Applied and Environmental Microbiology* **72**, 3756–3758.

Ensign, S. A., Small, F. J., Allen, J. R. & Sluis, M. K. (1998). New roles for CO_2 in the microbial metabolism of aliphatic epoxides and ketones. *Archives of Microbiology* **169**, 179–187.

Fetzner, S. (2002). Oxygenases without requirement for cofactors or metal ions. *Applied Microbiology and Biotechnology* **60**, 243–257.

Fujihara, H., Yoshida, H., Matsunaga, T., Goto, M. & Furukawa, K. (2006). Cross-regulation of biphenyl- and salicylate-catabolic genes by two regulatory systems in *Pseudomonas pseudoalcaligenes* KF707. *Journal of Bacteriology* **188**, 4690–4697.

Funhoff, E. G., Bauer, U., Garcia-Rubio, I., Witholt, B. & van Beilen, J. B. (2006). CYP153A6, a soluble P450 oxygenase catalyzing terminal-alkane hydroxylation. *Journal of Bacteriology* **188**, 5220–5227.

Gescher, J., Zaar, A., Mohamed, M., Schagger, H. & Fuchs, G. (2002). Genes coding for a new pathway of aerobic benzoate metabolism in *Azoarcus evansii*. *Journal of Bacteriology* **184**, 6301–6315.

Harwood, C. S. & Parales, R. E. (1996). The beta-ketoadipate pathway and the biology of self-identity. *Annual Review of Microbiology* **50**, 553–590.

Itoh, S. (2006). Mononuclear copper active-oxygen complexes. *Current Opinion in Chemical Biology* **10**, 115–122.

Jimenez, J. I., Minambres, B., Garcia, J. L. & Diaz, E. (2002). Genomic analysis of the aromatic catabolic pathways from *Pseudomonas putida* KT2440. *Environmental Microbiology* **4**, 824–841.

Johnson, E. L. & Hyman, M. R. (2006). Propane and *n*-butane oxidation by *Pseudomonas putida* GPo1. *Applied and Environmental Microbiology* **72**, 950–952.

Leahy, J. G., Batchelor, P. J. & Morcomb, S. M. (2003). Evolution of the soluble diiron monooxygenases. *FEMS Microbiology Reviews* **27**, 449–479.

Mattevi, A. (2006). To be or not to be an oxidase: challenging the oxygen reactivity of flavoenzymes. *Trends in Biochemical Sciences* **31**, 276–283.

Rosenberg, M. (2006). Microbial adhesion to hydrocarbons: twenty-five years of doing MATH. *FEMS Microbiology Letters* **262**, 129–134.

van Beilen, J. B., Neuenschwander, M., Smits, T. H. M., Roth, C., Balada, S. B. & Witholt, B. (2002). Rubredoxins involved in alkane oxidation. *Journal of Bacteriology* **184**, 1722–1732.

van Berkel, W. J. H., Kamerbeek, N. M. & Fraaije, M. W. (2006). Flavoprotein monooxygenases, a diverse class of oxidative biocatalysts. *Journal of Biotechnology* **124**, 670–689.

van Hamme, J. D., Singh, A. & Ward, O. P. (2003). Recent advances in petroleum microbiology. *Microbiology and Molecular Biology Reviews* **67**, 503–549.

Vangnai, A. S., Sayavedra-Soto, L. A. & Arp, D. J. (2002). Roles for the two 1-butanol dehydrogenases of *Pseudomonas butanovora* in butane and 1-butanol metabolism. *Journal of Bacteriology* **184**, 4343–4350.

Incomplete oxidation

Adachi, O., Moonmangmee, D., Toyama, H., Yamada, M., Shinagawa, E. & Matsushita, K. (2003). New developments in oxidative fermentation. *Applied Microbiology and Biotechnology* **60**, 643–653.

Holscher, T. & Gorisch, H. (2006). Knockout and overexpression of pyrrolo-quinoline quinone biosynthetic genes in *Gluconobacter oxydans* 621H. *Journal of Bacteriology* **188**, 7668–7676.

Keliang, G. & Dongzhi, W. (2006). Asymmetric oxidation by *Gluconobacter oxydans*. *Applied Microbiology and Biotechnology* **70**, 135–139.

Methyltrophy

Chistoserdova, L., Chen, S. W., Lapidus, A. & Lidstrom, M. E. (2003). Methylotrophy in *Methylobacterium extorquens* AM1 from a genomic point of view. *Journal of Bacteriology* **185**, 2980–2987.

Chistoserdova, L., Rasche, M. E. & Lidstrom, M. E. (2005). Novel dephosphote-trahydromethanopterin biosynthesis genes discovered via mutagenesis in *Methylobacterium extorquens* AM1. *Journal of Bacteriology* **187**, 2508–2512.

Choi, D. W., Kunz, R. C., Boyd, E. S., Semrau, J. D., Antholine, W. E., Han, J.-I., Zahn, J. A., Boyd, J. M., dela Mora, A. M. & Dispirito, A. A. (2003). The membrane-associated methane monooxygenase (pMMO) and pMMO-NADH:quinone oxidoreductase complex from *Methylococcus capsulatus* Bath. *Journal of Bacteriology* **185**, 5755–5764.

Dedysh, S. N., Smirnova, K. V., Khmelenina, V. N., Suzina, N. E., Liesack, W. & Trotsenko, Y. A. (2005). Methylotrophic autotrophy in *Beijerinckia mobilis*. *Journal of Bacteriology* **187**, 3884–3888.

Kelly, D. P. & Murrell, J. C. (1999). Microbial metabolism of methanesulfonic acid. *Archives of Microbiology* **172**, 341–348.

Kelly, D. P., Anthony, C. & Murrell, J. C. (2005). Insights into the obligate methanotroph *Methylococcus capsulatus*. *Trends in Microbiology* **13**, 195–198.

McDonald, I. R., Miguez, C. B., Rogge, G., Bourque, D., Wendlandt, K. D., Groleau, D. & Murrell, J. C. (2006). Diversity of soluble methane monooxygenase-containing methanotrophs isolated from polluted environments. *FEMS Microbiology Letters* **255**, 225–232.

Murrell, J. C. (1992). Genetics and molecular biology of methanotrophs. *FEMS Microbiology Reviews* **88**, 233–248.

Orita, I., Yurimoto, H., Hirai, R., Kawarabayasi, Y., Sakai, Y. & Kato, N. (2005). The archaeon *Pyrococcus horikoshii* possesses a bifunctional enzyme for formaldehyde fixation via the ribulose monophosphate pathway. *Journal of Bacteriology* **187**, 3636–3642.

Orita, I., Sato, T., Yurimoto, H., Kato, N., Atomi, H., Imanaka, T. & Sakai, Y. (2006). The ribulose monophosphate pathway substitutes for the missing pentose phosphate pathway in the archaeon *Thermococcus kodakaraensis*. *Journal of Bacteriology* **188**, 4698–4704.

Schubert, C. J., Coolen, M. J. L., Neretin, L. N., Schippers, A., Abbas, B., Durisch-Kaiser, E., Wehrli, B., Hopmans, E. C., Damste, J. S. S., Wakeham, S. & Kuypers, M. M. M. (2006). Aerobic and anaerobic methanotrophs in the Black Sea water column. *Environmental Microbiology* **8**, 1844–1856.

Theisen, A. R., Ali, M. H., Radajewski, S., Dumont, M. G., Dunfield, P. F., McDonald, I. R., Dedysh, S. N., Miguez, C. B. & Murrell, J. C. (2005). Regulation of methane oxidation in the facultative methanotroph *Methylocella silvestris* BL2. *Molecular Microbiology* **58**, 682–692.

Vorholt, J. A. (2002). Cofactor-dependent pathways of formaldehyde oxidation in methylotrophic bacteria. *Archives of Microbiology* **178**, 239–249.

Wood, A. P., Aurikko, J. P. & Kelly, D. P. (2004). A challenge for 21st century molecular biology and biochemistry: what are the causes of obligate autotrophy and methanotrophy? *FEMS Microbiology Reviews* **28**, 335–352.

Other metabolisms

Carterson, A. J., Morici, L. A., Jackson, D. W., Frisk, A., Lizewski, S. E., Jupiter, R., Simpson, K., Kunz, D. A., Davis, S. H., Schurr, J. R., Hassett, D. J. & Schurr, M. J. (2004). The transcriptional regulator AlgR controls cyanide production in *Pseudomonas aeruginosa*. *Journal of Bacteriology* **186**, 6837–6844.

Charoenpanich, J., Tani, A., Moriwaki, N., Kimbara, K. & Kawai, F. (2006). Dual regulation of a polyethylene glycol degradative operon by AraC-type and GalR-type regulators in *Sphingopyxis macrogoltabida* strain 103. *Microbiology-UK* **152**, 3025–3034.

Ciferri, O. (1999). Microbial degradation of paintings. *Applied and Environmental Microbiology* **65**, 879–885.

Coleman, N. V. & Spain, J. C. (2003). Epoxyalkane:coenzyme M transferase in the ethene and vinyl chloride biodegradation pathways of *Mycobacterium* strain JS60. *Journal of Bacteriology* **185**, 5536–5545.

Ebbs, S. (2004). Biological degradation of cyanide compounds. *Current Opinion in Biotechnology* **15**, 231–236.

Hawari, J., Beaudet, S., Halasz, A., Thiboutot, S. & Ampleman, G. (2000). Microbial degradation of explosives: biotransformation versus mineralization. *Applied Microbiology and Biotechnology* **54**, 605–618.

Hirota-Mamoto, R., Nagai, R., Tachibana, S., Yasuda, M., Tani, A., Kimbara, K. & Kawai, F. (2006). Cloning and expression of the gene for periplasmic poly(vinyl alcohol) dehydrogenase from *Sphingomonas* sp. strain 113P3, a novel-type quinohaemoprotein alcohol dehydrogenase. *Microbiology-UK* **152**, 1941–1949.

Jendrossek, D. & Handrick, R. (2002). Microbial degradation of polyhydroxyalkanoates. *Annual Review of Microbiology* **56**, 403–432.

Karpouzas, D. G. & Singh, B. K. (2006). Microbial degradation of organophosphorus xenobiotics: metabolic pathways and molecular basis. *Advances in Microbial Physiology* **51**, 119–225.

Kutsu-Shigeno, Y., Adachi, Y., Yamada, C., Toyoshima, K., Nomura, N., Uchiyama, H. & Nakajima-Kambe, T. (2006). Isolation of a bacterium that degrades urethane compounds and characterization of its urethane hydrolase. *Applied Microbiology and Biotechnology* **70**, 422–429.

Mooney, A., Ward, P. G. & O'Connor, K. E. (2006). Microbial degradation of styrene: biochemistry, molecular genetics, and perspectives for biotechnological applications. *Applied Microbiology and Biotechnology* **72**, 1–10.

Ohta, T., Tani, A., Kimbara, K. & Kawai, F. (2005). A novel nicotinoprotein aldehyde dehydrogenase involved in polyethylene glycol degradation. *Applied Microbiology and Biotechnology* **68**, 639–646.

Pessi, G. & Haas, D. (2000). Transcriptional control of the hydrogen cyanide biosynthetic genes *hcnABC* by the anaerobic regulator ANR and the quorum-sensing regulators LasR and RhlR in *Pseudomonas aeruginosa*. *Journal of Bacteriology* **182**, 6940–6949.

Shimao, M. (2001). Biodegradation of plastics. *Current Opinion in Biotechnology* **12**, 242–247.

Urgun-Demirtas, M., Stark, B. & Pagilla, K. (2006). Use of genetically engineered microorganisms (GEMs) for the bioremediation of contaminants. *Critical Reviews in Biotechnology* **26**, 145–164.

Vaillancourt, F. H., Haro, M. A., Drouin, N. M., Karim, Z., Maaroufi, H. & Eltis, L. D. (2003). Characterization of extradiol dioxygenases from a polychlorinated biphenyl-degrading strain that possesses higher specificities for chlorinated metabolites. *Journal of Bacteriology* **185**, 1253–1260.

Vimr, E. R., Kalivoda, K. A., Deszo, E. L. & Steenbergen, S. M. (2004). Diversity of microbial sialic acid metabolism. *Microbiology and Molecular Biology Reviews* **68**, 132–153.

8

Anaerobic fermentation

Anaerobic conditions are maintained in some ecosystems where the rate of oxygen supply is lower than that of consumption. Organic compounds are removed from anaerobic ecosystems through the concerted action of fermentative and anaerobic respiratory microorganisms. In microbiology, the term 'fermentation' can be used to describe either microbial processes that produce useful products or a form of anaerobic microbial growth using internally supplied electron acceptors and generating ATP mainly through substrate-level phosphorylation (SLP).

8.1 | Electron acceptors used in anaerobic metabolism

8.1.1 Fermentation and anaerobic respiration

Respiration refers to the reduction of oxygen by electrons from the electron transport chains coupled to the generation of a proton motive force through electron transport phosphorylation (ETP; Section 5.8). Under anaerobic conditions, some microorganisms grow using an ETP process with externally supplied oxidized compounds other than oxygen as the terminal electron acceptor. This type of growth is referred to as anaerobic respiration. In a fermentative process, ATP is generated through SLP with the oxidation of electron donors coupled to the reduction of electron carriers such as $NAD(P)^+$ or flavin adenine dinucleotide (FAD). The reduced electron carriers are reoxidized reducing the metabolic intermediate.

This chapter describes the fermentation processes carried out by various anaerobic prokaryotes. In fermentation, ATP is generated not only through SLP but also by other mechanisms such as the reactions catalyzed by fumarate reductase and Na^+-dependent decarboxylase, and lactate/H^+ symport as described earlier (Section 5.8.6).

8.1.2 Hydrogen in fermentation

The product formed/substrate consumed ratio is constant in some fermentations such as the ethanol (Section 8.3) and homolactate

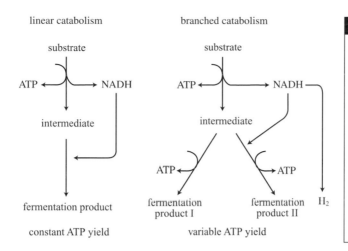

Figure 8.1 Linear and branched pathways in anaerobic fermentation.

(Bacteriol. Rev. 41:100–180, 1977)

Ethanol fermentation is an example of a linear fermentative pathway, and more than one product is formed in the acetate-butyrate fermentation by clostridia. In a branched fermentative pathway, more ATP can be generated when protons are used as the electron acceptor.

(Section 8.4) fermentations while it is variable in others such as the clostridial fermentation (Section 8.5). Fermentation processes with a constant ratio are referred to as linear pathways, while the others are referred to as branched fermentative pathways (Figure 8.1). A branched pathway yields more ATP and more oxidized products than a linear pathway. To produce more oxidized products, a proportion of the reduced electron carriers such as NAD(P)H should be oxidized coupled to the reduction of H^+ to H_2. The formation of products in a branched pathway is dictated by the environmental growth conditions, especially the hydrogen partial pressure.

8.2 | Molecular oxygen and anaerobes

As discussed in Chapter 2 (Section 2.2.2), microbes are classified according to their response to molecular oxygen (O_2) into aerobes, facultative anaerobes and obligate anaerobes. Aerotolerant obligate anaerobes are distinguished from strict anaerobes among the obligate anaerobes. O_2 comprises about 20% of air, and air-saturated liquid media contains about 7–8 mg/l O_2 at ambient temperature. Microaerophiles use O_2 as their electron acceptor at a low dissolved O_2 concentration of 0.1–3 mg/l O_2, but they cannot grow above this concentration. Strict anaerobes and microaerophiles are inhibited by molecular oxygen or its metabolites. Several hypotheses have been proposed to explain the inhibitory mechanism.

Molecular oxygen reacts with reduced flavoproteins, Fe-S proteins and cytochromes to be reduced to hydrogen peroxide (H_2O_2) or superoxide ($O_2^{\cdot -}$). These are very strong oxidants with high redox potentials $[E^{0'}(O_2^-/H_2O_2) = +0.98\,V,\ E^{0'}(H_2O_2/H_2O) = +1.35\,V]$ and destroy cellular polymers such as DNA, RNA, proteins and other essential components. Aerobes and facultative

Table 8.1. | *Superoxide dismutase and catalase activities in selected prokaryotes*

Organisms	Specific activity (U/mg)	
	Superoxide dismutase	Catalase
Aerobes and facultative anaerobes		
Escherichia coli	1.8	6.1
Salmonella typhimurium	1.4	2.4
Rhizobium japonicum	2.6	0.7
Micrococcus radiodurans	7.0	289.0
Pseudomonas species	2.0	22.5
Aerotolerant anaerobes		
Eubacterium limosum	11.6	0
Enterococcus (Streptococcus) faecalis	0.8	0
Lactococcus (Streptococcus) lactis	1.4	0
Clostridium oroticum	0.6	0
Lactobacillus plantarum	0	0
Strict anaerobes		
Veillonella alcalescens	0	0
Clostridium pasteurianum	0	0
Clostridium butyricum[a]	1.4	0
Clostridium sticklandii	0	0
Butyrivibrio fibrisolvens	0	0.1
Methanosarcina barkeri	+[b]	40.0
Desulfovibrio gigas	3.4	52.6

[a] A strong peroxidase activity (5.8 U/mg) is found in this bacterium.
[b] Superoxide dismutase has been purified in this archaeon.

anaerobes therefore possess enzymes which detoxify them. These enzymes are superoxide dismutase (SOD) and catalase (Table 8.1). Peroxidase is another enzyme that removes hydrogen peroxide. The reactions are:

$$2O_2^{\cdot-} + 2H^+ \xrightarrow{\text{SOD}} H_2O_2 + O_2$$

$$2H_2O_2 \xrightarrow{\text{catalase}} 2H_2O + O_2$$

$$H_2O_2 + RH_2 \xrightarrow{\text{peroxidase}} 2H_2O + R$$

Strict anaerobes have been thought to be sensitive against O_2 because they do not possess SOD and catalase. This is true in some cases, but these enzyme activities have been identified in some strict anaerobes, and genes for these enzymes and related proteins have been found in their genomes. For example, SOD was found in *Clostridium butyricum*. Methanogenic archaea are one of the most O_2-sensitive groups of organisms. *Methanosarcina barkeri* possesses SOD and catalase activities, and the SOD gene has been identified in *Methanobacterium bryantii* and *Methanobacterium thermoautotrophicum*. Sulfate-reducing bacteria are less sensitive to O_2 than methanogens, and *Desulfovibrio gigas*, *Desulfovibrio vulgaris* and other *Desulfovibrio* species have SOD and catalase activities.

In addition to SOD and catalase, species of *Desulfovibrio* have proteins that detoxify reactive oxygen species including superoxide. These are Fe-containing electron carriers such as rubredoxin, desulfoferredoxin, neelaredoxin and rubrerythrin. A similar protein to neelaredoxin in an archaeon, *Archaeoglobus fulgidus*, functions as a SOD, and another similar protein, superoxide reductase, destroys superoxide directly to water, and not through O_2, in *Pyrococcus furiosus*.

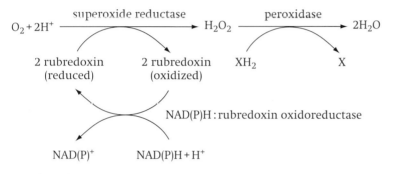

Though some strict anaerobes have SOD and catalase activities (Table 8.1), they do not grow under aerobic conditions in the laboratory. This may be due to the nature of the molecular oxygen. Dissolved oxygen increases the redox potential of the solution and a high redox potential inhibits the growth of some strict anaerobes. Methanogens grow at a redox potential lower than $-0.3\,V$. Sulfide is an essential component of some enzymes, and molecular oxygen oxidizes this to form a disulfide. The organisms might not be able to grow with such inactivated enzymes. One hypothesis proposes that growth is impossible due to a lack of reducing equivalents for biosynthesis since electrons are exhausted to reduce oxygen. It is most likely that multiple mechanisms are responsible for growth inhibition by oxygen.

8.3 | Ethanol fermentation

Saccharomyces cerevisiae ferments carbohydrates through the EMP pathway to ethanol, and the ED pathway is used by *Zymomonas mobilis*. Pyruvate is decarboxylated to acetaldehyde, which is used as the electron acceptor. Acetaldehyde is reduced to ethanol, which consumes the electrons generated during the glycolytic process where ATP is generated through SLP (Figure 8.2). *Saccharomyces cerevisiae*

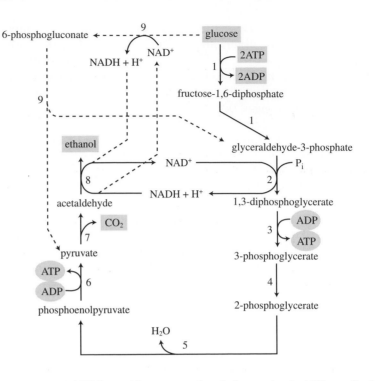

Figure 8.2 Ethanol fermentation by *Saccharomyces cerevisiae* and *Zymomonas mobilis*.

1–6, EMP pathway (in solid lines); 7, pyruvate decarboxylase; 8, alcohol dehydrogenase; 9, ED pathway (in dotted lines).

generates 2 ATP from 1 hexose molecule but a single ATP results from 1 hexose molecule in *Zymomonas mobilis*.

$$CH_3-CO-COOH \xrightarrow{\text{pyruvate decarboxylase}} CH_3-CHO + CO_2$$

$$CH_3-CHO + NADH + H^+ \xrightarrow{\substack{\text{alcohol} \\ \text{dehydrogenase}}} CH_3-CH_2OH + NAD^+$$

Pyruvate decarboxylase has thiamine pyrophosphate as a prosthetic group as in pyruvate dehydrogenase. Pyruvate decarboxylase is known mainly in eukaryotes. In addition to *Zymomonas mobilis*, this enzyme is found in a facultative anaerobe, *Erwinia amylovora*, and in a strictly anaerobic acidophile, *Sarcina ventriculi*. Pyruvate decarboxylase is a key enzyme of ethanol fermentation.

It should be noted that ethanol is produced through different reactions in saccharolytic clostridia, heterofermentative lactic acid bacteria and enteric bacteria. These bacteria oxidize pyruvate to acetyl-CoA before reducing it to ethanol. They do not possess pyruvate decarboxylase. Ethanol production in clostridia is catalyzed by the following reactions:

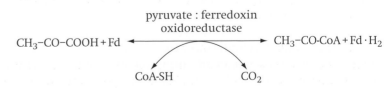

$$\text{CH}_3\text{-CO-CoA} + \text{NADH} + \text{H}^+ \underset{\text{dehydrogenase}}{\overset{\text{aldehyde}}{\rightleftharpoons}} \text{CH}_3\text{-CHO} + \text{NAD}^+$$

$$\text{CH}_3\text{-CHO} + \text{NADH} + \text{H}^+ \overset{\text{alcohol dehydrogenase}}{\rightleftharpoons} \text{CH}_3\text{-CH}_2\text{OH} + \text{NAD}^+$$

Ethanol fermentation through pyruvate decarboxylase in a linear fermentative pathway does not produce any by-products except CO_2 and water, while ethanol fermentation through acetyl-CoA is a branched fermentative pathway and produces various fermentation products such as lactate, acetate and H_2. Thermophilic anaerobes ferment various carbohydrates including cellulose and pentoses through acetyl-CoA to ethanol. Among them are *Thermoanaerobium brockii*, *Thermoanaerobacter ethanolicus* and *Clostridium thermocellum* (Section 8.5.1.5).

8.4 | Lactate fermentation

Lactate is a common fermentation product in many facultative and obligate anaerobes. Some bacteria produce lactate as a major fermentation product and these are referred to as lactic acid bacteria (LAB). Most LAB have a limited ability to synthesize monomers for biosynthesis and vitamins which are needed as growth factors. LAB are regarded as obligate anaerobes but they can use oxygen, synthesizing cytochromes when hemin is provided in the medium. Some LAB produce only lactate from sugars while others produce acetate and ethanol in addition to lactate (Table 8.2). The former are referred to as homofermentative and the latter heterofermentative LAB. Homofermentative LAB ferment sugars through the EMP pathway and heterofermentative LAB ferment sugars through the phosphoketolase pathway (Section 4.5).

8.4.1 Homolactate fermentation

Homofermentative LAB include most species of *Lactobacillus*, *Sporolactobacillus*, *Pediococcus*, *Enterococcus* and *Lactococcus*. They use hexoses through the EMP pathway to generate ATP. Lactate dehydrogenase reoxidizes the NADH reduced during the EMP pathway using pyruvate as the electron acceptor (Figure 8.3). As fermentation proceeds, lactate is accumulated lowering the intracellular pH. Lactate dehydrogenase is active in acidic conditions producing lactate as the major product. Under alkaline conditions, homofermentative LAB produce large quantities of acetate and ethanol.

8.4.2 Heterolactate fermentation

Species of *Leuconostoc* and *Bifidobacterium* produce ethanol and acetate in addition to lactate. They employ a unique glycolytic pathway

	Fermentation mode	
Strain	Homofermentative	Heterofermentative
Lactobacillus		
L. delbrueckii	+	−
L. lactis	+	−
L. bulgaricus	+	−
L. casei	+	−
L. curvantus	+	−
L. plantarum	+	−
L. brevis	−	+
L. fermentum	−	+
Sporolactobacillus		
S. inulinus	+	−
Enterococcus		
E. faecalis	+	−
Lactococcus		
L. cremoris	−	+
L. lactis	+	−
Leuconostoc		
L. mesenteroides	−	+
L. dextranicum	−	+
Pediococcus		
P. damnosus	+	−
Bifidobacterium[a]		
B. bifidum	−	+

Table 8.2. *Representative LAB and their fermentation mode*

[a] Phylogenetically, these bacteria are classified apart from lactic acid bacteria.

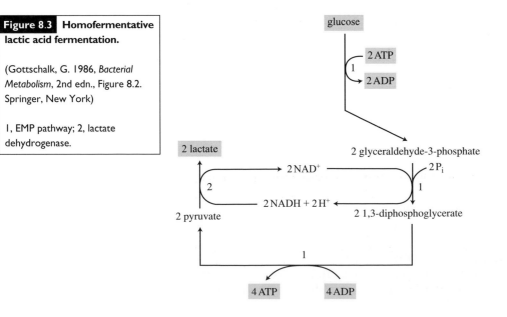

Figure 8.3 **Homofermentative lactic acid fermentation.**

(Gottschalk, G. 1986, *Bacterial Metabolism*, 2nd edn., Figure 8.2. Springer, New York)

1, EMP pathway; 2, lactate dehydrogenase.

known as the phosphoketolase pathway (Section 4.5). As shown in Figure 4.9, heterofermentative LAB like *Leuconostoc mesenteroides* oxidize glucose-6-phosphate to ribulose-5-phosphate. Epimerase converts ribulose-5-phosphate to xylulose-5-phosphate, before cleavage to glyceraldehyde-3-phosphate and acetyl-phosphate by the action of phosphoketolase. Glyceraldehyde-3-phosphate is metabolized to lactate as in the homolactate fermentation generating ATP. Acetyl-phosphate is reduced to ethanol acting as the electron acceptor to oxidize the NADH reduced in the glucose-6-phosphate oxidation process. One ATP per hexose is available from this fermentation.

Pentoses are converted to xylulose-5-phosphate without reducing NAD^+. In this case, acetyl-phosphate is not used as the electron acceptor but is used to synthesize ATP. *Leuconostoc mesenteroides* synthesizes 1 ATP from a molecule of hexose and 2 ATP from a molecule of pentose.

Bifidobacterium bifidum ferments 2 molecules of hexose to 2 molecules of lactate and 3 molecules of acetate employing two separate phosphoketolases, one active on fructose-6-phosphate and the other on xylulose-5-phosphate (Figure 4.10, Section 4.5). This bacterium synthesizes 5 ATP from 2 molecules of glucose. Since hexose-6-phosphate is not metabolized through a reductive process, acetyl-phosphate is used to synthesize ATP as in pentose metabolism by *Leuconostoc mesenteroides*.

8.4.3 Biosynthesis in latic acid bacteria (LAB)

LAB require many amino acids, vitamins, nucleic acid bases and other substances as growth factors because they cannot synthesize them. However, they can synthesize a few monomers and polymers for biosynthesis from acetyl-CoA. Acetyl-CoA or acetyl-phosphate are not intermediates in homolactate fermentation. Pyruvate is oxidized to acetyl-CoA through different routes depending on the strain. The pyruvate dehydrogenase multienzyme complex oxidizes pyruvate to acetyl-CoA in most species of *Lactococcus* and *Enterococcus* as in aerobic bacteria. *Enterococcus faecalis*, *Bifidobacterium bifidum* and *Lactobacillus casei* rely on pyruvate:formate lyase, an enzyme found in anaerobic metabolism in facultative anaerobic enteric bacteria (Section 8.6):

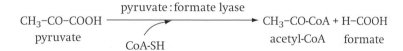

$$CH_3-CO-COOH \xrightarrow[\text{CoA-SH}]{\text{pyruvate : formate lyase}} CH_3-CO\text{-}CoA + H\text{-}COOH$$

pyruvate → acetyl-CoA formate

Pyruvate oxidase and phosphotransacetylase convert pyruvate to acetyl-CoA in other LAB including *Lactobacillus delbrueckii* and *Lactobacillus plantarum*:

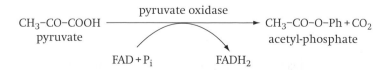

$$CH_3-CO-COOH \xrightarrow[\text{FAD} + P_i \quad \text{FADH}_2]{\text{pyruvate oxidase}} CH_3-CO-O-Ph + CO_2$$

pyruvate → acetyl-phosphate

$$CH_3\text{-}CO\text{-}O\text{-}Ph + CoA\text{-}SH \xrightarrow{\text{phosphotransacetylase}} CH_3\text{-}CO\text{-}CoA + P_i$$

$FADH_2$ is reoxidized by lactate dehydrogenase:

$$\text{pyruvate} + FADH_2 \xrightarrow[\text{dehydrogenase}]{\text{lactate}} \text{lactate} + FAD$$

8.4.4 Oxygen metabolism in LAB

LAB are aerotolerant obligate anaerobes and possess superoxide dismutase and peroxidase to detoxify superoxide ($O_2^{\cdot-}$) and hydrogen peroxide (H_2O_2), respectively. Peroxidase generates sulfite radicals from sulfite which are lethal to cells. Sulfite can therefore be used for the selective elimination of LAB from peroxidase-negative aerobic organisms.

$$H_2O_2 + XH_2 \, (NADH + H^+) \xrightarrow{\text{peroxidase}} 2H_2O + X \, (NAD^+)$$

Catalase activity is negative but becomes positive in some LAB such as *Enterococcus faecalis*, *Lactobacillus brevis* and *Lactobacillus plantarum* when hemin is provided. These organisms have the genetic information to synthesize the catalase apoprotein but not hemin. The $Y_{glucose}$ of *Enterococcus faecalis* is 22 g under anaerobic conditions but increases up to 52 g under aerobic conditions. This is because *Enterococcus faecalis* can synthesize ATP through an electron transport phosphorylation (ETP) process using O_2 as an electron acceptor.

In addition to ETP, peroxidase (or oxidase) increases the cell yield, oxidizing NADH with O_2 as electron acceptor. When (per)oxidase oxidizes NADH in heterofermentative LAB, extra ATP can be synthesized through substrate-level phosphorylation from acetyl-phosphate that is not needed as an electron acceptor. This reaction is catalyzed by acetate kinase.

Similarly, homofermentative *Lactococcus lactis* oxidizes pyruvate to acetate or condenses it to acetoin instead of using it as an electron acceptor under aerobic conditions when H_2O-forming NADH oxidase oxidizes NADH to NAD^+ reducing O_2.

8.4.5 Lactate/H^+ symport

When the intracellular lactate concentration is higher than that of the medium, lactate is exported with $2H^+$ generating a proton motive force in *Lactococcus cremoris* (Figure 5.28, Section 5.8.6.3).

8.4.6 LAB in fermented food

Dairy products and fermented vegetables, such as *kimchi*, in far eastern countries are typical LAB fermented foods. In addition to lactate production, LAB produce flavours specific for each food.

Citrate is produced in the initial phase of soybean sauce fermentation and converted to acetate in the later stages by LAB. *Pediococcus halophilus*, isolated from maturing soybean sauce, ferments citrate to acetate and formate (Figure 8.4). In this fermentation, citrate lyase

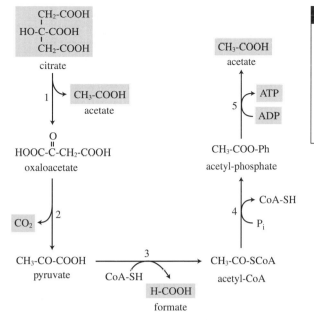

cleaves citrate to acetate and oxaloacetate (OAA). OAA is decarboxylated to pyruvate. It is not known if energy is conserved during the decarboxylation reaction of OAA in this bacterium, and this reaction is not coupled to proton (sodium) motive force generation as in other LAB. Since pyruvate is not needed to dispose of electrons, it is converted to acetyl-CoA through a reaction catalyzed by pyruvate:formate lyase. The acetyl-CoA is used to synthesize ATP. Some enteric bacteria including *Salmonella typhimurium* and *Klebsiella pneumoniae* ferment citrate through a similar metabolism (Figure 8.14, Section 8.6.3). Citrate fermentation by *Pediococcus halophilus* can be summarized as:

$$\text{citrate} + \text{ADP} + P_i \rightarrow 2\ \text{acetate} + \text{formate} + \text{ATP}$$

Milk contains about 1.5 g/l citrate, which is converted to diacetyl, acetoin, acetate and lactate during the butter fermentation by LAB such as *Lactococcus lactis* subsp. *diacetylactis*, *Leuconostoc cremoris*, *Leuconostoc oenos* and *Leuconostoc mesenteroides* (Figure 8.5). Citrate is metabolized to acetate and pyruvate as in *Pediococcus halophilus*. When the NADH/NAD^+ ratio is high, pyruvate is reduced to lactate, and acetoin is produced through 2-acetolactate when the ratio is low. A similar reaction is found in some *Bacillus* species (Figure 7.40, Section 7.10.2) and in enteric bacteria (Figure 8.12, Section 8.6.2). ATP is synthesized by acetate kinase during this fermentation, but energy is not conserved in the decarboxylation of OAA. A proton motive force is generated through citrate/lactate exchange, which is a similar transport system to the malate/lactate exchange (Section 5.8.6).

Lactobacillus plantarum, *Lactobacillus casei*, *Leuconostoc mesenteroides*, *Leuconostoc oenos* and *Lactococcus lactis* convert malate to lactate in a process known as the malolactic fermentation:

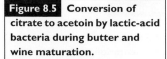

Figure 8.5 Conversion of citrate to acetoin by lactic-acid bacteria during butter and wine maturation.

1, citrate lyase; 2, oxaloacetate decarboxylase; 3, lactate dehydrogenase; 4, 2-acetolactate synthase; 5, 2-acetolactate decarboxylase; 6, spontaneous chemical reaction; 7, acetoin dehydrogenase; 8, pyruvate dehydrogenase complex; 9, phosphotransacetylase; 10, acetate kinase.

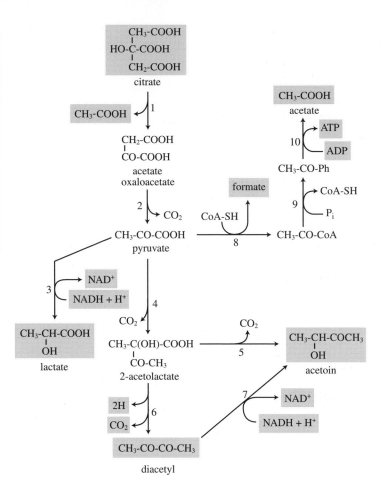

$$malate^{2-} \longrightarrow lactate^- + CO_2$$

Since these bacteria cannot use malate as their carbon and energy source, the malolactic fermentation is possible in the presence of fermentable substrates such as glucose. A proton motive force is generated in this fermentation through H^+ consumption in the carboxylation of malate to lactate and through $malate^{2-}/lactate^-$ exchange (Section 5.8.6).

Glycerol is dehydrated to 3-hydroxypropionaldehyde that serves as an electron acceptor in heterofermentative *Lactobacillus* species. Since these bacteria preferentially use this electron acceptor, pyruvate is oxidized to acetate, synthesizing ATP, through the reaction catalyzed by acetate kinase.

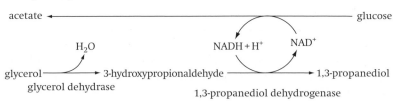

Consequently, the growth yield of heterofermentative LAB is higher on glucose with glycerol than on glucose alone. A similar fermentation is found in the metabolism of glycerol by enteric bacteria (Figure 8.13, Section 8.6.2).

8.5 | Butyrate and acetone–butanol–ethanol fermentations

Spore-forming anaerobic Gram-positive bacteria without sulfate-reducing ability are classified as the genus *Clostridium*. They are divided into saccharolytic and proteolytic clostridia according to their preferred electron donor. Saccharolytic clostridia ferment carbohydrates to butyrate and acetate, and proteinaceous compounds are fermented by proteolytic clostridia. The latter organisms are mostly pathogenic. Clostridial fermentation is a typical branched fermentative pathway. In addition to clostridia, species belonging to the genera *Butyrivibrio*, *Eubacterium* and *Fusobacterium* also produce butyrate (Table 8.3). All of these are obligate anaerobes.

8.5.1 Butyrate fermentation

Clostridium butyricum transports glucose by group translocation before metabolizing it to pyruvate. Pyruvate is oxidized to acetyl-CoA through a reaction known as a phosphoroclastic reaction catalyzed by pyruvate:ferredoxin oxidoreductase. This is different from the reaction catalyzed by pyruvate dehydrogenase (Figure 5.1). Hydrogenase can oxidize the reduced ferredoxin in this reaction to produce H_2:

$$\text{ferredoxin} \cdot H_2 \xrightleftharpoons[]{\text{hydrogenase}} \text{ferredoxin} + H_2$$

8.5.1.1 Phosphoroclastic reaction

Pyruvate:ferredoxin oxidoreductase contains thiamine pyrophosphate as a prosthetic group, like pyruvate dehydrogenase (Section 5.1), and catalyzes the following reactions:

$$CH_3\text{–}CO\text{–}COOH + \text{TPP-enzyme} \rightleftharpoons CH_3\text{–}CHOH\text{–}\text{TPP-enzyme} + CO_2$$

$$CH_3\text{–}CHOH\text{–}\text{TPP-enzyme} + \text{ferredoxin} + \text{CoA-SH} \rightleftharpoons$$
$$\text{TPP-enzyme} + CH_3\text{–}CO\text{-CoA} + \text{ferredoxin} \cdot H_2$$

As shown above, the phosphoroclastic reaction is reversible, and this enzyme can reduce acetyl-CoA to pyruvate while pyruvate dehydrogenase does not catalyze the reduction of acetyl-CoA. Ferredoxin used in the phosphoroclastic reaction has a redox potential similar to that of the acetyl-CoA $+ CO_2$/pyruvate half reaction, while the pyridine nucleotide (NAD^+/$NADH + H^+$) has a higher redox potential:

Table 8.3.	Anaerobes producing butyrate as their main fermentation product

Butyribacterium methylotrophicum
Butyrivibrio fibrisolvens
Clostridium butyricum
C. kluyveri
C. pasteurianum
Eubacterium limosum
Fusobacterium nucleatum

$2H^+/H_2$	$E^{0'} = -0.41\,V$
Fd(oxidized)/Fd(reduced)	$E^{0'} = -0.41\,V$
$NAD^+/NADH + H^+$	$E^{0'} = -0.32\,V$
Acetyl-CoA + CO_2/pyruvate	$E^{0'} = -0.42\,V$

Since ferredoxin has a redox potential similar to that of the $2H^+/H_2$ half reaction, hydrogenase activity to oxidize the reduced ferredoxin depends on the hydrogen partial pressure, and so does the phosphoroclastic reaction. Ferredoxin is a [Fe-S] protein (Figure 5.17, Section 5.8.2.2).

8.5.1.2 Butyrate formation

Acetyl-CoA produced from the phosphoroclastic reaction is metabolized either to acetate through acetyl-phosphate or to butyrate through acetoacetyl-CoA (Figure 8.6). The NADH reduced during glycolysis is oxidized, reducing acetoacetyl-CoA to butyrate. When the hydrogen partial pressure is low, hydrogenase oxidizes reduced ferredoxin producing H_2. Under this condition, the ferredoxin (oxidized)/ferredoxin(reduced) ratio is high, and NADH: ferredoxin oxidoreductase couples NADH oxidation to ferredoxin reduction. Since NADH is oxidized through these reactions, acetyl-CoA is not needed as the electron acceptor, and it is converted to acetyl-phosphate on which acetate kinase reacts to synthesize ATP.

$$NADH + H^+ \underset{\text{oxidoreductase}}{\overset{\text{NADH:ferredoxin}}{\rightleftharpoons}} Fd \cdot H_2 \overset{\text{hydrogenase}}{\rightleftharpoons} H_2$$

In an undisturbed culture of Clostridium butyricum, 100 mol glucose is fermented to 76 mol butyrate and 42 mol acetate. However, when H_2 is continuously removed by shaking, the butyrate/acetate ratio becomes 1. This is due to the fact that the equilibrium shifts to the right in the above reactions. Since kinases synthesize ATP from acetyl-phosphate

Table 8.4. | *Fermentation balance in* Clostridium butyricum *(mmol/100 mmol substrate consumed)*

	Lactate + acetate	Pyruvate	Glucose
Acetate	-32^a	33	42
Butyrate	65	33	76
CO_2	100	93	188
H_2	59	30	235

[a] – means consumption.

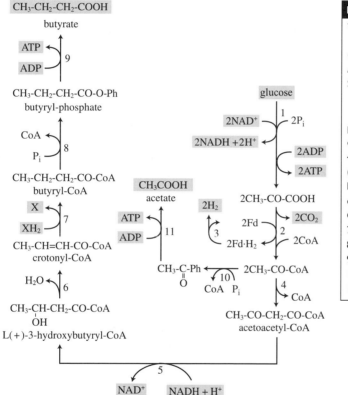

Figure 8.6 **Butyrate-acetate fermentation of glucose.**

(Gottschalk, G. 1986, *Bacterial Metabolism*, 2nd edn., Figure 8.9. Springer, New York)

1, EMP pathway; 2, pyruvate:ferredoxin oxidoreductase; 3, hydrogenase; 4, acetyl-CoA-acetyltransferase (thiolase); 5, L($+$)-3-hydroxybutyryl-CoA dehydrogenase; 6, 3-hydroxyacyl-CoA hydrolase (crotonase); 7, butyryl-CoA dehydrogenase; 8, phosphotransbutyrylase; 9, butyrate kinase; 10, phosphotransacetylase; 11, acetate kinase.

and butyryl-phosphate, 4 ATP are synthesized from glucose fermentation to acetate and 3 ATP from the butyrate fermentation.

8.5.1.3 Lactate fermentation by *Clostridium butyricum*

Silage is pasturage fermented by LAB. Stale silage contains butyrate, which can be formed from lactate and acetate by *Clostridium butyricum* and *Clostridium tyrobutyricum*. They do not ferment lactate alone, but produce butyrate from lactate and acetate (Figure 8.7). As shown in Table 8.4, 65 mmol butyrate and 59 mmol H_2 are produced from 100 mmol lactate and 32 mmol acetate.

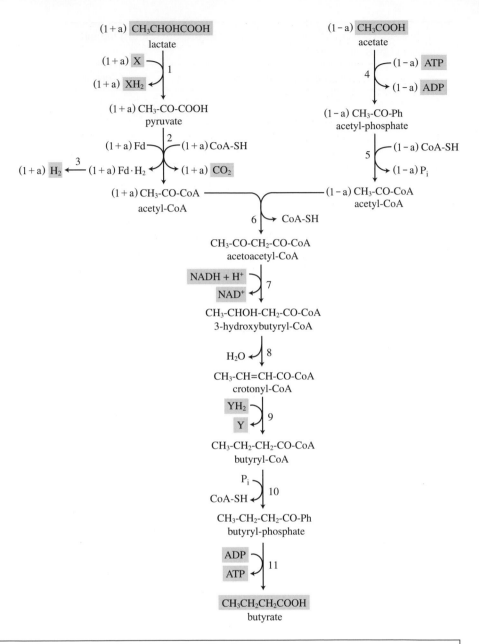

Figure 8.7 **Lactate-acetate fermentation to butyrate by *Clostridium butyricum*.**

When $(1+a)$ lactate is oxidized to $(1+a)$ acetyl-CoA, $(1-a)$ acetate are activated to $(1-a)$ acetyl-CoA consuming $(1-a)$ ATP. Butyrate (1) is produced consuming $[(1+a)+(1-a)=2]$ acetyl-CoA and 4 H with generation of 1 ATP. The net ATP generated is $\{[1-(1-a)]=a\}$, with 2a H_2 evolution.

1, lactate dehydrogenase; 2, pyruvate:ferredoxin oxidoreductase; 3, hydrogenase; 4, acetate kinase; 5, phosphotransacetylase; 6–11, butyrate fermentation pathway.

As shown in the reactions below, when lactate and acetate are consumed in a 1:1 ratio, a net ATP gain is not possible:

$$2 \text{ lactate} + \text{ADP} \rightarrow \text{butyrate} + \text{ATP} + 4\text{H}$$
$$\underline{+) \; 2 \text{ acetate} + \text{ATP} + 4\text{H} \rightarrow \text{butyrate} + \text{ADP}}$$
$$2 \text{ lactate} + 2 \text{ acetate} \rightarrow 2 \text{ butyrate}$$

A net gain of ATP is only possible when more lactate is consumed than acetate, and this becomes feasible when some of the electrons from lactate oxidation are consumed to produce H_2. Lactate dehydrogenase and pyruvate:ferredoxin oxidoreductase are involved in the lactate oxidation process. The redox potential of the pyruvate/lactate half-reaction ($E^{0'} = -0.19\,\text{V}$) is too high for lactate oxidation to be coupled to H_2 evolution ($E^{0'}$ of $2\text{H}^+/\text{H}_2 = -0.41\text{V}$). These electrons are consumed in the reduction of acetoacetyl-CoA to butyrate. Some of the electrons of the reduced ferredoxin from pyruvate oxidation are used for H_2 evolution. From the fermentation balance it can be seen that 59/2 mmol butyrate are produced solely from lactate and 59/2 mmol ATP are synthesized from 100 mmol lactate consumed since four electrons can be generated in the conceptual conversion of lactate to butyrate. The lactate–acetate fermentation by *Clostridium butyricum* is summarized as:

$$\text{acetate} + 3 \text{ lactate} + \text{ADP} + P_i \longrightarrow 2 \text{ butyrate} + 2\text{H}_2 + \text{ATP}$$

Similarly, *Clostridium kluyveri* ferments ethanol and acetate to butyrate and caproate as in the following reaction:

$$6 \text{ ethanol} + 3 \text{ acetate} \longrightarrow 3 \text{ butyrate} + \text{caproate} + 2\text{H}_2 + 4\text{H}_2\text{O} + \text{H}^+$$

Just as *Clostridium butyricum* ferments lactate with acetate, *Clostridium kluyveri* ferments ethanol in the presence of acetate. Those anaerobes fermenting lactate and ethanol with acetate should obtain all the carbon skeletons needed for biosynthesis from these substrates. Pyruvate is provided by lactate dehydrogenase in *Clostridium butyricum*, and through the phosphoroclastic reaction catalyzed by pyruvate:ferredoxin oxidoreductase from acetyl-CoA in *Clostridium kluyveri*:

$$\text{CH}_3\text{–CO–CoA} + \text{Fd·H}_2 + \text{CO}_2 \xrightleftharpoons[]{\substack{\text{pyruvate:ferredoxin} \\ \text{oxidoreductase}}} \text{CH}_3\text{–CO–COOH} + \text{CoA–SH} + \text{Fd}$$

Glucose-6-phosphate is synthesized from pyruvate through gluconeogenesis, and pyruvate carboxylase converts pyruvate to oxaloacetate to obtain TCA cycle intermediates through the incomplete TCA fork (Figure 8.8 and Figure 5.6, Section 5.4.1).

$$\text{pyruvate} + \text{HCO}_3^- + \text{ATP} \xrightleftharpoons[]{\text{pyruvate carboxylase}} \text{oxaloacetate} + \text{ADP} + P_i$$

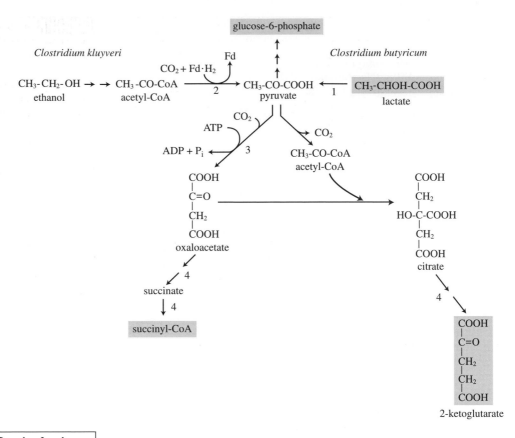

Figure 8.8 Supply of carbon skeletons through the incomplete TCA fork from lactate and ethanol in *Clostridium butyricum* and *Clostridium kluyveri*.

1, lactate dehydrogenase; 2, pyruvate:ferredoxin oxidoreductase; 3, pyruvate carboxylase; 4, incomplete TCA fork.

8.5.1.4 *Clostridium butyricum* as a probiotic

Saccharolytic clostridia and lactic acid bacteria produce acetate, butyrate and lactate in high concentrations. Their pK_a values are 4.82, 4.75 and 3.86, respectively. As the fermentation proceeds, the acidic fermentation products accumulate with a decrease in pH to values near or lower than the pK_a. At these conditions, undissociated forms of the acids accumulate. These are toxic to cells since the undissociated acids are hydrophobic and permeable to the cytoplasmic membrane, dissipating the proton motive force. However, the producing organisms do have some mechanisms of resistance to such acids in their undissociated forms (Section 8.5.2). *Clostridium butyricum* spores have been used as a probiotic for over half a century in the Far East. It is believed that the fatty acids produced by this bacterium can control undesirable bacteria in the intestine in a similar manner to the lactate produced by probiotic lactic acid bacteria.

8.5.1.5 Non-butyrate clostridial fermentation

Not all saccharolytic clostridia produce butyrate as their fermentation product. As described earlier (Section 8.3) some anaerobic fermentative bacteria ferment carbohydrate to acetate, ethanol, lactate, CO_2 and H_2. These include saccharolytic clostridia such

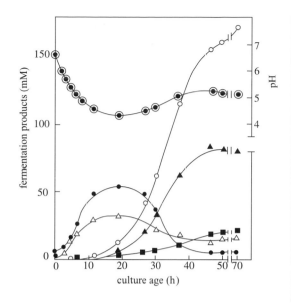

Figure 8.9 **Butanol fermentation by _Clostridium acetobutylicum._**

(Gottschalk, G. 1986, _Bacterial Metabolism_, 2nd edn., Figure 8.10. Springer, New York)

The bacterium ferments sugar to butyrate and acetate at the initial stage of growth. When the acids accumulate with a decrease in pH, the bacterium switches its metabolism to produce solvents such as butanol, acetone and ethanol, thus consuming the acids.

○, butanol; ▲, acetone; ■, ethanol; ●, butyrate; △, acetate, ◉, pH.

as _Clostridium sphenoides_, and cellulolytic clostridia such as _Clostridium thermocellum_, _Clostridium cellulolyticum_, _Clostridium josui_, and _Clostridium cellulovorans_.

8.5.2 Acetone–butanol–ethanol fermentation

Most butyrate-producing saccharolytic clostridia form small amounts of butanol, and a few strains such as _Clostridium acetobutylicum_, _Clostridium beijerinckii_, _Clostridium saccharobutylicum_, and _Clostridium saccharoperbutylacetonicum_ produce butanol in high concentration with acetone (or isopropanol) and ethanol. At the beginning of the fermentation they produce butyrate and acetate, disposing of the excess electrons to reduce H^+ to H_2 just like _Clostridium butyricum_. The solvent production starts when the acidic products accumulate (Figure 8.9).

During the solventogenic phase, sugars are fermented directly to solvents, and the acidic products are also converted to solvents. Acetate and butyrate are activated to acetyl-CoA and butyryl-CoA through the reactions catalyzed by acetoacetyl-CoA:acetate coenzyme A transferase or kinase and phosphotransacetylase. The acyl-CoAs are reduced to ethanol and butanol by aldehyde dehydrogenase and alcohol dehydrogenase. Acetoacetate is decarboxylated to acetone. Acetone is further reduced to isopropanol in _Clostridium beijerinckii_, since its alcohol dehydrogenase is active not only on aldehydes but also on ketones (Figure 8.10).

For the onset of solventogenesis, electron flux as well as carbon flux should be diverted. More electrons are needed to produce butanol and ethanol than the acidic products. The electrons used to reduce H^+ to H_2 during the acidogenic phase are used by aldehyde dehydrogenase and alcohol dehydrogenase during the solventogenic phase. In the butanol-producing clostridia, NAD(P)$^+$:ferredoxin oxidoreductase is

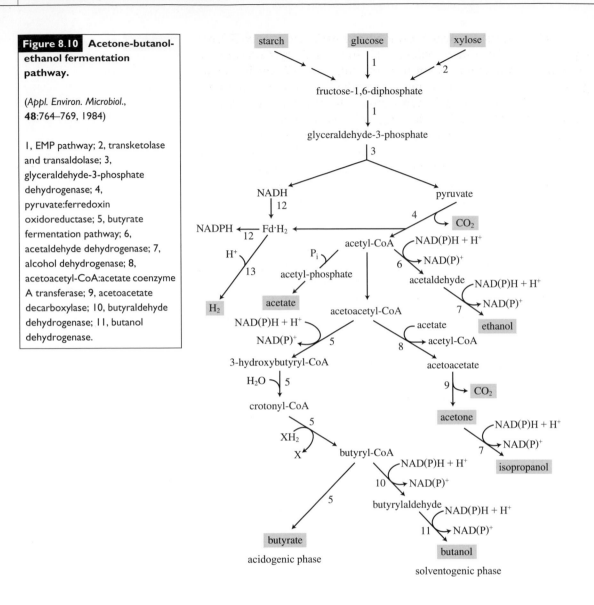

Figure 8.10 **Acetone-butanol-ethanol fermentation pathway.**

(*Appl. Environ. Microbiol.*, **48**:764–769, 1984)

1, EMP pathway; 2, transketolase and transaldolase; 3, glyceraldehyde-3-phosphate dehydrogenase; 4, pyruvate:ferredoxin oxidoreductase; 5, butyrate fermentation pathway; 6, acetaldehyde dehydrogenase; 7, alcohol dehydrogenase; 8, acetoacetyl-CoA:acetate coenzyme A transferase; 9, acetoacetate decarboxylase; 10, butyraldehyde dehydrogenase; 11, butanol dehydrogenase.

active, exchanging electrons between $NAD(P)^+$ and ferredoxin. During the acidogenic phase this enzyme is active to reduce ferredoxin, oxidizing $NAD(P)H$, and catalyzes the reverse reaction during solventogenesis. H_2 produced during acidogenesis is taken up by the bacteria for solvent production. Solventogenic clostridia have a H_2-producing hydrogenase as well as an uptake hydrogenase.

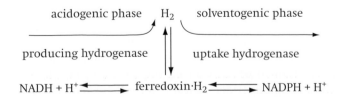

This is a typical branched fermentative pathway. Electrons from glycolysis are disposed of, reducing H^+ to H_2, and acyl-CoAs are not used as electron acceptors but used to synthesize ATP during acidogenesis through acyl-phosphate. Their reduction to solvent does not involve ATP synthesis. It is not known why and how the bacteria switch from acidogenesis with a high energy conservation efficiency to the less energy conserving solventogenesis. Solventogenesis might be a protection mechanism from the acidic products. The concentration of guanosine-3'-diphosphate-5'-triphosphate (pppGpp) increases during solventogenesis in *Clostridium acetobutylicum*, suggesting that genes for solventogenic enzymes are under stringent control (Section 12.2.1). A plasmid carries some genes of the solventogenic metabolism in *Clostridium acetobutylicum*.

Acidogenic reactions from acyl-CoAs to acids are reversible. When the acidic products accumulate, the acyl-CoA concentration increases. Under these conditions the free coenzyme A concentration becomes too low for efficient metabolism. Solventogenesis can be explained as a coenzyme A recovery mechanism. Solventogenesis is also accompanied by sporulation. For this reason solventogenesis can be considered to be a secondary metabolism.

Clostridium acetobutylicum does not ferment lactate only, but uses it with carbohydrate, increasing butyrate and butanol production with a decrease in acetone production. Lactate oxidation to pyruvate ($E^{0'} = -0.19\,V$) is coupled with the reduction of crotonyl-CoA to butyryl-CoA ($E^{0'} = +0.19\,V$).

8.5.3 Fermentation balance

Unlike linear fermentative pathways such as the ethanol and homofermentative lactic acid fermentations, saccharolytic clostridial fermentation produces a variety of products in a branched fermentative pathway (Table 8.5).

In a fermentation, electrons generated during ATP-synthesizing metabolism should be properly disposed of using the metabolic

Table 8.5. | *Sugar fermentation by selected* Clostridium *species (mmol product/100 mmol sugar consumed)*

Product	C. butyricum	C. perfringens	C. acetobutylicum
Butyrate	76	34	4
Acetate	42	60	14
Lactate	–	33	–
CO_2	188	176	221
H_2	235	214	135
Ethanol	–	26	7
Butanol	–	–	56
Acetone	–	–	22

–, not produed.

Table 8.6. *Carbon and electron recoveries and oxidation–reduction balance in the acetone–butanol–ethanol fermentation*

Substrate and product	Mol/100 mol substrate	Mol carbon	O/R balance		Balance of available hydrogen	
			O/R value	O/R value (mol/100 mol)	Available H	Available H (mol/100 mol)
Glucose	100	600	0	–	24	2400
Butyrate	4	16	−2	−8	20	80
Acetate	14	28	0	–	8	112
CO_2	221	221	+2	+442	0	–
H_2	135	–	−1	−135	2	270
Ethanol	7	14	−2	−14	12	1344
Butanol	56	224	−4	−224	24	1344
Acetone	22	66	−2	−44	16	352
Total		569		−425, +442		2242

Carbon recovery: $569/600 \times 100 = 94.8\%$.
O/R balance $= 425/442 = 1.04$.
H recovery: $2242/2400 \times 100 = 93\%$.
–, not applicable.

intermediates as electron acceptors. In complex fermentations, such as that carried out by *Clostridium acetobutylicum*, it is difficult to judge if the electron balance is even. To make it simple, the oxidation/reduction (O/R) balance can be calculated in complex fermentations as shown in Table 8.6. Arbitrarily, the O/R values of formaldehyde (CH_2O) and its multiples, for example hexoses and pentoses, are taken as zero. Each 2H in excess is expressed as −1, and +1 for a lack of 2H. For example, ethanol with a formula of C_2H_6O added to H_2O gives $C_2H_8O_2$. In comparison with $C_2H_4O_2$ (acetate), 4H are in excess to give an O/R value of −2. Similarly, acetate ($C_2H_4O_2$) has an O/R value of 0, and carbon dioxide, C(-H_2O)O gives a value of +2 (Table 8.6).

Since the cell yield in a fermentation is low and the cell mass has an O/R value of about zero, the O/R value of the fermentation products should be similar to that of the substrate.

8.6 | Mixed acid and butanediol fermentation

8.6.1 Mixed acid fermentation

Some Gram-negative facultative anaerobic bacteria ferment glucose, producing various products including lactate, acetate, succinate, formate, CO_2 and H_2. These include species of *Escherichia*, *Salmonella*, *Shigella* and *Enterobacter* (Figure 8.11).

Glucose is metabolized through the EMP pathway. Phosphoenolpyruvate (PEP) carboxylase synthesizes oxaloacetate from PEP before being reduced to succinate. Pyruvate is either reduced to lactate

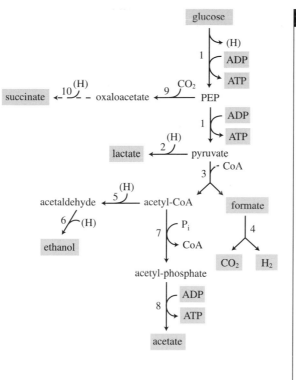

Figure 8.11 Mixed acid fermentation by some Gram-negative facultative anaerobic bacteria.

(Gottschalk, G. 1986, *Bacterial Metabolism*, 2nd edn., Figure 8.15. Springer, New York)

Facultative anaerobes belonging to the genera *Escherichia*, *Salmonella*, *Shigella*, *Enterobacter* and others ferment sugars to lactate, acetate, formate, succinate and ethanol in the absence of electron acceptors.

1, EMP pathway; 2, lactate dehydrogenase; 3, pyruvate:formate lyase; 4, formate:hydrogen lyase; 5, acetaldehyde dehydrogenase; 6, alcohol dehydrogenase; 7, phosphotransacetylase; 8, acetate kinase; 9, phosphoenolpyruvate (PEP) carboxylase; 10, enzymes of the TCA cycle.

by lactate dehydrogenase, or cleaved to acetyl-CoA and formate by pyruvate:formate lyase. According to the availability of electrons, acetyl-CoA is either reduced to ethanol or used to synthesize ATP.

Strictly anaerobic bacteria such as *Anaerobiospirillum succiniciproducens* and *Actinobacillus succinogenes* ferment carbohydrate mainly to succinate through a similar metabolism. In this case the succinate yield is as high as the amount of carbohydrate fermented. PEP carboxylase (reaction 9, Figure 8.11) fixes a large amount of CO_2 in this fermentation.

8.6.2 Butanediol fermentation

Some *Erwinia*, *Klebsiella* and *Serratia* species produce 2,3-butanediol in addition to lactate and ethanol from pyruvate, the EMP pathway product. Pyruvate is the substrate for one of three enzymes in these bacteria. These are lactate dehydrogenase, pyruvate:formate lyase and 2-acetolactate synthase (Figure 8.12). The reactions catalyzed by these enzymes are similar to those of the mixed acid fermentation except for 2-acetolactate synthase. This enzyme condenses two molecules of pyruvate to 2-acetolactate that is further decarboxylated and reduced to 2,3-butanediol. A similar metabolism is found in *Bacillus polymyxa* during vegetative growth (Section 7.10.2) and in lactic acid bacteria fermenting citrate (Section 8.4.6).

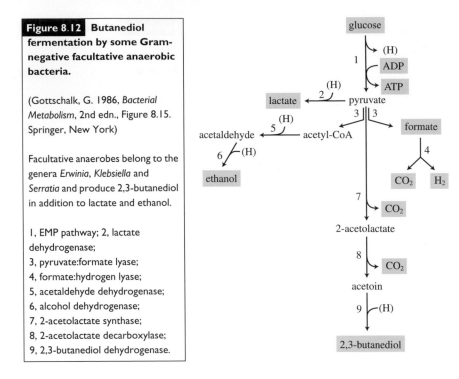

Figure 8.12 **Butanediol fermentation by some Gram-negative facultative anaerobic bacteria.**

(Gottschalk, G. 1986, *Bacterial Metabolism*, 2nd edn., Figure 8.15. Springer, New York)

Facultative anaerobes belong to the genera *Erwinia*, *Klebsiella* and *Serratia* and produce 2,3-butanediol in addition to lactate and ethanol.

1, EMP pathway; 2, lactate dehydrogenase;
3, pyruvate:formate lyase;
4, formate:hydrogen lyase;
5, acetaldehyde dehydrogenase;
6, alcohol dehydrogenase;
7, 2-acetolactate synthase;
8, 2-acetolactate decarboxylase;
9, 2,3-butanediol dehydrogenase.

The first enzyme of this metabolism, 2-acetolactate synthase, is best characterized in Gram-negative facultative bacteria. This enzyme has thiamine pyrophosphate as a cofactor to catalyze the following reactions:

$$CH_3-CO-COOH + E-TPP \longrightarrow CH_3-CHOH-TPP-E + CO_2$$

$$CH_3-CHOH-TPP-E + CH_3-CO-COOH \longrightarrow \begin{array}{l} CH_3-C=O \\ CH_3-COH-COOH + E-TPP \\ \text{2-acetolactate} \end{array}$$

Under anaerobic conditions, 2,3-butanediol-producing facultative anaerobes produce acidic products, lowering the external and intracellular pH. 2-acetolactate synthase, which catalyzes the first reaction to produce 2,3-butanediol, has an optimum at pH 6.0. When the intracellular pH drops, this enzyme becomes active to divert carbon flux from acid production to the neutral solvent. An enzyme catalyzing the same reaction catalyzes the first reaction of valine synthesis from pyruvate (Figure 6.15, Section 6.4.1). The enzyme involved in valine synthesis has an optimum at pH 8.0, and also catalyzes the condensing reaction of 2-ketobutyrate and pyruvate to synthesize isoleucine. This enzyme is referred to as the pH 8.0 enzyme while the enzyme involved in 2,3-butanediol synthesis is referred to as the pH 6.0 enzyme.

Klebsiella pneumoniae, *Klebsiella oxytoca* and *Enterobacter aerogenes* ferment glycerol to various products including 2,3-butanediol (Figure 8.13). They oxidize a part of glycerol to pyruvate, and dispose

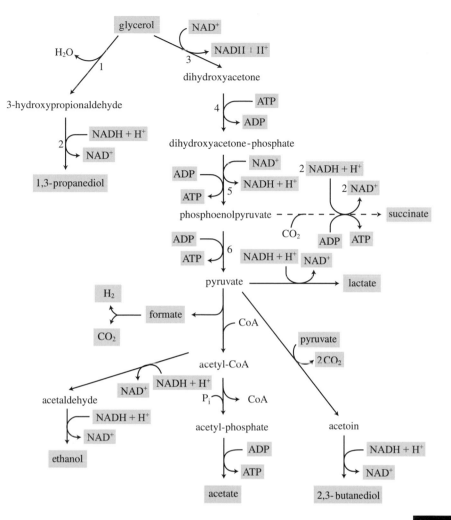

of the resulting electrons to reduce the remaining glycerol to 1,3-propanediol. Pyruvate is metabolized as in the 2,3-butanediol fermentation. Glycerol is reduced to 1,3-propanediol by lactic acid bacteria while oxidizing carbohydrate (Section 8.4.6). These diols are important petrochemical intermediates.

8.6.3 Citrate fermentation by facultative anaerobes
Since a TCA cycle enzyme, 2-ketoglutarate dehydrogenase, is not expressed in facultative anaerobes under fermentative conditions, they cannot oxidize citrate through the TCA cycle. However, citrate is metabolized in several different pathways depending on the organism (Figure 8.14). *Salmonella typhimurium* and *Klebsiella pneumoniae* cleave citrate to acetate and oxaloacetate through the action of citrate lyase. Oxaloacetate is decarboxylated to pyruvate (Figure 8.14a). Pyruvate is oxidized as in glucose metabolism, synthesizing ATP through the action of acetate kinase. This reaction is similar to citrate oxidation by *Pediococcus halophilus* (Figure 8.4). Oxaloacetate

Figure 8.13 **Glycerol fermentation by *Klebsiella pneumoniae*.**

(*Appl. Microbiol. Biotechnol.* **50**:24–29, 1998)

Glycerol is metabolized though a similar pathway as in the butanediol fermentation as well as being used as an electron acceptor to be reduced to 1,3-propanediol.

1, glycerol dehydratase;
2, 1,3-propanediol dehydrogenase;
3, glycerol dehydrogenase;
4, dihydroxyacetone kinase;
5, 6, enzymes in Figure 8.14.

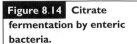

Figure 8.14 **Citrate fermentation by enteric bacteria.**

(*Arch. Microbiol.* 167:78–88, 1997)

(a) In the absence of electron acceptors, *Klebsiella pneumoniae* and *Salmonella typhimurium* ferment citrate to acetate, as in *Pediococcus halophilus* (Figure 8.4).
(b) *Escherichia coli* reduces citrate as an electron acceptor with a fermentable substrate.
(c) *Providencia rettgeri* metabolizes citrate to succinate in a similar metabolism to that of the incomplete TCA fork.

1, citrate lyase; 2, oxaloacetate decarboxylase; 3, pyruvate:formate lyase; 4, phosphotransacetylase; 5, acetate kinase; 6, malate dehydrogenase; 7, fumarase; 8, fumarate reductase; 9, aconitase; 10, isocitrate dehydrogenase; 11, 2-ketoglutarate dehydrogenase complex; 12, succinate thiokinase.

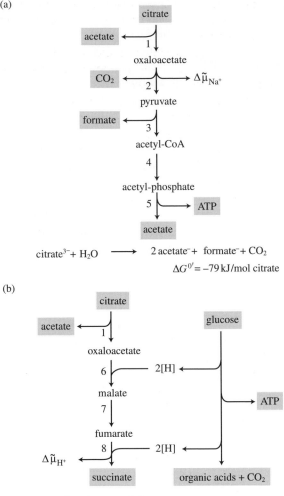

(a)

$$citrate^{3-} + H_2O \longrightarrow 2\,acetate^- + formate^- + CO_2$$
$$\Delta G^{0\prime} = -79\,kJ/mol\ citrate$$

(b)

$$citrate^{3-} + glucose + H_2O \longrightarrow$$
$$succinate^{2-} + 3\,acetate^- + 2\,CO_2 + 2\,H_2 + 2\,H^+$$
$$\Delta G^{0\prime} = -344\ kJ/reaction$$

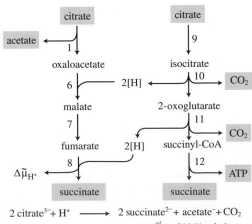

(c)

$$2\,citrate^{3-} + H^+ \longrightarrow 2\,succinate^{2-} + acetate^- + CO_2$$
$$\Delta G^{0\prime} = -81\,kJ/mol\ citrate$$

decarboxylase in these bacteria is a Na^+-dependent enzyme and generates a sodium gradient, while the same reaction in the lactic acid bacteria is not coupled to energy conservation.

Citrate is used as an electron acceptor to oxidize glucose by certain strains of *Escherichia coli* that have plasmids carrying the gene for the citrate transporter (Figure 8.14b).

Providencia rettgeri metabolizes citrate to succinate. Genes for all the TCA cycle enzymes are expressed in this bacterium under fermentative conditions, but citrate is not oxidized through the functional TCA cycle since electrons from the TCA cycle in the form of NAD(P)H cannot be disposed of under the given conditions (Figure 8.14c). Instead, this bacterium metabolizes citrate to succinate partly through the forward reaction of the TCA cycle and also through the reverse reaction to balance the oxidation and reduction. Energy is conserved in this metabolism through the action of fumarate reductase in the form of a proton motive force.

8.6.4 Anaerobic enzymes

Enteric bacteria oxidize pyruvate in a reaction catalyzed by the pyruvate dehydrogenase complex under aerobic conditions. This enzyme is not expressed under anaerobic conditions, and NADH inhibits its activity. On the other hand, pyruvate:formate lyase functions only under anaerobic conditions since it is expressed under fermentative conditions and irreversibly inactivated by molecular oxygen. This enzyme can oxidize pyruvate to acetyl-CoA required in anabolism without reducing NAD^+ under fermentative conditions.

Formate:hydrogen lyase cleaves the formate produced by pyruvate:formate lyase to CO_2 and H_2 under fermentative conditions. This enzyme is a complex of formate dehydrogenase II (FDH_{II}) and hydrogenase. When electron acceptors such as nitrate or fumarate are present, FDH_I oxidizes formate to CO_2 transferring the electrons to nitrate reductase or fumarate reductase via NADH for energy conservation through anaerobic respiration. Formate metabolism under anaerobic conditions is shown in Figure 8.15. These enzymes are present in species of *Escherichia* and *Enterobacter*, but not in species of *Shigella* and *Erwinia*.

$$HCOO^- + H^+ \xrightarrow{\text{formate : hydrogen lyase}} CO_2 + H_2$$

$$HCOO^- + NAD^+ \xrightarrow{\text{formate dehydrogenase I}} CO_2 + NADH$$

Enzymes expressed only under anaerobic conditions in facultative anaerobes are referred to as anaerobic enzymes. They include pyruvate:formate lyase, hydrogenase, nitrate reductase and fumarate reductase. Their expression is controlled by the FNR protein (Section 12.2.4).

Figure 8.15 **Formate metabolism in facultative anaerobes.**

(Gottschalk, G. 1986, *Bacterial Metabolism*, 2nd edn., Figure 8.16. Springer, New York)

Facultative anaerobic enteric bacteria oxidize pyruvate in a reaction catalyzed by pyruvate:formate lyase without reducing pyridine nucleotides. Formate dehydrogenase oxidizes formate to reduce nitrate or fumarate through anaerobic respiration (a) or to reduce protons (b).

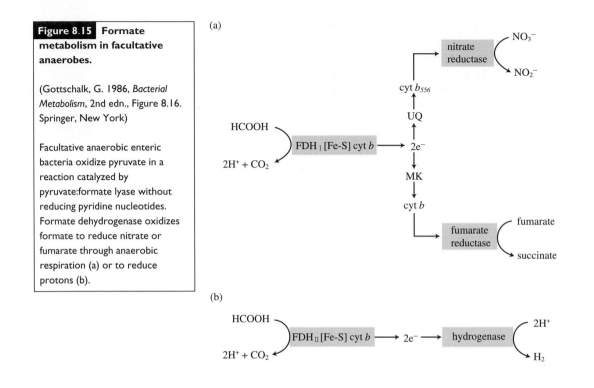

8.7 | Propionate fermentation

Species of the genera *Propionibacterium*, *Clostridium propionicum* and *Megasphaera elsdenii* ferment carbohydrate or lactate to propionate, acetate, and CO_2:

$$3 \text{ glucose} \longrightarrow 4 \text{ propionate} + 2 \text{ acetate} + 2CO_2$$

$$3 \text{ lactate} \longrightarrow 2 \text{ propionate} + \text{acetate} + CO_2$$

Lactate is the preferred substrate over carbohydrate in most propionate producers. They ferment glucose or lactate to propionate through either the acrylate pathway or the succinate–propionate pathway. Spore-forming *Propionispora vibrioides* ferments sugar alcohols such as mannitol, sorbitol and xylitol to propionate and acetate through an unknown pathway. *Propionispora vibrioides* ferments the aliphatic polyester poly(propylene adipate) to propionate.

8.7.1 Succinate–propionate pathway

Species belonging to the genus *Propionibacterium* ferment lactate to propionate via succinate through this pathway (Figure 8.16). Lactate dehydrogenase with flavin oxidizes lactate to pyruvate. Two molecules of pyruvate are reduced to propionate as the electron acceptor while one pyruvate is oxidized to acetate through acetyl-CoA synthesizing 1 ATP. Some enzymes of this pathway are of interest. They are transcarboxylase requiring biotin, fumarate reductase

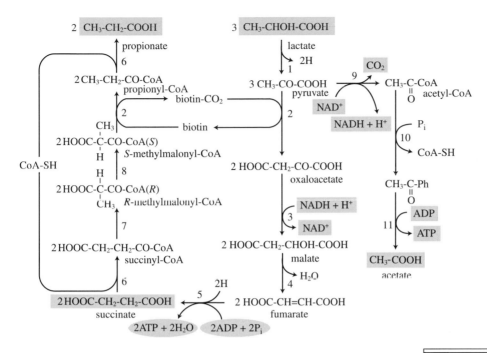

Figure 8.16 Succinate–propionate pathway in the genus *Propionibacterium*, which ferments lactate to propionate.

(Gottschalk, G. 1986, *Bacterial Metabolism*, 2nd edn., Figure 8.18. Springer, New York)

1, lactate dehydrogenase; 2, methylmalonyl-CoA:pyruvate transcarboxylase; 3, malate dehydrogenase; 4, fumarase; 5, fumarate reductase; 6, coenzyme A transferase; 7, methylmalonyl-CoA mutase; 8, methylmalonyl-CoA racemase; 9, pyruvate dehydrogenase; 10, phospho-transacetylase; 11, acetate kinase.

coupled to ATP synthesis and methylmalonyl-CoA mutase that uses coenzyme B_{12}. In this fermentation 3 ATP are synthesized fermenting 3 lactate, one by acetate kinase and two by fumarate reductase.

Methylmalonyl-CoA:pyruvate transcarboxylase transfers a carboxyl group from methylmalonyl-CoA to pyruvate to form propionyl-CoA and oxaloacetate as in reaction 2 of Figure 8.16. Biotin is involved in this reaction as a cofactor.

Fumarate reductase couples the reduction of fumarate, using NADH as the electron donor, to the formation of a proton motive force as described previously (Section 5.8.6.1).

When heme is provided, *Bacteroides fragilis* ferments carbohydrates to propionate and acetate through the succinate–propionate pathway with functional cytochromes. However, without heme, lactate and acetate are produced from the carbohydrate fermentation because fumarate reductase cannot function in the absence of functional cytochromes. As in lactic acid bacteria of the *Lactococcus* genus, this bacterium synthesizes the apoprotein of cytochromes.

Methylmalonyl-CoA mutase catalyzes the carbon-rearranging reaction to convert succinyl-CoA to methylmalonyl-CoA. This reaction requires coenzyme B_{12}. Coenzyme B_{12} has the structure of 5'-deoxyadenosylcobalamin (Figure 8.17). This coenzyme is required for various enzyme reactions involving carbon rearrangement. Vitamin B_{12} is produced commercially using *Propionibacterium shermanii* and related bacteria since they synthesize coenzyme B_{12} in large concentrations for methylmalonyl-CoA mutase.

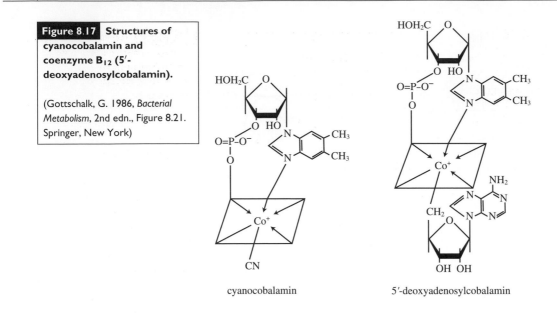

Figure 8.17 **Structures of cyanocobalamin and coenzyme B$_{12}$ (5′-deoxyadenosylcobalamin).**

(Gottschalk, G. 1986, *Bacterial Metabolism*, 2nd edn., Figure 8.21. Springer, New York)

cyanocobalamin   5′-deoxyadenosylcobalamin

Propionate producers through the succinate–propionate pathway excrete a small amount of succinate into the medium. In this case, methylmalonyl-CoA:pyruvate transcarboxylase cannot produce propionyl-CoA because the concentration of methylmalonyl-CoA becomes too low. To replace the excreted succinate, pyruvate or phosphoenolpyruvate are carboxylated to oxaloacetate which is reduced to succinate via malate:

$$\text{pyruvate} + \text{ATP} + \text{P}_i \xrightarrow{\substack{\text{pyruvate orthophosphate} \\ \text{dikinase}}} \text{PEP} + \text{AMP} + \text{PP}_i$$

$$\text{PEP} + \text{CO}_2 + \text{ADP} \xrightarrow{\text{PEP carboxykinase}} \text{oxaloacetate} + \text{ATP}$$

$$\text{PEP} + \text{CO}_2 + \text{P}_i \xrightarrow{\text{PEP carboxytransphosphorylase}} \text{oxaloacetate} + \text{PP}_i$$

Since pyrophosphatase activity is low in species of *Propionibacterium*, the pyrophosphate produced in the above reactions is used to phosphorylate sugars, thus conserving the energy carried by this inorganic compound.

8.7.2 Acrylate pathway

The majority of the propionate producers ferment carbohydrate or lactate through the succinate–propionate pathway, but *Clostridium propionicum* and *Megasphaera elsdenii* ferment lactate to propionate via lactyl-CoA and acrylyl-CoA. Electrons for this reductive metabolism are supplied from the oxidation of lactate to acetate (Figure 8.18). The ATP yield of the acrylate pathway is 1 ATP/3 lactate, much lower than the 1 ATP/lactate in the succinate–propionate pathway.

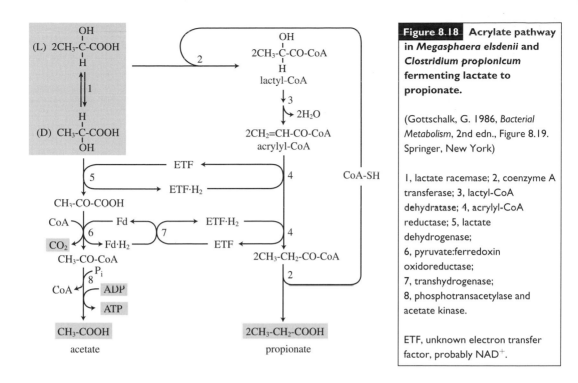

Figure 8.18 Acrylate pathway in *Megasphaera elsdenii* and *Clostridium propionicum* fermenting lactate to propionate.

(Gottschalk, G. 1986, *Bacterial Metabolism*, 2nd edn., Figure 8.19. Springer, New York)

1, lactate racemase; 2, coenzyme A transferase; 3, lactyl-CoA dehydratase; 4, acrylyl-CoA reductase; 5, lactate dehydrogenase; 6, pyruvate:ferredoxin oxidoreductase; 7, transhydrogenase; 8, phosphotransacetylase and acetate kinase.

ETF, unknown electron transfer factor, probably NAD^+.

8.8 | Fermentation of amino acids and nucleic acid bases

In addition to carbohydrates and organic acids, amino acids and nucleic acid bases are fermented under anaerobic conditions. Many strains of *Clostridium* ferment these nitrogenous compounds. They are distinguished as proteolytic clostridia from the saccharolytic clostridia that ferment carbohydrates. The proteolytic clostridia are mainly pathogenic to animals including humans.

Many anaerobic bacteria, including proteolytic clostridia, ferment individual amino acids but a mixture is a better substrate in many cases. When amino acid mixtures are fermented, some amino acids are oxidized with ATP synthesis using other amino acids as the electron acceptors. The fermentation of amino acid mixtures is referred to as the Stickland reaction.

8.8.1 Fermentation of individual amino acids
Amino acids are deaminated during their fermentation as in oxidation under aerobic conditions. Reductive deamination is the commonest anaerobic amino acid metabolism (shown in the reaction below) although some amino acids are deaminated through oxidative deamination or transamination reactions depending on the organism.

$$R\text{–}CHNH_2\text{–}COOH + 2H \rightarrow RCH_2\text{–}COOH + NH_3$$

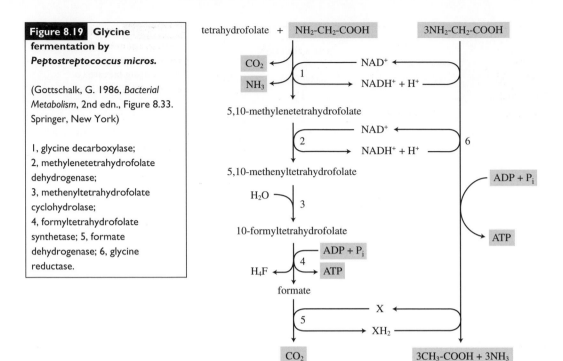

Figure 8.19 Glycine fermentation by *Peptostreptococcus micros.*

(Gottschalk, G. 1986, *Bacterial Metabolism*, 2nd edn., Figure 8.33. Springer, New York)

1, glycine decarboxylase; 2, methylenetetrahydrofolate dehydrogenase; 3, methenyltetrahydrofolate cyclohydrolase; 4, formyltetrahydrofolate synthetase; 5, formate dehydrogenase; 6, glycine reductase.

Clostridium propionicum deaminates alanine to pyruvate that is fermented to propionate and acetate as in the acrylate pathway (Figure 8.18).

Clostridium litoralis, Clostridium sticklandii, Eubacterium acidaminophilum and *Peptostreptococcus micros* ferment glycine according to the following stoichiometry:

$$4 \text{ glycine} + 2H_2O \rightarrow 3 \text{ acetate} + 2CO_2 + 4NH_3$$

Glycine is oxidatively decarboxylated forming methylenetetrahydrofolate (methylene-H_4F). Methylene-H_4F is further oxidized to CO_2 and the resulting electrons are transferred to thioredoxin before being consumed, reducing glycine in a reaction catalyzed by glycine reductase (Figure 8.19). Glycine reductase is a selenium-containing protein reducing glycine to acetyl-phosphate using reduced thioredoxin as the electron donor. ATP is synthesized through substrate-level phosphorylation in the reactions catalyzed by formyltetrahydrofolate synthetase and acetate kinase.

Threonine is a substrate for fermentative growth of *Clostridium propionicum* and *Peptostreptococcus prevotii*. Threonine dehydratase deaminates threonine to 2-ketobutyrate before being oxidized to propionyl-CoA in a reaction catalyzed by 2-ketobutyrate:ferredoxin oxidoreductase, which is a similar reaction to pyruvate oxidation by pyruvate:ferredoxin oxidoreductase. Hydrogenase oxidizes the reduced ferredoxin producing hydrogen, and propionyl-CoA is converted to propionate producing ATP.

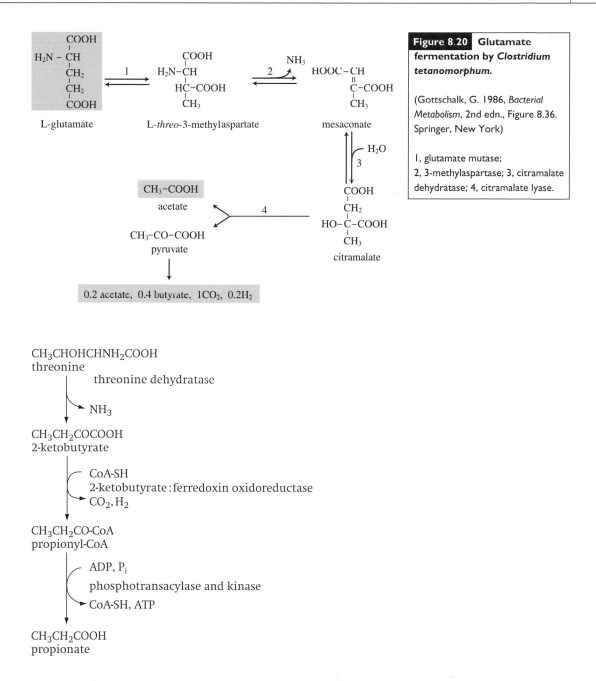

Figure 8.20 **Glutamate fermentation by *Clostridium tetanomorphum*.**

(Gottschalk, G. 1986, *Bacterial Metabolism*, 2nd edn., Figure 8.36. Springer, New York)

1, glutamate mutase;
2, 3-methylaspartase; 3, citramalate dehydratase; 4, citramalate lyase.

Glutamate is fermented through different pathways depending on the bacterial strain. *Clostridium tetanomorphum* splits glutamate to acetate and pyruvate through a series of reactions initiated by coenzyme B_{12}-containing glutamate mutase (Figure 8.20). Pyruvate is metabolized to acetate and butyrate as in the saccharolytic clostridial butyrate fermentation with ATP synthesis by kinases (Figure 8.6). On the other hand, *Acidaminococcus fermentans* and *Peptostreptococcus asaccharolyticus* ferment glutamate to acetate and butyrate via glutaconyl-CoA (Figure 8.21). The Na^+-dependent glutaconyl-CoA

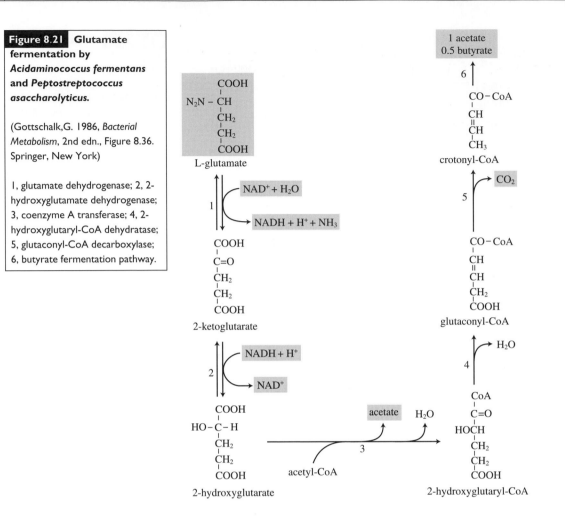

Figure 8.21 Glutamate fermentation by *Acidaminococcus fermentans* and *Peptostreptococcus asaccharolyticus.*

(Gottschalk, G. 1986, *Bacterial Metabolism*, 2nd edn., Figure 8.36. Springer, New York)

1, glutamate dehydrogenase; 2, 2-hydroxyglutamate dehydrogenase; 3, coenzyme A transferase; 4, 2-hydroxyglutaryl-CoA dehydratase; 5, glutaconyl-CoA decarboxylase; 6, butyrate fermentation pathway.

decarboxylase generates a sodium motive force (Section 5.8.6.2), and the kinases synthesize ATP.

Many obligate and facultative anaerobes, including *Clostridium novyi* and *Bacteroides melaninogenicus*, ferment aspartate to acetate and succinate. The former converts aspartate to alanine in a reaction catalyzed by aspartate decarboxylase. Alanine is fermented as shown above. *Bacteroides melaninogenicus* deaminates aspartate to fumarate, one third of which is oxidized to acetate and two-thirds are reduced to succinate, consuming the electrons generated from the oxidation reactions. This bacterium possesses menaquinone and cytochrome *b* for the generation of a proton motive force by fumarate reductase as in propionate producers (Section 8.7.1).

Lysine is fermented to acetate and butyrate by *Clostridium sticklandii* and *Clostridium subterminale* (Figure 8.22). Lysine is converted to 3-keto-5-aminohexanoate before being split to acetoacetate and 3-aminobutyryl-CoA with the addition of acetyl-CoA. Acetoacetate is converted to acetate with the synthesis of ATP and 3-aminobutyryl-CoA is reduced to butyrate.

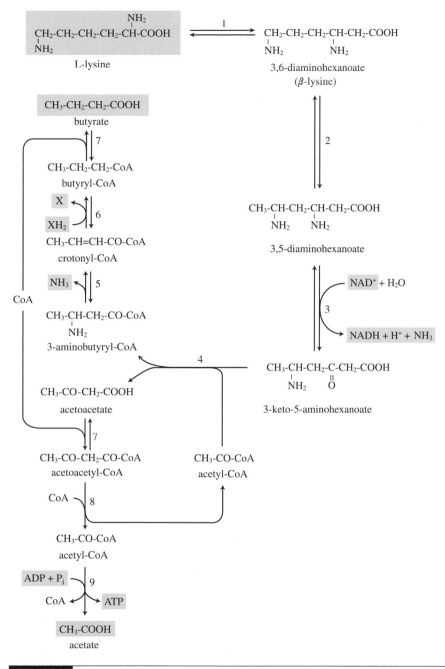

Figure 8.22 Lysine fermentation by *Clostridium sticklandii.*

(Gottschalk, G. 1986, *Bacterial Metabolism*, 2nd edn., Figure 8.37. Springer, New York)

1, lysine-2,3-aminomutase; 2, β-lysine mutase; 3, 3,5-diaminohexanoate dehydrogenase; 4, 3-keto-5-aminohexanoate cleavage enzyme; 5, 3-aminobutyryl-CoA deaminase; 6, butyryl-CoA dehydrogenase; 7, coenzyme A transferase; 8, β-ketothiolase; 9, phosphotransacetylase and acetate kinase.

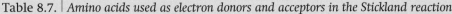

| Table 8.7. | Amino acids used as electron donors and acceptors in the Stickland reaction | | | |

Electron donor			Electron acceptor	
Amino acid	Products		Amino acid	Products
Alanine	acetate + NH_3 + CO_2		Glycine	acetate + NH_3
Leucine	3-methylbutyrate + NH_3 + CO_2		Proline	5-aminovalerate
Isoleucine	2-methylbutyrate + NH_3 + CO_2		Phenylalanine	phenylpropionate + NH_3
Valine	2-methylpropionate + NH_3 + CO_2		Tryptophan	indolepropionate + NH_3
Phenylalanine	phenylacetate + NH_3 + CO_2		Ornithine	5-aminovalerate + NH_3
Tryptophan	indoleacetate + NH_3 + CO_2		Leucine	4-methylvalerate + NH_3
Histidine	glutamate + NH_3 + CO_2		Betaine	acetate + trimethylamine
			Sarcosine	acetate + monomethylamine

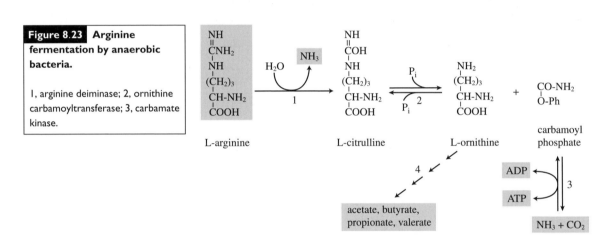

Figure 8.23 **Arginine fermentation by anaerobic bacteria.**

1, arginine deiminase; 2, ornithine carbamoyltransferase; 3, carbamate kinase.

Arginine is fermented as shown in Figure 8.23. This fermentation was first discovered in *Mycoplasma* spp. and later in *Clostridium sticklandii*, *Clostridium botulinum*, and *Halobacterium salinarium*. In this fermentation, arginine deiminase converts arginine to citrulline, which is split to ornithine and carbamoyl phosphate. Ornithine is metabolized to acetate and butyrate, and carbamate kinase synthesizes ATP from carbamoyl phosphate. A similar fermentation by heterofermentative bacteria such as *Leuconostoc oenos* produces flavour from arginine during wine maturation.

In addition to those discussed above, all natural amino acids are fermented by many anaerobic bacteria including proteolytic clostridia.

8.8.2 Stickland reaction

Many anaerobic bacteria ferment single amino acids but grow better with amino acid mixtures. In this case, some amino acids serve as electron donors while others serve as electron acceptors. Alanine and valine are used as electron donors, and glycine and proline are used as electron acceptors. Depending on the mixture, leucine and aromatic amino acids can be used as electron donors or acceptors (Table 8.7).

Generally, amino acids used as electron donors are oxidatively deaminated and the resulting 2-ketoacids are further oxidized by 2-ketoacid:ferredoxin oxidoreductase to acyl-CoAs. ATP is synthesized from acyl-CoA through substrate-level phosphorylation (SLP). Electrons from the oxidation reactions are disposed of, reducing the electron-accepting amino acids.

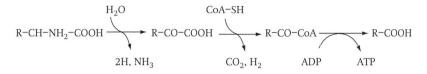

During the reduction of the electron acceptors in Stickland reactions, ATP may be synthesized through SLP, such as in the reaction catalyzed by glycine reductase, but not in all cases. Some results have suggested electron-transport phosphorylation from the reduction of proline and phenylalanine as electron acceptors.

8.8.3 Fermentation of purine and pyrimidine bases

Clostridium acidiurici, *Clostridium cylindrosporum Clostridium purinolyticum* and other species ferment purine bases to acetate, CO_2 and NH_3, and pyrimidine bases are fermented by *Clostridium glycolicum* and *Clostridium oroticum*.

Purine bases, adenine, guanine and urate, are converted to formiminoglycine (Figure 8.24) and to acetate with ATP synthesis (Figure 8.25)

Many strains of *Clostridium* ferment pyrimidine bases though they are less common than purine fermenters. *Clostridium glycolicum* ferments uracil to β-alanine, CO_2 and NH_3, and orotate is fermented by *Clostridium oroticum* to acetate, CO_2 and NH_3.

8.9 | Fermentation of dicarboxylic acids

Some anaerobes ferment dicarboxylic acids, conserving free energy from the decarboxylation reactions in the form of a sodium motive force. *Succinispira mobilis* oxidizes succinate to acetate. Malonate and oxalate are fermented by *Malonomonas rubra* and *Oxalobacter formigenes*, respectively. *Oxalobacter formigenes* removes oxalate in the human gut preventing the development of kidney stones.

8.10 | Hyperthermophilic archaeal fermentation

Many prokaryotes have been isolated from hot springs and underwater hydrothermal vents. Some of them can grow at a temperature over 100 °C, and most of these are archaea. Some of them are aerobic, but the majority are anaerobes. Many of these organisms use reduced sulfur compounds for their energy metabolism. These

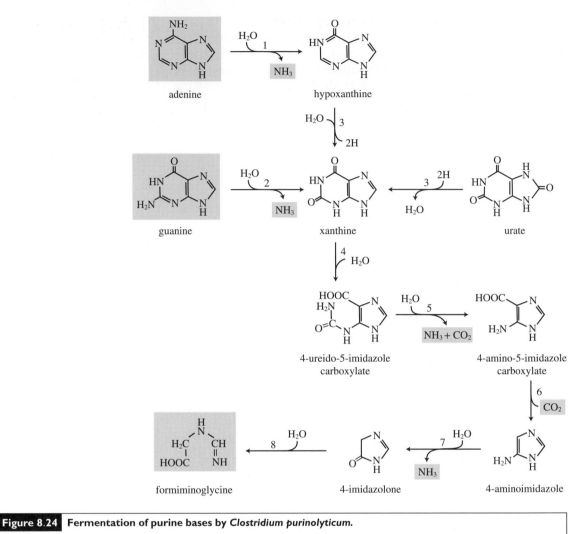

Figure 8.24 Fermentation of purine bases by *Clostridium purinolyticum.*

(Gottschalk, G. 1986, *Bacterial Metabolism*, 2nd edn., Figure 8.38. Springer, New York)

1, adenine deaminase; 2, guanine deaminase; 3, xanthine dehydrogenase; 4, xanthine amidohydrolase; 5, 4-ureido-5-imidazole carboxylate amidohydrolase; 6, 4-amino-5-imidazole carboxylate decarboxylase; 7, 4-aminoimidazole deaminase; 8, 4-imidazolonase.

are referred to as sulfur-dependent archaea. Among them, strains of the genus *Archeaoglobus* conserve energy through anaerobic respiration using sulfate as the electron acceptor. Others reduce sulfur not as an electron acceptor, but as an electron sink (Table 9.5, Section 9.3). They use sulfur to dispose of electrons without conserving energy. This process is referred to as fermentative sulfur reduction.

Some hyperthermophilic archaea, including *Pyrococcus furiosus*, *Pyrococcus woesei* and *Thermococcus litoralis*, ferment protein and carbohydrates in the absence of sulfur as the electron sink. *Pyrococcus furiosus* ferments carbohydrates in a modified EMP pathway (Figure 4.3,

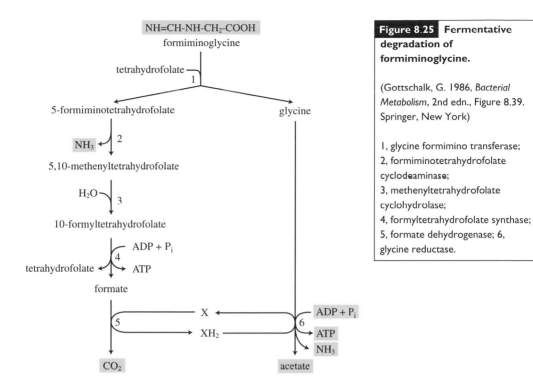

Figure 8.25 Fermentative degradation of formiminoglycine.

(Gottschalk, G. 1986, *Bacterial Metabolism*, 2nd edn., Figure 8.39. Springer, New York)

1, glycine formimino transferase;
2, formiminotetrahydrofolate cyclodeaminase;
3, methenyltetrahydrofolate cyclohydrolase;
4, formyltetrahydrofolate synthase;
5, formate dehydrogenase; 6, glycine reductase.

Section 4.1.3.2). In this pathway, ferredoxin oxidoreductase oxidizes glyceraldehyde-3-phosphate and pyruvate, reducing ferredoxin. Electrons from the reduced ferredoxin are transferred to NADPH before being consumed to produce H_2S, H_2 and alanine (Figure 8.26). This archaeon possesses a unique enzyme, sulfhydrogenase, that can reduce either H^+ or S^0. This organism uses pyruvate as an electron acceptor in a two-step reaction different from that carried out in bacteria. Alanine aminotransferase catalyzes the amino group transfer reaction from glutamate to pyruvate, producing 2-ketoglutarate and alanine. 2-ketoglutarate is reduced by glutamate dehydrogenase consuming electrons in the form of NADPH.

Pyruvate:ferredoxin oxidoreductase oxidizes pyruvate to acetyl-CoA. Acetyl-CoA synthetase synthesizes ATP from acetyl-CoA. In this reaction acetyl-phosphate is not involved, unlike in most bacteria:

$$\text{acetyl-CoA} + \text{ADP} + P_i \xrightarrow{\text{acetyl-CoA synthetase}} \text{acetate} + \text{CoA-SH} + \text{ATP}$$

8.11 | Degradation of xenobiotics under fermentative conditions

The petrochemical industry produces many anthropogenic compounds unknown in natural ecosystems. These are resistant to microbial metabolism and referred to as xenobiotics. Generally they

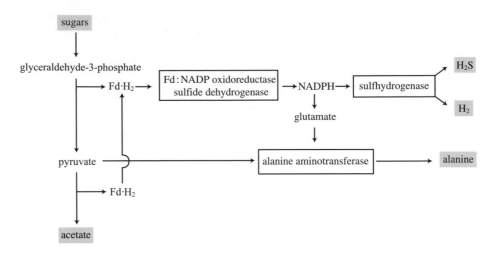

Figure 8.26 **Sugar fermentation by *Pyrococcus furiosus*.**

(*FEMS Microbiol. Rev.* **18**:119–137, 1996)

This archaeon metabolizes carbohydrates through a modified EMP pathway and has an enzyme sulfhydrogenase that reduces H^+ and S^0. Pyruvate or H^+ is used as electron acceptors. S^0 is used as an electron sink when available. Pyruvate is reduced as an electron acceptor in a two-step reaction. Alanine aminotransferase reduces pyruvate to alanine transferring the amino group from glutamate, and the resulting 2-ketoglutarate is reduced to glutamate oxidizing NADPH in a reaction catalyzed by glutamate dehydrogenase.

---

are better substrates for aerobic metabolism but some of them are susceptible to anaerobic respiratory metabolism (Section 9.9). Some xenobiotics can be fermented. For example, *Pelobacter venetianus* ferments polyethylene glycol to ethanol and acetate.

FURTHER READING

Oxygen toxicity

Atack, J. M. & Kelly, D. J. (2006). Structure, mechanism and physiological roles of bacterial cytochrome *c* peroxidases. *Advances in Microbial Physiology* **52**, 73–106.

Brioukhanov, A. L., Thauer, R. K. & Netrusov, A. I. (2002). Catalase and superoxide dismutase in the cells of strictly anaerobic microorganisms. *Microbiology-Moscow* **71**, 281–285.

Brioukhanov, A. L., Netrusov, A. I. & Eggen, R. I. L. (2006). The catalase and superoxide dismutase genes are transcriptionally up-regulated upon oxidative stress in the strictly anaerobic archaeon *Methanosarcina barkeri*. *Microbiology-UK* **152**, 1671–1677.

Chen, L., Sharma, P., Le Gall, J., Mariano, A. M., Teixeira, M. & Xavier, A. V. (1994). A blue non-heme iron protein from *Desulfovibrio gigas*. *European Journal of Biochemistry* **226**, 613–618.

Coulter, E. D. & Kurtz, D. M. (2001). A role for rubredoxin in oxidative stress protection in *Desulfovibrio vulgaris*: catalytic electron transfer to rubrerythrin and two-iron superoxide reductase. *Archives of Biochemistry and Biophysics* **394**, 76–86.

Davydova, M. N. & Tarasova, N. B. (2005). Carbon monoxide inhibits superoxide dismutase and stimulates reactive oxygen species production by *Desulfovibrio desulfuricans* 1388. *Anaerobe* **11**, 335–338.

Dolla, A., Fournier, M. & Dermoun, Z. (2006). Oxygen defense in sulfate-reducing bacteria. *Journal of Biotechnology* **126**, 87–100.

Fournier, M., Zhang, Y., Wildschut, J. D., Dolla, A., Voordouw, J. K., Schriemer, D. C. & Voordouw, G. (2003). Function of oxygen resistance proteins in the anaerobic, sulfate-reducing bacterium *Desulfovibrio vulgaris* Hildenborough. *Journal of Bacteriology* **185**, 71–79.

Frey, A. D. & Kallio, P. T. (2003). Bacterial hemoglobins and flavohemoglobins: versatile proteins and their impact on microbiology and biotechnology. *FEMS Microbiology Reviews* **27**, 525–545.

Grunden, A. M., Jenney, F. E., Jr., Ma, K., Ji, M., Weinberg, M. V. & Adams, M. W. W. (2005). *In vitro* reconstitution of an NADPH-dependent superoxide reduction pathway from *Pyrococcus furiosus*. *Applied and Environmental Microbiology* **71**, 1522–1530.

Horsburgh, M. J., Wharton, S. J., Karavolos, M. & Foster, S. J. (2002). Manganese: elemental defence for a life with oxygen. *Trends in Microbiology* **10**, 496–501.

Imlay, J. A. (2002). How oxygen damages microbes: oxygen tolerance and obligate anaerobiosis. *Advances in Microbial Physiology* **46**, 111–153.

Imlay, J. A. (2006). Iron-sulphur clusters and the problem with oxygen. *Molecular Microbiology* **59**, 1073–1082.

Jacobson, M. D. (1996). Reactive oxygen species and programmed cell death. *Trends in Biochemical Sciences* **21**, 83–86.

Jean, D., Briolat, V. & Reysset, G. (2004). Oxidative stress response in *Clostridium perfringens*. *Microbiology-UK* **150**, 1649–1659.

Jovanovic, T., Ascenso, C., Hazlett, K. R. O., Sikkink, R., Kerbs, C., Litwiller, R., Benson, L. M., Moura, I., Moura, J. J. G., Radolf, J. D., Huynh, B. H., Naylor, S. & Rusnak, F. (2000). Neelaredoxin, an iron-binding protein from the syphilis spirochete, *Treponema pallidum*, is a superoxide reductase. *Journal of Biological Chemistry* **275** , 28439–28448.

Kawasaki, S., Watamura, Y., Ono, M., Watanabe, T., Takeda, K. & Niimura, Y. (2005). Adaptive responses to oxygen stress in obligatory anaerobes *Clostridium acetobutylicum* and *Clostridium aminovalericum*. *Applied and Environmental Microbiology* **71**, 8442–8450.

Kawasaki, S., Mimura, T., Satoh, T., Takeda, K. & Niimura, Y. (2006). Response of the microaerophilic *Bifidobacterium* species, *B. boum* and *B. thermophilum*, to oxygen. *Applied and Environmental Microbiology* **72**, 6854–6858.

Kim, J. H. & Suh, K. H. (2005). Light-dependent expression of superoxide dismutase from cyanobacterium *Synechocystis* sp. strain PCC 6803. *Archives of Microbiology* **183**, 218–223.

Kitamura, M., Nakanishi, T., Kojima, S., Kumagai, I. & Inoue, H. (2001). Cloning and expression of the catalase gene from the anaerobic bacterium *Desulfovibrio vulgaris* (Miyazaki F). *Journal of Biochemistry* **129**, 357–364.

Kjeldsen, K. U., Joulian, C. & Ingvorsen, K. (2004). Oxygen tolerance of sulfate-reducing bacteria in activated sludge. *Environmental Science and Technology* **38**, 2038–2043.

Le Gall, J. & Xavier, A. V. (1996). Anaerobes response to oxygen: the sulfate-reducing bacteria. *Anaerobe* **2**, 1–9.

Lee, J. H., Yeo, W. S. & Roe, J. H. (2004). Induction of the *sufA* operon encoding Fe-S assembly proteins by superoxide generators and hydrogen peroxide: involvement of OxyR, IHF and an unidentified oxidant-responsive factor. *Molecular Microbiology* **51**, 1745–1755.

Ludwig, R. A. (2004). Microaerophilic bacteria transduce energy via oxidative metabolic gearing. *Research in Microbiology* **155**, 61–70.

Lumppio, H. L., Shenvi, N. V., Summers, A. O., Voordouw, G. & Kurtz, D. M. (2001). Rubrerythrin and rubredoxin oxidoreductase in *Desulfovibrio vulgaris*: a novel oxidative stress protection system. *Journal of Bacteriology* **183**, 101–108.

Mongkolsuk, S. & Helmann, J. D. (2002). Regulation of inducible peroxide stress responses. *Molecular Microbiology* **45**, 9–15.

Podkopaeva, D. A., Yu Grabovich, M. & Dubinina, G. A. (2003). Oxidative stress and antioxidant cell protection systems in the microaerophilic bacterium *Spirillum winogradskii*. *Microbiology-Moscow* **72**, 534–542.

Romao, C. V., Liu, M. Y., Le Gall, J., Gomes, C. M., Braga, V., Pacheco, I., Xavier, A. V. & Teixeira, M. (1999). The superoxide dismutase activity of desulfoferrodoxin from *Desulfovibrio desulfuricans* ATCC 27774. *European Journal of Biochemistry* **261**, 438–443.

Sawers, G. (1999). The aerobic/anaerobic interface. *Current Opinion in Microbiology* **2**, 181–187.

Silva, G., Legall, J., Xavier, A. V., Teixeira, M. & Rodrigues-Pousada, C. (2001). Molecular characterization of *Desulfovibrio gigas* neelaredoxin, a protein involved in oxygen detoxification in anaerobes. *Journal of Bacteriology* **183**, 4413–4420.

Storz, G. & Imlay, J. A. (1999). Oxidative stress. *Current Opinion in Microbiology* **2**, 188–194.

Wildschut, J. D., Lang, R. M., Voordouw, J. K. & Voordouw, G. (2006). Rubredoxin:oxygen oxidoreductase enhances survival of *Desulfovibrio vulgaris* Hildenborough under microaerophilic conditions. *Journal of Bacteriology* **188**, 6253–6260.

Lactic acid

Asanuma, N., Yoshii, T. & Hino, T. (2004). Molecular characteristics and transcription of the gene encoding a multifunctional alcohol dehydrogenase in relation to the deactivation of pyruvate formate-lyase in the ruminal bacterium *Streptococcus bovis*. *Archives of Microbiology* **181**, 122–128.

Cocaign-Bousquet, M., Even, S., Lindley, N. D. & Loubiere, P. (2002). Anaerobic sugar catabolism in *Lactococcus lactis*: genetic regulation and enzyme control over pathway flux. *Applied Microbiology and Biotechnology* **60**, 24–32.

de Vos, W. M., Bron, P. A. & Kleerebezem, M. (2004). Post-genomics of lactic acid bacteria and other food-grade bacteria to discover gut functionality. *Current Opinion in Biotechnology* **15**, 86–93.

Diaz-Muniz, I. & Steele, J. L. (2006). Conditions required for citrate utilization during growth of *Lactobacillus casei* ATCC334 in chemically defined medium and Cheddar cheese extract. *Antonie Van Leeuwenhoek* **90**, 233–243.

Drider, D., Bekal, S. & Prevost, H. (2004). Genetic organization and expression of citrate permease in lactic acid bacteria. *Genetics and Molecular Research* **3**, 273–281.

Garrigues, C., Mercade, M., Cocaign-Bousquet, M., Lindley, N. D. & Loubiere, P. (2001). Regulation of pyruvate metabolism in *Lactococcus lactis* depends on the imbalance between catabolism and anabolism. *Biotechnology and Bioengineering* **74**, 108–115.

Klijn, A., Mercenier, A. & Arigoni, F. (2005). Lessons from the genomes of bifidobacteria. *FEMS Microbiology Reviews* **29**, 491–509.

Liu, S. Q. (2002). Malolactic fermentation in wine: beyond deacidification. *Journal of Applied Microbiology* **92**, 589.

Martin, M. G., Sender, P. D., Peiru, S., de Mendoza, D. & Magn, C. (2004). Acid-inducible transcription of the operon encoding the citrate lyase complex of *Lactococcus lactis* biovar diacetylactis CRL264. *Journal of Bacteriology* **186**, 5649–5660.

Ouwehand, A. C., Salminen, S. & Isolauri, E. (2002). Probiotics: an overview of beneficial effects. *Antonie van Leeuwenhoek* **82**, 279–289.

Sarantinopoulos, P., Kalantzopoulos, G. & Tsakalidou, E. (2001). Citrate metabolism by *Enterococcus faecalis* FAIR-E 229. *Applied and Environmental Microbiology* **67**, 5482–5487.

Smit, G., Smit, B. A. & Engels, W. J. M. (2005). Flavour formation by lactic acid bacteria and biochemical flavour profiling of cheese products. *FEMS Microbiology Reviews* **29**, 591–610.

Vido, K., le Bars, D., Mistou, M. Y., Anglade, P., Gruss, A. & Gaudu, P. (2004). Proteome analyses of heme-dependent respiration in *Lactococcus lactis*: involvement of the proteolytic system. *Journal of Bacteriology* **186**, 1648–1657.

Zaunmueller, T., Eichert, M., Richter, H. & Unden, G. (2006). Variations in the energy metabolism of biotechnologically relevant heterofermentative lactic acid bacteria during growth on sugars and organic acids. *Applied Microbiology and Biotechnology* **72**, 421–429.

Ethanol

Kalnenieks, U. (2006). Physiology of *Zymomonas mobilis*: some unanswered questions. *Advances in Microbial Physiology* **51**, 73–117.

Kalnenieks, U., Toma, M. M., Galinina, N. & Poole, R. K. (2003). The paradoxical cyanide-stimulated respiration of *Zymomonas mobilis*: cyanide sensitivity of alcohol dehydrogenase (ADH II). *Microbiology-UK* **149**, 1739–1744.

Roustan, J. L. & Sablayrolles, J. M. (2002). Impact of the addition of electron acceptors on the by-products of alcoholic fermentation. *Enzyme and Microbial Technology* **31**, 142–152.

Butyrate and butanol

Adams, C. J., Redmond, M. C. & Valentine, D. L. (2006). Pure-culture growth of fermentative bacteria, facilitated by H_2 removal: bioenergetics and H_2 production. *Applied and Environmental Microbiology* **72**, 1079–1085.

Alsaker, K. V. & Papoutsakis, E. T. (2005). Transcriptional program of early sporulation and stationary-phase events in *Clostridium acetobutylicum*. *Journal of Bacteriology* **187**, 7103–7118.

Armstrong, F. A. (2004). Hydrogenases: active site puzzles and progress. *Current Opinion in Chemical Biology* **8**, 133–140.

Bourriaud, C., Robins, R. J., Martin, L., Kozlowski, F., Tenailleau, E., Cherbut, C. & Michel, C. (2005). Lactate is mainly fermented to butyrate by human intestinal microfloras but inter-individual variation is evident. *Journal of Applied Microbiology* **99**, 201–212.

Buckel, W. & Golding, B. T. (2006). Radical enzymes in anaerobes. *Annual Review of Microbiology* **60**, 27–49.

Ceccarelli, E. A., Arakaki, A. K., Cortez, N. & Carrillo, N. (2004). Functional plasticity and catalytic efficiency in plant and bacterial ferredoxin-NADP(H) reductases. *Biochimica et Biophysica Acta – Proteins & Proteomics* **1698**, 155–165.

Colin, T., Bories, A., Lavigne, C. & Moulin, G. (2001). Effects of acetate and butyrate during glycerol fermentation by *Clostridium butyricum*. *Current Microbiology* **43**, 238–243.

Duncan, S. H., Barcenilla, A., Stewart, C. S., Pryde, S. E. & Flint, H. J. (2002). Acetate utilization and butyryl coenzyme A (CoA):acetate-CoA transferase in butyrate-producing bacteria from the human large intestine. *Applied and Environmental Microbiology* **68**, 5186–5190.

Duncan, S. H., Louis, P. & Flint, H. J. (2004). Lactate-utilizing bacteria, isolated from human feces, that produce butyrate as a major fermentation product. *Applied and Environmental Microbiology* **70**, 5810–5817.

Gonzalez-Pajuelo, M., Meynial-Salles, I., Mendes, F., Soucaille, P. & Vasconcelos, I. (2006). Microbial conversion of glycerol to 1,3-propanediol: physiological comparison of a natural producer, *Clostridium butyricum* VPI 3266, and an engineered strain, *Clostridium acetobutylicum* DG1(pSPD5). *Applied and Environmental Microbiology* **72**, 96–101.

Guedon, E. & Petitdemange, H. (2001). Identification of the gene encoding NADH-rubredoxin oxidoreductase in *Clostridium acetobutylicum*. *Biochemical and Biophysical Research Communications* **285**, 496–502.

Hillmann, F., Fischer, R. J. & Bahl, H. (2006). The rubrerythrin-like protein Hsp21 of *Clostridium acetobutylicum* is a general stress protein. *Archives of Microbiology* **185**, 270–276.

Kutty, R. & Bennett, G. (2005). Biochemical characterization of trinitrotoluene transforming oxygen-insensitive nitroreductases from *Clostridium acetobutylicum* ATCC 824. *Archives of Microbiology* **184**, 158–167.

Malaoui, H. & Marczak, R. (2001). Influence of glucose on glycerol metabolism by wild-type and mutant strains of *Clostridium butyricum* E5 grown in chemostat culture. *Applied Microbiology and Biotechnology* **55**, 226–233.

Malaoui, H. & Marczak, R. (2001). Separation and characterization of the 1,3-propanediol and glycerol dehydrogenase activities from *Clostridium butyricum* E5 wild-type and mutant D. *Journal of Applied Microbiology* **90**, 1006–1014.

May, A., Hillmann, F., Riebe, O., Fischer, R. J. & Bahl, H. (2004). A rubrerythrin-like oxidative stress protein of *Clostridium acetobutylicum* is encoded by a duplicated gene and identical to the heat shock protein Hsp21. *FEMS Microbiology Letters* **238**, 249–254.

Pryde, S. E., Duncan, S. H., Hold, G. L., Stewart, C. S. & Flint, H. J. (2002). The microbiology of butyrate formation in the human colon. *FEMS Microbiology Letters* **217**, 133–139.

Saint-Amans, S., Girbal, L., Andrade, J., Ahrens, K. & Soucaille, P. (2001). Regulation of carbon and electron flow in *Clostridium butyricum* VPI 3266 grown on glucose-glycerol mixtures. *Journal of Bacteriology* **183**, 1748–1754.

Thormann, K., Feustel, L., Lorenz, K., Nakotte, S. & Duerre, P. (2002). Control of butanol formation in *Clostridium acetobutylicum* by transcriptional activation. *Journal of Bacteriology* **184**, 1966–1973.

Tummala, S. B., Junne, S. G. & Papoutsakis, E. T. (2003). Antisense RNA downregulation of coenzyme A transferase combined with alcohol-aldehyde dehydrogenase overexpression leads to predominantly alcohologenic *Clostridium acetobutylicum* fermentations. *Journal of Bacteriology* **185**, 3644–3653.

Vignais, P. M., Billoud, B. & Meyer, J. (2001). Classification and phylogeny of hydrogenases. *FEMS Microbiology Reviews* **25**, 455–501.

Zhao, Y., Hindorff, L. A., Chuang, A., Monroe-Augustus, M., Lyristis, M., Harrison, M. L., Rudolph, F. B. & Bennett, G. N. (2003). Expression of a cloned cyclopropane fatty acid synthase gene reduces solvent formation in *Clostridium acetobutylicum* ATCC 824. *Applied and Environmental Microbiology* **69**, 2831–2841.

Zhao, Y., Tomas, C. A., Rudolph, F. B., Papoutsakis, E. T. & Bennett, G. N. (2005). Intracellular butyryl phosphate and acetyl phosphate concentrations in *Clostridium acetobutylicum* and their implication. *Applied and Environmental Microbiology* **71**, 530–537.

Zverlov, V., Berezina, O., Velikodvorskaya, G. & Schwarz, W. (2006). Bacterial acetone and butanol production by industrial fermentation in the Soviet Union: use of hydrolyzed agricultural waste for biorefinery. *Applied Microbiology and Biotechnology* **71**, 587–597.

Mixed acid fermentation

Altaras, N. E., Etzel, M. R. & Cameron, D. C. (2001). Conversion of sugars to 1,2-propanediol by *Thermoanaerobacterium thermosaccharolyticum* HG-8. *Biotechnology Progress* **17**, 52–56.

Bagramyan, K., Galstyan, A. & Trchounian, A. (2000). Redox potential is a determinant in the *Escherichia coli* anaerobic fermentative growth and survival: effects of impermeable oxidant. *Bioelectrochemistry* **51**, 151–156.

Berrios-Rivera, S. J., San, K.-Y. & Bennett, G. N. (2003). The effect of carbon sources and lactate dehydrogenase deletion on 1,2-propanediol production in *Escherichia coli*. *Journal of Industrial Microbiology and Biotechnology* **30**, 34–40.

Carlier, J. P., Marchandin, H., Jumas-Bilak, E., Lorin, V., Henry, C., Carriere, C. & Jean-Pierre, H. (2002). *Anaeroglobus geminatus* gen. nov., sp. nov., a novel member of the family *Veillonellaceae*. *International Journal of Systematic and Evolutionary Microbiology* **52**, 983–986.

Cecchini, G., Schroder, I., Gunsalus, R. P. & Maklashina, E. (2002). Succinate dehydrogenase and fumarate reductase from *Escherichia coli*. *Biochimica et Biophysica Acta – Bioenergetics* **1553**, 140–157.

Chen, X., Zhang, D.-J., Qi, W.-T., Gao, S.-J., Xiu, Z.-L. & Xu, P. (2003). Microbial fed-batch production of 1,3-propanediol by *Klebsiella pneumoniae* under micro-aerobic conditions. *Applied Microbiology and Biotechnology* **63**, 143–146.

Garnova, E. S. & Krasil'nikova, E. N. (2003). Carbohydrate metabolism of the saccharolytic alkaliphilic anaerobes *Halonatronum saccharophilum*, *Amphibacillus fermentum*, and *Amphibacillus tropicus*. *Microbiology-Moscow* **72**, 558–563.

Hasona, A., Kim, Y., Healy, F. G., Ingram, L. O. & Shanmugam, K. T. (2004). Pyruvate formate lyase and acetate kinase are essential for anaerobic growth of *Escherichia coli* on xylose. *Journal of Bacteriology* **186**, 7593–7600.

Kim, P., Laivenieks, M., Vieille, C. & Zeikus, J. G. (2004). Effect of overexpression of *Actinobacillus succinogenes* phosphoenolpyruvate carboxykinase on succinate production in *Escherichia coli*. *Applied and Environmental Microbiology* **70**, 1238–1241.

Laurinavichene, T. V., Zorin, N. A. & Tsygankov, A. A. (2002). Effect of redox potential on activity of hydrogenase 1 and hydrogenase 2 in *Escherichia coli*. *Archives of Microbiology* **178**, 437–442.

Lee, P. C., Lee, S. Y., Hong, S. H. & Chang, H. N. (2002). Isolation and characterization of a new succinic acid-producing bacterium, *Mannheimia succiniciproducens* MBEL55E, from bovine rumen. *Applied Microbiology and Biotechnology* **58**, 663–668.

Nakamura, C. E. & Whited, G. M. (2003). Metabolic engineering for the microbial production of 1,3-propanediol. *Current Opinion in Biotechnology* **14**, 454–459.

Skibinski, D. A. G., Golby, P., Chang, Y. S., Sargent, F., Hoffman, R., Harper, R., Guest, J. R., Attwood, M. M., Berks, B. C. & Andrews, S. C. (2002). Regulation of the hydrogenase-4 operon of *Escherichia coli* by the σ^{54}-dependent transcriptional activators FhlA and HyfR. *Journal of Bacteriology* **184**, 6642–6653.

van Houdt, R., Moons, P., Hueso Buj, M. & Michiels, C. W. (2006). *N*-acyl-L-homoserine lactone quorum sensing controls butanediol fermentation in *Serratia plymuthica* RVH1 and *Serratia marcescens* MG1. *Journal of Bacteriology* **188**, 4570–4572.

Wang, W., Sun, J., Hartlep, M., Deckwer, W. D. & Zeng, A. P. (2003). Combined use of proteomic analysis and enzyme activity assays for metabolic pathway analysis of glycerol fermentation by *Klebsiella pneumoniae*. *Biotechnology and Bioengineering* **83**, 525–536.

Propionate

Abou-Zeid, D. M., Biebl, H., Sproer, C. & Muller, R. J. (2004). *Propionispora hippei* sp. nov., a novel Gram-negative, spore-forming anaerobe that produces propionic acid. *International Journal of Systematic and Evolutionary Microbiology* **54**, 951–954.

Biebl, H., Schwab-Hanisch, H., Sproer, C. & Lunsdorf, H. (2000). *Propionispora vibrioides*, nov. gen., nov. sp., a new Gram-negative, spore-forming anaerobe that ferments sugar alcohols. *Archives of Microbiology* **174**, 239–247.

Deborde, C. & Boyaval, P. (2000). Interactions between pyruvate and lactate metabolism in *Propionibacterium freudenreichii* subsp shermanii: in vivo C-13 nuclear magnetic resonance studies. *Applied and Environmental Microbiology* **66**, 2012–2020.

Janssen, P. H. (1998). Pathway of glucose catabolism by strain VeGlc2, an anaerobe belonging to the Verrucomicrobiales lineage of bacterial descent. *Applied and Environmental Microbiology* **64**, 4830–4833.

Janssen, P. H. & Liesack, W. (1995). Succinate decarboxylation by *Propionigenium maris* sp. nov., a new anaerobic bacterium from an estuarine sediment. *Archives of Microbiology* **164**, 29–35.

Kiatpapan, P. & Murooka, Y. (2002). Genetic manipulation system in propionibacteria. *Journal of Bioscience and Bioengineering* **93**, 1–8.

Koussemon, M., Combet-Blanc, Y. & Ollivier, B. (2003). Glucose fermentation by *Propionibacterium microaerophilum*: effect of pH on metabolism and bioenergetics. *Current Microbiology* **46**, 141–145.

Seeliger, S., Janssen, P. H. & Schink, B. (2002). Energetics and kinetics of lactate fermentation to acetate and propionate via methylmalonyl-CoA or acrylyl-CoA. *FEMS Microbiology Letters* **211**, 65–70.

Quesada-Chanto, A., Silveira, M. M., Schmidmeyer, A. C., Schroeder, A. G., Dacosta, J. P. C. L., Lopez, J., Carvalhojonas, M. F., Artolozaga, M. J. & Jonas, R. (1998). Effect of oxygen supply on pattern of growth, and corrinoid and organic acid production of *Propionibacterium shermanii*. *Applied Microbiology and Biotechnology* **49**, 732–736.

Tholozan, J. L., Grivet, J. P. & Vallet, C. (1994). Metabolic pathway to propionate of *Pectinatus frisingensis*, a strictly anaerobic beer-spoilage bacterium. *Archives of Microbiology* **162**, 401–408.

Ye, K. M., Shijo, M., Miyano, K. & Shimizu, K. (1999). Metabolic pathway of *Propionibacterium* growing with oxygen: enzymes, C-13 NMR analysis, and its application for vitamin B-12 production with periodic fermentation. *Biotechnology Progress* **15**, 201–207.

Fermentation of amino acids

Buckel, W. (2001). Unusual enzymes involved in five pathways of glutamate fermentation. *Applied Microbiology and Biotechnology* **57**, 263–273.

Lan, J. & Newman, E. B. (2003). A requirement for anaerobically induced redox functions during aerobic growth of *Escherichia coli* with serine, glycine and leucine as carbon source. *Research in Microbiology* **154**, 191–197.

Menes, R. J. & Muxi, L. (2002). *Anaerobaculum mobile* sp. nov., a novel anaerobic, moderately thermophilic, peptide-fermenting bacterium that uses crotonate as an electron acceptor, and emended description of the genus *Anaerobaculum*. *International Journal of Systematic and Evolutionary Microbiology* **52**, 157–164.

Wallace, R. J., McKain, N., McEwan, N. R., Miyagawa, E., Chaudhary, L. C., King, T. P., Walker, N. D., Apajalahti, J. H. A. & Newbold, C. J. (2003). *Eubacterium pyruvativorans* sp. nov., a novel non-saccharolytic anaerobe from the rumen that ferments pyruvate and amino acids, forms caproate and utilizes acetate and propionate. *International Journal of Systematic and Evolutionary Microbiology* **53**, 965–970.

Fermentation of dicarboxylic acids

Duncan, S. H., Richardson, A. J., Kaul, P., Holmes, R. P., Allison, M. J. & Stewart, C. S. (2002). *Oxalobacter formigenes* and its potential role in human health. *Applied and Environmental Microbiology* **68**, 3841–3847.

Janssen, P. H. & Hugenholtz, P. (2003). Fermentation of glycolate by a pure culture of a strictly anaerobic Gram-positive bacterium belonging to the family *Lachnospiraceae*. *Archives of Microbiology* **179**, 321–328.

Sahin, N. (2003). Oxalotrophic bacteria. *Research in Microbiology* **154**, 399–407.

Sidhu, H., Hoppe, B., Hesse, A., Tenbrock, K., Bromme, S., Rietschel, E. & Peck, A. B. (1998). Absence of *Oxalobacter formigenes* in cystic fibrosis patients: a risk factor for hyperoxaluria. *Lancet* **352**, 1026–1029.

Stewart, C. S., Duncan, S. H. & Cave, D. R. (2004). *Oxalobacter formigenes* and its role in oxalate metabolism in the human gut. *FEMS Microbiology Letters* **230**, 1–7.

Ye, L., Jia, Z., Jung, T. & Maloney, P. C. (2001). Topology of OxlT, the oxalate transporter of *Oxalobacter formigenes*, determined by site-directed fluorescence labeling. *Journal of Bacteriology* **183**, 2490–2496.

Anaerobic respiration

In the previous chapter, respiration was defined as an energy conservation process achieved through electron transport phosphorylation (ETP) using externally supplied electron acceptors. Electron acceptors used in anaerobic respiration include oxidized sulfur and nitrogen compounds, metal ions, organic halogens and carbon dioxide. Other oxidized compounds reduced under anaerobic conditions include iodate, (per)chlorate, and phosphate. There is evidence to suggest that these compounds are used as electron acceptors in anaerobic ecosystems but there are some exceptions. ATP synthesis mechanisms dependent on a proton motive force are known in some fermentative bacteria. These include Na^+-dependent decarboxylation, fumarate reduction and product/proton symport, as described earlier (Section 5.8.6). Sulfidogenesis and methanogenesis are described as fermentations in some cases since a small amount of energy is conserved in these anaerobic processes. However, in these processes ATP is generated mainly through the proton motive force and they can therefore be classified as anaerobic respiration.

Many ecosystems become anaerobic when oxygen consumption is greater than its supply. Even under anaerobic conditions, natural organic compounds are continuously recycled. Anaerobic respiratory microbes convert organic materials to carbon dioxide and methane under anaerobic conditions in conjunction with fermentative microbes.

Energy is required for all forms of life. At any given conditions, those organisms utilizing energy sources more efficiently will become dominant over the others. Among the anaerobic respiratory prokaryotes, denitrifiers conserve more energy than other groups. For this reason sulfidogenesis and methanogenesis are inhibited in the presence of nitrate, and sulfate inhibits methanogenesis. Ferric iron is ubiquitous on Earth, and has a redox potential higher than sulfate (Table 9.1). Because of its availability, ferric iron is a more important electron acceptor than nitrate. It has been estimated that more than half of the degradation of organic compounds under anaerobic conditions is coupled to the reduction of ferric iron. Halogenated hydrocarbons are generally toxic and recalcitrant to

Table 9.1. | *Free energy from NADH oxidation coupled to electron acceptors used by prokaryotes*

Reduction half reaction	$\Delta G^{0'} (kJ/2e^-)$
$\frac{1}{2} O_2 \rightarrow 2H_2O$	-219.07
$2NO_3^- \rightarrow N_2$	-206.12
$Fe^{3+} \rightarrow Fe^{2+}$	-209.46
$CH_3Cl \rightarrow CH_4 + HCl$	-135.08
$MnO_2 \rightarrow Mn^{2+}$	-134.52
$Se(VI) \rightarrow Se(IV)$	-129.96
$Cr(VI) \rightarrow Cr(III)$	-90.04
$As(V) \rightarrow As(0)$	-46.11
$SO_4^{2-} \rightarrow HS^-$	-20.24
$CO_2 \rightarrow CH_4$	-14.58

Source: FEMS Microbiology Reviews **23**, 615–627, 1999, and other sources.

degradation under aerobic conditions but can serve as electron acceptors under anaerobic conditions.

9.1 | Denitrification

Denitrification is an economically and environmentally important microbial process. Nitrogen fertilizer is lost from farmland through this process, which is also exploited to remove nitrogen from wastewater treatment plants before discharge to prevent eutrophication. Many facultative anaerobes use nitrate and nitrite as electron acceptors under oxygen-limited conditions (Table 9.2). Strains of *Paracoccus*, *Ralstonia* (*Alcaligenes*) and *Pseudomonas* are the best-known denitrifiers. These metabolize carbohydrates through glycolysis and the TCA cycle as do the aerobes. The electrons from these metabolic pathways are consumed reducing nitrate and nitrite.

9.1.1 Biochemistry of denitrification

Denitrification is defined as a microbial process reducing nitrate (NO_3^-) or nitrite (NO_2^-) to generate a proton motive force under anaerobic conditions. Gaseous nitrogen (N_2) is the main product with small amounts of NO (nitric oxide) and N_2O (nitrous oxide). The denitrifiers have a similar electron transport system to aerobic respiratory organisms. Reduced coenzyme Q provides electrons for nitrate reductase, and the other enzymes oxidize reduced cytochrome *c* (Figure 9.1).

Nitrate reductase (NaR) reduces NO_3^- to NO_2^- coupled to the oxidation of quinol. Subsequently nitrite reductase (NiR) oxidizes reduced cytochrome *c* to reduce NO_2^- to nitric oxide (NO). Two molecules of NO are reduced further to nitrous oxide (N_2O) by nitric oxide reductase (Figure 9.2). Finally, nitrous oxide reductase reduces

Table 9.2. *Typical denitrifying prokaryotes*

Organism	Electron donor	Electron acceptor
Alcaligenes cycloclastes	organics	NO_2^{-a}
Alcaligenes faecalis	organics	NO_3^-
Ralstonia eutropha (Alcaligenes eutrophus)	H_2	NO_3^-
Bacillus licheniformis	organics	NO_3^-
Bacillus azotoformans	organics	NO_2^-
Hyphomicrobium vulgare	CH_4	NO_3^-
Paracoccus denitrificans	organics, H_2	NO_3^-
Propionibacterium pentosaceum	organics	NO_3^-
Pseudomonas fluorescens	organics	NO_3^-
Thiobacillus denitrificans	S^{2-}	NO_3^-
Pyrobaculum aerophilum[b]	organics, H_2	NO_3^-

[a] NO_3^- is not reduced.
[b] archaeon.

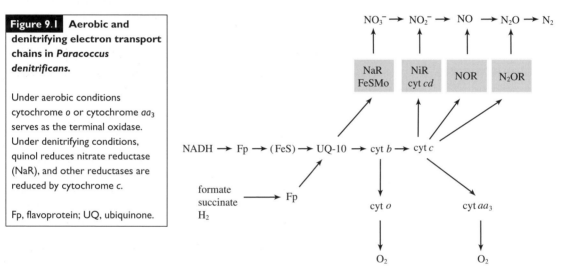

Figure 9.1 **Aerobic and denitrifying electron transport chains in *Paracoccus denitrificans*.**

Under aerobic conditions cytochrome o or cytochrome aa_3 serves as the terminal oxidase. Under denitrifying conditions, quinol reduces nitrate reductase (NaR), and other reductases are reduced by cytochrome c.

Fp, flavoprotein; UQ, ubiquinone.

N_2O to N_2. In *Paracoccus denitrificans*, nitrate reductase and nitric oxide reductase are cytoplasmic membrane proteins while nitrite reductase is located in the periplasm.

9.1.1.1 Nitrate reductase

Nitrate reductase is a complex protein consisting of α, β and γ subunits. This enzyme is an [Fe-S] protein containing molybdenum in addition to iron. The [Fe-S] centres of the β subunits participate in electron transfer within the molecule, and the b-type cytochromes of the γ subunit are involved in electron transfer from quinol to the [Fe-S] clusters (Figure 9.3). The α subunit containing molybdopterin and a [4Fe-4S] cluster catalyzes the reductive reaction:

$$NO_3^- + 2H^+ + 2e^- \rightarrow NO_2^- + H_2O$$

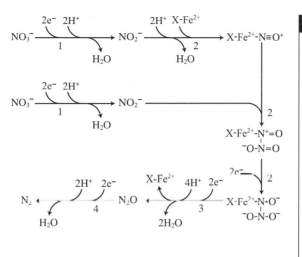

Figure 9.2 **Dissimilatory nitrate reduction.**

(Gottschalk, G. 1986, *Bacterial Metabolism*, 2nd edn., Figure 5.18. Springer, New York)

Nitrate is reduced to nitrite by nitrate reductase (1) that oxidizes quinol. Nitrite reductase (2) reduces nitrite to nitric oxide. Two molecules of nitric oxide are reductively condensed to nitrous oxide by nitric oxide reductase (3). Nitrous oxide reductase (4) reduces its substrate to gaseous nitrogen. Reduced cytochrome *c* supplies electrons for the last three reducing steps. These enzymes function either at the cytoplasmic membrane or in the periplasm.

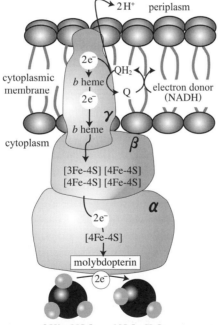

Figure 9.3 **Model of dissimilatory nitrate reductase in *Escherichia coli*.**

(*BBA-Gene Struct. Express.* **1446**:1–23, 1999)

Nitrate reductase is a complex protein consisting of α, β and γ subunits. The *b*-type cytochrome associated with the γ subunit is reduced, oxidizing quinol and liberating protons. The [Fe-S] cluster of the β subunit transfers electrons from the γ subunit to the α subunit where nitrate is reduced to nitrite.

The β subunit connects the α subunit at the cytoplasmic side and the γ subunit embedded in the membrane. The γ subunit takes electrons using *b*-type cytochromes from the reduced quinone and transfers them to the α subunit through the β subunit involving a [3Fe-3S] cluster and three [4Fe-4S] clusters. The *b*-type cytochromes export protons during the electron transfer reactions. *Paracoccus denitrificans* has a separate nitrate reductase in the periplasm of unknown function.

The nitrate reductase that initiates denitrification is different from the assimilatory nitrate reductase that is involved in the use of nitrate as a nitrogen source (Section 6.2.2). This enzyme is referred to as the dissimilatory nitrate reductase and is inhibited by oxygen but not by NH_3. On the other hand, NH_3 inhibits the assimilatory enzyme but O_2 does not. Energy in the form of the proton motive force is not conserved by the assimilatory enzyme.

9.1.1.2 Nitrite reductase

Nitrite produced by nitrate reductase is excreted to the periplasmic region by a specific transporter to be reduced by nitrite reductase. Two different nitrite reductases are known, one containing *c*-type and *d*-type cytochromes (cdNiR) and the other a copper protein (CuNiR). CuNiR is widely distributed in prokaryotes including Gram-positive denitrifiers such as species of the genus *Bacillus*, Gram-negative bacteria such as *Pseudomonas aureofaciens* and in archaea (species of the genus *Haloarcula*): cdNiR is found only in Gram-negative bacteria.

Cytochrome *c* or a Cu-containing small protein, pseudoazurin, provides electrons to the homodimeric cdNiR. Cytochrome *c* of the cdNiR takes electrons and nitrite is reduced to nitric oxide (NO) at cytochrome *d*. cdNiR reduces not only nitrite but also nitric oxide. The gene for nitric oxide reductase has been identified and its mutant cannot grow under denitrifying conditions. The function of nitric oxide reducing activity in cdNiR is not known.

This dissimilatory nitrite reductase is different from the assimilatory nitrite reductase, which is a cytoplasmic enzyme catalyzing the reduction of NO_2^- directly to NH_4^+ (Section 6.2.2).

$$NO_2^- + 3NADH + 5H^+ \rightarrow NH_4^+ + 3NAD^+ + 2H_2O$$

9.1.1.3 Nitric oxide reductase and nitrous oxide reductase

Nitric oxide produced by nitrite reductase is toxic and is reduced by nitric oxide reductase to nitrous oxide as soon as it is produced. Nitric oxide reductase is a complex enzyme consisting of a small subunit containing *c*-type cytochromes and two large subunits containing *b*-type cytochromes. This enzyme reduces two molecules of NO to N_2O using the electrons available from reduced cytochrome *c*. Protons are not translocated during this reaction.

N_2O is reduced to N_2 by nitrous oxide reductase, generating a proton motive force through consuming $2H^+$ in the cytoplasm. Reduced cytochrome *c* provides electrons for this reaction.

9.1.2 ATP synthesis in denitrification

During aerobic respiration, the H^+/O ratio is approximately 10 in prokaryotes depending on the electron carriers involved in ETP (Section 5.8.4). The redox potentials of the denitrification reactions are shown in Table 9.3. The redox potentials of the first two reactions

Table 9.3.	Redox potentials and ATP yield of denitrification reactions	
Half reaction	$E^{0'}$ (mV)	$P/2e^-$
$\frac{1}{2} O_2 / H_2O$	+815	3
NO_3^- / NO_2^-	+421	2
NO_2^-/NO	+337	2
$NO/\frac{1}{2}N_2O$	+1180	3
N_2O/ N_2	+1350	3

are lower but those of the other two reactions are much higher than that of O_2. Assuming that the $P/2e^-$ (ATP synthesized/$2e^-$ transported) ratio is 3 in aerobic respiration (Section 5.8.4), the ratio can be 2 in the first two reactions and 3 in the other reactions.

9.1.3 Regulation of denitrification

Denitrification is an alternative respiratory metabolism in facultative anaerobes under O_2-limited conditions. Since less energy is conserved in denitrification than in aerobic respiration, denitrification is strongly inhibited by O_2 with few exceptions. The expression of the genes for denitrification is regulated, and so are the enzyme activities after they are expressed. The enzyme activities appear 4–120 minutes after the culture becomes anaerobic, and their expression is stimulated by nitrate, indicating that their expression is repressed by O_2. When the culture is transferred from anaerobic to aerobic conditions, the enzymes are slowly irreversibly inactivated. The enzymes become inactive under aerobic conditions because their affinity for reduced coenzyme Q and cytochrome c is lower than that for aerobic respiratory enzymes.

Many denitrifiers can use various other electron acceptors including oxygen, nitrate, dimethyl sulfoxide (DMSO, $+0.15$ V), dimethyl sulfide (DMS) and fumarate ($+0.03$ V). Organisms utilizing more than one electron acceptor have elaborate regulatory mechanisms to conserve more energy under given growth conditions. Electron acceptors with a higher redox potential are preferentially used over those with a lower redox potential. Under aerobic conditions, a two-component system, ArcA/ArcB, stimulates expression of the genes for aerobic respiration, and FNR activates the expression of anaerobic enzymes under anaerobic conditions (Section 12.2.4). Through these mechanisms, oxygen represses the expression of anaerobic respiratory enzymes. Enzymes that use electron acceptors of lower redox potential such as DMSO and fumarate are repressed by nitrate.

In FNR activation, dimeric FNR binds to the FNR box that is upstream of the anaerobic enzyme genes, increasing the affinity of the promoter for RNA polymerase. The expression of the nitrate reductase gene is further stimulated by a two-component system, NarX/NarL, in the presence of nitrate (Section 12.2.10). The membrane sensor protein NarX is phosphorylated consuming ATP in the presence of nitrate, and transfers phosphate to the regulatory protein NarL. The phosphorylated NarL is an activator of nitrate reductase gene expression. Another two-component system, NarQ/NarP, controls expression of the other enzymes of denitrification including nitrite reductase, nitric oxide reductase and nitrous oxide reductase. Regulation by two-component systems is discussed later (Section 12.2).

9.1.4 Denitrifiers other than facultatively anaerobic chemoorganotrophs

Denitrification has been regarded as a purely prokaryotic metabolism occurring mainly in facultative anaerobes and using organic electron donors. However, this process has now been identified in many other organisms including fungi such as *Fusarium oxysporum* and related strains, *Cylindrocarpon tonkinense*, *Fusarium solani*, *Gibberella fujikuroi*, *Talaromyces flavus*, *Trichoderma hamatum* and *Trichosporon cutaneum*. These reduce nitrate and nitrite to N_2 or N_2O.

Some chemolithotrophs can use nitrate as their electron acceptor. *Thiobacillus denitrificans* and *Thiomicrospira denitrificans* use sulfur as the electron donor and nitrate as the electron acceptor. This anaerobic chemolithotrophic metabolism is ubiquitous in freshwater and marine environments. *Thiobacillus denitrificans* can couple uranium(IV) oxidation to nitrate reduction. Some alkalophilic bacteria can also use reduced sulfur as the electron donor and nitrate or nitrite as electron acceptors. *Thioalkalivibrio denitrificans* and *Thioalkalivibrio nitratireducens* oxidize thiosulfate reducing nitrate, and *Thioalkalivibrio thiocyanodenitrificans* grows chemolithotrophically oxidizing thiocyanate and thiosulfate to sulfate coupled to the reduction of nitrate. These are facultative anaerobes (Section 10.3.1).

A carboxydobacterium (Section 10.6), *Pseudomonas carboxydoflava*, uses nitrate under anaerobic conditions with carbon monoxide as its electron donor. Physiologically diverse H_2-oxidizing bacteria and archaea can also use nitrate as their electron acceptor (Section 10.5). A thermophilic chemolithotrophic hydrogen bacterium, *Hydrogenobacter thermophilus*, uses O_2 as well as nitrate as electron acceptors (Section 10.5.1). A group of thermophilic bacteria isolated from hydrothermal vents grow chemolithotrophically using H_2 as the electron donor reducing nitrate to ammonia. These include *Caminibacter* spp., *Desulfurobacterium crinifex*, *Thermovibrio ruber*, *Thermovibrio ammonificans* and *Hydrogenomonas thermophila*. They reduce elemental sulfur to hydrogen sulfide as an alternative electron acceptor (Section 9.3). Species of *Caminibacter* and *Hydrogenomonas thermophila* are microaerophiles, and therefore use O_2 as their electron acceptor. *Ferroglobus placidus*, a strictly anaerobic archaeon, grows chemoautotrophically oxidizing hydrogen coupled to nitrate reduction. There is some evidence for some anaerobes using iron (Fe^0) or ferrous iron (Fe^{2+}) as their electron donors under denitrifying conditions, but pure cultures have not yet been isolated.

Another group of chemolithotrophs uses NH_4^+ as an electron donor reducing nitrite:

$$NH_4^+ + NO_2^- \rightarrow N_2 + 2H_2O \ (\Delta G^{0'} = -358 \text{ kJ/mol } NH_4^+)$$

These have not been isolated in pure culture but have been enriched in a reactor treating wastewater containing a high concentration of NH_4^+ in a process known as the anaerobic ammonia oxidation (ANAMMOX) process. Based on 16S ribosomal RNA gene sequences, these organisms have been named as *Brocadia anammoxidans*, *Kuenenia stuttgartiensis*, *Scalindua brodae* and *Scalindua wagneri*. Similar clones are widely distributed in natural and artificial ecosystems. A nitrifier, *Nitromonas europaea*, oxidizes ammonia under anaerobic conditions using nitrogen dioxide (NO_2) as the electron acceptor.

Some strictly anaerobic bacteria grow chemoheterotrophically reducing nitrate to NH_4^+. These include *Denitrovibrio acetiphilus*, *Thauera selenatis*, *Wolinella succinogenes* and *Desulfovibrio gigas*. Since this process generates a proton motive force and is not inhibited by NH_4^+, the reduction of nitrate to NH_4^+ is regarded as dissimilatory nitrate reduction. This process may also be referred to as nitrate ammonification. A similar reaction is found in *Bacillus subtilis* and *Moorella thermoacetica* (formerly *Clostridium thermoaceticum*). Homoacetogenic (Section 9.5) *Moorella thermoacetica* ferments glucose to three molecules of acetate, two arising from pyruvate from glycolysis and the third from CO_2 produced by pyruvate:ferredoxin oxidoreductase. A culture grown with nitrate produces only two molecules of acetate since the electrons used for the synthesis of the third acetate on glucose are consumed to reduce nitrate. The electron transport system in nitrate ammonification has not yet been established.

Filamentous sulfur bacteria of the genera *Thioploca* and *Beggiatoa* and a related bacterium, *Thiomargarita namibiensis*, reduce nitrate to ammonia (Section 10.3.1). They accumulate nitrate within intracellular vacuoles. In these bacteria, nitrate is used as an electron acceptor and reduced to ammonia with sulfide or sulfur as the electron donors when the oxygen supply is limited.

Clostridium perfringens and propionate-producing anaerobes reduce nitrate to NH_4^+ as in nitrate ammonification, but this reaction in these bacteria is not coupled to the generation of a proton motive force. Since electrons are disposed of in nitrate reduction, the electron acceptor, acetyl-CoA, is converted to acetate synthesizing ATP. The fermentative propionate-producers do not produce propionate in the presence of nitrate, but produce only acetate from lactate (Figures 8.16 and 8.18, and Section 8.7). This process is referred to as fermentative nitrate reduction, and nitrate in this case is termed an electron sink.

Though it is well known that O_2 represses expression of the denitrification enzyme genes, aerobic nitrate reduction has also been known for a long time. A facultative anaerobic bacterium isolated from an anaerobic digester, *Microvigula aerodenitrificans*, can reduce nitrate under air-saturated conditions. This bacterium can use O_2 and nitrate as electron acceptors simultaneously. A similar property is found in some other bacteria including *Alcaligenes faecalis*, *Citrobacter diversus*, *Pseudomonas nautica* and *Thiospaera pantotropha*.

9.1.5 Oxidation of xenobiotics under denitrifying conditions

Many xenobiotics can be oxidized under denitrifying conditions including aromatic compounds such as benzene, toluene, alkylbenzene, xylene, phenol and resorcinol, and chlorinated hydrocarbons such as atrazine and 3-chlorobenzoate (Section 9.9).

9.2 | Metal reduction

Some chemolithotrophs oxidize various metal ions as electron donors (Sections 10.4 and 10.7) while oxidized metal ions can serve as electron acceptors in anaerobic respiration. Microbes play important roles in the cycling of metals and metalloids such as iron, manganese, selenium, and arsenic (as well as all the elements comprising cellular constituents such as carbon, nitrogen, sulfur, phosphorus, etc.). Microbes reduce metal ions as electron acceptors (dissimilatory metal reduction) and for biosynthetic purposes (assimilatory metal reduction), both of which may reduce metal toxicity. During the reduction of the metals and metalloids listed in Table 9.4, free energy is conserved in the form of a proton motive force. Some metal reductions, for example Au(III) to Au(0), are not, however, coupled to energy conservation. Table 9.4 shows a partial list of metal-reducing bacteria.

Before an Fe(III)-reducing bacterium, *Shewanella oneidensis*, was characterized, metal reduction in anaerobic ecosystems was regarded as a chemical process coupled to the oxidation of sulfide (HS^-) produced through sulfidogenesis (Section 9.3). Fe(III) is virtually insoluble in water and this electron acceptor is reduced on the cell surface. Cr(VI), As(V), U(VI) and Tc(VII) are water soluble and toxic. Dissimilatory reducers of these metal species can reduce toxicity and remove them from the aqueous phase.

9.2.1 Fe(III) and Mn(IV) reduction

Most Fe(III) and Mn(IV) reducers use fermentation products such as acetate and lactate as their electron donors, though some use carbohydrates and glycerol (Table 9.4). Carbohydrate and glycerol metabolism in these organisms is not fully elucidated.

Shewanella alga and *Shewanella oneidensis* oxidize lactate and pyruvate to acetate, and the resulting electrons are consumed in reducing the metal ions. They cannot oxidize acetate further and are referred to as incomplete oxidizers similar to some strains of the sulfate-reducing *Desulfovibrio* genus.

$$\text{lactate}^- + 4Fe^{3+}\,(2Mn^{4+}) + 2H_2O \longrightarrow \text{acetate}^- + 4Fe^{2+}\,(2Mn^{2+}) + HCO_3^- + 5H^+$$

$$\text{pyruvate}^- + 2Fe^{3+}\,(Mn^{4+}) + 2H_2O \longrightarrow \text{acetate}^- + 2Fe^{2+}\,(Mn^{2+}) + HCO_3^- + 3H^+$$

Table 9.4. | *Anaerobes capable of using oxidized metal ions as electron acceptors*

Electron acceptor	Organism	Electron donor
Fe(III)	*Acidiphilum cryptum*	carbohydrate
	Aeromonas hydrophila	glycerol
	Deferribacter thermophilus	acetate, amino acids
	Desulfuromonas acetoxidans	acetate
	Ferroglobus placidus[a]	acetate
	Ferroplasma acidarmanus[a]	carbohydrate
	Geobacter chapellei	acetate
	Geobacter hydrogenophilus	acetate, propionate, H_2
	Geobacter metallireducens	acetate
	Geobacter sulfurreducens	acetate
	Geoglobus ahangari	acetate
	Geothrix fermentans	acetate
	Geovibrio ferrireducens	acetate
	Pantoea agglomerans	acetate
	Pelobacter carbinolicus	ethanol
	Pyrobaculum islandicum[a]	H_2
	Shewanella alga	lactate
	Shewanella frigidimarina	carbohydrate
	Shewanella saccharophilia	carbohydrate
	Shewanella oneidensis	lactate
	Thermoterrabacterium ferrireducens	carbohydrate, glycerol, amino acids
Mn(IV)	*Desulfurimonas acetoxidans*	acetate
	Geobacter metallireducens	acetate
	Geobacter sulfurreducens	acetate
	Pyrobaculum islandicum[a]	H_2
	Shewanella oneidensis	lactate
U(VI)	*Desulfovibrio desulfuricans*	lactate
	Geobacter metallireducens	acetate
	Pyrobaculum islandicum[a]	H_2
	Shewanella alga	lactate
Selenate [Se(VI)]	*Aeromonas hydrophila*	glycerol
	Bacillus arsenicoselenatis	lactate
	Bacillus selenitireducens	amino acids
	Desulfotomaculum auripigmentum	lactate
	Enterobacter cloacae	amino acids
	Geospirillum barnesii	amino acids
	Pyrobaculum aerophilum[a]	H_2
	Selenihalanaerobacter shriftii	acetate, glucose
	Thauera selenatis	acetate
	Wolinella succinogenes	formate, hydrogen
Chromate [Cr(VI)]	*Bacterium dechromaticans*	lactate
	Enterobacter cloacae	amino acids
	Pyrobaculum islandicum	H_2
	Shewanella oneidensis	lactate
Co(III)	*Aeromonas hydrophila*	glycerol
	Pyrobaculum islandicum[a]	H_2

Table 9.4. (cont.)		
Electron acceptor	Organism	Electron donor
Arsenate [As(V)]	*Bacillus arsenicoselenatis*	lactate
	Chrysiogenes arsenatis	acetate
	Pyrobaculum arsenaticum[a]	H_2
	Sulfurospirillum arsenophilum	lactate
	Sulfurospirillum barnesii	lactate
Pertechnetate [Tc(VII)]	*Geobacter metallireducens*	acetate
	Shewanella oneidensis	lactate
Vanadate [V(V)]	*Shewanella oneidensis*	lactate

[a] Archaea.

Geobacter metallireducens, *Geobacter sulfurreducens* and *Desulfuromonas acetoxidans* oxidize acetate completely to CO_2 through the modified TCA cycle described later (Figure 9.5).

$$CH_3COO^- + 8Fe^{3+}(4Mn^{4+}) + 4H_2O \longrightarrow 8Fe^{2+}(4Mn^{2+}) + 2HCO_3^- + 9H^+$$

In addition to acetate, *Geobacter metallireducens* can use aromatic compounds such as phenol, benzoate and toluene as their electron donors. Other xenobiotics, including benzene, toluene, ethylbenzene, xylene, and trinitrotoluene, can also be removed under Fe(III)-reducing conditions (Section 9.9).

Fe(III) and Mn(IV) are practically insoluble in water under physiological conditions. Bacteria using these insoluble electron acceptors need to export electrons to reduce them extracellularly as they cannot import these metal species. The electrons are transferred from the bacterial cell to the insoluble electron acceptor either through direct contact between them and a mineral surface or facilitated by soluble mediators such as humic acid.

Cells of *Shewanella oneidensis* and *Geobacter sulfurreducens* are electrochemically active due to cell surface *c*-type cytochromes. Ferric reductase is an outer membrane enzyme in these bacteria. The ferric reductase of *Geobacter sulfurreducens* is a complex protein containing *c*-type cytochromes and flavin adenine dinucleotide (FAD). The electron transport system linking electron donor oxidation and electron acceptor reduction is not fully elucidated. The whole genome has been sequenced in *Shewanella oneidensis* and *Geobacter sulfurreducens* and more than 40 genes encoding cytochromes have been identified in these bacteria. More than four *c*-type cytochromes are present on the outer membrane of *Shewanella oneidensis*.

Many Fe(III)-reducers are electrochemically active and capable of exchanging electrons with an electrode. They metabolize electron donors with the electrode being used as an electron acceptor or electron sink in a fuel-cell-type electrochemical device. Similar devices are used to enrich microbial consortia oxidizing organic contaminants in wastewater with concomitant electricity generation.

Since Fe(III)-reducers metabolize a wide range of electron donors, including many xenobiotics, and Fe(III) is a major constituent of the Earth's crust, these organisms have considerable potential for the bioremediation of contaminated soil. They are also responsible for reduced methanogenesis in freshwater ecosystems such as paddy fields, since they can outcompete methanogens.

9.2.2 Microbial reduction of other metals

In addition to Fe(III) and Mn(IV), many other metal ions are reduced by prokaryotes for biosynthetic purposes, as electron acceptors or in detoxification mechanisms. Selenium (a metalloid) is a component of some enzymes including glycine reductase and formate dehydrogenase in the form of selenocysteine. Selenate is reduced in an assimilatory process, and also reduced as an electron acceptor. Other metal ions used as electron acceptors include As(V), Cr(VI), Mo(VI), Se(VI), U(VI) and V(V) (Table 9.4). Many microbes, including fermentative bacteria, reduce toxic As(V), Cr(VI), Se(VI) and V(V) without conserving energy available from the oxidation–reduction. These reactions result in detoxification. It is not known if energy is conserved during the microbial reduction of mercury(II), palladium(II), tellurium(VI), neptunium(V) and technetium(VII).

As(V), Cr(VI), Mo(VI), Se(VI) and U(VI) are rapidly reduced in anaerobic ecosystems, and various bacteria and archaea have been isolated based on their abilities to use these metal ions as their electron acceptor. These organisms are phylogenetically diverse and found widely in nature. Though many prokaryotes have been identified as metal reducers, their carbon metabolism and electron transport processes have been less studied.

The reduction of metals as electron acceptors is different from their reduction for biosynthetic purposes. As in nitrate reduction and sulfate reduction, metal reduction using the metal as an electron acceptor is referred to as dissimilatory metal reduction, while biosynthetic metal reduction is referred to as assimilatory reduction. The arsenate reductase of *Chrysiogenes arsenatis* is a [Fe-S] protein containing molybdenum and zinc. Various respiratory inhibitors (Section 5.8.2.3) inhibit Cr(VI) and V(V) reduction by *Shewanella oneidensis*, indicating the involvement of quinone and cytochromes in this electron transport process.

Metal reducers offer potential for treating wastewater containing toxic metals and metalloids including As(V), Cr(VI), Mo(VI), Se(VI), U(VI), Np(V), Tc(VII) and V(V). For example, a radiation-resistant bacterium, *Deinococcus radiodurans*, reduces water-soluble U(VI) and Tc(VII) to insoluble U(IV) and Tc(IV) respectively, in effluents from nuclear power plants.

9.2.3 Metal reduction and the environment

Before molecular oxygen accumulated in the atmosphere during the evolution of the Earth, Fe(III) and Mn(IV) were the most important electron acceptors in the carbon cycle since they are widely distributed in the Earth's crust, and chemolithotrophs used reduced metal ions as electron donors. Through these microbial activities, metal ions were subject to solubilization–immobilization cycles, thereby concentrating metal ores upon which some of the mining industries depend.

Among the electron acceptors used by microbes under anaerobic conditions, Fe(III) and Mn(IV) are most commonly used, since nitrate is not widely distributed. These metal ions have a lower redox potential than other electron donors and their reduction conserves more energy than sulfidogenesis and methanogenesis. It is estimated that most organic compound oxidization under anaerobic conditions is coupled to reduction of these metal ions. This has an important environmental impact as metal reducers can oxidize xenobiotics in contaminated soil (Section 9.9). However, Fe(III) reduction to soluble Fe(II) can be accompanied by the mobilization of other toxic metals and phosphate, thereby deteriorating subsurface water quality. Toxic metals, Fe(III) and phosphate can form stable complexes.

9.3 | Sulfidogenesis

Many bacteria and archaea use sulfate (SO_4^{2-}) and elemental sulfur (S^0) as their electron acceptors. All of these organisms are obligate anaerobes (Table 9.5). Sulfidogens are grouped into mesophilic Gram-negative bacteria, spore-forming Gram-positive bacteria, thermophilic bacteria and hyperthermophilic archaea. In Table 9.5 they are listed as sulfate reducers and sulfur reducers, with each of these groups being further divided into complete oxidizers and incomplete oxidizers.

The main habitat of sulfidogens is sediments rich in organic electron donors and sulfate. They cause corrosion of underground and underwater structures and are especially troublesome in petroleum refineries and sewage works, causing great economic loss.

Sulfur can be reduced by some metal reducers including species of *Wolinella*, *Shewanella*, *Sulfurospirillum* and *Geobacter*.

Electron donors used by sulfidogens are mainly fatty acids and alcohols produced by fermentation, with a few exceptions. Incomplete oxidizers metabolize ethanol and lactate to acetate, while acetate is completely oxidized by the complete oxidizers. Sulfidogens couple electron transport to sulfate or sulfur with the generation of a proton motive force in most cases, but incomplete oxidizing sulfur-reducing archaea do not. They dispose of electrons from fermentative metabolism in a process referred to as fermentative sulfidogenesis. A similar metabolism is found among the denitrifiers (Section 9.1.4).

As shown in Table 9.5, some sulfidogenic bacteria and archaea use H_2 as their electron donor. Some of them are facultative chemolithotrophs and others are heterotrophs using organic compounds as their major electron donor. A group of thermophilic obligately chemolithotrophic bacteria have been isolated from hydrothermal vents

Table 9.5. *Sulfidogens*		
Organism (genus or species)	Electron donor	Acetate catabolism
Sulfate reducers		
Incomplete oxidizer		
Eubacteria		
Desulfovibrio	lactate, ethanol, malate, H_2, methanol[a], glycerol[a]	–
Desulfomonas	lactate	–
Desulfomicrobium	lactate, ethanol, malate, succinate, fumarate, H_2	–
Desulfobulbus	lactate, ethanol, H_2, propionate	–
Desulfobotulus	lactate, long-chain fatty acids	–
Desulfotomaculum nigrificans	lactate, ethanol, H_2	–
Thermodesulfobacterium	lactate, H_2	–
Complete oxidizer		
Eubacteria		
Desulfoarculus	acetate, propionate, long-chain fatty acids	CO[b]
Desulfobacca acetoxidans	acetate, propionate, butyrate, ethanol, propanol	CO
Desulfobacter	acetate, H_2[a], ethanol[a]	TCA[c]
Desulfobacterium	lactate, acetate, ethanol	CO
Desulfococcus	lactate, ethanol, propionate, long-chain fatty acids, benzoate	CO
Desulfosarcina	lactate, ethanol, H_2, acetate, propionate, long-chain fatty acids, benzoate	CO
Desulfonema	lactate, acetate, H_2, succinate, fumarate, malate, propionate, formate, benzoate, long-chain fatty acids	CO
Desulfosarcina	lactate, ethanol, H_2, acetate, propionate, long-chain fatty acids, benzoate	CO
Desulfotomaculum acetoxidans	ethanol, butanol, H_2, acetate, butyrate, long-chain fatty acids	CO
Desulfovibrio baarsii	acetate, propionate, butyrate, long-chain fatty acids	CO
Desulfovirga adipica	acetate, propionate, long-chain fatty acids, alcohols	?[d]
Archaeon		
Archaeoglobus fulgidus	lactate, malate, H_2, succinate[a], fumarate[a], glucose[a]	CO
Sulfur reducers		
Incomplete oxidizer		
Eubacteria		
Desulfovibrio gigas	lactate, ethanol, malate	–
Desulfomicrobium	lactate	–
Wolinella	lactate	–
Shewanella	lactate	–
Sulfurospirillum arcachonense (microaerophile)	lactate	–

Table 9.5.	(cont.)		
Organism (genus or species)	Electron donor	Acetate catabolism
Archaea		
Thermoproteus	H$_2$, yeast extract	– (fermentative)
Pyrobaculum	H$_2$, yeast extract	– (fermentative)
Desulfurococcus	H$_2$, yeast extract	– (fermentative)
Complete oxidizer		
Eubacteria		
Desulfuromonas	lactate[a], ethanol, acetate, propionate, succinate, glutamate	TCA
Desulfurella	acetate	TCA
Desulfuromusa	acetate, propionate, yeast extract, peptone	?[d]
Geobacter	acetate	TCA

[a] Used by some species.
[b] Carbon monoxide dehydrogenase (acetyl-CoA or Wood–Ljungdahl) pathway.
[c] Modified TCA cycle.
[d] Not known.

using H$_2$ as the electron donor and reducing elemental sulfur to hydrogen sulfide (Section 10.5.3). These include *Balnearium lithotrophicum*, *Caminibacter* spp., *Desulfurobacterium crinifex*, *Thermovibrio ruber*, *Thermovibrio ammonificans* and *Hydrogenomonas thermophila*. They all reduce nitrate to ammonia except for *Balnearium lithotrophicum* (Section 9.1.4). *Hydrogenomonas thermophila* and species of *Caminibacter* are microaerophiles while the others are strict anaerobes.

9.3.1 Biochemistry of sulfidogenesis

9.3.1.1 Reduction of sulfate and sulfur

Sulfate is activated to adenosine-5′-phosphosulfate (APS) consuming two high energy phosphate bonds of ATP. This reaction is similar to the initial reaction of assimilatory sulfate reduction (Section 6.3) but APS is not activated further to adenosine-3′-phosphate-5′-phosphosulfate (PAPS). The redox potential of the HSO$_4^-$/HSO$_3^-$ half reaction is −0.454 V, which is much lower than that of any biological electron carriers including ferredoxin (−0.41 V). The activation process makes the electron transfer to the electron acceptor a downhill reaction.

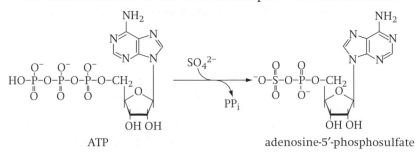

Pyrophosphate (PP$_i$) produced by ATP sulfurylase is hydrolyzed to two P$_i$ by the action of inorganic phosphatase in most cases, but strains of *Desulfotomaculum* use PP$_i$ to activate acetate to acetyl-phosphate through the reaction catalyzed by acetate-pyrophosphate kinase.

APS reductase liberates AMP on reducing APS to sulfite that is further reduced to sulfide (HS$^-$) by sulfite reductase consuming 6e$^-$ (Figure 9.4).

Membrane-bound sulfur reductase reduces water-insoluble elemental sulfur in sulfur reducers. Electrons are transferred to sulfur reductase from electron donors through menaquinone and cytochrome *c*. Sulfur reducers grow attached to sulfur granules for efficient utilization of this water-insoluble electron acceptor. Sulfide in solution reacts with granular sulfur to form polysulfides such as tetrasulfide and pentasulfide. Polysulfides are highly soluble and might be the actual electron acceptors of sulfur reductase.

9.3.1.2 Carbon metabolism

Incomplete oxidizers provide electrons for the reduction of sulfate and sulfur from the oxidation of lactate and ethanol to acetate. Lactate dehydrogenase oxidizes lactate to pyruvate that is further oxidized to acetyl-CoA through the phosphoroclastic reaction catalyzed by pyruvate:ferredoxin oxidoreductase (Section 8.5.1). The electron carrier of lactate dehydrogenase is not known. Acetyl-CoA is converted to acetate, synthesizing ATP through substrate-level phosphorylation (Figure 9.4). Alcohol dehydrogenase and acetaldehyde dehydrogenase oxidize ethanol to acetyl-CoA:

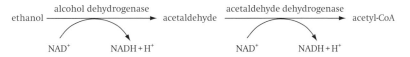

Acetate is metabolized through the acetyl-CoA pathway (Sections 9.5.2 and 10.8.3) in most complete oxidizers except in strains of *Desulfobacter*, *Desulfuromonas* and *Desulfurella* that oxidize acetate through a modified TCA cycle (Table 9.5). Each of these organisms has its own form of modified TCA cycle (Figure 9.5) with the differences occurring in the reactions of acetyl-CoA and citrate synthesis. Acetate is activated to acetyl-CoA by succinyl-CoA:acetate CoA-SH transferase in strains of *Desulfobacter* and *Desulfuromonas*, while acetate kinase and phosphotransacetylase catalyze the reactions in strains of *Desulfurella* consuming ATP. ATP:citrate lyase synthesizes citrate and ATP from acetyl-CoA and oxaloacetate in strains of *Desulfobacter*, and citrate synthase condenses acetyl-CoA and oxaloacetate into citrate without ATP synthesis in *Desulfuromonas* and *Desulfurella* strains.

Ferredoxin is reduced coupled to the oxidation of 2-ketoglutarate in a reaction catalyzed by 2-ketoglutarate:ferredoxin oxidoreductase; succinate oxidation is coupled to the reduction of menaquinone. Electrons from the reduced menaquinone undergo reverse electron transport to be used in reducing sulfur (Figure 9.8b, and Section 10.1).

Figure 9.4 **Hydrogen cycling in sulfate reduction by *Desulfovibrio* species.**

(Gottschalk, G. 1986, *Bacterial Metabolism*, 2nd edn., Figure 8.29. Springer, New York)

Incomplete oxidizing sulfate-reducers, including *Desulfovibrio* species, oxidize two lactate to acetate to reduce one sulfate. Through substrate-level phosphorylation (reaction 4) two ATPs are produced, and two high energy phosphate bonds are consumed (reaction 5) per sulfate reduced. Extra ATP needs to be generated for bacterial growth and this is achieved by hydrogen cycling. Cytoplasmic hydrogenase produces H_2 and oxidizes the reduced electron carriers that result during lactate oxidation. H_2 diffuses to the periplasm where another hydrogenase reduces cytochrome c_3 leaving protons. The reduced cytochrome c_3 supplies electrons for sulfate reduction across the membrane.

1, lactate dehydrogenase; 2, pyruvate: ferredoxin oxidoreductase; 3, phosphotransacetylase; 4, acetate kinase; 5, ATP sulfurylase; 6, pyrophosphatase; 7, APS reductase; 8, sulfite reductase; 9, cytoplasmic hydrogenase; 10, periplasmic hydrogenase.

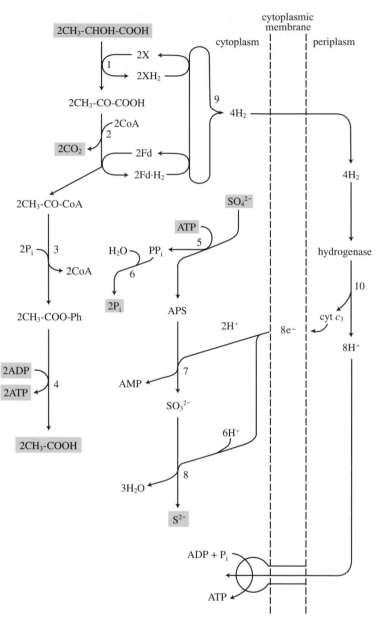

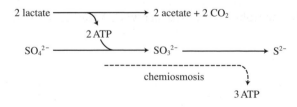

ATP balance

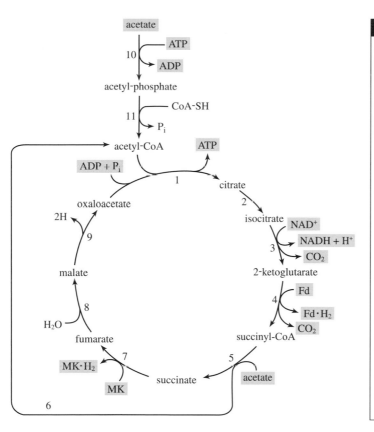

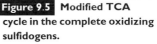

Figure 9.5 Modified TCA cycle in the complete oxidizing sulfidogens.

Acetate is activated in a reaction catalyzed by succinyl-CoA:acetate CoA-SH transferase (6) in *Desulfobacter* and *Desulfuromonas*, and by acetate kinase (10) and phosphotransacetylase (11) in *Desulfurella*. Citrate is formed by ATP:citrate lyase in *Desulfobacter*, and by citrate synthase in *Desulfuromonas* and *Desulfurella*.

1, ATP:citrate lyase (*Desulfobacter*) or citrate synthase (*Desulfuromonas* and *Desulfurella*); 2, 3, 8, 9 as in the TCA cycle; 4, 2-ketoglutarate:ferredoxin oxidoreductase; 5, succinyl-CoA synthetase (*Desulfurella*); 6, succinyl-CoA:acetate CoA transferase (*Desulfobacter* and *Desulfuromonas*); 7, succinate:menaquinone oxidoreductase; 10, acetate kinase; 11, phosphotransacetylase.

All complete oxidizing sulfate reducers, except strains of *Desulfobacter*, oxidize acetate through the acetyl-CoA pathway which is not known in sulfur reducers. (Figure 9.6). Since the key enzyme of this pathway is carbon monoxide dehydrogenase (CODH), this pathway is also referred to as the CODH pathway or the Wood–Ljungdahl pathway to honour the individuals who discovered this metabolism. CODH splits acetyl-CoA to form methyl-tetrahydrofolate (CH_3-H_4F) and enzyme-bound [CO]. The enzyme-bound [CO] is oxidized to CO_2 reducing ferredoxin. CH_3-H_4F is oxidized to CO_2. In the archaeon *Archaeoglobus fulgidus*, the archaeal C1-carriers tetrahydromethanopterin (H_4MPT) and methanofuran (MF) replace tetrahydrofolate.

Since all reactions of the acetyl-CoA pathway are reversible, some chemolithotrophic anaerobes synthesize acetyl-CoA for biosynthetic purposes from CO_2 and H_2 through this pathway (Section 10.8.3). These include methanogens, homoacetogens and chemolithotrophic sulfidogens.

Strains of *Desulfobulbus*, incomplete oxidizers, oxidize propionate to acetate through a similar pathway as the succinate–propionate pathway in strains of *Propionibacterium* (Section 8.7.1, Figure 8.16). These bacteria oxidize ethanol and lactate to acetate via acetyl-CoA and pyruvate, respectively (Figure 9.7).

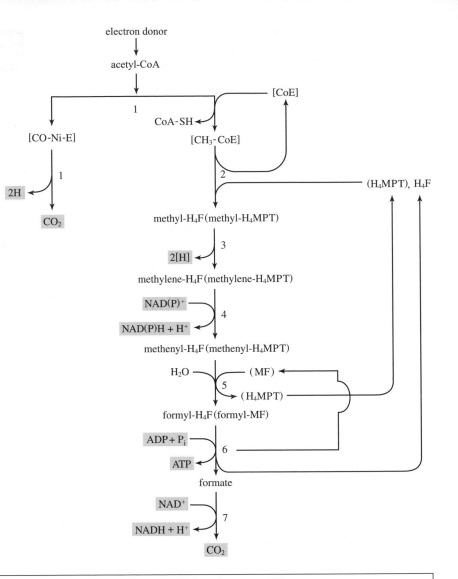

Figure 9.6 **Complete oxidation of acetate through the acetyl-CoA pathway in sulfidogens.**

All complete oxidizing sulfate-reducers, except *Desulfobacter*, employ the acetyl-CoA pathway involving carbon monoxide dehydrogenase (1, or acetyl-CoA synthase). This pathway is the reverse of acetate production from $H_2 + CO_2$ in homoacetogens. H_4F (tetrahydrofolate) carries C_1 compounds in bacteria: the archaeon *Archaeoglobus fulgidus* uses tetrahydromethanopterin (H_4MPT) and methanofuran (MF) for this purpose.

1, carbon monoxide dehydrogenase; 2, methyltransferase; 3, methyl-H_4F dehydrogenase; 4, methylene-H_4F dehydrogenase; 5, methenyl-H_4F cyclohydrolase; 6, formyl-H_4F synthetase; 7, formate dehydrogenase.

[CO-Ni-E]: enzyme-bound carbon monoxide; CoE, corrinoid enzyme.

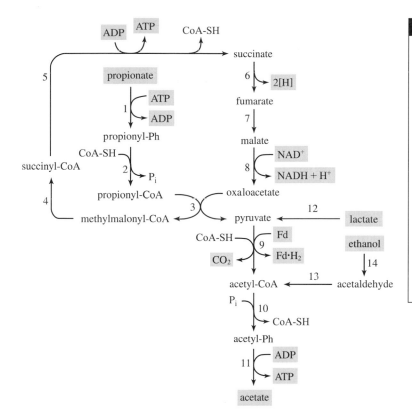

Figure 9.7 **Oxidation of propionate, lactate and ethanol to acetate by strains of Desulfobulbus.**

1, propionate kinase; 2, phosphotransacylase; 3, methylmalonyl-CoA:pyruvate transcarboxylase; 4, methylmalonyl-CoA mutase; 5, succinyl-CoA synthetase; 6, succinate dehydrogenase; 7, fumarase; 8, malate dehydrogenase; 9, pyruvate:ferredoxin oxidoreductase; 10, phosphotransacetylase; 11, acetate kinase; 12, lactate dehydrogenase; 13, aldehyde dehydrogenase, 14, alcohol dehydrogenase.

9.3.2 Electron transport and ATP yield in sulfidogens

9.3.2.1 Incomplete oxidizers

As shown below, incomplete oxidizing strains of the genus *Desulfovibrio* gain two ATP through substrate-level phosphorylation, and consume the same number of ATP molecules to reduce a molecule of sulfate, oxidizing two molecules of lactate to acetate:

$$SO_4^{2-} + 8H^+ + ATP \rightarrow S^{2-} + AMP + PP_i + 4H_2O$$

$$PP_i + H_2O \rightarrow 2P_i$$

$$2 \text{ lactate} + 2ADP + 2P_i \rightarrow 2 \text{ acetate} + 8H^+ + 2CO_2 + 2ATP$$

$$ATP + AMP \rightarrow 2ADP$$

$$\overline{\text{(sum) } SO_4^{2-} + 2 \text{ lactate} \rightarrow S^{2-} + 2 \text{ acetate} + 2H_2O + 2CO_2}$$

These bacteria need other mechanisms to synthesize ATP to grow on lactate using sulfate as the electron acceptor. They synthesize ATP through an electron transport phosphorylation (ETP) mechanism. Hydrogenases play an important role in generating a proton motive force which is coupled to sulfidogenesis in these bacteria. The cytoplasmic hydrogenase reduces $8H^+$ to produce $4H_2$. The H_2 diffuses into the periplasm where it is oxidized by periplasmic hydrogenase to reduce cytochrome c_3. Electrons are transferred back into the cytoplasm to reduce sulfate consuming another $8H^+$. These reactions result in the generation of a proton motive force and this process is referred to as the hydrogen cycling mechanism (Figure 9.4).

The hydrogen cycling mechanism is not known in strains of *Desulfotomaculum*. They use PP_i produced by ATP sulfurylase to phosphorylate acetate through the action of acetate-pyrophosphate kinase. These bacteria synthesize 2ATP from lactate oxidation to acetate and invest 1ATP in sulfate reduction. Pyrophosphatase activity is low in these bacteria, and PP_i is not hydrolyzed to $2P_i$. In many bacteria, polyphosphate is used to store energy (Section 13.2.4).

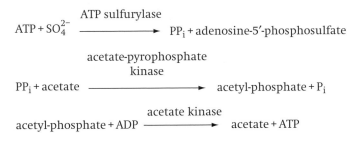

9.3.2.2 Complete oxidizers

Though ATP is synthesized through SLP in complete oxidizers (reaction 1 in Figure 9.5 and reaction 6 in Figure 9.6), the number of ATP molecules is less than that required for the activation of acetate and sulfate. As shown below, enough free energy is generated from acetate oxidation coupled to the reduction of sulfate or sulfur. This free energy is conserved in the form of a proton motive force.

$$acetate^- + SO_4^{2-} + 3H^+ \rightarrow 2CO_2 + H_2S + 2H_2O$$
$$(\Delta G^{0\prime} = -63\,\text{kJ/mol acetate})$$
$$acetate^- + H^+ + 4S^0 + 2H_2O \rightarrow 2CO_2 + 4H_2S$$
$$(\Delta G^{0\prime} = -39\,\text{kJ/mol acetate})$$

Acetate oxidation is coupled to the reduction of $NAD(P)^+$ and ferredoxin. Electrons from NAD(P)H and reduced ferredoxin are transported to menaquinone, exporting H^+ to generate a proton motive force for ATP synthesis (Figure 9.8). Since the redox potential of S^0/H_2S is lower than that of menaquinone, electrons of the reduced menaquinone with succinate oxidation (reaction 6 in Figure 9.7) undergo reverse electron transport (Section 10.1) to reduce S^0 in sulfur reducers employing a modified TCA cycle (Figure 9.8b).

The ATP yield of the complete oxidizers is 0.6 ATP/acetate in a sulfur reducer (*Desulfuromonas acetoxidans*) and 0.8 ATP/acetate in a sulfate reducer (*Desulfotomaculum acetoxidans*). These figures are calculated from cell yield, Y_{ATP} and the maintenance energy (Section 6.14.3).

9.3.3 Carbon skeleton supply in sulfidogens

Some complete oxidizing sulfate reducers can grow chemolithotrophically using H_2 as the electron donor (Table 9.5). Among these

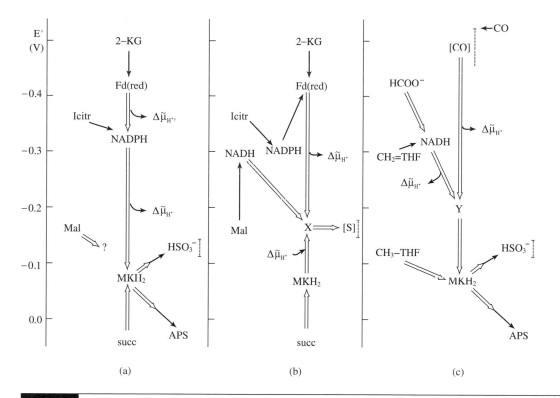

Figure 9.8 Electron metabolism in complete oxidizing sulfidogens.

(Widdel, F. & Bak, F. 1992. In *The Prokaryotes*, 2nd edn., pp. 3352–3378. Springer, New York)

Acetate oxidation is mediated through a modified TCA cycle to reduce sulfate in *Desulfobacter postgatei* (a) and to reduce sulfur in *Desulfuromonas acetoxidans* (b). *Desulfotomaculum acetoxidans* (c) oxidizes acetate through the acetyl-CoA pathway to reduce sulfate. The TCA cycle intermediate 2-ketoglutarate (2-KG) reduces ferredoxin (Fd), and isocitrate (Icitr) reduces $NADP^+$. Reduced ferredoxin and NADPH transfer electrons to menaquinone (MKH_2) in *Desulfobacter postgatei* (a) to an unknown electron carrier (X) in *Desulfuromonas acetoxidans* (b), generating a proton motive force. The electrons are finally consumed to reduce adenosine-5′-phosphosulfate (APS), sulfite (HSO_3^-) or sulfur ([S]). Another TCA cycle intermediate, malate (Mal), reduces NAD^+ in *Desulfuromonas acetoxidans* (b), but the electron carrier in *Desulfobacter postgatei* (a) is not known. Succinate (Suc) reduces menaquinone. Electrons from menaquinone are transferred to the unknown electron carrier (X) to reduce sulfur through a reverse electron transport process in *Desulfuromonas acetoxidans* (b). Electrons from the acetyl-CoA pathway are transferred to an unknown electron carrier (Y) with generation of a proton motive force. Reactions indicated with single arrows take place in the cytoplasm, and those with double arrows occur in the cytoplasmic membrane. Electrons from acetyl-CoA pathway intermediates, formate ($HCOO^-$) and methylene-tetrahydrofolate (CH_2=THF) are transferred to NAD^+, and those from methyl-tetrahydrofolate (CH_3–THF) to menquinone.

organisms, *Desulfobacter hydrogenophilus* fixes CO_2 through a reductive TCA cycle (Section 10.8.2) and the acetyl-CoA pathway (Section 10.8.3) is employed in others, including *Desulfosarcina variabilis*, *Desulfonema limicola*, and *Desulfotomaculum orientis*.

Acetyl-CoA synthesized from $H_2 + CO_2$ or ethanol is reduced to pyruvate by pyruvate:ferredoxin oxidoreductase. Pyruvate is metabolized through the incomplete TCA fork (Section 5.4.1) and gluconeogenesis (Section 4.2) to supply carbon skeletons for biosynthesis.

9.3.4 Oxidation of xenobiotics under sulfidogenic conditions

Lactate and acetate are common electron donors for most sulfidogens. However, it has been found that many xenobiotics can be oxidized under sulfidogenic conditions. They include aliphatic, aromatic, and halogenated hydrocarbons. Microbial consortia that include sulfidogens are generally responsible for xenobiotic oxidation. Methane is also oxidized under these conditions. Sulfate reducers have been isolated from soil contaminated with petroleum and related substances. Xenobiotics that can be used as electron donors in anaerobic respiratory processes are discussed later (Section 9.9).

9.4 | Methanogenesis

The conversion of organic materials to methane (CH_4) has been known for a long time in anaerobic ecosystems. This microbial process has been applied to treat wastewaters rich in organic content. Methanogens are strictly anaerobic archaea. They flourish in anaerobic organic-rich ecosystems with low sulfate concentrations, but they are one of the most tedious groups of microbes to cultivate in pure culture. CO_2 is the electron acceptor in methanogens. The common electron donors used by methanogens are formate, methanol, acetate, methylamines, carbon monoxide and H_2 to reduce CO_2. Some methanogens can use ethanol, 2-propanol, 2-butanol and ketones.

9.4.1 Methanogens

Methanogens occupy five phylogenetic orders in the archaea (Table 9.6). They are grouped according to the electron donors they use and comprise hydrogenotrophic, methylotrophic and aceticlastic methanogens.

9.4.1.1 Hydrogenotrophic methanogens

The majority of known methanogens grow on the free energy available from the reduction of CO_2 to CH_4 using H_2 as the electron donor. These are referred to as chemolithotrophic methanogens.

$$CO_2 + 4H_2 \rightarrow CH_4 + 2H_2O \quad (\Delta G^{0'} = -136 \text{ kJ/mol } CH_4)$$

They use formate and CO in addition to $CO_2 + H_2$. These are oxidized to CO_2 before being reduced to CH_4.

$$4HCOOH \rightarrow CH_4 + 3CO_2 + 2H_2O \quad (\Delta G^{0'} = -144 \text{ kJ/mol } CH_4)$$
$$4CO + 2H_2O \rightarrow CH_4 + 3CO_2 \quad (\Delta G^{0'} = -211 \text{ kJ/mol } CH_4)$$

Some of the chemolithotrophic methanogens oxidize ethanol, 2-propanol and 2-butanol completely or partially using CO_2 as the electron acceptor. These include strains of *Methanobacterium*, *Methanoculleus*, *Methanogenium*, *Methanolacinia*, *Methanospirillum*, and *Methanocorpusculum*.

Table 9.6. *Methanogens*

Methanogen	Electron donor	Characteristics
Order Methanobacteriales		
Methanobacterium	$H_2 + CO_2$, formate[a], CO[a] $2°$ alcohol $+ CO_2$[a]	pseudomurein
Methanobrevibacter	$H_2 + CO_2$, formate[a]	pseudomurein, S-layer
Methanosphaera	$H_2 +$ methanol	pseudomurein, acetate or CO_2 as carbon source in mammalian intestine
Methanothermobacter	$H_2 + CO_2$, formate	sulfur reducer
Methanothermus	$H_2 + CO_2$	pseudomurein, S-layer, thermophile, hot springs
Order Methanococcales		
Methanococcus	$H_2 + CO_2$, formate[a]	S-layer
Methanothermococcus	$H_2 + CO_2$, formate[a]	S-layer
Methanocaldococcus	$H_2 + CO_2$	S-layer, fast growth
Methanotorris	$H_2 + CO_2$	S-layer
Order Methanomicrobiales		
Methanomicrobium	$H_2 + CO_2$, formate[a]	acetate as carbon source
Methanoculleus	$H_2 + CO_2$, formate[a], $2°$ alcohol $+ CO_2$[a]	acetate as carbon source
Methanofollis	$H_2 + CO_2$, formate[a], $2°$ alcohol $+ CO_2$[a]	S-layer, acetate as carbon source
Methanogenium	$H_2 + CO_2$, formate[a], $2°$ alcohol $+ CO_2$[a]	S-layer, acetate as carbon source
Methanolacinia	$H_2 + CO_2$, $2°$ alcohol	acetate as carbon source
Methanoplanus	$H_2 + CO_2$, formate	sulfur reducer
Methanocorpusculum	$H_2 + CO_2$, formate, $2°$ alcohol $+ CO_2$[a]	yeast extract or peptone as carbon source
Methanospirillum	$H_2 + CO_2$, formate, $2°$ alcohol $+ CO_2$[a]	grows in chains
Methanocalculus	$H_2 + CO_2$	halophilic (12% NaCl), disc-shaped, S-layer
Order Methanosarcinales		
Methanosarcina	$H_2 + CO_2$[a], acetate[a], CO[a], methanol, methylamines	forms clumps
Methanococcoides	methanol, methylamines	halophilic (0.2 M NaCl), requires Mg^{2+}
Methanohalobium	methanol, methylamines	halophilic (2.6–5.1 M NaCl)
Methanohalophilus	methanol, methylamines	halophilic (1.0–2.5 M NaCl)
Methanolobus	methanol, methylamines, dimethyl sulfide[a]	S-layer
Methanosalsum	methanol, methylamines, dimethylsulfide	halophilic and alkalophilic, S-layer
Methanosaeta	acetate	grows in chains, halophilic (2.5 M NaCl)
Order Methanopyrales		
Methanopyrus	$H_2 + CO_2$	hyperthermophilic (110 °C), sulfur reducer, pseudomurein

[a] Dependent on strain.
$2°$, secondary.

9.4.1.2 Methylotrophic methanogens

Methyl compounds such as methanol and methylamines are used as the substrate by methylotrophic methanogens. These include strains of the order *Methanosarcinales* except those of the genus *Methanosaeta*. Strains of *Methanolobus* can use methyl sulfide.

$$4CH_3OH \rightarrow 3CH_4 + CO_2 + 2H_2O \quad (\Delta G^{0'} = -106 \text{ kJ/mol } CH_4)$$

Methyltrophic methanogens of the genus *Methanosphaera* use methanol only in the presence of CO_2 as the electron acceptor.

9.4.1.3 Aceticlastic methanogens

Limited numbers of methanogens can use acetate as their substrate. These include strains of *Methanosaeta* and *Methanosarcina*. The former use only acetate and grow in chains. Strains of *Methanosarcina* are most versatile in terms of their substrate. They can use all known methanogenic substrates except secondary alcohols.

$$CH_3COOH \rightarrow CH_4 + CO_2 \quad (\Delta G^{0'} = -37 \text{ kJ/mol } CH_4)$$

9.4.2 Coenzymes in methanogens

Methanogenic archaea employ some unique coenzymes. These coenzymes are known in some hyperthermophilic archaea and in a limited number of the eubacteria (Sections 4.3.4, 7.7 and 7.9.2). The presence of unique coenzymes in methanogenic archaea is one of the bases for the hypothesis of separate archaeal evolution. The presence of archaeal coenzymes in eubacteria is taken as evidence of recent lateral gene transfer.

Tetrahydrofolate (H_4F) and *S*-adenosylmethionine (SAM) are C1-carriers in eubacteria and eukaryotes (Section 6.6.2). Methanofuran (MF), 5,6,7,8-tetrahydromethanopterin (H_4MPT), and coenzyme M replace them in methanogenic archaea. MF is a formyl-carrier, H_4MPT is the archaeal analogue of tetrahydrofolate in eubacteria carrying formyl-, methenyl-, methylene- and methyl-groups, while coenzyme M is a methyl-group carrier. In addition to these C1-carriers, methanogenic archaea use coenzyme F_{420}, coenzyme F_{430}, 7-mercaptoheptanoylthreonine phosphate, and methanophenazine (Figure 9.9).

Coenzyme F_{420} is reduced by hydrogenase, formate dehydrogenase and carbon monoxide dehydrogenase, and oxidized in reducing CO_2 to CH_4. Coenzyme F_{420} is involved in the reactions catalyzed by $NADP^+$ reductase, pyruvate synthase and 2-ketoglutarate synthase in methanogenic archaea. Since coenzyme F_{420} is fluorescent, methanogenic archaea containing this coenzyme can be distinguished from eubacteria using fluorescence microscopy. The structure of F_{420} shown in Figure 9.9 is found in *Methanobacterium thermoautotrophicum* and is referred to as F_{420}-2 because it has a side chain consisting of two glutamates. In other strains, 4–7 glutamate residues form the F_{420} side chain. Coenzyme F_{430} is another methanogenic electron carrier involved in the reaction catalyzed by methyl-coenzyme M methylreductase.

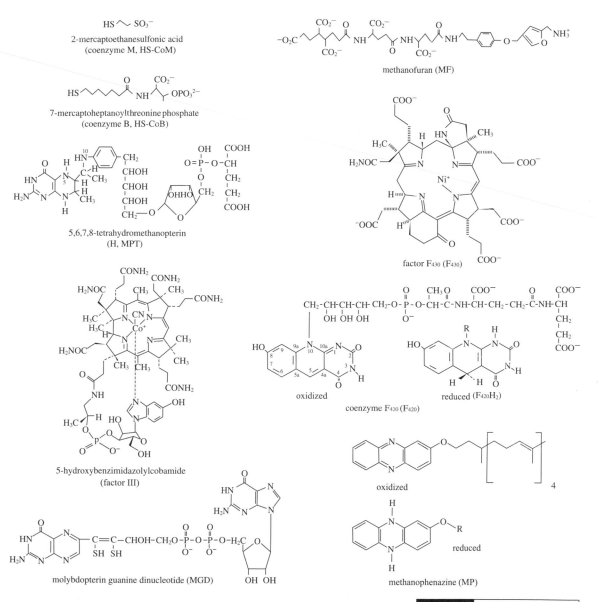

Figure 9.9 Methanogenic cofactors and their structures.

(*Microbiol. Mol. Biol. Rev.* **63**:570–620, 1999; *FEMS Microbiol. Rev.* **23**:13–38, 1999)

MF, H₄MPT and F₄₂₀ are known in some thermophilic archaea, e.g. *Archaeoglobus fulgidus* and *Solfolobus acidocaldarius*. Methanogenic coenzymes are also known in eubacteria. Coenzyme F₄₂₀ is used in strains of *Streptomyces* in reactions synthesizing tetracycline and lincomycin. F₄₂₀-dependent glucose-6-phosphate dehydrogenase catalyzes the first reaction of the HMP pathway in eubacteria, *Mycobacterium smegmatis* and some species of *Nocardia* (Section 4.3.4). Coenzyme M is also known in eubacteria. *Rhodococcus rhodochrous* uses Co-M in the oxidative metabolism of propylene (Section 7.7). A methylotroph, *Methylobacterium extorquens*, metabolizes C1-compounds not only bound to tetrahydrofolate but also to H₄MPT (Section 7.9.2).

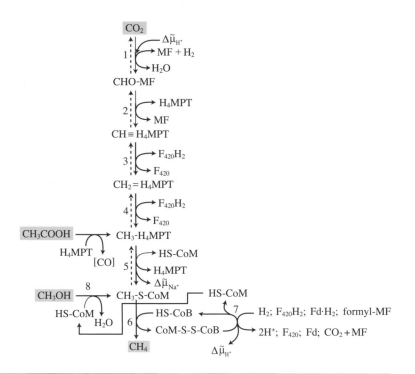

Figure 9.10 Reduction of methanol and carbon dioxide to methane.

(*Microbiol. Mol. Biol. Rev.* **63**:570–620, 1999)

Formyl-MF dehydrogenase (1) reduces carbon dioxide to form formyl-MF. The formyl group is transferred from formyl-MF to tetrahydromethanopterin (H_4MPT) to be reduced to methyl-H_4MPT. The methyl group is transferred to coenzyme M (CoM-SH) before being reduced to methane with the electrons supplied from the reduced forms of coenzyme F_{420} indirectly. Energy is conserved in the reactions catalyzed by methyl-H_4MPT:CoM-methyltransferase (5) in the form of a sodium motive force and by heterodisulfide reductase (7) in the form of a proton motive force. The proton motive force is consumed in the reaction catalyzed by formyl-MF dehydrogenase (1). Methanol is converted to methyl-CoM, part of which is oxidized (dotted line) to supply the reducing equivalents for the energy-conserving reaction catalyzed by heterodisulfide reductase (7).

1, formyl-MF dehydrogenase; 2, formyl-MF:H_4MPT formyltransferase and methenyl-H_4MPT cyclohydrolase; 3, F_{420}-dependent methylene-H_4MPT dehydrogenase; 4, F_{420}-dependent methylene-H_4MPT reductase; 5, methyl-H_4MPT:CoM-methyltransferase; 6, methyl-CoM reductase; 7, heterodisulfide reductase; 8, methyltransferase.

MF, methanofuran; H_4MPT, tetrahydromethanopterin; HS-CoB, 7-mercaptoheptanoylthreonine phosphate; $Fd \cdot H_2$, reduced ferredoxin (Figure 9.13).

9.4.3 Methanogenic pathways

Molecular oxygen not only inhibits the growth of methanogens, but also irreversibly inactivates many of the methanogenic enzymes. Due to such oxygen sensitivity, it can be difficult to study the reactions of methanogenesis. However, most of the enzymes of methanogenesis have been characterized and further information is now available from genomic analysis.

9.4.3.1 Hydrogenotrophic methanogenesis

CO_2 is reduced to formate and bound to methanofuran (MF) by the action of a membrane enzyme, formyl-MF dehydrogenase, which

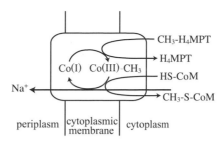

Figure 9.11 Sodium motive force generation by the methyl-H$_4$MPT:CoM-methyltransferase complex.

(*Microbiol. Mol. Biol. Rev.* **63**:570–620, 1999)

The methyl-H$_4$MPT:CoM-methyltransferase complex is a Na$^+$-dependent enzyme present in the cytoplasmic membrane. The prosthetic group cobamide mediates methyl group transfer from methyl-H$_4$MPT to coenzyme M. During this process, Na$^+$ is exported generating a sodium motive force. When methanol is oxidized to CO$_2$, the sodium motive force is consumed in the reverse reaction.

oxidizes H$_2$. In the next reaction, the formyl group is transferred from formyl-MF to H$_4$MPT before being reduced to the methyl group. Methyl-H$_4$MPT:CoM-transferase forms methyl-CoM by transferring the methyl group. Methyl-CoM is finally reduced to CH$_4$ in a reaction catalyzed by methyl-CoM reductase forming a heterodisulfide CoM-S-S-CoB. Methyl-CoM reductase contains coenzyme F$_{430}$ as a prosthetic group. The heterodisulfide is reduced to CoM-SH and CoB-SH (Figure 9.10). The reduced F$_{420}$ provides electrons for all these reactions except the first reaction which is catalyzed by formyl-MF dehydrogenase. The cytoplasmic F$_{420}$-reducing hydrogenase reduces F$_{420}$ oxidizing H$_2$.

Hydrogenotrophic methanogens synthesize carbon skeletons for biosynthesis from methyl-CoM. Carbon monoxide dehydrogenase synthesizes acetyl-CoA from methyl-CoM and enzyme-bound [CO] (Figure 9.16).

Methanogens require Na$^+$ for growth since this cation is used in an energy conservation process in a reaction catalyzed by a membrane-bound enzyme, methyl-H$_4$MPT:CoM-methyltransferase. This enzyme exports Na$^+$ in the methyl-CoM-forming reaction to generate a sodium motive force (Figure 9.11). Methyl-H$_4$MPT:CoM-methyltransferase consists of 6–8 subunits depending on the strain, and contains a cobalamin (vitamin B$_{12}$) derivative, 5-hydroxybenzimidazolyl cobamide, and [4S-4Fe] clusters.

Methyl-CoM reductase catalyzes the last reaction of methanogenesis and forms CH$_4$ and CoM-S-S-CoB heterodisulfide from methyl-CoM and reduced 7-mercaptoheptanoylthreonine phosphate (HS-CoB). This enzyme contains coenzyme F$_{430}$ as a prosthetic group. Heterodisulfide reductase reduces CoM-S-S-CoB heterodisulfide to HS-CoM and HS-CoB using electrons through the electron transport system from H$_2$. A proton motive force is generated in this reaction (Figure 9.12a).

9.4.3.2 Methylotrophic methanogenesis

Methyl transferase forms methyl-CoM from methanol and methylamine (Figure 9.10). As in hydrogenotrophic methanogens, methyl-CoM is reduced to CH$_4$ forming heterodisulfide (Figure 9.10). To supply electrons for the reaction catalyzed by heterodisulfide reductase, a part of methyl-CoM is oxidized to CO$_2$ through the reverse reactions of CO$_2$ reduction to methyl-CoM (Figure 9.10) in the absence of other electron donors such as H$_2$. During the oxidation of methyl-CoM, the sodium motive force is consumed in the reaction catalyzed

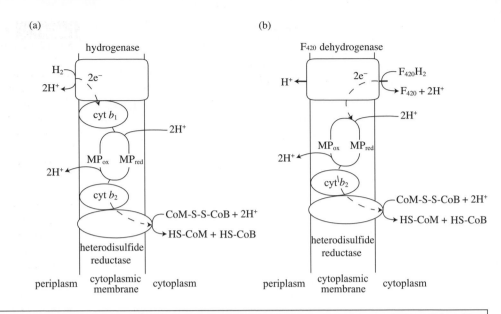

(a)   (b)

Figure 9.12 Proton motive force generation by heterodisulfide reductase during methanogensis from $CO_2 + H_2$, and from methanol in *Methanosarcina mazei*.

(*Microbiol. Mol. Biol. Rev.* **63**:570–620, 1999)

(a) With $CO_2 + H_2$, the cytoplasmic membrane hydrogenase oxidizes hydrogen, leaving protons in the periplasmic region. The electrons are transferred to the heterodisulfide reductase through cytochrome b_1, methanophenazine (MP), and cytochrome b_2. During electron transport, protons are exported and the heterodisulfide reductase consumes protons. (b) When methanol is used, F_{420} dehydrogenase transfers electrons from reduced $F_{420}H_2$ with substrate oxidation to methanophenazine (MP) accompanied by proton export.

by methyl-H_4MPT:CoM-methyltransferase (Figure 9.11). It is not known what electron carrier formyl-MF dehydrogenase reduces or if the enzyme reaction is coupled to the generation of a proton motive force. The reaction catalyzed by heterodisulfide reductase in methylotrophic methanogens is different from that in hydrogenotrophic methanogens. The latter use hydrogen as the electron donor of the reaction (Figure 9.12a) while the former uses reduced F_{420} (Figure 9.12b). A proton motive force is generated in this reaction.

9.4.3.3 Aceticlastic methanogenesis

Strains of *Methanosarcina* and *Methanosaeta* use acetate to produce CH_4. Species of the genus *Methanosaeta* have a higher affinity for acetate and activate it to acetyl-CoA in a one-step reaction catalyzed by acetyl-CoA synthetase:

$$\text{acetate} + \text{CoA-SH} + \text{ATP} \xrightarrow{\text{acetyl-CoA synthetase}} \text{acetyl-CoA} + \text{AMP} + \text{PP}_i$$
$$(\Delta G^{0'} = -6 \text{ kJ/mol acetate})$$

However, in species of *Methanosarcina* that have a low affinity for acetate, acetyl-CoA is formed through a two-step process catalyzed by acetate kinase and phosphotransacetylase:

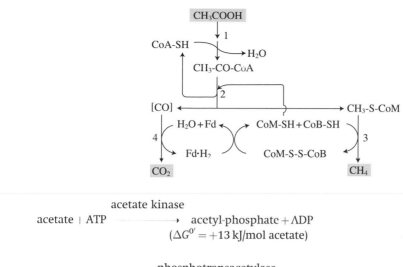

Figure 9.13 **Acetate degradation to methane.**

Acetate is activated (1) by acetate kinase/phosphotransacetylase (in *Methanosarcina*) or by acetyl-CoA synthetase (in *Methanosaeta*). Carbon monoxide dehydrogenase (2) transfers methyl groups to form methyl-S-CoM, oxidizing the carbonyl group to carbon dioxide to reduce ferredoxin. Methyl-CoM methylreductase (3) reduces methyl-CoM to methane, oxidizing HS-CoB to form CoM-S-S-CoB heterodisulfide. Heterodisulfide reductase (4) reduces the heterodisulfide, oxidizing the reduced ferredoxin in a reaction similar to that shown in Figure 9.12(a) involving a membrane protein, ferredoxin: methanophenazine oxidoreductase, with cytochrome *b*.

$$\text{acetate} + \text{ATP} \xrightarrow{\text{acetate kinase}} \text{acetyl-phosphate} + \text{ADP}$$
$$(\Delta G^{0'} = +13 \,\text{kJ/mol acetate})$$

$$\text{acetyl-phosphate} + \text{CoA-SH} \xrightarrow{\text{phosphotransacetylase}} \text{acetyl-CoA} + P_i$$
$$(\Delta G^{0'} = -9 \,\text{kJ/mol acetyl-phosphate})$$

Carbon monoxide dehydrogenase splits acetyl-CoA to form CH_3-H_4MPT and enzyme-bound [CO]. The methyl group of CH_3-H_4MPT is transferred to methyl-CoM, and [CO] is oxidized to CO_2, reducing ferredoxin (Figure 9.13). In *Methanosarcina barkeri*, H_4MPT is replaced by tetrahydrosarcinapterin that has a similar structure to H_4MPT. Electrons from the reduced ferredoxin are transferred to heterodisulfide reductase through ferredoxin:methanophenazine oxidoreductase that contains cytochrome *b*. This reaction is similar to that of the methylotrophic methanogens that use reduced ferredoxin in place of F_{420}.

9.4.4 Energy conservation in methanogenesis

The free energy change in each step of methanogenesis (Table 9.7) is less than the phosphorylation potential (-51.8 kJ/mol, Section 5.6.3) needed for ATP synthesis through substrate-level phosphorylation (SLP). The enzymes of SLP are not known in the methanogenic process. The methanogenic archaea synthesize ATP through electron transport phosphorylation (ETP) with the generation of a proton motive force. Experiments using uncouplers and ATPase inhibitors support this hypothesis (Figure 9.14).

With the addition of methanol and H_2, cell suspensions of *Methanosarcina barkeri* produce CH_4 with concomitant generation of a proton motive force (Δp) and an increase in ATP concentration (Figure 9.14). Δp increases with the decrease in ATP concentration when an ATPase inhibitor, N,N'-dicyclohexylcarbodiimide (DCCD) is added. DCCD inhibits ATP synthesis and Δp consumption. The high level of Δp inhibits methanogenesis, the Δp generation process. A rapid decrease in Δp and ATP concentration is observed on addition of uncouplers such as carbonylcyanide m-chlorophenylhydrazone (CCCP) and carbonylcyanide p-trifluoromethoxy-phenylhydrazone

Table 9.7. | *Changes in free energy in each step of methanogenesis*

Reaction	$\Delta G^{0'}$ (kJ/mol reactants)
$CO_2 + H_2 + MF \rightarrow HCO\text{-}MF + H_2O$	$+16$
$HCO\text{-}MF + H_4MPT \rightarrow HCO\text{-}H_4MPT + MF$	-5
$HCO\text{-}H_4MPT + H^+ \rightarrow CH\equiv H_4MPT^+ + H_2O$	-2
$CH\equiv H_4MPT^+ + F_{420}H_2 \rightarrow CH=H_4MPT + F_{420} + H^+$	$+6.5$
$CH_2=H_4MPT + F_{420}H_2 \rightarrow CH_3\text{-}H_4MPT + F_{420}$	-5
$CH_3\text{-}H_4MPT + HS\text{-}CoM \rightarrow CH_3\text{-}S\text{-}CoM + H_4MPT$	-29
$CH_3\text{-}S\text{-}CoM + HS\text{-}CoB \rightarrow CH_4 + CoM\text{-}S\text{-}S\text{-}CoB$	-43
$CoM\text{-}S\text{-}S\text{-}CoB + H_2 \rightarrow CH_3\text{-}S\text{-}CoM + HS\text{-}CoB$	-42
$CH_3OH + HS\text{-}CoM \rightarrow H\text{-}S\text{-}CoM + H_2O$	-27.5
$CH_3\text{-}CO\text{-}S\text{-}CoA + H_4MPT \rightarrow CH_3\text{-}H_4MPT + [CO] + HS\text{-}CoA$	$+62$
$CoM\text{-}S\text{-}S\text{-}CoB + F_{420}H_2 \rightarrow CH_3\text{-}S\text{-}CoM + HS\text{-}CoB + F_{420}$	-29

Source: Antonie van Leeuwenhoek, **66**, 187–208, 1994.

Figure 9.14 **Proton motive force generation during methanol reduction to methane by *Methanosarcina barkeri*.**

(Gottschalk, G. 1986, *Bacterial Metabolism*, 2nd edn., Figure 8.26. Springer, New York)

When a cell suspension of *Methanosarcina barkeri* is supplied with methanol and H_2, methane is generated, increasing the proton motive force and ATP concentration. At this point ATPase inhibitors inhibit methane generation with a slight increase in the proton motive force and a decrease in ATP concentration to the background level. With the addition of an uncoupler, methane generation resumes, but the proton motive force and ATP concentration remain low. This shows that the proton motive force is generated during methanogenesis before ATP is synthesized.

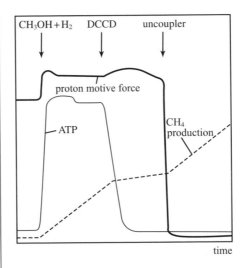

(FCCP). Uncouplers dissipate Δp. At a low Δp level, ATP cannot be synthesized, and methanogenesis is accelerated.

As described previously, a sodium motive force is generated in the methyl-CoM-forming reaction catalyzed by methyl-H_4MPT:CoM-methyltransferase. The sodium motive force is used in various energy consuming reactions including transport, motility and ATP synthesis. The sodium motive force is used for ATP synthesis either through the proton motive force by the action of Na^+/H^+ antiport or directly with the Na^+-ATPase. A Na^+-ATPase is known in many prokaryotes and has a V-type ATPase structure (Section 5.8.4.1).

9.4.5 Biosynthesis in methanogens

The hydrogenotrophic methanogens use CO_2 as their carbon source, and simple carbon compounds such as formate, methanol, methylamines and acetate are used by others. They metabolize these carbon

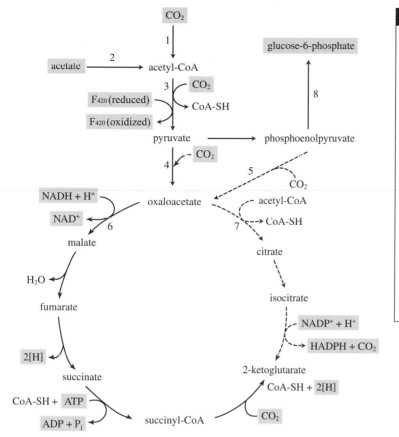

Figure 9.15 Supply of carbon skeletons for biosynthesis in methanogens.

(*Microbiol. Rev.* **51**:135–177, 1987)

Pyruvate synthase (3) produces pyruvate from the reduction of acetyl-CoA that is produced either from the acetyl-CoA pathway (1) or through acetate activation (2). PEP carboxylase (5) starts an oxidative TCA fork (7, dotted line) in species of *Methanosarcina*, while species of *Methanobacterium* and *Methanococcus* produce oxaloacetate catalyzed by pyruvate carboxylase (4) for the reductive TCA fork (6). Gluconeogenesis (8) is used to provide glucose-6-phosphate.

sources to acetyl-CoA to supply carbon skeletons for biosynthesis. Aceticlastic methanogens activate acetate as in reaction 1 in Figure 9.13. Other methanogens use carbon monoxide dehydrogenase in the synthesis of acetyl-CoA from methyl-H_4MPT and enzyme-bound [CO] as in the acetyl-CoA pathway in reaction 2 in Figure 9.13. The acetyl-CoA pathway is the reverse of the acetate oxidation pathway (Figure 9.6) and this pathway is also known as the carbon monoxide dehydrogenase pathway or the Wood–Ljungdahl pathway (Section 9.3.1.2). This pathway is known in some complete oxidizing sulfidogens (Table 9.5) and in homoacetogens (Section 9.5.2) in addition to methanogens. Tetrahydrofolate is used as the C1-carrier in the bacterial acetyl-CoA pathway while archaea use methanofuran (MF) and tetrahydromethanopterin (H_4MPT) for the same purpose.

Pyruvate synthase reduces acetyl-CoA to pyruvate, oxidizing reduced coenzyme F_{420}. Pyruvate is carboxylated to oxaloacetate (OAA) either directly or through phosphoenolpyruvate (PEP). OAA is metabolized to 2-ketoglutarate through an incomplete TCA fork (Section 5.4.1). Species of *Methanobacterium* and *Methanococcus* that lack citrate synthase employ the reductive branch of the incomplete TCA fork while species of *Methanosarcina* metabolize PEP via the oxidative branch (Figure 9.15).

Gluconeogenesis has not been firmly established in methanogens but some EMP pathway enzymes, including fructose-1,6-diphosphate aldolase, have been identified, suggesting that normal gluconeogenesis (Section 4.2) occurs in methanogens (Figure 9.15).

9.5 | Homoacetogenesis

Homoacetogens are a group of strictly anaerobic bacteria growing on the free energy available from the reduction of CO_2 to acetate. In addition to $CO_2 + H_2$, they metabolize sugars and methanol to acetate. Saccharolytic clostridia and enteric bacteria ferment carbohydrates to acetate with other fermentation products such as butyrate, lactate, succinate and others. Homoacetogens ferment a molecule of glucose to three molecules of acetate, two via pyruvate and acetyl-CoA. The third acetate is produced from two CO_2, NADH and reduced ferredoxin. Pyruvate oxidation to acetyl-CoA produces CO_2 and reduces ferredoxin, and NADH is from glycolysis. CO_2, CO and methanol are converted to acetate by a similar mechanism. Homoacetogens are phylogenetically diverse, and all of them have carbon monoxide dehydrogenase. Some studies have investigated the use of homoacetogens to produce acetate from CO-containing resources such as synthetic gas.

9.5.1 Homoacetogens

Since *Clostridium aceticum* was discovered, many other homoacetogens have been isolated (Table 9.8). All homoacetogens known to date reduce CO_2 to acetate using H_2, and ferment sugars except for strains of *Sporomusa*. *Clostridium formicoaceticum* ferments fructose but not glucose. Most of the homoacetogens can use methanol and CO. Some of them hydrolyze the methyl esters in lignin and its degradation products to produce acetate. Many aromatic compounds are metabolized by homoacetogens and halogenated hydrocarbons such as CH_2Cl_2 are dehalogenated. Species of *Sporomusa* were first identified as Gram-negative spore-formers but reclassified as firmicutes.

9.5.2 Carbon metabolism in homoacetogens

9.5.2.1 Sugar metabolism

Homoacetogens metabolize hexoses through the EMP pathway in most cases, and the modified ED pathway (Section 4.4.3) is the glycolytic pathway in *Clostridium aceticum*. The resulting pyruvate is oxidized to acetyl-CoA through the phosphoroclastic reaction. Phosphotransacetylase and acetate kinase cleave acetyl-CoA to acetate with ATP synthesis. NADH and reduced ferredoxin are oxidized coupled with the reduction of CO_2 to acetate (Figure 9.16).

Formate dehydrogenase reduces part of the CO_2 produced from the phosphoroclastic reaction to formate before being bound to tetrahydrofolate (H_4F) to be reduced to methyl-H_4F. The remaining CO_2 is reduced to an enzyme-bound form of [CO] by carbon monoxide

Table 9.8. | *Homoacetogens and their electron donors*

Organism	Gram stain	Spores	Electron donor[a]
Acetitomaculum ruminis	+	−	CO
Acetobacterium carbinolicum	+	−	methanol, 2,3-butanediol, ethylene glycol, phenylmethylether
Acetobacterium malicum	+	−	malate, 2-methoxyethanol
Acetobacterium wieringae	+	−	phenylmethylether
Acetobacterium woodii	+	−	CO
Butyribacterium methylotrophicum[b]	+	−	CO, methanol
Clostridium aceticum	+	+	sugars
Clostridium formicoaceticum	+	+	methanol, phenylmethylether
Clostridium methoxybenzovorans	+	+	methanol, sugars, lactate, methoxylated aromatics
Clostridium scatologenes	+	+	ethanol, butanol, glycerol, sugars, methoxylated aromatics
Eubacterium limosum[b]	+	−	CO, methanol, betaine, phenylmethylether
Halophaga foetida	−	−	trihydroxybenzenes, trimethoxybenzoate, pyruvate
Moorella glycerini	+	+	glycerol, sugars, pyruvate
Moorella thermoacetica[c]	+	+	CO, methanol, phenylmethylether
Moorella thermoautotrophica[d]	+	+	CO, phenylmethylether
Ruminococcus hydrogenotrophicus	+	−	
Ruminococcus productus[e]	+	−	CO, phenylmethylether
Sporomusa malonica[f]	+	+	malonate
Sporomusa silvacetica[f]	+	+	vanillate, sugars, fumarate, ethanol
Sporomusa sphaeroides[f]	+	+	
Sporomusa termitida[f]	+	+	CO, phenylmethylether
Syntrophococcus sucromutans[b]	+	−	phenylmethylether
Thermoanaerobacter kivui[g]	+	−	

[a] All grow on $CO_2 + H_2$, and sugars are used except for species of *Sporomusa* that do not use sugars as their electron donor.

[b] Butyrate is produced in addition to acetate by *Butyribacterium methylotrophicum*, *Eubacterium limosum* and *Syntrophococcus sucromutans*.

[c] Formerly *Clostridium thermoaceticum*.

[d] Formerly *Clostridium thermoautotrophicum*.

[e] Formerly *Peptostreptococcus productus*.

[f] Formerly known as Gram-negative.

dehydrogenase (CODH). CODH catalyzes acetyl-CoA synthesis from methyl-H_4F and [CO]. CODH is a dual function enzyme catalyzing the reversible reaction of CO oxidation and acetyl-CoA synthesis. This enzyme is referred to either as CO dehydrogenase or as acetyl-CoA synthase.

This CO_2-fixation acetyl-CoA pathway or CO dehydrogenase pathway is found in autotrophic methanogens (Section 9.4.5) and sulfidogens (Section 9.3.3) for the supply of carbon skeletons for biosynthesis, as well as in homoacetogens (Section 10.8.3).

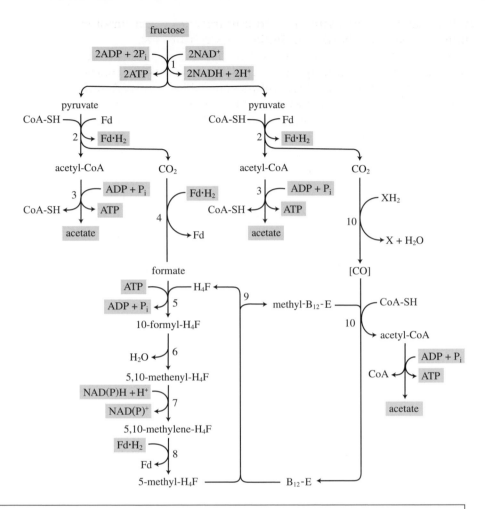

Figure 9.16 Hexose metabolism by homoacetogens.

(Gottschalk, G. 1986, *Bacterial Metabolism*, 2nd edn., Figure 8.23. Springer, New York)

Homoacetogens ferment glucose to produce two acetate and synthesize the third acetate from CO_2 through the acetyl-CoA pathway. 1, glycolysis; 2, pyruvate:ferredoxin oxidoreductase; 3, phosphotransacetylase and acetate kinase; 4, formate dehydrogenase; 5, formyl tetrahydrofolate synthetase; 6, methenyl-tetrahydrofolate cyclohydrolase; 7, methylene-tetrahydrofolate dehydrogenase; 8, methylene-tetrahydrofolate reductase; 9, tetrahydrofolate:B_{12} methyltransferase; 10, carbon monoxide dehydrogenase (acetyl-CoA synthase).

[CO], enzyme-bound carbon monoxide.

A tungsten-containing aldehyde oxidoreductase (AOR) is known in *Clostridium formicoaceticum* and *Moorela thermoacetica* (*Clostridium thermoaceticum*). This enzyme oxidizes aldehyde directly to the corresponding acid. Formate dehydrogenase is another enzyme containing tungsten. *Clostridium formicoaceticum* has a molybdenum-containing AOR in addition to the enzyme containing tungsten. A similar enzyme, glyceraldehyde:ferredoxin oxidoreductase, is known in the hyperthermophilic archaea *Sulfolobus acidocaldarius* and *Thermoproteus tenax* that metabolize sugars without phosphorylation (Section 4.4.3.2, Figure 4.8).

It is interesting to note that AOR in homoacetogens oxidizes aldehyde without conserving free energy from the aldehyde oxidation to the corresponding acids. Aldehydes are oxidized through acyl-CoA and acyl-phosphate in most anaerobic metabolic pathways such as sulfidogenesis, synthesizing ATP through the action of kinases (Section 9.3.1). AOR oxidizes various aldehydes, including aromatic aldehydes derived from lignin, to corresponding acids to provide electrons for the reduction of CO_2 to acetate.

All reactions of the acetyl-CoA pathway are reversible. Acetate is used as the electron donor in aceticlastic methanogens and in some of the complete oxidizing sulfidogens, and anaerobic chemolithotrophs fix CO_2 through the same pathway. The syntrophic bacteria *Clostridium ultunense* and *Thermoacetogenium phaeum* oxidize acetate to CO_2 and H_2 in association with H_2-consuming sulfidogens or methanogens. The oxidation of acetate to CO_2 and H_2 is an endergonic reaction ($\Delta G^{0'} = +107.1$ kJ/mol acetate) and possible under very low hydrogen partial pressures (Section 9.8.2). These organisms grow chemolithotrophically, reducing CO_2 to acetate using H_2. They are referred to as reversibacteria.

Moorella thermoacetica produces two acetate from one hexose in the presence of nitrate. This bacterium uses methanol and CO as electron donors in the nitrate ammonification process (Section 9.1.4). Nitrate is the preferred electron acceptor over CO_2 in this bacterium.

CO is oxidized to CO_2 by CO dehydrogenase in aerobic CO-utilizing carboxydobacteria, reducing ubiquinone or cytochrome b (Section 10.6). The carboxydobacterial CO dehydrogenase does not catalyze the reverse reaction because the redox potential of the electron acceptor used by this enzyme is too high:

$$HCO_3^-/HCOO^- \qquad (E^{0'} = -0.43 \text{ V})$$
$$CO_2/CO \qquad (E^{0'} = -0.54 \text{ V})$$
$$Q/QH_2 \qquad (E^{0'} = 0 \text{ V})$$
$$NAD^+/NADH + H^+ \quad (E^{0'} = -0.32 \text{ V})$$
$$Fd/\ Fd\cdot H_2 \qquad (E^{0'} = -0.41 \text{ V})$$

The homoacetogenic CO dehydrogenase that uses ferredoxin as its electron carrier catalyzes the reverse reaction. Due to a similar reason, the formate dehydrogenase of homoacetogens catalyzes the reverse reaction while a similar enzyme in enteric bacteria cannot reduce CO_2 to formate.

9.5.2.2 Synthesis of carbon skeletons for biosynthesis in homoacetogens

Homoacetogens synthesize acetyl-CoA from $CO_2 + H_2$. Acetyl-CoA is metabolized through gluconeogenesis to supply carbon skeletons for biosynthesis as in methanogens (Section 9.4.5).

9.5.3 Energy conservation in homoacetogens

ATP is synthesized through SLP in glycolysis and in the reaction catalyzed by acetate kinase when homoacetogens metabolize sugars. During chemolithotrophic growth on $CO_2 + H_2$, one ATP is produced through SLP in the reaction catalyzed by acetate kinase (reaction 3, Figure 9.16) and consumed in the reaction catalyzed by formyl-H_4F synthetase (reaction 5, Figure 9.16). For chemolithotrophic growth, homoacetogens should produce ATP through ETP.

$$4H_2 + 2CO_2 \rightarrow CH_3COOH + 2H_2O \quad (\Delta G^{0'} = -107.1 \, kJ/mol \, acetate)$$

$$4CO + 2H_2O \rightarrow CH_3COOH + 2CO_2 \quad (\Delta G^{0'} = -484.4 \, kJ/mol \, acetate)$$

$$4CH_3OH + CO_2 \rightarrow 3CH_3COOH + 2H_2O \quad (\Delta G^{0'} = -235.4 \, kJ/mol \, acetate)$$

$$C_6H_{12}O_6 \rightarrow 3CH_3COOH \quad (\Delta G^{0'} = -103.5 \, kJ/mol \, acetate)$$

As shown in the reactions above, chemolithotrophic homoacetogenesis generates enough free energy for ATP synthesis. Enough free energy is available for the generation of a proton motive or sodium motive force at the reactions catalyzed by formyltetrahydrofolate synthase (-22 kJ/mol formate, reaction 5, Figure 9.16) and tetrahydrofolate:B_{12} methyltransferase (-38 kJ/mol 5-methyl-tetrahydrofolate, reaction 9, Figure 9.16). The exact mechanisms are not known as to how the free energy changes in homoacetogenesis are coupled to generation of the proton motive or sodium motive force. In the bacteria known as H^+-dependent acetogens, a proton motive force is generated, while a sodium motive force is generated in Na^+-dependent acetogens. Methylene-H_4F reductase generates a sodium motive force but it is not known if energy is conserved at the reaction catalyzed by tetrahydrofolate:B_{12} methyltransferase. Cytochromes and coenzyme Q that might participate in the energy conservation processes are known in H^+-dependent acetogens including *Moorella thermoacetica* and *Moorella thermoautotrophica*. Cytochromes are not found in Na^+-dependent acetogens, including *Acetobacterium woodii*, *Thermoanaerobacter kivui* and *Ruminococcus productus*, which possess membrane-bound corrinoids. Corrinoid is known to participate in the generation of a sodium motive force in methanogens (Figure 9.11). The sodium motive force is directly used for ATP synthesis through the action of the Na^+-ATPase. Since the redox potential of the CO_2/CO half reaction is much lower that that of ferredoxin, the reduction of CO_2 to [CO] (reaction 10, Figure 9.16) consumes the proton motive/sodium motive force.

9.6 | Dehalorespiration

Marine algae synthesize various halogenated compounds, and synthetic halogen compounds are widely used as solvents, agricultural chemicals and for other industrial purposes. Some of them are toxic, carcinogenic and recalcitrant. They are dehalogenated at a higher rate under anaerobic conditions than under aerobic conditions. Enzyme cofactors found in homoacetogens, methanogens and

Table 9.9. | *Dehalorespiratory bacteria*

Strain	Electron acceptor	Electron donor	Remark[a]
Gram-positive			
Desulfitobacterium chlororespirans	2,4,6-TCP	hydrogen, formate	ortho
Desulfitobacterium dehalogenans	PCE, 2,4,6-TCP	hydrogen, formate	ortho
Desulfitobacterium frappieri	2,4,6-TCP	pyruvate	ortho
			meta
			para
Desulfitobacterium hafniense	PCP	pyruvate	ortho
			meta
Dehalobacter restrictus	PCE, TCE	hydrogen	–
Gram-negative			
Desulfomonile tiedjei	PCE, TCE, 3-CB	hydrogen, formate	meta
Desulfomonile limimaris	3-CB, sulfate, nitrate	lactate	meta
Dehalospirillum multivorans	PCE, TCE	hydrogen, formate	–
Desulfuromonas chloroethenica	PCE, TCE	pyruvate, acetate	–
Dehalococcoides ethenogenes	PCE, TCE	hydrogen	–
Enterobacter agglomerans	PCE, TCE, DCE, chloroethene	hydrogen	–
Desulfovibrio dechloracetivorans[b]	2-CP, 2,6-DCP	acetate	ortho

2,4,6-TCP, 2,4,6-trichlorophenol; 2-CP, 2-chlorophenol; 2,6-DCP, 2,6-dichlorophenol; 3-CB, 3-chlorobenzoate; PCE, tetrachloroethene; TCE, trichloroethene; PCP, pentachlorophenol; DCE, dichloroethene.
[a] Dehalogenation position.
[b] Non-sulfidogenic.

other anaerobic bacteria remove halogens reductively in aqueous solution. These include corrinoids, iron porphyrins and F_{430}. These non-specific reactions are slower than those that occur when halogenated compounds are used as electron acceptors. The redox potential of halogenated compounds ranges from $+250$ to -600 mV.

$$CH_3Cl + NADH + H^+ \rightarrow CH_4 + HCl \quad (\Delta G^{0'} = -135.08 \text{ kJ/mol } CH_3Cl)$$

9.6.1 Dehalorespiratory organisms

A few pure cultures have been identified that use chlorinated benzoate, phenol and ethane as electron acceptors (Table 9.9) and enrichment cultures have been made which dehalogenate polychlorinated biphenyls (PCBs), chlorinated benzene derivatives and others.

Species of *Desulfomonile* and *Desulfitobacterium* use not only halogenated compounds but also sulfite and thiosulfate as their electron acceptors. Dehalogenating enzymes are induced in these bacteria by these electron acceptors. The transcription and activity of dehalogenating enzymes is repressed by sulfite and thiosulfate in *Desulfomonile tiedjei*. Halogenated compounds are used as electron sinks in sulfur-reducing *Desulfuromonas chloroethenica* coupled to the oxidation of acetate through the modified TCA cycle.

Dehalospirillum multivorans is closely related to the metal-reducing bacterium *Geospirillum barnesii* and uses arsenate and selenate as well as halogenated compounds as electron acceptors. The metalloid oxyanions inhibit transcription and activity of the dehalogenating enzymes in this bacterium. *Dehalococcoides ethenogenes* is the only known pure culture capable of complete dehalogenation of tetrachloroethene to ethane. *Enterobacter agglomerans* is the only known dehalogenating facultative anaerobe to date.

9.6.2 Energy conservation in dehalorespiration

Electron transport and energy conservation are not well understood in dehalorespiration. Most of the organisms listed in Table 9.9 can grow with hydrogen as the sole electron donor, which shows that dehalogenation with hydrogen is coupled to ATP generation. Under dehalogenating conditions, a *c*-type cytochrome is induced with the dehalogenating enzymes in *Desulfomonile tiedjei*. This bacterium has hydrogenase and formate dehydrogenase in the periplasmic region. Inhibitors of quinone function eliminate the dehalogenation capacity of this bacterium. Menaquinone and cytochromes *b* and *c* are known in dehalogenating organisms including *Desulfitobacterium*, *Dehalobacter restrictus* and *Dehalospirillum multivorans*.

9.7 | Miscellaneous electron acceptors

In addition to the electron acceptors discussed above, certain organic and inorganic compounds are used as electron acceptors by anaerobic bacteria (Table 9.10). Humic acid derived from lignin is used as an electron acceptor in a metal-reducing species of *Geobacter*. Reduced humic acid chemically reduces Fe(III) and is diffusible through the membrane: this compound may mediate electron transfer from the bacterial cells to the insoluble electron acceptor.

Perchlorate and chlorate are strong oxidizing agents and toxic, but some bacteria use them as electron acceptors. A similar oxygenated halogen, iodate, is present in seawater and can be used as an electron acceptor by some bacteria including *Shewanella putrefaciens*.

Table 9.10. *Electron acceptors used in anaerobic respiration*

Electron acceptor	Organism	Characteristics
Humic acid	*Geobacter sulfurreducens*	reoxidized in a chemical reaction reducing Fe(III)
(Per)chlorate	*Dechloromonas agitata*	
	Dechlorosoma suillum	chemolithotroph, Fe(II) as electron donor
	Dechloromonas sp.	chemolithotroph, H_2 as electron donor
	Ideonella dechloratans	
Iodate	*Shewanella putrefaciens*	
	Desulfovibrio desulfuricans	
Organic sulfonate	*Bilophila wadsworthia*	electron sink(?)
Organic nitro compounds	*Denitrobacterium detoxificans*	

Taurine, a natural sulfonate compound, is reduced to sulfide as an electron acceptor by bacteria. Nitropropanol and nitropropionate are toxic compounds found in some plants. They are reduced as electron acceptors in the rumen, thus protecting the host animals.

Some anaerobic bacteria including *Lactobacillus casei*, *Streptococcus lactis*, *Clostridium acetobutylicum*, *Aerobacter polymyxa* and sulfidogens are known to reduce phosphate to phosphine. However, the redox potential of these reactions is too low for them to be used as electron acceptors.

$$H_2PO_4^- \xrightarrow{\;E^{0\prime}=-0.65\,V\;} HPO_3^{2-} \xrightarrow{\;-0.74\,V\;} H_2PO_2^- \xrightarrow{\;-0.66\,V\;} PH_3$$

9.8 | Syntrophic associations

Organic materials introduced into the anaerobic environment are fermented to various alcohols and acids before being completely oxidized through anaerobic respiration coupled to the reduction of nitrate, metal ions, sulfate and others. Under anaerobic conditions where these electron acceptors are not available, methanogens are the final scavenger. Methanogens convert a limited range of substrates to methane (Table 9.6). The common fermentation products that cannot be utilized directly by methanogens are not accumulated under methanogenic conditions. These include ethanol, propionate and butyrate. These are oxidized to acetate and carbon dioxide by certain bacteria in a syntrophic association with methanogens. These bacteria are referred to as syntrophic bacteria or obligate proton-reducing acetogens (Table 9.11).

9.8.1 Syntrophic bacteria
Syntrophic bacteria catalyze the following reactions in syntrophic association with methanogens or sulfidogens:

ethanol + $H_2O \rightarrow$ acetate + $H^+ + 2H_2$ ($\Delta G^{0\prime} = +9.6$ kJ/mol ethanol)

butyrate + $H_2O \rightarrow$ 2 acetate + $H^+ + 2H_2$ ($\Delta G^{0\prime} = +48.1$ kJ/mol butyrate)

propionate + $H_2O \rightarrow$ acetate + $HCO_3^- + H^+ + 3H_2$ ($\Delta G^{0\prime} = +76.1$ kJ/mol propionate)

These endergonic reactions cannot take place, and therefore cannot support growth of the syntrophic bacteria under standard conditions. However, these reactions can be exergonic when the reaction product concentrations are kept very low (Section 5.5.1). Methanogens and sulfidogens remove hydrogen efficiently, which keeps its partial pressure low as these reactions become exergonic. Coculture of syntrophic bacteria and methanogens on butyrate and propionate is well documented. These mixed cultures are referred to as syntrophic associations, or described as interspecies hydrogen transfer.

Table 9.11. *Syntrophic bacteria*

Strains	Gram stain	Substrate used
Clostridium ultunense[a]	+	acetate
Smithella propionica[b]	−	propionate
Syntrophobacter fumaroxidans[c]	−	propionate
Syntrophobacter wolinii	−	propionate
Syntrophobotulus glycolicus	+	glycolate
Syntrophomonas sapovorans	−	butyrate and fatty acids (up to C_{18})
Syntrophomonas wolfei	−	butyrate and fatty acids (up to C_8)
Syntrophospora bryantii	+	butyrate and fatty acids (up to C_{11})
Syntrophothermus lipocalidus[d]	−	butyrate and fatty acids (up to C_{10})
Syntrophus aciditrophicus[e]	−	benzoate, cyclohexane carboxylate
Syntrophus buswellii[e]	−	benzoate, 3-phenylbenzoate
Syntrophus gentianae	−	benzoate
Thermoacetogenium phaeum[a]	+	acetate
Thermosyntropha lipolytica[b]	−	butyrate and fatty acids (up to C_{18})

[a] Pure culture on crotonate fermentation.
[b] Pure culture on fumarate fermentation and on anaerobic respiration using sulfate or fumarate.
[c] Pure culture on fumarate fermention and on anaerobic respiration using nitrate, sulfate or fumarate.
[d] Pure culture on crotonate fermentation or on benzoate + crotonate.
[e] Reversibacteria growing on $H_2 + CO_2$ producing acetate, and acetate oxidation under low partial pressure of H_2.

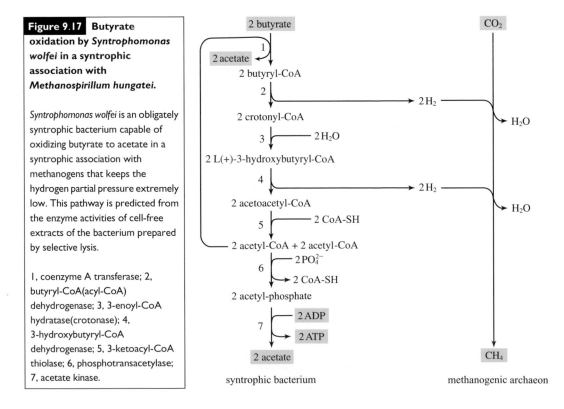

Figure 9.17 **Butyrate oxidation by *Syntrophomonas wolfei* in a syntrophic association with *Methanospirillum hungatei*.**

Syntrophomonas wolfei is an obligately syntrophic bacterium capable of oxidizing butyrate to acetate in a syntrophic association with methanogens that keeps the hydrogen partial pressure extremely low. This pathway is predicted from the enzyme activities of cell-free extracts of the bacterium prepared by selective lysis.

1, coenzyme A transferase; 2, butyryl-CoA(acyl-CoA) dehydrogenase; 3, 3-enoyl-CoA hydratase(crotonase); 4, 3-hydroxybutyryl-CoA dehydrogenase; 5, 3-ketoacyl-CoA thiolase; 6, phosphotransacetylase; 7, acetate kinase.

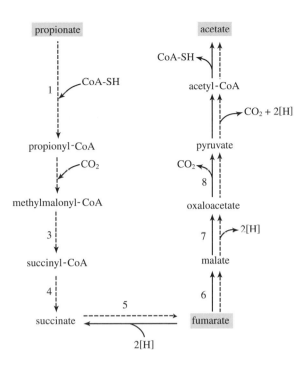

Figure 9.18 Metabolism of fumarate by *Syntrophobacter wolinii* (solid line) and the deduced propionate oxidation pathway (dotted line).

The oxidation of propionate to acetate is an endergonic reaction requiring the consumption of a large amount of free energy under standard conditions. *Syntrophobacter wolinii* disproportionates fumarate to succinate and acetate. A bacterial cell-free extract of the syntrophic consortium showed the enzyme activities shown in the figure. It is most probable that *Syntrophobacter wolinii* oxidizes propionate as shown (dotted lines).

1, propionate kinase/ phosphotransacylase; 2, methylmalonyl-CoA:pyruvate transcarboxylase; 3, methylmalonyl-CoA mutase; 4, succinyl-CoA synthetase; 5, succinate dehydrogenase; 6, fumarase; 7, malate dehydrogenase; 8, oxaloacetate decarboxylase.

9.8.2 Carbon metabolism in syntrophic bacteria

Syntrophobacter wolinii and *Syntrophomonas wolfei* grow fermentatively on propionate and butyrate under low hydrogen partial pressures. These are obligate syntrophs growing in coculture with hydrogen- and acetate-consuming sulfidogens or methanogens. Since they cannot be cultivated in pure culture, their carbon metabolism has been studied indirectly.

When a coculture of a Gram-negative syntrophic bacterium *Syntrophomonas wolfei* and an archaeon *Methanospirillum hungatei* is treated with lysozyme, the bacterial cells are lysed but not the archaeal cells. Enzyme activities were measured using this cell-free extract to establish butyrate catabolism to acetate as shown in Figure 9.17.

Syntrophomonas wolfei grows on crotonate in pure culture. This bacterium activates crotonate to crotonyl-CoA, and oxidizes half of it to 2 acetate synthesizing ATP and reduces the other half to butyrate. Other syntrophic bacteria including *Smithella propionica*, *Syntrophus aciditrophicus* and *Syntrophus buswellii* can use crotonate fermentatively. *Syntrophus aciditrophicus* and *Syntrophus buswellii* oxidize benzoate fermentatively reducing crotonate to butyrate (Table 9.11).

Fumarate is fermented in pure culture to acetate and succinate by *Syntrophobacter wolinii* that oxidizes propionate in a syntrophic association. In this fermentation, a part of fumarate is oxidized to acetate with ATP synthesis, and the remaining fumarate is reduced as an electron acceptor. From this fumarate fermentation, the propionate oxidation reactions are predicted as shown in Figure 9.18. This metabolism is similar to propionate oxidation by an incomplete oxidizing sulfidogen, *Desulfobulus propionicus* (see Figure 9.7). Another propionate

Figure 9.19 **Facultative syntrophic association between *Ruminococcus albus* and methanogens through interspecies hydrogen transfer.**

(Gottschalk, G. 1986, *Bacterial Metabolism*, 2nd edn., Figure 8.31. Springer, New York)

The cell yield of the rumen bacterium *Ruminococcus albus* increases in association with the hydrogen-consuming methanogens that keep the hydrogen partial pressure low. Under low hydrogen partial pressure, this bacterium synthesizes ATP from acetyl-CoA metabolism to acetate. With high hydrogen partial pressures, acetyl-CoA is reduced to ethanol or butyrate without ATP synthesis. This relationship is referred to as a facultative syntrophic association.

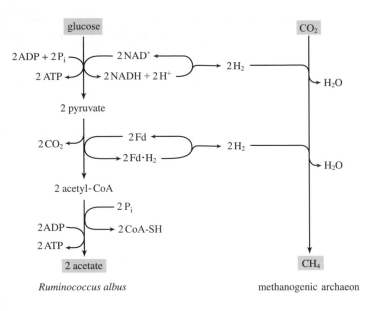

Ruminococcus albus   methanogenic archaeon

oxidizing syntrophic bacterium, *Syntrophobacter fumaroxidans*, ferments fumarate in pure culture, and the fatty acid oxidizing *Syntrophothermus lipocalidus* not only ferments fumarate but also uses fumarate, nitrate and sulfate as electron acceptors to oxidize fatty acids to acetate (Table 9.11).

The term disproportionation is used to describe a metabolism where a part of the substrate is used as electron donor and the other part is used as electron acceptor.

Clostridium ultunense and *Thermoacetogenium phaeum* grow syntrophically with methanogens oxidizing acetate to CO_2. In pure culture they grow on $H_2 + CO_2$ as homoacetogens, and can be referred to as reversibacteria (Section 9.5.2).

9.8.3 Facultative syntrophic associations

A rumen bacterium, *Ruminococcus albus*, ferments various carbohydrates including cellulose. Under normal conditions, this bacterium reduces acetyl-CoA to ethanol or butyrate. When the hydrogen partial pressure is low, electrons are consumed to reduce protons to hydrogen and acetyl-CoA is metabolized to acetate with ATP synthesis. With extra ATP, the growth yield of the bacterium is higher under low hydrogen partial pressures. Similarly the bacterium grows better in coculture with a hydrogen-consuming partner (Figure 9.19). This coculture is referred to as facultative syntrophy.

9.9 | Element cycling under anaerobic conditions

Various organic compounds are removed as CO_2 and CH_4, and various electron acceptors are reduced under anaerobic conditions. Through these oxidation–reduction reactions many compounds are transformed and these have been discussed in several previous sections. Many xenobiotics can be degraded under anaerobic conditions (Table 9.12).

Table 9.12. | *Examples of xenobiotics metabolized under anaerobic conditions*

Reaction	Xenobiotics
Denitrification	phenol, p-cresol, hydroxybenzoate, benzoate, toluene, resorcinol, phenylacetate, phthalate polychlorobenzoate, polychlorinated biphenyls, aminobenzoate, vanillate, protocatechuate
Metal reduction	benzoate, toluene, phenol, p-cresol hydroxybenzoates, phthalate
Sulfidogenesis	chlorophenols, benzoate, toluene, xylene, hydroxybenzoate, phenol, indole, phenylacetate, trinitrotoluene, 3-aminobenzoate, catechol
Methanogenesis[a]	benzene, toluene, o-xylene, benzoate, tetrachloroethanes, 1,2-dichloroethane, CFC-11 dichloromethane, chlorophenol, benzaldehyde 3-aminobenzoate, phenylpropionate, catechol, vanillate, methylbenzoate, chlorobenzene
Homoacetogenesis	ethylene glycol, phenylmethylether, dichloromethane
Fermentation	3-hydroxybenzoate, polyethylene glycol, hydroxybenzoate, polyphenol

[a] Mixed culture.

9.9.1 Oxidation of aromatic hydrocarbons under anaerobic conditions

For a long time, hydrocarbons were believed to be oxidized only under aerobic conditions by the action of oxygenases utilizing molecular oxygen (Section 7.8.3). Hydrocarbons can also be oxidized by different mechanisms under anaerobic conditions using a variety of electron acceptors. These reactions are thermodynamically feasible. The free energy changes for toluene oxidation under anaerobic respiratory conditions are:

Denitrification:

$$C_6H_5(CH_3) + 7.2NO_3^- + 0.2H^+ \rightarrow 7HCO_3^- + 3.6N_2 + 0.6H_2O \quad (\Delta G^{0\prime} = -493.6 \text{ kJ/mol } NO_3^-)$$

Nitrate ammonification:

$$C_6H_5(CH_3) + 4.5NO_3^- + 0.2H^+ + 7.5H_2O \rightarrow 7HCO_3^- + 4.5NH_4^+ \quad (\Delta G^{0\prime} = -493.1 \text{ kJ/mol } NO_3^-)$$

Ferric iron reduction:

$$C_6H_5(CH_3) + 36Fe(OH)_3 + 29HCO_3^- + 29H^+ \rightarrow 36FeCO_3 + 87H_2O \quad (\Delta G^{0\prime} = -39.1 \text{ kJ/mol Fe})$$

Sulfate reduction:

$$C_6H_5(CH_3) + 4.5SO_4^{2-} + 2H^+ + 3H_2O \rightarrow 7HCO_3^- + 4.5H_2S \quad (\Delta G^{0\prime} = -45.6 \text{ kJ/mol } SO_4^{2-})$$

Methanogenesis:

$$C_6H_5(CH_3) + 7.5H_2O \rightarrow 2.5HCO_3^- + 4.5CH_4 + 2.5H^+ \quad (\Delta G^{0\prime} = -29.1 \text{ kJ/mol } CH_4)$$

(*FEMS Microbiol. Rev.* **22**, 459–473, 1998)

Hydrocarbon oxidation under anaerobic conditions is best known with enriched mixed cultures. Toluene is oxidized under anaerobic conditions by denitrifying *Thauera aromatica* and *Azoarcus tolulyticus*, metal-reducing *Geobacter metallireducens* and *Geobacter grbiciae*, and sulfidogenic *Desulfobacula toluolica* and *Desulfobacter centonicum*. Benzene is less

Figure 9.20 The initial activation reactions during anaerobic degradation of saturated aliphatic and aromatic hydrocarbons.

(*Curr. Opin. Biotechnol.* 12:259–276, 2001)

Saturated aliphatic hydrocarbons (a), toluene and xylene (b), 2-methylnaphthalene (c), and ethylbenzene (d) are activated through binding with fumarate. Some denitrifying bacteria dehydrogenate ethylbenzene and propylbenzene (e). Polyaromatic hydrocarbons (PAH) including naphthalene are carboxylated (f).

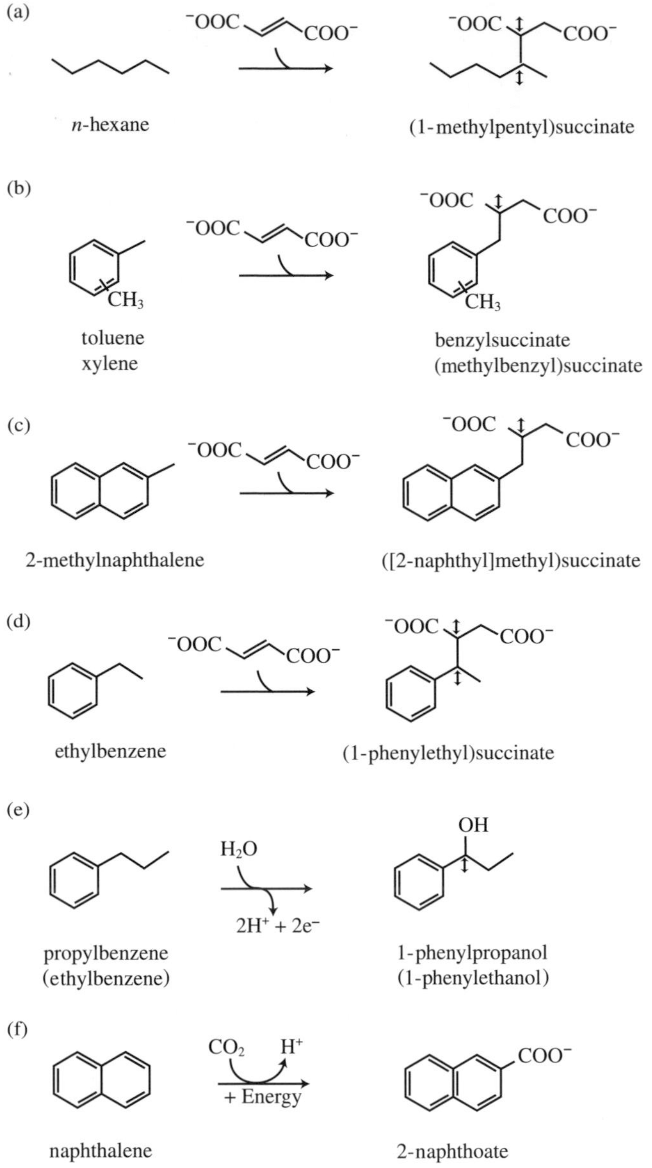

(a)

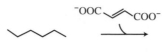

n-hexane

(1-methylpentyl)succinate

(b)

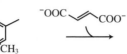

toluene
xylene

benzylsuccinate
(methylbenzyl)succinate

(c)

2-methylnaphthalene

([2-naphthyl]methyl)succinate

(d)

ethylbenzene

(1-phenylethyl)succinate

(e)

H₂O

2H⁺ + 2e⁻

propylbenzene
(ethylbenzene)

1-phenylpropanol
(1-phenylethanol)

(f)

CO_2 H^+

+ Energy

naphthalene

2-naphthoate

degradable than toluene and ethylbenzene. Species of *Dechloromonas* oxidize toluene under denitrifying conditions. Pure cultures have not been isolated that can oxidize saturated aliphatic hydrocarbons.

Aromatic hydrocarbons can serve as electron donors for denitrification, metal reduction, sulfidogenesis and methanogenesis. They include toluene, benzene, and aromatic acids such as benzoate and gentisate. The anaerobic degradation is initiated by one of three reactions (Figure 9.20):

(1) binding with fumarate
(2) dehydrogenation
(3) carboxylation.

Fumarate binding is the initial reaction in the anaerobic degradation of saturated aliphatic hydrocarbons, and aromatic hydrocarbons

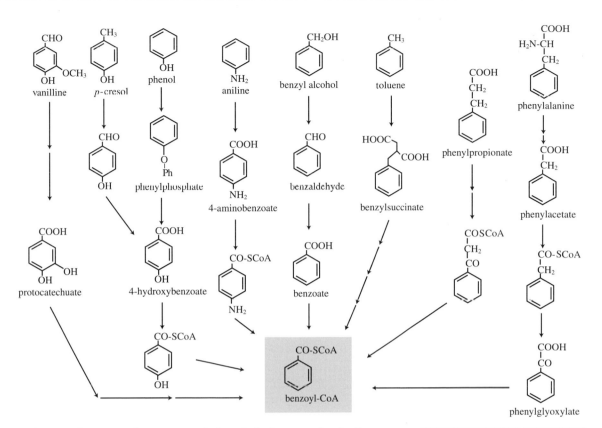

Figure 9.21 Aromatic compounds metabolized through benzoyl-CoA under anaerobic conditions.

(*FEMS Microbiol. Rev.* **22**:439–458, 1998)

Different aromatic substrates are converted into a few central aromatic intermediates including benzoyl-CoA, resorcinol, phloroglucinol, hydroxyhydroquinone and possibly others before further degradation. Among them, benzoyl-CoA is the best known intermediate during their degradation under anaerobic conditions.

such as toluene, xylene, 2-methylnaphthalene and ethylbenzene under all electron-accepting conditions. Denitrifiers dehydrogenate ethylbenzene and propylbenzene to initiate anaerobic degradation. Polyaromatic hydrocarbons (PAHs) are carboxylated before anaerobic degradation. Benzene is degraded via phenol and benzoate, but the exact mechanisms are not yet fully established.

After the initial activation reactions, aromatic compounds are further metabolized via benzoyl-CoA, resorcinol, phloroglucinol or hydroxyhydroquinone. Benzoyl-CoA is the best known intermediate, as shown in Figure 9.21. Based on enzyme activities, the benzoyl-CoA pathway has been deduced in photosynthetic *Rhodopseudomonas palustris*, denitrifying *Thauera aromatica* and *Azoarcus tolulyticus*, and syntrophic *Syntrophus gentianae* (Figure 9.22). Benzoyl-CoA reductase reduces the benzene ring of benzoyl-CoA. At this reaction 2ATP is consumed to make the substrate reduction ($E^{0'}-1.8$ volt) thermodynamically possible coupling the oxidation of reduced ferredoxin ($E^{0'}-0.42\,V$). The resulting cyclic dienoyl-CoA is dehydrated and metabolized to acetyl-CoA through 3-hydroxypimelyl-CoA.

9.9.2 Transformation of xenobiotics under anaerobic conditions

As discussed above, many xenobiotics are transformed under anaerobic conditions using various electron acceptors including

Figure 9.22 **Proposed metabolic routes of the benzoyl-CoA pathway.**

(*FEMS Microbiol. Rev.* **22**:439–458, 1998)

Benzoate is activated to benzoyl-CoA consuming ATP before being reduced to cyclic dienoyl-CoA by benzoyl-CoA reductase. Since the redox potential ($E^{0'}$) of the benzoyl-CoA/cyclic dienoyl-CoA half reaction is − 1.8 V, much lower than that of ferredoxin, 2ATP is consumed in the reaction. The resulting cyclic dienoyl-CoA is further metabolized to 3– hydroxypimelyl-CoA through different reactions depending on the strains, and finally to acetyl-CoA.

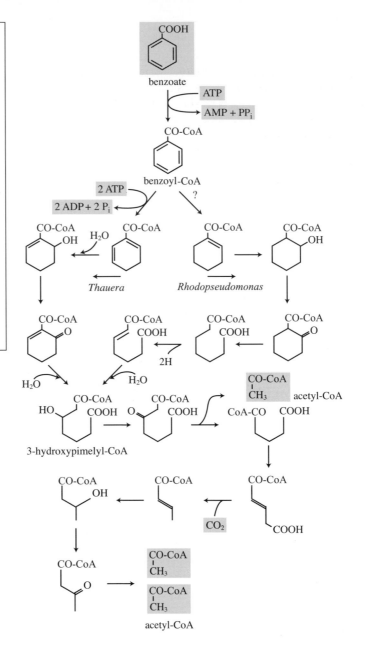

oxidized sulfur and nitrogen compounds, and metal ions. These microbial activities play an important role in the cycling of elements including degradation of natural and anthropogenic xenobiotics. Such properties can be exploited to clean up contaminated water and soil in a process known as bioremediation. Not only are organic xenobiotics oxidized, but toxic metals can also be immobilized, extracted from soil or removed from aqueous solution.

FURTHER READING

General

Croal, L. R., Gralnick, J. A., Malasarn, D. & Newman, D. K. (2004). The genetics of geochemistry. *Annual Review of Genetics* **38**, 175–202.

Ehrlich, H. L. (2002). *Geomicrobiology*. New York: Marcel Dekker.

Gal'chenko, V. F. (2004). On the problem of anaerobic methane oxidation. *Microbiology-Moscow* **73**, 599–608.

Strous, M. & Jetten, M. S. M. (2004). Anaerobic oxidation of methane and ammonium. *Annual Review of Microbiology* **58**, 99–117.

Teske, A. P. (2005). The deep subsurface biosphere is alive and well. *Trends in Microbiology* **13**, 402–404.

Warren, L. A. & Kauffman, M. E. (2003). Geoscience: microbial geoengineers. *Science* **299**, 1027–1029.

Denitrification

Baker, S. C., Ferguson, S. J., Ludwig, B., Page, M. D., Richter, O. M. H. & van Spanning, R. J. M. (1998). Molecular genetics of the genus *Paracoccus*: metabolically versatile bacteria with bioenergetic flexibility. *Microbiology and Molecular Biology Reviews* **62**, 1046–1078.

Blasco, F., Guigliarelli, B., Magalon, A., Asso, M., Giordano, G. & Rothery, R. A. (2001). The coordination and function of the redox centres of the membrane-bound nitrate reductases. *Cellular and Molecular Life Sciences* **58**, 179–193.

Cabello, P., Roldan, M. D. & Moreno-Vivian, C. (2004). Nitrate reduction and the nitrogen cycle in archaea. *Microbiology-UK* **150**, 3527–3546.

Cole, J. A. (1996). Nitrate reduction to ammonia by enteric bacteria: redundancy, or a strategy for survival during oxygen starvation? *FEMS Microbiology Letters* **136**, 1–11.

Ferguson, S. J. (1994). Denitrification and its control. *Antonie van Leeuwenhoek* **66**, 89–110.

Fritz, G., Einsle, O., Rudolf, M., Schiffer, A. & Kroneck, P. M. H. (2005). Key bacterial multi-centered metal enzymes involved in nitrate and sulfate respiration. *Journal of Molecular Microbiology and Biotechnology* **10**, 223–233.

Gregory, K. B., Bond, D. R. & Lovley, D. R. (2004). Graphite electrodes as electron donors for anaerobic respiration. *Environmental Microbiology* **6**, 596–604.

Hendriks, J., Oubrie, A., Castresana, J., Urbani, A., Gemeinhardt, S. & Saraste, M. (2000). Nitric oxide reductases in bacteria. *Biochimica et Biophysica Acta – Bioenergetics* **1459**, 266–273.

Jetten, M. S. M., Logemann, S., Muyzer, G., Robertson, L. A., Devries, S., Van Loosdrecht, M. C. M. & Kuenen, J. G. (1997). Novel principles in the microbial conversion of nitrogen compounds. *Antonie van Leeuwenhoek* **71**, 75–93.

Jetten, M. S. M., Strous, M., van de Pas-Schoonen, K. T., Schalk, J., van Dongen, U., van de Graaf, A. A., Logemann, S., Muyzer, G., van Loosdrecht, M. C. M., & Kuenen, J. G. (1999). The anaerobic oxidation of ammonium. *FEMS Microbiology Reviews* **22**, 421–437.

Moura, I., Bursakov, S., Costa, C. & Moura, J. J. G. (1997). Nitrate and nitrite utilization in sulfate-reducing bacteria. *Anaerobe* **3**, 279–290.

Park, H. I., Kim, J. S., Kim, D. K., Choi, Y. J. & Pak, D. (2006). Nitrate-reducing bacterial community in a biofilm-electrode reactor. *Enzyme and Microbial Technology* **39**, 453–458.

Philippot, L. (2002). Denitrifying genes in bacterial and archaeal genomes. *Biochimica et Biophysica Acta – Gene Structure and Expression* **1577**, 355–376.

Philippot, L. (2005). Denitrification in pathogenic bacteria: for better or worse? *Trends in Microbiology* **13**, 191–192.

Philippot, L. & Hojberg, O. (1999). Dissimilatory nitrate reductases in bacteria. *Biochimica et Biophysica Acta – Gene Structure and Expression* **1446**, 1–23.

Potter, L., Angove, H., Richardson, D. & Cole, J. (2001). Nitrate reduction in the periplasm of Gram-negative bacteria. *Advances in Microbial Physiology* **45**, 51–86.

Richardson, D. J. & Watmough, N. J. (1999). Inorganic nitrogen metabolism in bacteria. *Current Opinion in Chemical Biology* **3**, 207–219.

Simon, J. (2002). Enzymology and bioenergetics of respiratory nitrite ammonification. *FEMS Microbiology Reviews* **26**, 285–309.

Stouthamer, A. H., Deboer, A. P. N., Vanderoost, J. & Vanspanning, R. J. M. (1997). Emerging principles of inorganic nitrogen metabolism in *Paracoccus denitrificans* and related bacteria. *Antonie van Leeuwenhoek* **71**, 33–41.

Takaya, N. (2002). Dissimilatory nitrate reduction metabolisms and their control in fungi. *Journal of Bioscience and Bioengineering* **94**, 506–510.

van Niftrik, L. A., Fuerst, J. A., Damste, J. S. S., Kuenen, J. G., Jetten, M. S. M. & Strous, M. (2004). The anammoxosome: an intracytoplasmic compartment in anammox bacteria. *FEMS Microbiology Letters* **233**, 7–13.

Ward, B. B. (2003). Significance of anaerobic ammonium oxidation in the ocean. *Trends in Microbiology* **11**, 408–410.

Zumft, W. G. (2005). Biogenesis of the bacterial respiratory CuA, Cu-S enzyme nitrous oxide reductase. *Journal of Molecular Microbiology and Biotechnology* **10**, 154–166.

Zumft, W. G. & Kroneck, P. M. H. (2006). Respiratory transformation of nitrous oxide (N_2O) to dinitrogen by bacteria and archaea. *Advances in Microbial Physiology* **52**, 107–227.

Metal reduction

Barkay, T. & Schaefer, J. (2001). Metal and radionuclide bioremediation: issues, considerations and potentials. *Current Opinion in Microbiology* **4**, 318–323.

Carmona, M. & Diaz, E. (2005). Iron-reducing bacteria unravel novel strategies for the anaerobic catabolism of aromatic compounds. *Molecular Microbiology* **58**, 1210–1215.

Cervantes, C., Campos-Garcia, J., Devars, S., Gutierrez-Corona, F., Loza-Tavera, H., Torres-Guzman, J. C. & Moreno-Sanchez, R. (2001). Interactions of chromium with microorganisms and plants. *FEMS Microbiology Reviews* **25**, 335–347.

Chang, I. S., Moon, H., Bretschger, O., Jang, J. K., Park, H. I., Nealson, K. H. & Kim, B. H. (2006). Electrochemically active bacteria (EAB) and mediator-less microbial fuel cells. *Journal of Microbiology and Biotechnology* **16**, 163–177.

Kim, B. H., Kim, H. J., Hyun, M. S. & Park, D. H. (1999). Direct electrode reaction of an Fe(III)-reducing bacterium, *Shewanella putrefaciens*. *Journal of Microbiology and Biotechnology* **9**, 127–131.

Kim, B. H., Park, H. S., Kim, H. J., Kim, G. T., Chang, I. S., Lee, J. & Phung, N. T. (2004). Enrichment of microbial community generating electricity using a fuel-cell-type electrochemical cell. *Applied Microbiology and Biotechnology* **63**, 672–681.

Landa, E. R. (2005). Microbial biogeochemistry of uranium mill tailings. *Advances in Applied Microbiology* **57**, 113–130.

Lee, A. K. & Newman, D. K. (2003). Microbial iron respiration: impacts on corrosion processes. *Applied Microbiology and Biotechnology* **62**, 134–139.

Lloyd, J. R. (2003). Microbial reduction of metals and radionuclides. *FEMS Microbiology Reviews* **27**, 411–425.

Lovley, D. R., Holmes, D. E. & Nevin, K. P. (2004). Dissimilatory Fe(III) and Mn(IV) reduction. *Advances in Microbial Physiology* **49**, 219–286.

Messens, J. & Silver, S. (2006). Arsenate reduction: thiol cascade chemistry with convergent evolution. *Journal of Molecular Biology* **362**, 1–17.

Mukhopadhyay, R., Rosen, B. P., Phung, L. T. & Silver, S. (2002). Microbial arsenic: from geocycles to genes and enzymes. *FEMS Microbiology Reviews* **26**, 311–325.

Nealson, K. & Cox, B. (2002). Microbial metal-ion reduction and Mars: extraterrestrial expectations? *Current Opinion in Microbiology* **5**, 296–300.

Newman, D. K. (2001). Microbiology: how bacteria respire minerals. *Science* **292**, 1312–1313.

Oremland, R. S., Stolz, J. F. & Hollibaugh, J. T. (2004). The microbial arsenic cycle in Mono Lake, California. *FEMS Microbiology Ecology* **48**, 15–27.

Schroder, I., Johnson, E. & de Vries, S. (2003). Microbial ferric iron reductases. *FEMS Microbiology Reviews* **27**, 427–447.

Slobodkin, A. (2005). Thermophilic microbial metal reduction. *Microbiology-Moscow* **74**, 501–514.

Stolz, J. F., Basu, P., Santini, J. M. & Oremland, R. S. (2006). Arsenic and selenium in microbial metabolism. *Annual Review of Microbiology* **60**, 107–130.

Valls, M. & de Lorenzo, V. (2002). Exploiting the genetic and biochemical capacities of bacteria for the remediation of heavy metal pollution. *FEMS Microbiology Reviews* **26**, 327–338.

Wall, J. D. & Krumholz, L. R. (2006). Uranium reduction. *Annual Review of Microbiology* **60**, 149–166.

Weber, K. A., Achenbach, L. A. & Coates, J. D. (2006). Microorganisms pumping iron: anaerobic microbial iron oxidation and reduction. *Nature Reviews Microbiology* **4**, 752–764.

Wilkins, M., Livens, F., Vaughan, D. & Lloyd, J. (2006). The impact of Fe(III)-reducing bacteria on uranium mobility. *Biogeochemistry* **78**, 125–150.

Sulfidogenesis

Angell, P. (1999). Understanding microbially influenced corrosion as biofilm-mediated changes in surface chemistry. *Current Opinion in Biotechnology* **10**, 269–272.

Castro, H. F., Williams, N. H. & Ogram, A. (2000). Phylogeny of sulfate-reducing bacteria. *FEMS Microbiology Ecology* **31**, 1–9.

Colwell, F. S., Onstott, T. C., Delwiche, M. E., Chandler, D., Fredrickson, J. K., Yao, Q. J., McKinley, J. P., Boone, D., Griffiths, R., Phelps, T. J., Ringelberg, D., White, D. C., LaFreniere, L., Balkwill, D., Lehman, R. M., Konisky, J. & Long, P. E. (1997). Microorganisms from deep, high temperature sandstones: constraints on microbial colonization. *FEMS Microbiology Reviews* **20**, 425–435.

Cypionka, H. (2000). Oxygen respiration by *Desulfovibrio* species. *Annual Review of Microbiology* **54**, 827–848.

Hansen, T. A. (1994). Metabolism of sulfate-reducing prokaryotes. *Antonie van Leeuwenhoek* **66**, 165–185.

Hedderich, R., Klimmek, O., Kroger, A., Dirmeier, R., Keller, M. & Stetter, K. O. (1999). Anaerobic respiration with elemental sulfur and with disulfides. *FEMS Microbiology Reviews* **22**, 353–381.

Hockin, S. and Gadd, G. M. (2003). Linked redox-precipitation of sulfur and selenium under anaerobic conditions by sulfate-reducing bacterial biofilms. *Applied and Environmental Microbiology* **69**, 7063–7072.

Hockin, S. and Gadd, G. M. (2006). Removal of selenate from sulphate-containing media by sulphate-reducing bacterial biofilms. *Environmental Microbiology* **8**, 816–826.

Hockin, S. and Gadd, G. M. (2007). Bioremediation of metals by precipitation and cellular binding. In *Sulphate-reducing Bacteria: Environmental and Engineered Systems*, ed. L. L. Barton and W. A. Hamilton, pp. 405–434. Cambridge: Cambridge University Press.

Holmer, M. & Storkholm, P. (2001). Sulphate reduction and sulphur cycling in lake sediments: a review. *Freshwater Biology* **46**, 431–451.

Le Gall, J. & Xavier, A. V. (1996). Anaerobes response to oxygen: the sulfate-reducing bacteria. *Anaerobe* **2**, 1–9.

Lie, T. J., Leadbetter, J. R. & Leadbetter, E. R. (1998). Metabolism of sulfonic acids and other organosulfur compounds by sulfate-reducing bacteria. *Geomicrobiology Journal* **15**, 135–149.

Rueter, P., Rabus, R., Wilkes, H., Aeckersberg, F., Rainey, F. A., Jannasch, H. W. & Widdel, F. (1994). Anaerobic oxidation of hydrocarbons in crude oil by new types of sulphate-reducing bacteria. *Nature* **372**, 455–458.

Schauder, R. & Kroger, A. (1993). Bacterial sulphur respiration. *Archives of Microbiology* **159**, 491–497.

Villemur, R., Lanthier, M., Beaudet, R. & Lepine, F. (2006). The *Desulfitobacterium* genus. *FEMS Microbiology Reviews* **30**, 706–733.

White, C. and Gadd, G. M. (1998). Accumulation and effects of cadmium on sulphate-reducing bacterial biofilms. *Microbiology–UK* **144**, 1407–1415.

White, C., Dennis, J. S. and Gadd, G. M. (2003). A mathematical process model for cadmium precipitation by sulphate-reducing bacterial biofilms. *Biodegradation* **14**, 139–151.

Methanogenesis

Blaut, M. (1994). Metabolism of methanogens. *Antonie van Leeuwenhoek* **66**, 187–208.

Deppenmeier, U., Lienard, T. & Gottschalk, G. (1999). Novel reactions involved in energy conservation by methanogenic archaea. *FEBS Letters* **457**, 291–297.

Dybas, M. & Konisky, J. (1992). Energy transduction in the methanogen *Methanococcus voltae* is based on a sodium current. *Journal of Bacteriology* **174**, 5575–5583.

Ferry, J. G. (1992). Biochemistry of methanogenesis. *Critical Reviews in Biochemistry and Molecular Biology* **27**, 473–503.

Ferry, J. G. (1999). Enzymology of one-carbon metabolism in methanogenic pathways. *FEMS Microbiology Reviews* **23**, 13–38.

Lin, W. C., Yang, Y.-L. & Whitman, W. B. (2003). The anabolic pyruvate oxidor-eductase from *Methanococcus maripaludis*. *Archives of Microbiology* **179**, 444–456.

Macario, A. J. L., Lange, M., Ahring, B. K. & De Macario, E. C. (1999). Stress genes and proteins in the archaea. *Microbiology and Molecular Biology Reviews* **63**, 923–967.

Maden, B. E. H. (2000). Tetrahydrofolate and tetrahydromethanopterin compared: functionally distinct carriers in C-1 metabolism. *Biochemical Journal* **350**, 609–629.

Reeve, J. N., Nolling, J., Morgan, R. M. & Smith, D. R. (1997). Methanogenesis: genes, genomes, and who's on first? *Journal of Bacteriology* **179**, 5975–5986.

Schaefer, G., Engelhard, M. & Mueller, V. (1999). Bioenergetics of the Archaea. *Microbiology and Molecular Biology Reviews* **63**, 570–620.

Schink, B. (1997). Energetics of syntrophic cooperation in methanogenic degradation. *Microbiology and Molecular Biology Reviews* **61**, 262–280.

Shima, S. & Thauer, R. K. (2005). Methyl-coenzyme M reductase and the anaerobic oxidation of methane in methanotrophic Archaea. *Current Opinion in Microbiology* **8**, 643–648.

Thauer, R. K. (1998). Biochemistry of methanogenesis: a tribute to Marjory Stephenson. *Microbiology-UK* **144**, 2377–2406.

Valentine, D. L. & Reeburgh, W. S. (2000). New perspectives on anaerobic methane oxidation. *Environmental Microbiology* **2**, 477–484.

Homoacetogenesis

Bacher, A., Rieder, C., Eichinger, D., Arigoni, D., Fuchs, G. & Eisenreich, W. (1998). Elucidation of novel biosynthetic pathways and metabolite flux patterns by retrobiosynthetic NMR analysis. *FEMS Microbiology Reviews* **22**, 567–598.

Detkova, E. & Pusheva, M. (2006). Energy metabolism in halophilic and alkaliphilic acetogenic bacteria. *Microbiology-Moscow* **75**, 1–11.

Diekert, G. & Wohlfarth, G. (1994). Metabolism of homoacetogens. *Antonie van Leeuwenhoek* **66**, 209–221.

Drake, H. L. & Daniel, S. L. (2004). Physiology of the thermophilic acetogen *Moorella thermoacetica*. *Research in Microbiology* **155**, 422–436.

Ferry, J. G. (1995). CO dehydrogenase. *Annual Review of Microbiology* **49**, 305–333.

Grahame, D. A. (2003). Acetate C-C bond formation and decomposition in the anaerobic world: the structure of a central enzyme and its key active-site metal cluster. *Trends in Biochemical Sciences* **28**, 221–224.

Hansen, T. A. (1994). Metabolism of sulfate-reducing prokaryotes. *Antonie van Leeuwenhoek* **66**, 165–185.

Mueller, V. (2003). Energy conservation in acetogenic bacteria. *Applied and Environmental Microbiology* **69**, 6345–6353.

Ragsdale, S. (2004). Life with carbon monoxide. *Critical Reviews in Biochemistry and Molecular Biology* **39**, 165–195.

Russell, M. J. & Martin, W. (2004). The rocky roots of the acetyl-CoA pathway. *Trends in Biochemical Sciences* **29**, 358–363.

Siebers, B. & Schonheit, P. (2005). Unusual pathways and enzymes of central carbohydrate metabolism in Archaea. *Current Opinion in Microbiology* **8**, 695–705.

Sipma, J., Henstra, A. M., Parshina, S. N., Lens, P. N. L., Lettinga, G. & Stams, A. J. M. (2006). Microbial CO conversions with applications in synthesis gas purification and bio-desulfurization. *Critical Reviews in Biotechnology* **26**, 41–65.

Dehalorespiration

Abraham, W., Nogales, B., Golyshin, P., Pieper, D. & Timmis, K. (2002). Polychlorinated biphenyl-degrading microbial communities in soils and sediments. *Current Opinion in Microbiology* **5**, 246–253.

Borja, J., Taleon, D.M., Auresenia, J. & Gallardo, S. (2005). Polychlorinated biphenyls and their biodegradation. *Process Biochemistry* **40**, 1999–2013.

Chen, G. (2004). Reductive dehalogenation of tetrachloroethylene by microorganisms: current knowledge and application strategies. *Applied Microbiology and Biotechnology* **63**, 373–377.

El Fantroussi, S., Naveau, H. & Agathos, S.N. (1998). Anaerobic dechlorinating bacteria. *Biotechnology Progress* **14**, 167–188.

Fetzner, S. (1998). Bacterial dehalogenation. *Applied Microbiology and Biotechnology* **50**, 633–657.

Fetzner, S. & Lingens, F. (1994). Bacterial dehalogenases: biochemistry, genetics, and biotechnological applications. *Microbiological Reviews* **58**, 641–685.

Furukawa, K. (2000). Biochemical and genetic bases of microbial degradation of polychlorinated biphenyls (PCBs). *Journal of General and Applied Microbiology* **46**, 283–296.

Hoehener, P., Werner, D., Balsiger, C. & Pasteris, G. (2003). Worldwide occurrence and fate of chlorofluorocarbons in groundwater. *Critical Reviews in Environmental Science and Technology* **33**, 1–29.

Holliger, C. & Schumacher, W. (1994). Reductive dehalogenation as a respiratory process. *Antonie van Leeuwenhoek* **66**, 239–246.

Holliger, C., Wohlfarth, G. & Diekert, G. (1999). Reductive dechlorination in the energy metabolism of anaerobic bacteria. *FEMS Microbiology Reviews* **22**, 383–398.

Janssen, D.B. (2004). Evolving haloalkane dehalogenases. *Current Opinion in Chemical Biology* **8**, 150–159.

Janssen, D.B., Oppentocht, J.E. & Poelarends, G.J. (2001). Microbial dehalogenation. *Current Opinion in Biotechnology* **12**, 254–258.

Janssen, D.B., Dinkla, J.T., Poelarends, G.J. & Terpstra, P. (2005). Bacterial degradation of xenobiotic compounds: evolution and distribution of novel enzyme activities. *Environmental Microbiology* **7**, 1868–1882.

McCarty, P.L. (1997). Microbiology: breathing with chlorinated solvents. *Science* **276**, 1521–1522.

Smidt, H. & de Vos, W.M. (2004). Anaerobic microbial dehalogenation. *Annual Review of Microbiology* **58**, 43–73.

Vlieg, J.E.T.V., Poelarends, G.J., Mars, A.E. & Janssen, D.B. (2000). Detoxification of reactive intermediates during microbial metabolism of halogenated compounds. *Current Opinion in Microbiology* **3**, 257–262.

Wiegel, J. & Wu, Q.Z. (2000). Microbial reductive dehalogenation of polychlorinated biphenyls. *FEMS Microbiology Ecology* **32**, 1–15.

Anaerobic respiration on miscellaneous electron acceptors

Arkhipova, O. & Akimenko, V. (2005). Unsaturated organic acids as terminal electron acceptors for reductase chains of anaerobic bacteria. *Microbiology-Moscow* **74**, 629–639.

Coates, J.D. & Achenbach, L.A. (2004). Microbial perchlorate reduction: rocket-fueled metabolism. *Nature Reviews Microbiology* **2**, 569–580.

Councell, T.B., Landa, E.R. & Lovley, D.R. (1997). Microbial reduction of iodate. *Water, Air and Soil Pollution* **100**, 99–106.

Geng, J., Jin, X., Wang, Q., Niu, X., Wang, X., Edwards, M. & Glindemann, D. (2005). Matrix bound phosphine formation and depletion in eutrophic lake sediment fermentation: simulation of different environmental factors. *Anaerobe* **11**, 273–279.

Kroger, A., Geisler, V., Lemma, E., Theis, F. & Lenger, R. (1992). Bacterial fumarate respiration. *Archives of Microbiology* **158**, 311–314.

Richardson, D.J. (2000). Bacterial respiration: a flexible process for a changing environment. *Microbiology-UK* **146**, 551–571.

Roels, J. & Verstraete, W. (2001). Biological formation of volatile phosphorus compounds. *Bioresource Technology* **79**, 243–250.

Slobodkin, A.I., Zavarzina, D.G., Sokolova, T.G. & Bonch-Osmolovskaya, E.A. (1999). Dissimilatory reduction of inorganic electron acceptors by thermophilic anaerobic prokaryotes. *Microbiology-Moscow* **68**, 522–542.

Syntrophic associations

Cao, X., Liu, X. & Dong, X. (2003). *Alkaliphilus crotonatoxidans* sp. nov., a strictly anaerobic, crotonate-dismutating bacterium isolated from a methanogenic environment. *International Journal of Systematic and Evolutionary Microbiology* **53**, 971–975.

de Bok, F.A.M., Stams, A.J.M., Dijkema, C. & Boone, D.R. (2001). Pathway of propionate oxidation by a syntrophic culture of *Smithella propionica* and *Methanospirillum hungatei*. *Applied and Environmental Microbiology* **67**, 1800–1804.

de Bok, F.A.M., Luijten, M.L.G.C. & Stams, A.J.M. (2002). Biochemical evidence for formate transfer in syntrophic propionate-oxidizing cocultures of *Syntrophobacter fumaroxidans* and *Methanospirillum hungatei*. *Applied and Environmental Microbiology* **68**, 4247–4252.

de Bok, F.A.M., Plugge, C.M. & Stams, A.J.M. (2004). Interspecies electron transfer in methanogenic propionate degrading consortia. *Water Research* **38**, 1368–1375.

Grabowski, A., Blanchet, D. & Jeanthon, C. (2005). Characterization of long-chain fatty-acid-degrading syntrophic associations from a biodegraded oil reservoir. *Research in Microbiology* **156**, 814–821.

Hattori, S., Galushko, A.S., Kamagata, Y. & Schink, B. (2005). Operation of the CO dehydrogenase/acetyl coenzyme A pathway in both acetate oxidation and acetate formation by the syntrophically acetate-oxidizing bacterium *Thermacetogenium phaeum*. *Journal of Bacteriology* **187**, 3471–3476.

Imachi, H., Sekiguchi, Y., Kamagata, Y., Hanada, S., Ohashi, A. & Harada, H. (2002). *Pelotomaculum thermopropionicum* gen. nov., sp. nov., an anaerobic, thermophilic, syntrophic propionate-oxidizing bacterium. *International Journal of Systematic and Evolutionary Microbiology* **52**, 1729–1735.

Imachi, H., Sekiguchi, Y., Qiu, Y.L., Hugenholtz, P., Kimura, N., Wagner, M., Ohashi, A. & Harada, H. (2006). Non-sulfate-reducing, syntrophic bacteria affiliated with *Desulfotomaculum* cluster I are widely distributed in methanogenic environments. *Applied and Environmental Microbiology* **72**, 2080–2091.

Ishii, S., Kosaka, T., Hori, K., Hotta, Y. & Watanabe, K. (2005). Coaggregation facilitates interspecies hydrogen transfer between *Pelotomaculum thermopropionicum* and *Methanothermobacter thermautotrophicus*. *Applied and Environmental Microbiology* **71**, 7838–7845.

Johnson, M. R., Conners, S. B., Montero, C. I., Chou, C. J., Shockley, K. R. & Kelly, R. M. (2006). The *Thermotoga maritima* phenotype is impacted by syntrophic interaction with *Methanococcus jannaschii* in hyperthermophilic coculture. *Applied and Environmental Microbiology* **72**, 811–818.

Kendall, M. M., Liu, Y. & Boone, D. R. (2006). Butyrate- and propionate-degrading syntrophs from permanently cold marine sediments in Skan Bay, Alaska, and description of *Algorimarina butyrica* gen. nov., sp. nov. *FEMS Microbiology Letters* **262**, 107–114.

Kosaka, T., Uchiyama, T., Ishii, S., Enoki, M., Imachi, H., Kamagata, Y., Ohasi, A., Harada, H., Ikenaga, H. & Watanabe, K. (2006). Reconstruction and regulation of the central catabolic pathway in the thermophilic propionate-oxidizing syntroph *Pelotomaculum thermopropionicum*. *Journal of Bacteriology* **188**, 202–210.

Plugge, C. M., Balk, M., Zoetendal, E. G. & Stams, A. J. M. (2002). *Gelria glutamica* gen. nov., sp. nov., a thermophilic, obligately syntrophic, glutamate-degrading anaerobe. *International Journal of Systematic and Evolutionary Microbiology* **52**, 401–407.

Qiu, Y.-L., Sekiguchi, Y., Imachi, H., Kamagata, Y., Tseng, I.-C., Cheng, S.-S., Ohashi, A. & Harada, H. (2003). *Sporotomaculum syntrophicum* sp. nov., a novel anaerobic, syntrophic benzoate-degrading bacterium isolated from methanogenic sludge treating wastewater from terephthalate manufacturing. *Archives of Microbiology* **179**, 242–249.

Qiu, Y. L., Sekiguchi, Y., Hanada, S., Imachi, H., Tseng, I. C., Cheng, S. S., Ohashi, A., Harada, H. & Kamagata, Y. (2006). *Pelotomaculum terephthalicum* sp. nov. and *Pelotomaculum isophthalicum* sp. nov.: two anaerobic bacteria that degrade phthalate isomers in syntrophic association with hydrogenotrophic methanogens. *Archives of Microbiology* **185**, 172–182.

Zhang, C., Liu, X. & Dong, X. (2004). *Syntrophomonas curvata* sp. nov., an anaerobe that degrades fatty acids in co-culture with methanogens. *International Journal of Systematic and Evolutionary Microbiology* **54**, 969–973.

Zhilina, T., Zavarzina, D., Kolganova, T., Tourova, T. & Zavarzin, G. (2005). *Candidatus 'Contubernalis alkalaceticum'*, an obligately syntrophic alkaliphilic bacterium capable of anaerobic acetate oxidation in a coculture with *Desulfonatronum cooperativum*. *Microbiology-Moscow* **74**, 695–703.

Degradation of xenobiotics under anaerobic conditions

Boll, M. (2005). Key enzymes in the anaerobic aromatic metabolism catalysing Birch-like reductions. *Biochimica et Biophysica Acta – Bioenergetics* **1707**, 34–50.

Boll, M., Fuchs, G. & Heider, J. (2002). Anaerobic oxidation of aromatic compounds and hydrocarbons. *Current Opinion in Chemical Biology* **6**, 604–611.

Carmona, M. & Diaz, E. (2005). Iron-reducing bacteria unravel novel strategies for the anaerobic catabolism of aromatic compounds. *Molecular Microbiology* **58**, 1210–1215.

Chakraborty, R. & Coates, J. D. (2004). Anaerobic degradation of monoaromatic hydrocarbons. *Applied Microbiology and Biotechnology* **64**, 437–446.

Coates, J. D., Chakraborty, R. & McInerney, M. J. (2002). Anaerobic benzene biodegradation: a new era. *Research in Microbiology* **153**, 621–628.

Corvini, P., Schaeffer, A. & Schlosser, D. (2006). Microbial degradation of nonylphenol and other alkylphenols: our evolving view. *Applied Microbiology and Biotechnology* **72**, 223–243.

Esteve-Nunez, A., Caballero, A. & Ramos, J. L. (2001). Biological degradation of 2,4,6-trinitrotoluene. *Microbiology and Molecular Biology Reviews* **65**, 335–352.

Eyers, L., George, I., Schuler, L., Stenuit, B., Agathos, S. N. & El Fantroussi, S. (2004). Environmental genomics: exploring the unmined richness of microbes to degrade xenobiotics. *Applied Microbiology and Biotechnology* **66**, 123–130.

Gibson, J. & Harwood, S. (2002). Metabolic diversity in anaerobic compound utilization by anaerobic microbes. *Annual Review of Microbiology* **56**, 345–369.

Meckenstock, R. U., Safinowski, M. & Griebler, C. (2004). Anaerobic degradation of polycyclic aromatic hydrocarbons. *FEMS Microbiology Ecology* **49**, 27–36.

Spormann, A. M. & Widdel, F. (2000). Metabolism of alkylbenzenes, alkanes, and other hydrocarbons in anaerobic bacteria. *Biodegradation* **11**, 85–105.

Veeresh, G. S., Kumar, P. & Mehrotra, I. (2005). Treatment of phenol and cresols in upflow anaerobic sludge blanket (UASB) process: a review. *Water Research* **39**, 154–170.

Zhang, C. & Bennett, G. N. (2005). Biodegradation of xenobiotics by anaerobic bacteria. *Applied Microbiology and Biotechnology* **67**, 600–618.

Chemolithotrophy

Some prokaryotes grow by using reduced inorganic compounds as their energy source and CO_2 as the carbon source. These are called chemolithotrophs. The electron donors used by chemolithotrophs include nitrogen and sulfur compounds, Fe(II), H_2, and CO. The Calvin cycle is the most common CO_2 fixation mechanism, and the reductive TCA cycle, acetyl-CoA pathway and 3-hydroxypropionate cycle are found in some chemolithotrophic prokaryotes. Some can use organic compounds as their carbon source while metabolizing an inorganic electron donor. This kind of bacterial metabolism is referred to as mixotrophy.

10.1 | Reverse electron transport

As with chemoorganotrophs, metabolism of chemolithotrophs requires ATP and NAD(P)H for carbon metabolism and biosynthetic processes. Some of the electron donors used by chemolithotrophs have a redox potential higher than that of $NAD(P)^+/NAD(P)H$ (Table 10.1). Electrons from these electron donors are transferred to coenzyme Q or to cytochromes. Some of the electrons are used to generate a proton motive force reducing O_2 while the remaining electrons reduce $NAD(P)^+$ to NAD(P)H through a reverse of the electron transport chain. The latter is an uphill reaction and coupled with the consumption of the proton motive force (Figure 10.1). This is referred to as reverse electron transport. In most cases, electron donors with a redox potential lower than $NAD(P)^+/NAD(P)H$ are oxidized and this is coupled with the reduction of coenzyme Q or cytochromes for the efficient utilization of the electron donors at low concentration. The energy consumed in reverse electron transport from cytochrome c to $NAD(P)^+$ is about five times the energy generated from the forward electron transport process.

Table 10.1. | *Redox potential of inorganic electron donors used by chemolithotrophs*

Electron donating reaction	Redox potential ($E^{0'}$, V)
$CO + H_2O \longrightarrow CO_2 + 2H^+ + 2e^-$	-0.54
$SO_3^{2-} + H_2O \longrightarrow SO_4^{2-} + 2H^+ + 2e^-$	-0.45
$H_2 \longrightarrow 2H^+ + 2e^-$	-0.41
$NAD(P)H + H^+ \longrightarrow NAD(P)^+ + 2H^+ + 2e^-$	-0.32
$H_2S \longrightarrow S^0 + 2H^+ + 2e^-$	-0.25
$S^0 + 3H_2O \longrightarrow SO_3^{2-} + 6H^+ + 4e^-$	$+0.05$
$NO_2^- + H_2O \longrightarrow NO_3^- + 2H^+ + 2e^-$	$+0.42$
$NH_4^+ + 2H_2O \longrightarrow NO_2^- + 8H^+ + 6e^-$	$+0.44$
$Fe^{2+} \longrightarrow Fe^{3+} + e^-$	$+0.78$
$O_2 + 4H^+ + 4e^- \longrightarrow 2H_2O$	$+0.86$

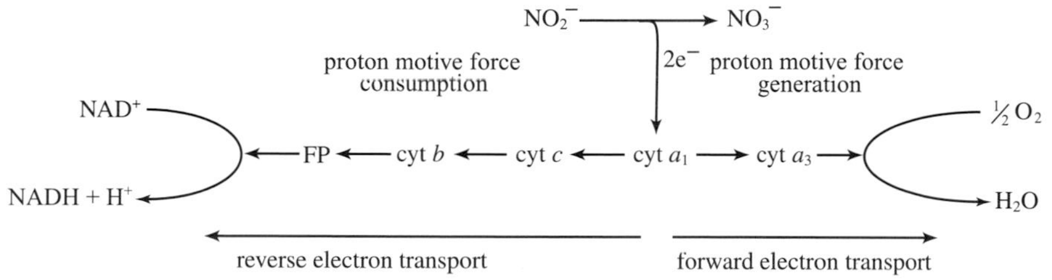

Figure 10.1 Reverse electron transport to reduce $NAD(P)^+$ in a nitrite oxidizer, *Nitrococcus mobilis.*

Nitrococcus mobilis oxidizes nitrite as the electron donor reducing cytochrome a_1. Some of the electrons are consumed to reduce O_2 and generate a proton motive force (forward electron transport). The remaining electrons are transferred to $NAD(P)^+$ to supply reducing power for biosynthesis. The latter process requires energy in the form of a proton motive force, and is referred to as reverse electron transport.

10.2 | Nitrification

A group of bacteria oxidize ammonia to nitrite that is further oxidized to nitrate by another group of bacteria in an energy generating process known as nitrification. They are all Gram-negative, mostly obligately chemolithotrophic, and have an extensive membrane structure within the cytoplasm except for *Nitrosospira tenuis*. A separate group of bacteria oxidize nitrite to nitrate. These organisms are referred to as nitrifying bacteria and are widely distributed in soil and water. The nitrogen cycle cannot be completed without them.

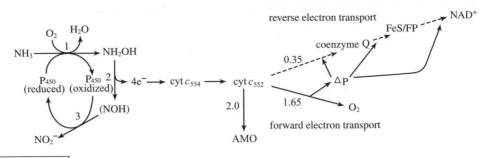

Figure 10.2 Ammonia oxidation to nitrite by denitrifiers.

(Modified from *Arch. Microbiol.* **178**:250–255, 2002)

Since the redox potential of NH_2OH/NH_3 (+0.899 V) is higher than that of $\frac{1}{2}O_2/H_2O$, ammonia monooxygenase (1) oxidizes ammonia, consuming $2e^-$ from the reduced form of cytochrome c_{552} mediated by P450. Hydroxylamine oxidation is coupled to the reduction of cytochrome c_{554}. Out of the four electrons released in the oxidation of NH_2OH by HAO, two electrons are consumed by AMO where they are used for the oxidation of ammonia. The other 1.65 electrons are routed to cytochrome oxidase to generate a proton motive force and the remaining 0.35 electrons pass to NAD^+ through reverse electron transport to supply reducing power for biosynthesis.

Ammonia and nitrite oxidizers fix CO_2 through the Calvin cycle (Section 10.8.1).

10.2.1 Ammonia oxidation

NH_3 is oxidized in a two-step reaction via hydroxylamine (NH_2OH) and nitroxyl (NOH) in reactions catalyzed by ammonia monooxygenase (AMO) and hydroxylamine oxidoreductase (HAO) (Figure 10.2).

$$2NH_3 + 3O_2 \longrightarrow 2NO_2^- + 2H^+ + 2H_2O \quad (\Delta G^{0\prime} = -272 \text{ kJ/mol } NH_3)$$

NH_3 is oxidized to hydroxylamine by AMO consuming two electrons available from the oxidation of hydroxylamine, probably through a membrane-bound cytochrome c_{552}. Since the redox potential of NH_2OH/NH_3 (+ 0.899 V) is higher than that of $\frac{1}{2}O_2/H_2O$, ammonia monooxygenase oxidizes ammonia consuming $2e^-$ from the reduced form of cytochrome c_{552} mediated by P450, as shown in Figure 10.2. Hydroxylamine oxidation to nitrite is coupled to the reduction of cytochrome c_{554}.

$$2NH_3 + O_2 \xrightarrow{\text{AMO}} 2NH_2OH \qquad (\Delta G^{0\prime} = +16 \text{ kJ/mol } NH_3)$$

$$\text{hydroxylamine} + H_2O \xrightarrow{\text{HAO}} NO_2^- + 5H^+ + 4e^-$$

HAO from *Nitrosomonas europaea* is a homotrimer with each subunit containing eight *c*-type hemes, giving a total of 24 hemes. Seven of the hemes in each subunit are covalently attached to the protein by two thioester linkages. The eighth heme, designated P-460, is an unusual prosthetic group, and has an additional covalent bond through a tyrosine residue. The P-460 heme is located at the active site. The function of the *c*-hemes is believed to be the transfer of electrons from the active site of P-460 to cytochrome c_{554}.

The four electrons released from the oxidation of NH_2OH by HAO in *Nitrosomonas europaea* are channelled through cytochrome c_{554} to a membrane-bound cytochrome c_{552}. Two of the electrons are routed back to AMO, where they are used for the oxidation of ammonia, while 1.65 electrons are used to generate a proton motive force through cytochrome oxidase and 0.35 are used to reduce $NAD(P)^+$ through reverse electron transport. Out of

3×2 electrons, $2 \times 2e^-$ are consumed in ammonia oxidation by the monooxygenase, and the remaining $2e^-$ are used to generate Δp through ETP and NAD(P)H through the reverse electron transport.

Although ammonia oxidizers are known as obligate chemolithotrophs, they can utilize a limited number of organic compounds including amino acids and organic acids. The complete genome sequence of *Nitrosomonas europaea* has revealed a potential fructose permease, and this bacterium metabolizes fructose and pyruvate mixotrophically.

10.2.2 Nitrite oxidation

Nitrite produced from the oxidation of ammonia is used by a separate group of bacteria as their energy source (Table 10.2):

$$2NO_2^- + O_2 \longrightarrow 2NO_3^- \quad (\Delta G^{0\prime} = -74.8 \, \text{kJ/mol NO}_2^-)$$

Nitrite oxidoreductase (NOR) oxidizes nitrite to nitrate, reducing cytochrome a_1. It is a membrane-associated iron-sulfur molybdoprotein, and is part of an electron transfer chain which channels electrons from nitrite to molecular oxygen. NOR in *Nitrobacter hamburgensis* is a heterodimer consisting of α and β subunits. The NOR of *Nitrobacter winogradskyi* is composed of three subunits as well as heme a_1, heme c, non-heme iron and molybdenum. This enzyme transfers electrons to the cytochrome c oxidase through a membrane-bound cytochrome c (Figure 10.3). Since the free energy change is small, only one H^+ is transported coupled to these reactions. The electron transfer is not well understood. Hydride ion (H^-) is transferred from NOR to cytochrome c consuming the inside negative membrane potential,

Table 10.2. *Representative nitrifying bacteria*	
Organism	Characteristics
$NH_3 \longrightarrow NO_2^-$	
Nitrosomonas europaea	soil, fresh water, seawater, sewage works
Nitrosospira (*Nitrosovibrio*) *tenuis*	soil
Nitrosococcus nitrosus	soil
Nitrosococcus oceanus	seawater
Nitrosospira briensis	soil
Nitrosolobus multiformis	soil
$NO_2^- \longrightarrow NO_3^-$	
Nitrococcus mobilis	seawater
Nitrobacter winogradskyi	soil, fresh water, seawater, facultative chemolithotroph
Nitrospina gracilis	seawater
Nitrospira marina	seawater

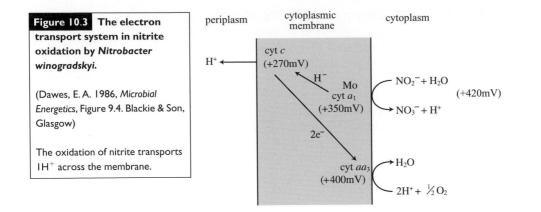

Figure 10.3 **The electron transport system in nitrite oxidation by *Nitrobacter winogradskyi*.**

(Dawes, E. A. 1986, *Microbial Energetics*, Figure 9.4. Blackie & Son, Glasgow)

The oxidation of nitrite transports 1H$^+$ across the membrane.

because electron transfer from NOR with a redox potential of $+420$ mV to cytochrome c ($+270$ V) is an uphill reaction.

Nitrite oxidizers are obligate chemolithotrophs with the exception of *Nitrobacter winogradskyi*, which is a facultative chemolithotroph.

10.2.3 Anaerobic nitrification

As discussed in Chapter 9 (Section 9.1.4) some bacteria of the phylum *Planctomycetes* oxidize ammonia under anaerobic conditions using nitrite as the electron acceptor. These are known as ANAMMOX bacteria. *Nitrosomonas europaea* oxidizes ammonia under anaerobic conditions using nitrogen dioxide (NO_2) as the electron acceptor.

10.3 | Sulfur bacteria and the oxidation of sulfur compounds

Certain prokaryotes can use inorganic sulfur compounds including sulfide (HS^-), elemental sulfur (S^0), thiosulfate ($HS_2O_3^-$), and sulfite (HSO_3^-) as their energy source. These are known as sulfur bacteria.

10.3.1 Sulfur bacteria

To distinguish them from photolithotrophic sulfur bacteria (Chapter 11), chemolithotrophic sulfur bacteria are referred to as colourless sulfur bacteria. These are phylogenetically diverse and include bacteria and archaea. They are grouped either according to the location of sulfur deposition after sulfide is oxidized (Table 10.3) or by their ability to use polythionate. Many can thrive at the aerobic–anaerobic interface where sulfide produced by sulfate-reducing bacteria diffuses from anaerobic regions. These organisms have to compete with molecular oxygen for sulfide, which is rapidly oxidized with molecular oxygen. Some other sulfur bacteria are acidophilic, oxidizing pyrite (FeS_2), and include species of *Thiobacillus*. The genus *Thiobacillus* is designated as small Gram-negative rod-shaped bacteria

| Table 10.3. | *Sulfur bacteria* |

1. Accumulating sulfur intracellularly

Gliding, filamentous cells	*Beggiatoa, Thiothrix, Thioploca*
Gliding, very large unicells	*Achromatium*
Immotile or motile with flagella, cocci or rod shaped	
immotile, rod	*Thiobacterium*
motile with flagella, rod	*Macromonas*
motile with flagella, cocci	*Thiovulum*
motile with flagella, vibrioid	*Thiospira*

2. Accumulating sulfur extracellularly

	Thiobacillus	*Thiomicrospira*	*Thioalkalimicrobium*	*Thioalkalivibrio*
Morphology	rod	vibrioid	rod to spirillum	rod to spirillum
Flagellum	+	+	+	+
DNA G + C (%)	34–70	48	61.0–65.6	48.0–51.2
Growth pH	1–8.5	5.0–8.5	7.5–10.6	7.5–10.6
Chemolithotrophy	facultative	obligate	obligate	obligate

3. Thermophilic

	Thermothrix	*Sulfolobus*	*Acidianus*
Classification	Gram-negative	archaeon	archaeon
Morphology	rod	cocci	cocci
DNA G + C content (%)	?	36–38	~31
Growth temperature ($^\circ$C)	40–80	50–85	45–95
Chemolithotrophy	facultative	facultative	facultative

Source: Applied and Environmental Microbiology, **67**, 2873–2882, 2001.

deriving energy from the oxidation of one or more reduced sulfur compounds including sulfides, thiosulfate, polythionate and thiocyanate. They fix CO_2 through the Calvin cycle. However, they are diverse phylogenetically in terms of 16S ribosomal RNA gene sequences, DNA G + C content and DNA homology in addition to physiological differences. Many species have been reclassified to *Paracoccus, Acidiphilium, Thiomonas, Thermithiobacillus, Acidithiobacillus* and *Halothiobacillus* but the classification of several species is still uncertain. For convenience, *Thiobacillus* is used here. In addition to species of *Beggiatoa, Thiobacillus* and *Thiomicrospira* within the sulfur bacteria, many other prokaryotes can oxidize sulfur compounds mixotrophically or chemolithotrophically. These include bacteria such as species of *Aquaspirillum, Aquifex, Bacillus, Paracoccus, Pseudomonas, Starkeya* and *Xanthobacter*, and archaea such as species of *Sulfolobus* and *Acidianus*. Species of the Gram-negative bacterial genus *Thermothrix*, and the archaea *Sulfolobus* and *Acidianus* are thermophiles. *Acidianus brierleyi* uses elemental sulfur not only as an electron donor under aerobic conditions, but also as an electron acceptor to reduce hydrogen under anaerobic conditions.

Species of *Thiobacillus* and *Thiomicrospira* can be isolated from diverse ecosystems including soil, freshwater and seawater. *Sulfolobus acidocaldarius* is an archaeon isolated from an acidic hot spring. In addition to reduced sulfur compounds, *Thiobacillus ferrooxidans* and *Sulfolobus acidocaldarius* oxidize Fe(II) to Fe(III) as their electron donor. *Thiobacillus ferrooxidans* and *Thiobacillus thiooxidans* grow optimally at around pH 2.0 and oxidize metal sulfides solubilizing the metals. Low quality ores are treated with these bacteria to recover metals such as Cu, U, and others. This process is referred to as bacterial leaching. This property can also be applied to remove sulfur (in pyrite, FeS_2) from coal. *Thiobacillus denitrificans* is an anaerobe which oxidizes sulfur compounds using nitrate as the electron acceptor.

Filamentous sulfur bacteria of the genera *Thioploca* and *Beggiatoa* accumulate nitrate in intracellular vacuoles. Nitrate, acting as electron acceptor, is reduced to ammonia with sulfide or sulfur as electron donors in these bacteria (Section 9.1.4). Many nitrate-accumulating sulfur bacteria inhabit the sediments of upwelling areas characterized by high sediment concentrations of soluble sulfide, and low levels of dissolved oxygen. The ecological implication of nitrate ammonification is that nitrogen is conserved within the ecosystem. *Thiomargarita namibiensis* is another sulfur bacterium with vacuoles containing nitrate. This bacterium has not been isolated in pure culture, but is found in sediments in the coastal waters of Namibia measuring up to 0.75 mm in size, which is about 100 times bigger than a normal sized bacterium.

In addition to the acidophilic and neutrophilic sulfur bacteria, alkalophilic sulfur bacteria thrive in alkaline soda lakes. These are species of *Thioalkalimicrobium* and *Thioalkalivibrio*. They accumulate elemental sulfur extracellularly before oxidizing it when sulfide is depleted. Members of the former genus are obligate aerobes, while some of the latter genus are facultative anaerobes using nitrate as electron acceptor (Section 9.1.4). An anaerobic Gram-negative bacterium is also known that can grow chemolithotrophically using HS^- as the electron donor and arsenate as the electron acceptor.

10.3.2 Biochemistry of sulfur compound oxidation

Sulfur compound oxidation mechanisms have not been clearly established, partly because sulfur chemistry is complicated and also because different organisms have different oxidative mechanisms that use different enzymes and coenzymes. Figure 10.4 outlines the sulfur compound oxidation pathways in bacteria and archaea. These oxidize HS^-, S^0, $HS_2O_3^-$ and HSO_3^- through a common pathway transferring electrons to the electron transport chain to generate a proton motive force.

Inorganic sulfur oxidation enzyme systems are best known in *Paracoccus tentotrophus*, a facultative chemolithotrophic Gram-negative

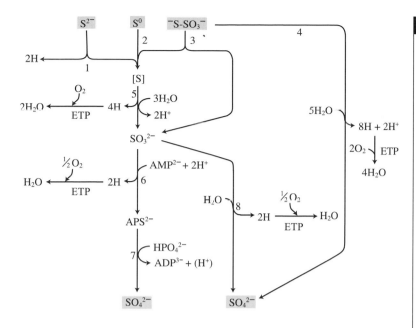

Figure 10.4 Oxidation of sulfur compounds by *Thiobacillus ferrooxidans*.

(Dawes, E. A. 1986, *Microbial Energetics*, Figure 9.5. Blackie & Son, Glasgow)

Sulfide and sulfur are converted to polysulfide before being oxidized further. Thiosulfate is metabolized to sulfite and polysulfide by rhodanese (3) or directly oxidized to sulfate by the thiosulfate-oxidizing multienzyme complex (4). Sulfur compound oxidation is coupled to the reduction of cytochrome *b* or *c*.

1, sulfide-oxidizing enzyme; 2, conversion to polysulfide; 3, rhodanese; 4, thiosulfate-oxidizing multienzyme complex; 5, sulfur-oxidizing enzyme; 6, APS reductase; 7, ADP sulfurylase; 8, sulfite cytochrome *c* reductase.

bacterium. The genes for sulfur oxidation (Sox) are encoded in the *sox* gene cluster consisting of 15 genes. The Sox system reconstituted from SoxB, SoxCD, SoxXA and SoxYZ oxidizes HS^-, S^0, HS_2O_3 and HSO_3 reducing cytochrome *c*, but each of the proteins is catalytically inactive in vitro. Other gene products include HS^--oxidizing enzyme, *c*-type cytochromes, regulator proteins and others. These proteins are believed to catalyze the reactions shown in Figure 10.4.

HS^- and S^0 are converted to polysulfide, [S], before being oxidized to sulfate. The reduced form of glutathione (GSH) is involved in the oxidation of sulfide to [S].

$$nS^- + GSH \longrightarrow GS_nSH + 2ne^-$$

Sulfur oxidase catalyzes [S] oxidation to sulfite. Sulfite is oxidized to sulfate either through the direct reaction catalyzed by sulfite cytochrome reductase (SCR) or through the reactions catalyzed by adenosine-5'-phosphosulfate (APS) reductase and ADP sulfurylase. Direct oxidation appears to be far more widespread than the APS reductase pathway. More energy is conserved in the latter reactions than in the former. The nature of SCR is different among the sulfur oxidizers. SCR of *Thiobacillus thioparus* is a cytoplasmic soluble enzyme containing molybdenum, and that of *Paracoccus* (*Thiobacillus*) *pantotrophicus* is a periplasmic enzyme containing Mo and *c*-type heme. It is not known if Mo is contained in the membrane SCR of *Thiobacillus thiooxidans*. The soluble and periplasmic SCRs oxidize sulfite coupled to the reduction of *c*-type cytochromes, while the membrane enzyme is coupled to the reduction of Fe(III). The electron transport chain is not known in detail, though genes of sulfur oxidizing enzymes have been characterized in many organisms. The mid-point redox potential of the sulfur compounds are:

$$S^0/H_2S = -0.25\,V$$
$$SO_3^{2-}/S = +0.5\,V$$
$$SO_4^{2-}/SO_3^{2-} = -0.454\,V$$

The cell yield on sulfide is higher in the anaerobic denitrifier *Thiobacillus denitrificans* than in *Thiobacillus thiooxidans*. This shows that some sulfur compounds are oxidized by oxidases directly reducing molecular oxygen in the aerobic bacterium without energy conservation, while the oxidative reactions are coupled to denitrification with energy conservation.

A sulfur bacterium *Paracoccus denitrificans* (*Thiosphaera pantotropha*) can grow chemolithotrophically oxidizing carbon disulfide (CS_2) or carbonyl sulfide (COS).

Thiobacillus ferrooxidans and *Sulfolobus acidocaldarius* can use Fe(III) as their electron acceptor and S^0 as the electron donor.

$$S^0 + 6Fe^{3+} + 4H_2O \longrightarrow HSO_4^- + 6Fe^{2+} + 7H^+ \quad (\Delta G^0 \text{ at pH } 2.0 = -314\,kJ/mol\ S^0)$$

10.3.3 Carbon metabolism in colourless sulfur bacteria

All species of the genus *Thiomicrospira* and some species of *Thiobacillus* and *Sulfolobus* are obligate chemolithotrophs. Other species are either facultative chemolithotrophs or mixotrophs using organic compounds. Obligately chemolithotrophic colourless sulfur bacteria fix CO_2 through the Calvin cycle, while the reductive TCA cycle (Section 10.8.2) and the 3-hydroxypropionate cycle (Section 10.8.4) are employed in the archaea *Sulfolobus acidocaldarius* and *Acidianus brierleyi*, respectively.

10.4 | Iron bacteria: ferrous iron oxidation

Many microorganisms can oxidize Fe(II) to Fe(III). Some of these are known as iron bacteria and use the free energy generated from the oxidation. Many heterotrophic bacteria also oxidize Fe(II), but the function of such ferrous iron oxidation is not known and they do not conserve the free energy.

Fe(II) is used as the electron donor in some Gram-negative bacteria including *Gallionella ferruginea*, *Thiobacillus ferrooxidans* and species of *Leptospirillum*, and the archaea *Sulfolobus acidocaldarius*, *Acidianus brierleyi* and species of *Ferroplasma*. *Thiobacillus ferrooxidans*, *Sulfolobus acidocaldarius* and *Acidianus brierleyi* use sulfur compounds as their electron donors, but the others do not use them. Since Fe(II) is chemically oxidized easily at neutral pH, iron bacteria growing at neutral pH are microaerophilic and prefer a medium redox potential around 200–320 mV at a slightly acidic pH of 6.0. Their growth can be observed in an agar gel with gradients of Fe(II) and O_2 in opposite directions. Species of

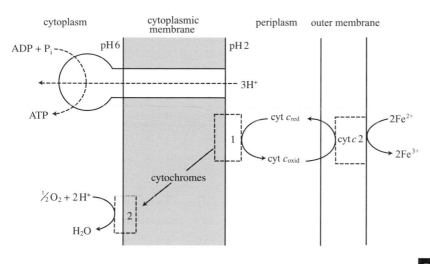

Figure 10.5. **ATP synthesis coupled to the oxidation of Fe²⁺ in *Thiobacillus ferrooxidans*.**

(Gottschalk, G. 1986, *Bacterial Metabolism*, 2nd edn, Figure 9.6. Springer, New York)

1, rusticyanin; 2, cytochrome oxidase.

Leptospirillum, *Thiobacillus ferrooxidans* and *Sulfolobus acidocaldarius* grow optimally around pH 2.0 where the chemical oxidation of Fe(II) is slow.

A water-insoluble mineral, pyrite, is the natural electron donor used by acidophilic iron bacteria including *Thiobacillus ferrooxidans*. They oxidize the electron donor at the cell surface, reducing an outer membrane *c*-type cytochrome which transfers electrons to the terminal oxidase located at the inner face of the cytoplasmic membrane (Figure 10.5). Electron transfer involves rusticyanin, a copper-containing blue protein. Rusticyanin is not involved in the electron transport chain that oxidizes sulfur compounds in this bacterium. The free energy change in Fe(II) oxidation is small since the redox potential of Fe(III)/Fe(II) is +0.78 V which is very similar to the +0.86 V of O_2/H_2O. The acidophilic Fe(II) oxidizers maintain their internal pH around neutrality with a H^+ gradient of 10^3-10^6 (Section 5.7.3). They maintain a low or inside positive membrane potential to compensate for the large potential generated by the H^+ gradient. Electron transfer from the periplasmic region to cytochrome oxidase contributes to the membrane potential and proton consumption by the oxidase contributes to the proton gradient part of the proton motive force.

The hyperthermophilic archaeon *Sulfolobus acidocaldarius* grows chemolithotrophically using sulfide and Fe(II) as its electron donors, like *Thiobacillus ferrooxidans*. The former is a facultative chemolithotroph while the latter is an obligate chemolithotroph. Species of *Leptospirillum*, *Gallionella ferruginea* and *Thiobacillus ferrooxidans* fix CO_2 through the Calvin cycle (Section 10.8.1), while the reductive TCA cycle (Section 10.8.2) and 3-hydroxypropionate cycle (Section 10.8.4) are employed by *Sulfolobus acidocaldarius* and *Acidianus brierleyi*, respectively.

An archeaon, *Ferroglobus placidus*, oxidizes Fe(II) under anaerobic conditions using nitrate as the electron acceptor. This is unusual since the redox potential of the electron donor ($E^{0'}$, Fe(III)/Fe(II) $= +0.78$ V) is higher than that of the electron acceptor ($E^{0'}$, $NO_3^-/NO_2^- = +0.42$ V). Several dissimilatory perchlorate-reducing bacteria including *Dechlorosoma suillum* use Fe(II) as their electron donor (Section 9.7).

In addition to Fe(II), As(III) and U(IV) can be used as electron donors by various prokaryotes. These are discussed later (Section 10.7).

10.5 | Hydrogen oxidation

10.5.1 Hydrogen-oxidizing bacteria
Various bacteria grow chemolithotrophcally on a $H_2 + CO_2$ mixture. With a few exceptions (e.g. *Hydrogenobacter thermophilus* and *Hydrogenovibrio marinus*), these are facultative chemolithotrophs (Table 10.4). They are phylogenetically diverse, and are grouped not on their chemolithotrophy but on their heterotrophic characteristics.

10.5.2 Hydrogenase
The bacteria listed in Table 10.4 use hydrogen as their electron donor to grow chemolithotrophically with the aid of hydrogenase. Most of them have a cytochrome-dependent particulate hydrogenase on their cytoplasmic membrane and *Ralstonia eutropha* (*Alcaligenes eutrophus*) and *Nocardia autotrophica* possess a NAD^+-dependent soluble hydrogenase in addition to the particulate enzyme (Figure 10.6). Only the soluble enzyme is found in the third group which includes *Alcaligenes denitrificans*, *Alcaligenes ruhlandii* and *Rhodococcus opacus*. The soluble hydrogenase reduces NAD^+, and the particulate enzyme transfers electrons from H_2 to coenzyme Q of the electron transport chain. The soluble hydrogenase gene in *Ralstonia eutropha* is plasmid encoded. Since the affinity of the soluble enzyme for the substrate is low, this enzyme cannot oxidize the substrate at low concentrations.

The particulate hydrogenase has a high affinity for the substrate, enabling the bacterium to use hydrogen at low concentrations. Organisms with only the particulate hydrogenase employ reverse electron transport to reduce $NAD(P)^+$. In most cases, the hydrogenase from hydrogen-oxidizing bacteria cannot produce hydrogen. These enzymes are referred to as uptake hydrogenases to differentiate them from those of anaerobic bacteria. The anaerobic bacterial enzymes are called evolution (or production) hydrogenases. The function of the evolution hydrogenase is to dispose of electrons generated from fermentative metabolism by reducing protons.

$$2H^+ + 2e^- \text{ (reduced ferredoxin)} \xrightleftharpoons{\text{evolution hydrogenase}} H_2$$

$$H_2 + \text{electron carrier (oxidized)} \xrightarrow{\text{uptake hydrogenase}} \text{electron carrier (reduced)}$$

Table 10.4. | *Hydrogen-oxidizing bacteria and their hydrogenase*

Organism	Hydrogenase	
	Soluble (NAD$^+$-dependent)	Particulate (cytochrome-dependent)
Facultative chemolithotroph		
Gram-negative		
Alcaligenes denitrificans	+	−
Ralstonia eutropha (Alcaligenes eutrophus)	+	+
Alcaligenes latus	−	+
Alcaligenes ruhlandii	+	
Aquaspirillum autotrophicum	−	+
Azospirillum lipoferum	−	+
Derxia gummosa	−	+
Flavobacterium autothermophilum	?	+
Microcyclus aquaticus	−	+
Paracoccus denitrificans	−	+
Pseudomonas facilis	−	+
Pseudomonas hydrogenovara	−	+
Ralstonia eutropha	+	+
Renobacter vacuolatum	−	+
Rhizobium japonicum	−	+
Xanthobacter flavus	−	+
Gram-positive		
Arthrobacter sp.	−	+
Bacillus schlegelii	−	+
Mycobacterium gordonae	?	+
Nocardia autotrophica	+	−
Rhodococcus opacus	+	−
Obligate chemolithotroph		
Hydrogenobacter thermophilus	−	+
Hydrogenovibrio marinus	−	+[a]

[a] The particulate hydrogenase in this bacterium is NAD$^+$-dependent and catalyzes the reverse reaction.

The particulate hydrogenase of the thermophilic chemolithotrophic *Hydrogenovibrio marinus* is NAD$^+$-dependent and catalyzes the reverse reaction.

10.5.3 Anaerobic H$_2$-oxidizers

It has been stated that some anaerobic respiratory prokaryotes can grow on $H_2 + CO_2$ with an appropriate electron acceptor. They include some sulfidogens (Section 9.3), methanogens (Section 9.4) and homoacetogens (Section 9.5). In addition to these, H$_2$-oxidizing anaerobic chemolithotrophs have been isolated from hydrothermal vents. *Desulfurobacterium crinifex*, *Thermovibrio ammonificans* and *Thermovibrio ruber* use H$_2$ as the electron donor reducing nitrate to ammonia, or S^0 to HS$^-$. These are strict anaerobes. Species of *Caminibacter* and

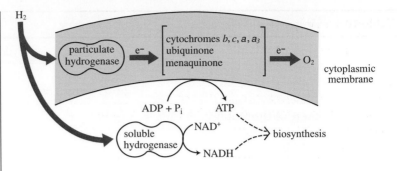

Figure 10.6 Hydrogen utilization by *Ralstonia eutropha*.

(*Alcaligenes eutrophus*) (Gottschalk, G. 1986, *Bacterial Metabolism*, 2nd edn., Figure 9.1. Springer, New York)

This bacterium has a NAD$^+$-dependent soluble hydrogenase in addition to the particulate enzyme. The soluble enzyme reduces pyridine nucleotide for biosynthesis, and the particulate hydrogenase channels electrons from hydrogen directly to the electron transport chain to generate a proton motive force.

Hydrogenomonas thermophila are microaerophilic H_2-oxidizers using O_2, nitrate or elemental sulfur as electron acceptors (Sections 9.1.4 and 9.3). Another strict anaerobe, *Balnearium lithotrophicum*, oxidizes H_2 reducing nitrate to ammonia, but cannot reduce sulfur. A *Dechloromonas* sp. isolated from a sewage works grows chemolithotrophically on H_2 and perchlorate. *Ferroglobus placidus*, a strictly anaerobic archaeon, grows chemoautotrophically oxidizing hydrogen coupled to nitrate reduction.

10.5.4 CO_2 fixation in H_2-oxidizers

The reductive TCA cycle (Section 10.8.2) is used to fix CO_2 in obligately chemolithotrophic *Hydrogenobacter thermophilus*, while all the other hydrogen bacteria tested to date fix CO_2 through the Calvin cycle (Section 10.8.1).

10.6 | Carbon monoxide oxidation: carboxydobacteria

Annually, about 1×10^9 tons of carbon monoxide (CO) are produced chemically and biologically as a by-product in photosynthesis, while a similar amount is oxidized so that the concentration remains around 0.1 ppm in the atmosphere. The concentration can, however, reach up to 100 ppm in industrial areas. CO is oxidized photochemically in the atmosphere, while bacterial oxidation is predominant in soil. Aerobic CO oxidizers are referred to as carboxydobacteria (Table 10.5). They are facultative chemolithotrophs like the hydrogen-oxidizing bacteria.

Most of the carboxydobacteria known to date are Gram-negative bacteria, and, except for *Alcaligenes carboxidus*, can use H_2, but not all H_2-oxidizing bacteria can use CO. H_2-oxidizing carboxydobacteria oxidize CO and H_2 simultaneously. Unexpectedly, many species of *Mycobacterium*, including *Mycobacterium tuberculosis*, can grow chemolithotrophically using CO as their sole carbon and energy source. *Pseudomonas carboxydoflava* uses nitrate under anaerobic conditions using carbon monoxide as electron donor (Section 9.1.4).

Carboxydobacteria possess carbon monoxide dehydrogenase. This aerobic enzyme is different from the anaerobic CO

Table 10.5.	*Typical carboxydobacteria*

Acinetobacter sp. Strain IC-1
Alcaligenes (*Carbophilus*) *carboxidus*
Comamonas (*Zavarzinia*) *compransoris*
Pseudomonas (*Oligotropha*) *carboxidovorans*
Pseudomonas carboxidoflava (*Hydrogenovibrio pseudoflava*)
Pseudomonas carboxidohydrogena
Pseudomonas gazotropha

dehydrogenases of methanogens and homoacetogens. The latter are dual function enzymes catalyzing CO oxidation and acetyl-CoA synthesis in both directions (Sections 9.4.3 and 9.5.2). They are soluble enzymes and use low redox potential electron carriers such as F_{420} and ferredoxin in methanogens and homoacetogens, respectively. The membrane-bound aerobic enzyme in the carboxydobacteria catalyzes CO oxidation only, reducing coenzyme Q:

$$CO + H_2O \xrightarrow{\text{CO dehydrogenase}} CO_2 + 2H^+ + 2e^-$$

As stated above, carboxydobacteria can use CO efficiently at low concentrations employing CO dehydrogenase to reduce coenzyme Q with a redox potential of around zero V, much higher than $-0.54\,V$, the redox potential of CO_2/CO. Carboxydobacteria reduce $NAD(P)^+$ through a reverse electron transport chain not directly coupled to CO oxidation. The Calvin cycle is the CO_2-fixing mechanism in carboxydobacteria.

Many methanogens and homoacetogens can use CO as their electron donor (Sections 9.4 and 9.5).

10.7 | Chemolithotrophs using other electron donors

In addition to the electron donors discussed above, other inorganic compounds can serve as an energy source in chemolithotrophic metabolism. *Pseudomonas arsenitoxidans* was isolated from a gold mine based on its ability to use arsenite as its electron donor:

$$2H_3AsO_3 + O_2 \longrightarrow HAsO_4^{2-} + H_2AsO_4^- + 3H^+ \quad (\Delta G^{0\prime} = -128\,kJ/mol\ H_3AsO_3)$$

Similar isolates include *Agrobacterium albertimagni* from fresh water, *Thiomonas* sp. from acid mine drainage, *Hydrogenobaculum* sp. from geothermal springs and several strains of α-proteobacteria isolated from a gold mine. A *Thermus* strain isolated from an arsenic-rich terrestrial geothermal environment rapidly oxidized inorganic As(III) to As(V) under aerobic conditions, but energy was not conserved. The same strain could use As(VI) as an electron acceptor in the absence of oxygen.

Mn(II) is oxidized by bacteria inhabiting a wide variety of environments, including marine and freshwater, soil, sediments, water pipes, Mn nodules, and hydrothermal vents. Phylogenetically, Mn(II)-oxidizing bacteria appear to be quite diverse, falling within the low $G + C$ Gram-positive bacteria, the Actinobacteria, or the Proteobacteria. The most well-characterized Mn(II)-oxidizing bacteria are a *Bacillus* sp., *Leptothrix discophora*, *Pedomicrobium manganicum* and *Pseudomonas putida*. They do not conserve energy from Mn(II) oxidation. Mn(II) oxidation is catalyzed by a member of the multicopper oxidase family that utilize multiple copper ions as cofactors. A quasi-photosynthetic bacterium (Section 11.1.3), *Erythrobacter* sp., grows better with Mn(II) oxidation and probably conserves energy from the oxidation. *Thiobacillus denitrificans* conserves energy from uranium(IV) oxidation coupled to nitrate reduction.

10.8 | CO$_2$ fixation pathways in chemolithotrophs

The Calvin cycle is the most common CO_2 fixation pathway in aerobic chemolithotrophs and in photolithotrophs, and some fix CO_2 through the reductive TCA cycle. The anaerobic chemolithotrophs, including methanogens, homoacetogens and sulfidogens, employ the acetyl-CoA pathway to fix CO_2. A fourth CO_2 fixation pathway, the 3-hydroxypropionate cycle, is known in some chemo- and photolithotrophs. The Calvin cycle is the only known CO_2-fixing metabolism in eukaryotes.

10.8.1 Calvin cycle

As shown in Figure 10.7, CO_2 is condensed to ribulose-1,5-bisphosphate to produce two molecules of 3-phosphoglycerate, which is reduced to glyceraldehyde-3-phosphate.

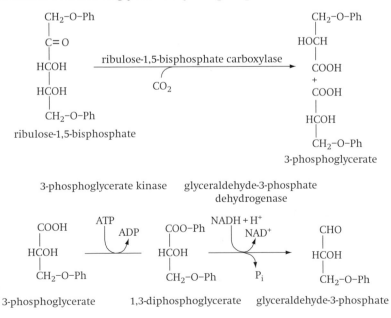

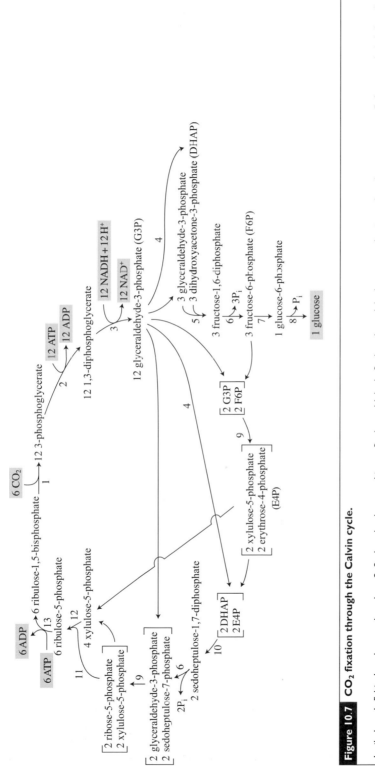

Figure 10.7 **CO₂ fixation through the Calvin cycle.**

1, ribulose-1, 5-bisphosphate carboxylase; 2, 3-phosphoglycerate kinase; 3, glyceraldehyde-3-phosphate dehydrogenase; 4, triose phosphate isomerase; 5, fructose-1, 6-diphosphate aldolase; 6, fructose-1,6-diphosphatase; 7, glucose-6-phosphate isomerase; 8, glucose-6-phosphatase; 9, transketolase; 10, sedoheptulose-1,7-diphosphate aldolase (the same as number 5); 11, ribose-5-phosphate isomerase; 12, ribose-5-phosphate-3-epimerase; 13, phosphoribulokinase.

A molecule of glyceraldehyde-3-phosphate is isomerized to dihydroxyacetone phosphate before being condensed to fructose-1,6-diphosphate with the second glyceraldehyde-3-phosphate molecule through the reverse reactions of the EMP pathway. Fructose-1,6-diphosphate is dephosphorylated to fructose-6-phosphate by the action of fructose-1,6-diphosphatase:

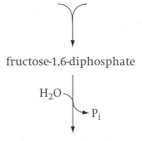

dihydroxyacetone phosphate + glyceraldehyde-3-phosphate

fructose-1,6-diphosphate

H_2O

P_i

fructose-6-phosphate

Through carbon rearrangement reactions similar to those of the HMP pathway, two molecules of fructose-6-phosphate ($2 \times C6$) and six molecules of glyceraldehyde-3-phosphate ($6 \times C3$) are converted to six molecules of ribulose-5-phosphate ($6 \times C5$) before being phosphorylated to ribulose-1,5-bisphosphate to begin the next round of reactions. The net result of these complex reactions is the synthesis of fructose-6-phosphate from $6CO_2$ consuming 18ATP and 12NAD(P)H:

$$6CO_2 + 18ATP + 12NAD(P)H + 12H^+ \longrightarrow \text{fructose-6-phosphate} + 18ADP + 18P_i + 12NAD(P)^+$$

ATP is consumed in the reactions catalyzed by 3-phosphoglycerate kinase and phosphoribulokinase, and glyceraldehyde-3-phosphate dehydrogenase oxidizes NAD(P)H.

10.8.1.1 Key enzymes of the Calvin cycle

Ribulose-1,5-bisphosphate carboxylase and phosphoribulokinase are key enzymes of the Calvin cycle, and are present only in the organisms fixing CO_2 through this highly energy-demanding pathway with a few exceptions (see below). Their activities are controlled at the transcriptional level and also after they are expressed. The enzymes are encoded by *cbb* genes organized in *cbb* operons differing in size and composition depending on the organism. In a facultative chemolithotroph, *Ralstonia eutropha*, the transcription of the operons, which may form regulons, is strictly controlled, being induced during chemolithotrophic growth but repressed to varying extents during heterotrophic growth. CbbR is a transcriptional regulator and the key activator protein of *cbb* operons. The *cbbR* gene is located adjacent to its cognate operon. The activating function of CbbR is modulated by metabolites which signal the nutritional state of the cell to the *cbb* system. Phosphoenolpyruvate is a negative effector of CbbR, whereas NADPH is a coactivator of the protein. In the photolithotrophs *Rhodobacter capsulatus* and *Rhodobacter sphaeroides*, a global two-component signal transduction system, RegBA, serves this function. Different *cbb* control systems have evolved in the diverse chemolithotrophs characterized by different metabolic capabilities.

Phosphoribulokinase activity is regulated by similar physiological signals. NADH activates enzyme activity, while AMP and phosphoenolpyruvate (PEP) are inhibitory. The increase in NADH concentration means the cells are ready to grow, activating the Calvin cycle. On the other hand, the biosynthetic pathway cannot be operated under a poor energy state with an increased AMP concentration. When a facultative chemolithotroph is provided with organic carbon, enzyme activity is inhibited by PEP. Similarly, 6-phosphogluconate inhibits ribulose-1,5-bisphosphate carboxylase activity.

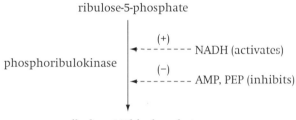

Ribulose-1,5-bisphosphate carboxylase is the most abundant single protein on Earth and is synthesized by all organisms fixing CO$_2$ through the Calvin cycle including plants. This enzyme is typically categorized into two forms. Type I, the most common form, consists of eight large and eight small subunits in a hexadecameric (L$_8$S$_8$) structure. This type is widely distributed in CO$_2$-fixing organisms, including all higher plants, algae, cyanobacteria, and many chemolithotrophic bacteria. The type II enzyme, on the other hand, consists of only large subunits (Lx), the number of which may be 2, 4 or 8 in different organisms. Type II ribulose-1,5-bisphosphate carboxylase is found in anaerobic purple photosynthetic bacteria and in some chemolithotrophs. Both types are found in some bacteria, especially in sulfur bacteria such as *Halothiobacillus neapolitanus* (formerly *Thiobacillus neapolitanus*), *Thiomonas intermedia* (formerly *Thiobacillus intermedius*) and *Thiobacillus denitrificans*, in photolithotrophs such as *Rhodobacter sphaeroides* and *Rhodobacter capsulatus*, and in the obligately chemolithotrophic hydrogen bacterium *Hydrogenovibrio marinus*. These have genes for both type I and type II enzymes.

In addition to type I and type II enzymes, two novel type III and type IV enzymes have been revealed by the complete genome sequences of some archaea and bacteria in which the Calvin cycle is not yet known. The type III enzyme is found in many thermophilic archaea including *Thermococcus* (formerly *Pyrococcus*) *kodakaraensis*, *Methanococcus jannaschii* and *Archaeoglobus fulgidus*. The gene for the enzyme from *Thermococcus kodakaraensis* has been cloned and expressed in *Escherichia coli*. The recombinant enzyme shows carboxylase activity with a decameric structure, but the function has not been elucidated. The existence of the type III enzyme in archaea is not limited to thermophilic organisms. The complete genomes of the mesophilic heterotrophic methanogens *Methanosarcina acetivorans*, *Methanosarcina mazei*, and *Methanosarcina barkeri* were also found to contain genes encoding putative ribulose-1,5-bisphosphate carboxylase proteins, but these were not found in *Methanobacterium thermoautotrophicum* and *Methanococcus maripaludis*. These methanogenic archaea fix CO_2 through the acetyl-CoA pathway. It is not known what the function of the carboxylase enzyme is in these organisms.

Microbial genome sequences have revealed open reading frames with a similar sequence to that of ribulose-1,5-bisphosphate carboxylase in *Bacillus subtilis* and green sulfur photosynthetic bacteria. This protein is referred to as type IV and is involved in oxidative stress response or sulfur metabolism. Green sulfur photosynthetic bacteria fix CO_2 through the reductive TCA cycle.

Ribulose-1,5-bisphosphate carboxylase is encapsulated in protein to form microcompartments named carboxysomes in many, but not all, chemolithotrophs. The rate of CO_2 fixation is much higher than expected from the CO_2 concentration in the cytoplasm and the low affinity of the enzyme for CO_2. The carboxysomes are believed to have a function in concentrating CO_2 to achieve the higher rate.

10.8.1.2 Photorespiration

In addition to carboxylase activity, ribulose-1,5-bisphosphate carboxylase has oxygenase activity under CO_2-limited conditions and a high O_2 concentration. Under these conditions the enzyme oxidizes ribulose-1,5-bisphosphate to 3-phosphoglycerate and phosphoglycolate. For this reason this enzyme is referred to as ribulose-1,5-bisphosphate carboxylase/oxygenase (RuBisCO).

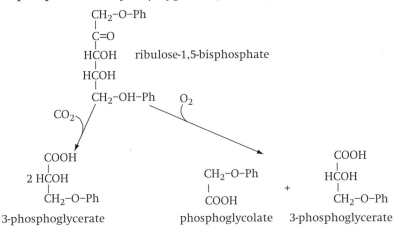

The enzyme efficiency is usually measured by the specificity factor (τ), which is the ratio of the rate constants for both carboxylase and oxygenase activities. The higher the RuBisCO τ value is, the better the RuBisCO can discern CO_2 from O_2. The τ value is generally

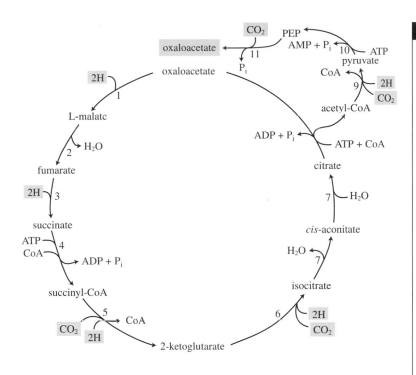

Figure 10.8 CO₂ fixation through the reductive TCA cycle.

Some bacteria and archaea reduce CO_2 to acetyl-CoA in the reverse direction of the TCA cycle. The TCA cycle enzymes unable to catalyze the reverse reaction, i.e. citrate synthase, 2-ketoglutarate dehydrogenase and succinate dehydrogenase, are replaced by ATP:citrate lyase (8), 2-ketoglutarate synthase (5) and fumarate reductase (3).

1, malate dehydrogenase; 2, fumarase; 3, fumarate reductase; 4, succinyl-CoA synthetase; 5, 2-ketoglutarate synthase; 6, isocitrate dehydrogenase; 7, aconitase; 8, ATP:citrate lyase; 9, pyruvate synthase; 10, PEP synthetase; 11, PEP carboxylase.

over 80 for type I RuBisCO in higher plants, between 25 and 75 for type I RuBisCO in bacteria, and under 20 for type II RuBisCO. The expression of the type I enzyme gene is activated at low CO₂ concentrations, and a high CO₂ concentration results in increased type II enzyme activity in organisms with both enzyme types including *Halothiobacillus neapolitanus*, *Hydrogenovibrio marinus* and *Rhodobacter sphaeroides*. It is hypothesized that photosynthetic bacteria cannot grow under aerobic conditions due to the low τ value of their RuBisCO (Section 11.1.2).

10.8.2 Reductive TCA cycle

The thermophilic H_2 bacterium *Hydrogenobacter thermophilus* is an obligate chemolithotroph. This bacterium does not have enzymes of the Calvin cycle, and fixes CO_2 through the reductive TCA cycle (Figure 10.8). This cyclic pathway is the CO₂ fixation process in some archaea including species of *Thermoproteus* and *Sulfolobus*, and in the photosynthetic green sulfur bacteria. This CO₂-fixing metabolism shares TCA cycle enzymes that catalyze the reverse reactions. ATP:citrate lyase, fumarate reductase and 2-ketoglutarate:ferredoxin oxidoreductase replace the TCA cycle enzymes that do not catalyze the reverse reactions, i.e. citrate synthase, succinate dehydrogenase and 2-ketoglutarate dehydrogenase. The reductive TCA cycle can be summarized as:

$$2CO_2 + 3NADH + 3H^+ + Fd \cdot H_2 + CoASH + 2ATP \longrightarrow$$
$$CH_3CO\text{-}CoA + 3NAD^+ + 3H_2O + Fd + 2ADP + 2P_i$$

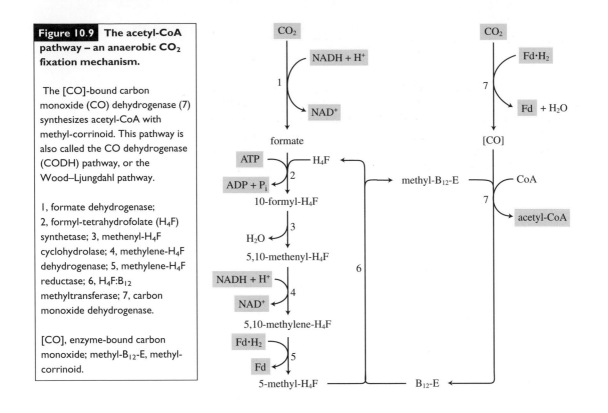

Figure 10.9 The acetyl-CoA pathway – an anaerobic CO_2 fixation mechanism.

The [CO]-bound carbon monoxide (CO) dehydrogenase (7) synthesizes acetyl-CoA with methyl-corrinoid. This pathway is also called the CO dehydrogenase (CODH) pathway, or the Wood–Ljungdahl pathway.

1, formate dehydrogenase; 2, formyl-tetrahydrofolate (H_4F) synthetase; 3, methenyl-H_4F cyclohydrolase; 4, methylene-H_4F dehydrogenase; 5, methylene-H_4F reductase; 6, $H_4F:B_{12}$ methyltransferase; 7, carbon monoxide dehydrogenase.

[CO], enzyme-bound carbon monoxide; methyl-B_{12}-E, methyl-corrinoid.

Acetyl-CoA is reduced to pyruvate by pyruvate:ferredoxin oxidoreductase. PEP synthetase converts pyruvate to PEP before entering gluconeogenesis to synthesize fructose-6-phosphate.

10.8.3 Anaerobic CO_2 fixation through the acetyl-CoA pathway

The acetyl-CoA pathway is employed for CO_2 fixation by anaerobic chemolithotrophs including sulfidogens, methanogens and homoacetogens (Figure 10.9).

Formate dehydrogenase reduces CO_2 to formate that is bound to the C1-carrier tetrahydrofolate (H_4F) to be reduced to methyl-H_4F. This methyl-group is transferred to coenzyme B_{12} (corrinoid). A second CO_2 molecule is reduced to the enzyme-bound form of [CO] by carbon monoxide dehydrogenase (CODH). [CO]-bound CODH synthesizes acetyl-CoA taking the methyl-group from methyl-corrinoid. CODH is a dual-function enzyme catalyzing CO oxidation/CO_2 reduction and acetyl-CoA synthesis/cleavage. This enzyme can also be called acetyl-CoA synthase.

This pathway has different names. 'Acetyl-CoA pathway' is commonly used since acetyl-CoA is the final product. 'CODH pathway' implies the pivotal role of the enzyme, while the 'Wood–Ljungdahl pathway' is another name to honour those who elucidated the pathway. The pathway can be summarized as:

$$2CO_2 + 2NADH + 2H^+ + 2Fd \cdot H_2 + CoASH + ATP \longrightarrow$$

$$CH_3CO\text{-}CoA + 2NAD^+ + 3H_2O + 2Fd + ADP + P_i$$

In comparison with the reductive TCA cycle, this pathway consumes one ATP less, and 2NADH and $2Fd \cdot H_2$ are oxidized while the reductive TCA cycle oxidizes 3NADH and $1Fd \cdot H_2$. Additional energy is conserved in the form of a sodium motive force at the reaction catalyzed by methylene-tetrahydrofolate reductase (Section 9.5.3). This comparison shows that the anaerobic process is more efficient than the reductive TCA cycle.

10.8.4 CO₂ fixation through the 3-hydroxypropionate cycle

CO_2 is reduced to glyoxylate in a green gliding bacterium, *Chloroflexus aurantiacus*, through a prokaryote-specific 3-hydroxypropionate cycle (Figure 10.10). A similar, but not identical, pathway appears to operate in CO_2 fixation by chemolithotrophic acidophilic archaea such as *Acidianus brierleyi*, *Acidianus ambivalens*, *Metallosphaera sedula*, and *Sulfolobus metallicus*.

Acetyl-CoA is carboxylated to malonyl-CoA by ATP-dependent acetyl-CoA carboxylase before being reduced to 3-hydroxypropionate. A bifunctional enzyme, malonyl-CoA reductase, catalyzes this two-step reductive reaction. 3-hydroxypropionate is reduced to propionyl-CoA. A single enzyme, propionyl-CoA synthase, catalyzes the three reactions from 3-hydroxypropionate to propionyl-CoA via 3-hydroxypropionyl-CoA and acrylyl-CoA consuming NADPH and ATP ($AMP + PP_i$). Propionyl-CoA is carboxylated to methylmalonyl-CoA followed by isomerization of methylmalonyl-CoA to succinyl-CoA. Succinyl-CoA is used for malate activation by CoA transfer, forming succinate and malyl-CoA; succinate in turn is oxidized to malate by the TCA cycle enzymes. Malyl-CoA is cleaved by malyl-CoA lyase with regeneration of the starting acetyl-CoA molecule and production of the first net CO_2 fixation product, glyoxylate.

In the second cycle, pyruvate is synthesized from glyoxylate fixing another molecule of CO_2. Acetyl-CoA is condensed with CO_2 and converted to propionyl-CoA as in the first cycle. Glyoxylate and propionyl-CoA are condensed to *erythro*-3-methylmalyl-CoA before being cleaved to acetyl-CoA and pyruvate via *erythro*-3-methylmalyl-CoA, mesaconyl-CoA, and citramalate. Pyruvate is the product of the second cycle through the reduction of glyoxylate and CO_2 with the regeneration of acetyl-CoA, the primary CO_2 acceptor molecule.

There are three unique processes in this CO_2 fixation pathway involving multifunctional enzymes that are not present in other chemolithotrophs. A bifunctional enzyme, malonyl-CoA reductase, catalyzes the two-step reduction of malonyl-CoA to 3-hydroxypropionate (reaction 2 in Figure 10.10). 3-hydroxypropionate is further metabolized to propionyl-CoA catalyzed by a trifunctional

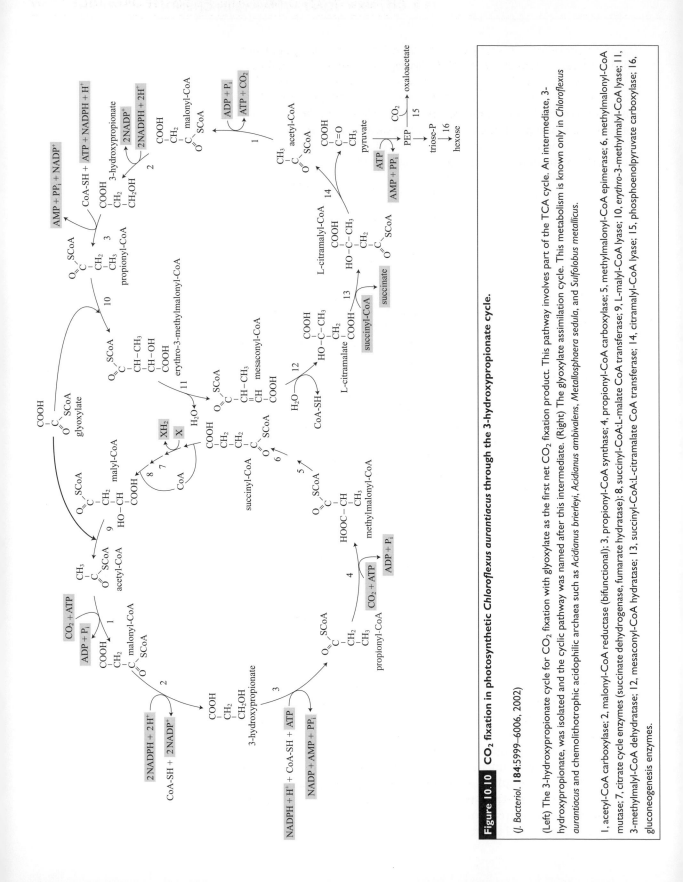

Figure 10.10 CO_2 fixation in photosynthetic *Chloroflexus aurantiacus* through the 3-hydroxypropionate cycle.

(*J. Bacteriol.* **184**:5999–6006, 2002)

(Left) The 3-hydroxypropionate cycle for CO_2 fixation with glyoxylate as the first net CO_2 fixation product. This pathway involves part of the TCA cycle. An intermediate, 3-hydroxypropionate, was isolated and the cyclic pathway was named after this intermediate. (Right) The glyoxylate assimilation cycle. This metabolism is known only in *Chloroflexus aurantiacus* and chemolithotrophic acidophilic archaea such as *Acidianus brierleyi*, *Acidianus ambivalens*, *Metallosphaera sedula*, and *Sulfolobus metallicus*.

1, acetyl-CoA carboxylase; 2, malonyl-CoA reductase (bifunctional); 3, propionyl-CoA synthase; 4, propionyl-CoA carboxylase; 5, methylmalonyl-CoA epimerase; 6, methylmalonyl-CoA mutase; 7, citrate cycle enzymes (succinate dehydrogenase, fumarate hydratase); 8, succinyl-CoA:L-malate CoA transferase; 9, L-malyl-CoA lyase; 10, *erythro*-3-methylmalyl-CoA lyase; 11, 3-methylmalyl-CoA dehydratase; 12, mesaconyl-CoA hydratase; 13, succinyl-CoA:L-citramalate CoA transferase; 14, citramalyl-CoA lyase; 15, phosphoenolpyruvate carboxylase; 16, gluconeogenesis enzymes.

enzyme, propionyl-CoA synthase (reaction 3 in Figure 10.10). Another bifunctional enzyme, malyl-CoA lyase/*erythro*-3-methylmalyl-CoA lyase, cleaves malyl-CoA to acetyl-CoA and glyoxylate (reaction 9 in Figure 10.10), and condenses glyoxylate and propionyl-CoA to *erythro*-3-methylmalyl-CoA (reaction 10 in Figure 10.10).

The 3-hydroxypropionate cycle can be summarized as:

$$3CO_2 + 6NADPH + 6H^+ + X + 5ATP \longrightarrow$$

$$pyruvate + 6NADP^+ + 3H_2O + XH_2 + 3ADP + 2AMP + 2PP_i + 3P_i$$

X is an unknown electron carrier reduced by succinate dehydrogenase.

10.8.5 Energy expenditure in CO_2 fixation

Assuming that the electrons consumed in CO_2 fixation processes are in the same energy state, efficiency can be compared in terms of ATP required for the synthesis of acetyl-CoA after normalization of the different products. Fructose-1,6-phosphate is produced in the Calvin cycle consuming 18ATP. When fructose-1,6-phosphate is metabolized through the EMP pathway to 2 pyruvate and then to 2 acetyl-CoA, 4ATP are generated. This gives a net 7ATP/acetyl-CoA from the Calvin cycle. The same parameters are 2ATP and 1ATP for the reductive TCA cycle and acetyl-CoA pathways, respectively. Pyruvate is produced consuming 5ATP to 3ADP and 2AMP through the 3-hydroxypropionate cycle. Assuming that PP_i is hydrolyzed to $2P_i$ without conserving energy, the ratio of ATP consumed/acetyl-CoA (pyruvate) is 7. From this comparison, it can be seen that energy efficiency varies from 1 to 7 ATP/acetyl-CoA in different CO_2 fixation mechanisms.

10.9 | Chemolithotrophs: what makes them unable to use organics?

Chemolithotrophs are divided into obligate and facultative chemolithotrophs. Obligate chemolithotrophs include a majority of the nitrifiers, some species of thiobacilli, and *Hydrogenobacter thermophilus* and *Hydrogenovibrio marinus* among the H_2-oxidizing bacteria. Ever since the discovery of chemolithotrophic bacteria over a century ago, a satisfactory explanation has not been made as to why the obligate chemolithotrophs cannot use organic electron donors and what the advantage is to be obligately chemolithotrophic. They have to supply 12-carbon compounds needed for biosynthesis (see Table 4.1) from CO_2 fixation products. Some of the obligate chemolithotrophs store polyglucose and other carbohydrates as carbon and energy reserves, so must be able to derive energy from their dissimilation.

Several hypotheses have been proposed to explain why obligate chemolithotrophs cannot use organic compounds as energy and carbon sources. These are:

(1) Obligate chemolithotrophs cannot use organic carbon because the organisms are unable to transport them into cells. As discussed above (Section 10.2.1), a potential fructose permease has been revealed in *Nitrosomonas europaea* through complete sequence determination of its genome, and this organism, as an obligate chemolithotroph, metabolizes fructose mixotrophically. However, amino acids and other monomers can diffuse into cells and are incorporated into biosynthesis in many chemolithotrophs.

(2) Chemolithotrophs are unable to synthesize ATP from NADH oxidation. With a few exceptions, obligate chemolithotrophs couple the oxidation of their electron donors to the reduction of quinone or cytochromes, and supply reducing equivalents for biosynthesis through reverse electron transport. The oxidation of organic compounds reduces NAD^+. Chemolithotrophs might lack enzymes for forward electron transport from NADH which makes them unable to develop a proton motive force via the electron transport chain from NADH. This hypothesis cannot explain the ATP levels sustained in many chemolithotrophs by the metabolism of organic storage materials, as stated above.

(3) All the obligate chemolithotrophs do not have a functional TCA cycle to assimilate organic substrates. They do not synthesize 2-ketoglutarate dehydrogenase, and supply oxaloacetate, succinate, 2-ketoglutarate, etc. through the incomplete TCA fork (Section 5.4.1). Pyridine nucleotides are not reduced in the incomplete TCA fork. 2-ketoglutarate dehydrogenase genes have been identified in many but not all of the chemolithotrophs examined. As the genes for the enzyme are repressed in facultative chemolithotrophs under the conditions for using an inorganic electron donor, the genes in obligate chemolithotrophs are regulated in such a way that they are repressed permanently.

(4) Obligate chemolithotrophs might not be able to use organic electron donors due to metabolic control. The facultative chemolithotroph *Ralstonia eutropha* (*Alcaligenes eutrophus*), metabolizes fructose through the ED pathway. When H_2 is supplied to the fructose culture, the bacterium stops growing due to the inability to use fructose as carbon source, because the enzymes for fructose metabolism are inhibited by H_2. When a mixture of $H_2/CO_2/O_2$ is supplied, growth is resumed.

These hypotheses do not provide completely plausible reasons for the inability of obligate chemolithotrophs to grow on organic electron donors. The mechanism might be much more complex than expected.

FURTHER READING

General

Douglas, S. & Beveridge, T. J. (1998). Mineral formation by bacteria in natural microbial communities. *FEMS Microbiology Ecology* **26**, 79–88.

Ehrlich, H. L. (1999). Microbes as geologic agents: their role in mineral formation. *Geomicrobiology Journal* **16**, 135–153.

Gadd, G. M., Semple, K. T. & Lappin-Scott, H. M. (2005). *Micro-organisms and Earth Systems: Advances in Geomicrobiology*. Cambridge: Cambridge University Press.

Heijnen, J. J. & Vandijken, J. P. (1992). In search of a thermodynamic description of biomass yields for the chemotrophic growth of microorganisms. *Biotechnology and Bioengineering* **39**, 833–858.

Maden, B. E. H. (1995). No soup for starters? Autotrophy and the origins of metabolism. *Trends in Biochemical Sciences* **20**, 337–341.

Mansch, R. & Beck, E. (1998). Biodeterioration of natural stone with special reference to nitrifying bacteria. *Biodegradation* **9**, 47–64.

Stevens, T. O. (1997). Lithoautotrophy in the subsurface. *FEMS Microbiology Reviews* **20**, 327–337.

Reverse electron transport

Elbehti, A., Brasseur, G. & Lemesle-Meunier, D. (2000). First evidence for existence of an uphill electron transfer through the bc_1 and NADH-Q oxidoreductase complexes of the acidophilic obligate chemolithotrophic ferrous ion-oxidizing bacterium *Thiobacillus ferrooxidans*. *Journal of Bacteriology* **182**, 3602–3606.

Jin, Q. & Bethke, C. M. (2003). A new rate law describing microbial respiration. *Applied and Environmental Microbiology* **69**, 2340–2348.

Nitrification

Arp, D. J., Sayavedra-Soto, L. A. & Hommes, N. G. (2002). Molecular biology and biochemistry of ammonia oxidation by *Nitrosomonas europaea*. *Archives of Microbiology*, **178**, 250–255.

Bothe, H., Jost, G., Schloter, M., Ward, B. B. & Witzel, K. P. (2000). Molecular analysis of ammonia oxidation and denitrification in natural environments. *FEMS Microbiology Reviews* **24**, 673–690.

Costa, E., Perez, J. & Kreft, J. U. (2006). Why is metabolic labour divided in nitrification? *Trends in Microbiology* **14**, 213–219.

Head, I. M., Hiorns, W. D., Embley, T. M., McCarthy, A. J. & Saunders, J. R. (1993). The phylogeny of autotrophic ammonia-oxidizing bacteria as determined by analysis of 16S ribosomal RNA gene sequences. *Journal of General Microbiology* **139**, 1147–1153.

Hermansson, A. & Lindgren, P. E. (2001). Quantification of ammonia-oxidizing bacteria in arable soil by real-time PCR. *Applied and Environmental Microbiology* **67**, 972–976.

Holmes, A. J., Costello, A., Lidstrom, M. E. & Murrell, J. C. (1995). Evidence that particulate methane monooxygenase and ammonia monooxygenase may be evolutionarily related. *FEMS Microbiology Letters* **132**, 203–208.

Hooper, A. B., Vannelli, T., Bergmann, D. J. & Arciero, D. M. (1997). Enzymology of the oxidation of ammonia to nitrite by bacteria. *Antonie van Leeuwenhoek* **71**, 59–67.

Ivanova, I. A., Stephen, J. R., Chang, Y. J., Bruggemann, J., Long, P. E., McKinley, J. P., Kowalchuk, G. A., White, D. C. & Macnaughton, S. J. (2000). A survey of 16 S rRNA and amoA genes related to autotrophic ammonia-oxidizing bacteria of the beta-subdivision of the class proteobacteria in contaminated groundwater. *Canadian Journal of Microbiology* **46**, 1012–1020.

Koch, G., Egli, K., van der Meer, J. R., Jr. & Siegrist, H. (2000). Mathematical modeling of autotrophic denitrification in a nitrifying biofilm of a rotating biological contactor. *Water Science and Technology* **41**, 191–198.

Kowalchuk, G. A. & Stephen, J. R. (2001). Ammonia-oxidizing bacteria: a model for molecular microbial ecology. *Annual Review of Microbiology* **55**, 485–529.

Nicol, G. W. & Schleper, C. (2006). Ammonia-oxidising Crenarchaeota: important players in the nitrogen cycle? *Trends in Microbiology* **14**, 207–212.

Richardson, D. J. & Watmough, N. J. (1999). Inorganic nitrogen metabolism in bacteria. *Current Opinion in Chemical Biology* **3**, 207–219.

Schramm, A., Larsen, L. H., Revsbech, N. P., Ramsing, N. B., Amann, R. & Schleifer, K. H. (1996). Structure and function of a nitrifying biofilm as determined by *in situ* hybridization and the use of microelectrodes. *Applied and Environmental Microbiology* **62**, 4641–4647.

Whittaker, M., Bergmann, D., Arciero, D. & Hooper, A. B. (2000). Electron transfer during the oxidation of ammonia by the chemolithotrophic bacterium *Nitrosomonas europaea*. *Biochimica et Biophysica Acta – Bioenergetics* **1459**, 346–355.

Ye, R. W. & Thomas, S. M. (2001). Microbial nitrogen cycles: physiology, genomics and applications. *Current Opinion in Microbiology* **4**, 307–312.

Colourless sulfur bacteria

Appia-Ayme, C., Guiliani, N., Ratouchniak, J. & Bonnefoy, V. (1999). Characterization of an operon encoding two *c*-type cytochromes, an aa_3-type cytochrome oxidase, and rusticyanin in *Thiobacillus ferrooxidans* ATCC 33020. *Applied and Environmental Microbiology* **65**, 4781–4787.

Bharathi, P. A. L., Nair, S. & Chandramohan, D. (1997). Anaerobic sulfide-oxidation in marine colorless sulfur-oxidizing bacteria. *Journal of Marine Biotechnology* **5**, 172–177.

Buonfiglio, V., Polidoro, M., Flora, L., Citro, G., Valenti, P. & Orsi, N. (1993). Identification of two outer membrane proteins involved in the oxidation of sulphur compounds in *Thiobacillus ferrooxidans*. *FEMS Microbiology Reviews* **11**, 43–50.

Friedrich, C. G. (1998). Physiology and genetics of sulfur-oxidizing bacteria. *Advances in Microbial Physiology*, **39**, 235–289.

Gevertz, D., Telang, A. J., Voordouw, G. & Jenneman, G. E. (2000). Isolation and characterization of strains CVO and FWKOB, two novel nitrate-reducing, sulfide-oxidizing bacteria isolated from oil field brine. *Applied and Environmental Microbiology* **66**, 2491–2501.

Howarth, R., Unz, R. F., Seviour, E. M., Seviour, R. J., Blackall, L. L., Pickup, R. W., Jones, J. G., Yaguchi, J. & Head, I. M. (1999). Phylogenetic relationships of filamentous sulfur bacteria (*Thiothrix* spp. and Eikelboom type 021 N bacteria) isolated from wastewater-treatment plants and description of *Thiothrix eikelboomii* sp. nov., *Thiothrix unzii* sp. nov., *Thiothrix fructosivorans* sp. *International Journal of Systematic Bacteriology* **49**, 1817–1827.

Johnson, D. B., Ghauri, M. A. & McGinness, S. (1993). Biogeochemical cycling of iron and sulphur in leaching environments. *FEMS Microbiology Reviews* **11**, 63–70.

Jorgensen, B. B. & Gallardo, V. A. (1999). *Thioploca* spp: filamentous sulfur bacteria with nitrate vacuoles. *FEMS Microbiology Ecology* **28**, 301–313.

Kelly, D. P. (1999). Thermodynamic aspects of energy conservation by chemolithotrophic sulfur bacteria in relation to the sulfur oxidation pathways. *Archives of Microbiology* **171**, 219–229.

Kelly, D. P., Shergill, J. K., Lu, W. P. & Wood, A. P. (1997). Oxidative metabolism of inorganic sulfur compounds by bacteria. *Antonie van Leeuwenhoek* **71**, 95–107.

Norris, P. R., Burton, N. P. & Foulis, N. A. M. (2000). Acidophiles in bioreactor mineral processing. *Extremophiles* **4**, 71–76.

Sand, W., Gerke, T., Hallmann, R. & Schippers, A. (1995). Sulfur chemistry, biofilm, and the (in)direct attack mechanism: a critical evaluation of bacterial leaching. *Applied Microbiology and Biotechnology* **43**, 961–966.

Schippers, A., Jozsa, P. G. & Sand, W. (1996). Sulfur chemistry in bacterial leaching of pyrite. *Applied and Environmental Microbiology* **62**, 3424–3431.

Sorokin, D. Y. (1994). Use of microorganisms in protection of environments from pollution by sulfur compounds. *Microbiology-Moscow* **63**, 533–547.

Vasquez, M. & Espejo, R. T. (1997). Chemolithotrophic bacteria in copper ores leached at high sulfuric acid concentration. *Applied and Environmental Microbiology* **63**, 332–334.

Ferrous iron and other metal oxides

Bacelar-Nicolau, P. & Johnson, D. B. (1999). Leaching of pyrite by acidophilic heterotrophic iron-oxidizing bacteria in pure and mixed cultures. *Applied and Environmental Microbiology* **65**, 585–590.

Blake, R. C., Shute, E. A., Greenwood, M. M., Spencer, G. H. & Ingledew, W. J. (1993). Enzymes of aerobic respiration on iron. *FEMS Microbiology Reviews* **11**, 9–18.

Briand, L., Thomas, H. & Donati, E. (1996). Vanadium(V) reduction in *Thiobacillus thiooxidans* cultures on elemental sulfur. *Biotechnology Letters* **18**, 505–508.

Buchholz-Cleven, B. E. E., Rattunde, B. & Straub, K. L. (1997). Screening for genetic diversity of isolates of anaerobic Fe(II)-oxidizing bacteria using DGGE and whole-cell hybridization. *Systematic and Applied Microbiology* **20**, 301–309.

Elbehti, A. & Lemeslemeunier, D. (1996). Identification of membrane-bound *c*-type cytochromes in an acidophilic ferrous ion oxidizing bacterium *Thiobacillus ferrooxidans*. *FEMS Microbiology Letters* **136**, 51–56.

Fuchs, T., Huber, H., Teiner, K., Burggraf, S. & Stetter, K. O. (1996). *Metallosphaera prunae*, sp nov, a novel metal-mobilizing, thermoacidophilic archaeum, isolated from a uranium mine in Germany. *Systematic and Applied Microbiology* **18**, 560–566.

Golyshina, O. V., Pivovarova, T. A., Karavaiko, G. I., Kondrat'eva, T. F., Moore, E. R. B., Abraham, W. R., Lunsdorf, H., Timmis, K. N., Yakimov, M.M. & Golyshin, P. N.

(2000). *Ferroplasma acidiphilum* gen. nov., sp nov., an acidophilic, autotrophic, ferrous-iron-oxidizing, cell-wall-lacking, mesophilic member of the *Ferroplasmaceae* fam. nov., comprising a distinct lineage of the Archaea. *International Journal of Systematic and Evolutionary Microbiology* **50**, 997–1006.

Hafenbradl, D., Keller, M., Dirmeier, R., Rachel, R., Rossnagel, P., Burggraf, S., Huber, H. & Stetter, K. O. (1996). *Ferroglobus placidus* gen. nov., sp. nov., a novel hyperthermophilic archaeum that oxidizes Fe^{2+} at neutral pH under anoxic conditions. *Archives of Microbiology* **166**, 308–314.

Johnson, D. B., Ghauri, M. A. & McGinness, S. (1993). Biogeochemical cycling of iron and sulphur in leaching environments. *FEMS Microbiology Reviews* **11**, 63–70.

Leduc, L. G. & Ferroni, G. D. (1994). The chemolithotrophic bacterium *Thiobacillus ferrooxidans*. *FEMS Microbiology Reviews* **14**, 103–119.

Rawlings, D. E. & Silver, S. (1995). Mining with microbes. *Biotechnology* **13**, 773–778.

Rawlings, D. E., Tributsch, H. & Hansford, G. S. (1999). Reasons why '*Leptospirillum*'-like species rather than *Thiobacillus ferrooxidans* are the dominant iron-oxidizing bacteria in many commercial processes for the biooxidation of pyrite and related ores. *Microbiology-UK* **145**, 5–13.

Santini, J. M., Sly, L. I., Schnagl, R. D. & Macy, J. M. (2000). A new chemolithoautotrophic arsenite-oxidizing bacterium isolated from a gold mine: phylogenetic, physiological, and preliminary biochemical studies. *Applied and Environmental Microbiology* **66**, 92–97.

Seeger, M. & Jerez, C. A. (1993). Response of *Thiobacillus ferrooxidans* to phosphate limitation. *FEMS Microbiology Reviews* **11**, 37–42.

Straub, K. L. & Buchholz-Cleven, B. E. (1998). Enumeration and detection of anaerobic ferrous iron-oxidizing, nitrate-reducing bacteria from diverse European sediments. *Applied and Environmental Microbiology* **64**, 4846–4856.

Straub, K. L., Benz, M., Schink, B. & Widdel, F. (1996). Anaerobic, nitrate-dependent microbial oxidation of ferrous iron. *Applied and Environmental Microbiology* **62**, 1458–1460.

Sugio, T., Hirayama, K., Inagaki, K., Tanaka, H. & Tano, T. (1992). Molybdenum oxidation by *Thiobacillus ferrooxidans*. *Applied and Environmental Microbiology* **58**, 1768–1771.

Yamanaka, T. & Fukumori, Y. (1995). Molecular aspects of the electron transfer system which participates in the oxidation of ferrous ion by *Thiobacillus ferrooxidans*. *FEMS Microbiology Reviews* **17**, 401–413.

Hydrogen oxidizers and carboxydobacteria

Aono, S., Kamachi, T. & Okura, I. (1993). Characterization and thermostability of a membrane-bound hydrogenase from a thermophilic hydrogen oxidizing bacterium, *Bacillus schlegelii*. *Bioscience, Biotechnology and Biochemistry* **57**, 1177–1179.

Dobbek, H., Gremer, L., Meyer, O. & Huber, R. (1999). Crystal structure and mechanism of CO dehydrogenase, a molybdo iron-sulfur flavoprotein containing S-selanylcysteine. *Proceedings of the National Academy of Sciences, USA* **96**, 8884–8889.

Gonzalez, J. M. & Robb, F. T. (2000). Genetic analysis of *Carboxydothermus hydrogenoformans* carbon monoxide dehydrogenase genes *cooF* and *cooS*. *FEMS Microbiology Letters* **191**, 243–247.

Gremer, L., Kellner, S., Dobbek, H., Huber, R. & Meyer, O. (2000). Binding of flavin adenine dinucleotide to molybdenum-containing carbon monoxide dehydrogenase from *Oligotropha carboxidovorans*: structural and functional analysis of a carbon monoxide dehydrogenase species in which the native flavoprotein has been replaced by its recombinant counterpart produced in *Escherichia coli*. *Journal of Biological Chemistry* **275**, 1864–1872.

Grzeszik, C., Lubbers, M., Reh, M. & Schlegel, H. G. (1997). Genes encoding the NAD-reducing hydrogenase of *Rhodococcus opacus* MR11. *Microbiology-UK* **143**, 1271–1286.

Grzeszik, C., Ross, K., Schneider, K., Reh, M. & Schlegel, H. G. (1997). Location, catalytic activity, and subunit composition of NAD-reducing hydrogenases of some *Alcaligenes* strains and *Rhodococcus opacus* MR22. *Archives of Microbiology* **167**, 172–176.

Hanzelmann, P. & Meyer, O. (1998). Effect of molybdate and tungstate on the biosynthesis of CO dehydrogenase and the molybdopterin cytosine-dinucleotide-type of molybdenum cofactor in *Hydrogenophaga pseudoflava*. *European Journal of Biochemistry* **255**, 755–765.

Nishihara, H., Yaguchi, T., Chung, S. Y., Suzuki, K., Yanagi, M., Yamasato, K., Kodama, T. & Igarashi, Y. (1998). Phylogenetic position of an obligately chemoautotrophic, marine hydrogen-oxidizing bacterium, *Hydrogenovibrio marinus*, on the basis of 16S rRNA gene sequences and two form I RuBisCO gene sequences. *Archives of Microbiology* **169**, 364–368.

Santiago, B. & Meyer, O. (1996). Characterization of hydrogenase activities associated with the molybdenum CO dehydrogenase from *Oligotropha carboxidovorans*. *FEMS Microbiology Letters* **136**, 157–162.

Suzuki, M., Cui, Z. J., Ishii, M. & Igarashi, Y. (2001). Nitrate respiratory metabolism in an obligately autotrophic hydrogen-oxidizing bacterium, *Hydrogenobacter thermophilus* TK-6. *Archives of Microbiology* **175**, 75–78.

Tachil, J. & Meyer, O. (1997). Redox state and activity of molybdopterin cytosine dinucleotide (MCD) of CO dehydrogenase from *Hydrogenophaga pseudoflava*. *FEMS Microbiology Letters* **148**, 203–208.

Yun, N. R., Arai, H., Ishii, M. & Igarashi, Y. (2001). The genes for anabolic 2-oxoglutarate:ferredoxin oxidoreductase from *Hydrogenobacter thermophilus* TK-6. *Biochemical and Biophysical Research Communications* **282**, 589–594.

Yurkova, N. A., Saveleva, N. D. & Lyalikova, N. N. (1993). Oxidation of molecular hydrogen and carbon monoxide by facultatively chemolithotrophic vanadate-reducing bacteria. *Microbiology-Moscow* **62**, 367–370.

CO$_2$ fixation

Beller, H. R., Letain, T. E., Chakicherla, A., Kane, S. R., Legler, T. C. & Coleman, M. A. (2006). Whole-genome transcriptional analysis of chemolithoautotrophic thiosulfate oxidation by *Thiobacillus denitrificans* under aerobic versus denitrifying conditions. *Journal of Bacteriology* **188**, 7005–7015.

Berg, I. A., Keppen, O. I., Krasil'nikova, E. N., Ugol'kova, N. V. & Ivanovsky, R. N. (2005). Carbon metabolism of filamentous anoxygenic phototrophic bacteria of the family *Oscillochloridaceae*. *Microbiology-Moscow* **74**, 258–264.

Cannon, G. C., Baker, S. H., Soyer, F., Johnson, D. R., Bradburne, C. E., Mehlman, J. L., Davies, P. S., Jiang, Q. L., Heinhorst, S. & Shively, J. M. (2003). Organization of carboxysome genes in the thiobacilli. *Current Microbiology* **46**, 115–119.

Eisenreich, W., Strauss, G., Werz, U., Fuchs, G. & Bacher, A. (1993). Retrobiosynthetic analysis of carbon fixation in the photosynthetic eubacterium *Chloroflexus aurantiacus*. *European Journal of Biochemistry* **215**, 619–632.

Finn, M. W. & Tabita, F. R. (2004). Modified pathway to synthesize ribulose 1,5-bisphosphate in methanogenic Archaea. *Journal of Bacteriology* **186**, 6360–6366.

Friedmann, S., Alber, B. E. & Fuchs, G. (2006). Properties of succinyl-coenzyme A:D-citramalate coenzyme A transferase and its role in the autotrophic 3-hydroxypropionate cycle of *Chloroflexus aurantiacus*. *Journal of Bacteriology* **188**, 6460–6468.

Friedmann, S., Steindorf, A., Alber, B. E. & Fuchs, G. (2006). Properties of succinyl-coenzyme A:L-malate coenzyme A transferase and its role in the autotrophic 3-hydroxypropionate cycle of *Chloroflexus aurantiacus*. *Journal of Bacteriology* **188**, 2646–2655.

Gibson, J. L. & Tabita, F. R. (1996). The molecular regulation of the reductive pentose phosphate pathway in Proteobacteria and cyanobacteria. *Archives of Microbiology* **166**, 141–150.

Hernandez, J. M., Baker, S. H., Lorbach, S. C., Shively, J. M. & Tabita, F. R. (1996). Deduced amino acid sequence, functional expression, and unique enzymatic properties of the form I and form II ribulose bisphosphate carboxylase oxygenase from the chemoautotrophic bacterium *Thiobacillus denitrificans*. *Journal of Bacteriology* **178**, 347–356.

Herter, S., Fuchs, G., Bacher, A. & Eisenreich, W. (2002). A bicyclic autotrophic CO_2 fixation pathway in *Chloroflexus aurantiacus*. *Journal of Biological Chemistry* **277**, 20277–20283.

Huegler, M., Huber, H., Stetter, K. O. & Fuchs, G. (2003). Autotrophic CO_2 fixation pathways in archaea (Crenarchaeota). *Archives of Microbiology* **179**, 160–173.

Hugler, M., Wirsen, C. O., Fuchs, G., Taylor, C. D. & Sievert, S. M. (2005). Evidence for autotrophic CO_2 fixation via the reductive tricarboxylic acid cycle by members of the ε-subdivision of Proteobacteria. *Journal of Bacteriology* **187**, 3020–3027.

Ishii, M., Miyake, T., Satoh, T., Sugiyama, H., Oshima, Y., Kodama, T. & Igarashi, Y. (1996). Autotrophic carbon dioxide fixation in *Acidianus brierleyi*. *Archives of Microbiology* **166**, 368–371.

Ishii, M., Chuakrut, S., Arai, H. & Igarashi, Y. (2004). Occurrence, biochemistry and possible biotechnological application of the 3-hydroxypropionate cycle. *Applied Microbiology and Biotechnology* **64**, 605–610.

Ivanovsky, R. N., Fal, Y. I., Berg, I. A., Ugolkova, N. V., Krasilnikova, E. N., Keppen, O. I., Zakharchuc, L. M. & Zyakun, A. M. (1999). Evidence for the presence of the reductive pentose phosphate cycle in a filamentous anoxygenic photosynthetic bacterium, *Oscillochloris trichoides* strain DG-6. *Microbiology-UK* **145**, 1743–1748.

Joshi, H. M. & Tabita, F. R. (2000). Induction of carbon monoxide dehydrogenase to facilitate redox balancing in a ribulose bisphosphate carboxylase/oxygenase-deficient mutant strain of *Rhodospirillum rubrum*. *Archives of Microbiology* **173**, 193–199.

Lelait, M. & Grivet, J. P. (1996). Carbon metabolism in *Eubacterium limosum*: a C-13 NMR study. *Anaerobe* **2**, 181–189.

Menon, S. & Ragsdale, S. W. (1999). The role of an iron-sulfur cluster in an enzymatic methylation reaction: methylation of CO dehydrogenase/acetyl-CoA synthase by the methylated corrinoid iron-sulfur protein. *Journal of Biological Chemistry* **274**, 11513–11518.

Qian, Y. L. & Tabita, F. R. (1996). A global signal transduction system regulates aerobic and anaerobic CO_2 fixation in *Rhodobacter sphaeroides*. *Journal of Bacteriology* **178**, 12–18.

Roberts, D. L., Zhao, S. Y., Doukov, T. & Ragsdale, S. W. (1994). The reductive acetyl coenzyme A pathway. Sequence and heterologous expression of active methyltetrahydrofolate: corrinoid/ iron-sulfur protein methyl-transferase from *Clostridium thermoaccticum*. *Journal of Bacteriology* **176**, 6127–6130.

Russell, M. J. & Martin, W. (2004). The rocky roots of the acetyl-CoA pathway. *Trends in Biochemical Sciences* **29**, 358–363.

Schouten, S., Strous, M., Kuypers, M. M. M., Rijpstra, W. I., Baas, M., Schubert, C. J., Jetten, M. S. M. & Sinninghe Damste, J. S. (2004). Stable carbon isotopic fractionations associated with inorganic carbon fixation by anaerobic ammonium-oxidizing bacteria. *Applied and Environmental Microbiology* **70**, 3785–3788.

Shively, J. M., Vankeulen, G. & Meijer, W. G. (1998). Something from almost nothing: carbon dioxide fixation in chemolithotrophs. *Annual Review of Microbiology* **52**, 191–230.

Tourova, T. P., Spiridonova, E. M., Berg, I. A., Kuznetsov, B. B. & Sorokin, D. Y. (2005). Phylogeny of ribulose-1,5-bisphosphate carboxylase/oxygenase genes in haloalkaliphilic obligately autotrophic sulfur-oxidizing bacteria of the genus *Thioalkalivibrio*. *Microbiology-Moscow* **74**, 321–328.

Tourova, T. P., Spiridonova, E. M., Berg, I. A., Kuznetsov, B. B. & Sorokin, D. Y. (2006). Occurrence, phylogeny and evolution of ribulose-1,5-bisphosphate carboxylase/oxygenase genes in obligately chemolithoautotrophic sulfur-oxidizing bacteria of the genera *Thiomicrospira* and *Thioalkalimicrobium*. *Microbiology-UK* **152**, 2159–2169.

Vorholt, J., Kunow, J., Stetter, K. O. & Thauer, R. K. (1995). Enzymes and coenzymes of the carbon monoxide dehydrogenase pathway for autotrophic CO_2 fixation in *Archaeoglobus lithotrophicus* and the lack of carbon monoxide dehydrogenase in the heterotrophic *A-profundus*. *Archives of Microbiology* **163**, 112–118.

Vorholt, J. A., Hafenbradl, D., Stetter, K. O. & Thauer, R. K. (1997). Pathways of autotrophic CO_2 fixation and of dissimilatory nitrate reduction to N_2O in *Ferroglobus placidus*. *Archives of Microbiology* **167**, 19–23.

Yoshizawa, Y., Toyoda, K., Arai, H., Ishii, M. & Igarashi, Y. (2004). CO_2-responsive expression and gene organization of three ribulose-1,5-bisphosphate carboxylase/oxygenase enzymes and carboxysomes in *Hydrogenovibrio marinus* strain MH-110. *Journal of Bacteriology* **186**, 5685–5691.

Photosynthesis

Photosynthetic organisms use light energy to fuel their biosynthetic processes. Oxygen is generated in oxygenic photosynthesis where water is used as the electron donor. In anoxygenic photosynthesis, organic or sulfur compounds are used as electron donors. Plants, algae and cyanobacteria carry out oxygenic photosynthesis, whereas the photosynthetic bacteria obtain energy from anoxygenic photosynthesis. Aerobic anoxygenic phototrophic bacteria use light energy in a similar way as the purple bacteria, and are a group of photosynthetic bacteria that grow under aerobic conditions.

Phototrophic organisms have a photosynthetic apparatus consisting of a reaction centre intimately associated with antenna molecules (or a light-harvesting complex). The antenna molecules and the reaction centre absorb light energy. The energy is concentrated at the reaction centre that is activated and initiates light-driven electron transport. Halophilic archaea convert light energy through a photophosphorylation process.

11.1 | Photosynthetic microorganisms

Microorganisms utilizing light energy include eukaryotic algae, and cyanobacteria, photosynthetic bacteria and aerobic anoxygenic phototrophic bacteria among the prokaryotes. The halophilic archaea synthesize ATP through photophosphorylation, but they are not considered to be photosynthetic organisms since they lack photosynthetic pigments.

Algae and cyanobacteria have similar photosynthetic processes, using chlorophyll, as plants. However, cyanobacteria are members of the proteobacteria according to their cell structure and ribosomal RNA sequences. Photosynthetic bacteria are different from other photosynthetic organisms. They have different photosynthetic pigments and do not use water as their electron donor. Some of them can grow chemoorganotrophically in the dark.

Table 11.1.	Cyanobacteria			
	Morphology	Cell division[a]	Heterocysts	N_2 fixation
Unicellular group	single cells	single	−	+[b]
Pleurocapsa group	single cells	multiple	−	+[b]
Oscillatoria group	filaments	single	−	−
Heterocystous group	filaments	single	+	+

[a] Cell divides into two daughter cells (single) or into more than two daughter cells (multiple).
[b] Several species do not fix N_2.

II.I.I Cyanobacteria

Cyanobacteria (also known as blue-green algae) grow photolithotrophically and fix CO_2 through the Calvin cycle. They do not generally require growth factors except for some that require vitamin B_{12}. They are classified into four groups according to their morphology. These are a unicellular group, the *Pleurocapsa* group, the *Oscillatoria* group and a heterocystous group (Table 11.1). N_2 is fixed by some members of the unicellular cyanobacteria and the *Oscillatoria* group, and by all members of the heterocystous group. Under N_2-fixing conditions, some of the cells within the filaments of heterocystous group cyanobacteria transform into heterocysts that lack photosystem II and this protects the nitrogenase from O_2 (Section 6.2.1.4).

II.I.2 Anaerobic photosynthetic bacteria

Photosynthetic bacteria are grouped according to their photosynthetic pigments and the electron donors used. These are purple bacteria, green bacteria and heliobacteria. Purple and green bacteria are further divided into purple non-sulfur and purple sulfur bacteria, and green sulfur and filamentous anoxygenic phototrophic bacteria (Table 11.2).

The purple sulfur bacteria include members of the *Chromatiaceae* and *Ectothiorhodospiraceae* within the γ-proteobacteria. The former accumulate sulfur granules intracellularly and the latter extracellularly. The purple non-sulfur bacteria are more diverse, belonging to α- and β-proteobacteria. They grow photosynthetically under anaerobic conditions, and many of them can grow chemoorganotrophically under aerobic conditions. The purple non-sulfur bacteria grow under all electron-accepting conditions (aerobic respiratory, anaerobic respiratory and fermentative conditions) in addition to anaerobic photosynthetic conditions. The purple sulfur and non-sulfur bacteria have pheophytin–quinone-type reaction centres (Section 11.3).

The photosynthetic green bacteria include two physiologically and phylogenetically distinct groups. These are the strictly anaerobic and obligately photolithotrophic green sulfur bacteria, and the filamentous anoxygenic photolithotrophic bacteria that are facultatively anaerobic. These have different reaction centres. The latter have the

Table 11.2.	Photosynthetic bacteria					
	Purple bacteria		Green bacteria			
Character	Non-sulfur	Sulfur	Sulfur	FAPB	Heliobacteria	AAPB
BCHL	a, b	a, b	a, c, d, e	a, c, d	g	a
H$_2$S as e$^-$ donor	±^{a}	+	+	+	−	−
S accumulation	−	intracellular^b	extracellular	−	−	−
H$_2$ as e$^-$ donor	+	+	+	+	−	−
Organics as e$^-$ donor	+	+	−	+	+	+
Carbon source	CO$_2$ organics	CO$_2$ organics	CO$_2$ organics	CO$_2$ organics	organics	organics
Aerobic respiration	+	−	−	+	−	+
CO$_2$ fixation	Calvin cycle	Calvin cycle	reductive TCA cycle	Calvin cycle^c	−	−

FAPB, filamentous anoxygenic phototrophic bacteria; AAPB, aerobic anoxygenic phototrophic bacteria; BCHL, bacteriochlorophyll.
^{a} Depending on the strain.
^{b} Members of the family *Ectothiorhodospiraceae* accumulate sulfur granules extracellularly.
^{c} Species of the genus *Chloroflexus* employ the 3-hydroxypropionate cycle.
Note: The halophilic archaea use light energy through photophosphorylation.

pheophytin–quinone type, while the former have the iron–sulfur type. The green sulfur bacteria cannot grow heterotrophically, while the filamentous anoxygenic phototrophic bacteria can grow heterotrophically under aerobic dark conditions. The latter, members of the *Chloroflexaceae*, belong to a deep-branching lineage of bacteria. These are also called photosynthetic flexibacteria. They stain Gram-negative but lack lipopolysaccharide.

The photoheterotrophic heliobacteria include three genera: *Heliobacterium*, *Heliobacillus* and *Heliophilum*. They do not grow aerobically in the dark, and fix N$_2$. They do not have photosynthetic organelles and the photosynthetic pigments, including the unique bacteriochlorophyll *g*, are located in the cytoplasmic membrane. They have an iron–sulfur-type reaction centre. Heliobacterial cells have several unusual features. They are extremely fragile and lyse when approaching the stationary phase. They stain Gram-negative but lack lipopolysaccharide like the filamentous anoxygenic phototrophic bacteria, and do not fix CO$_2$. Phylogenetically they belong to the low G + C Gram-positive bacteria.

11.1.3 Aerobic anoxygenic phototrophic bacteria

Photosynthetic bacteria utilize light energy under anaerobic conditions and their photosynthetic pigments are not synthesized under aerobic conditions. However, many bacteria are known to synthesize bacteriochlorophyll (BCHL) *a* and carotenoids under aerobic conditions. They can harvest light energy while they respire oxygen. These are referred to as quasi-photosynthetic bacteria or aerobic anoxygenic phototrophic bacteria. They comprise at least 11% of the total microbial community in the upper open ocean. They inhabit a variety of locations, including the extreme environments of acidic mine drainage waters, hot springs and deep ocean hydrothermal vent plumes.

Aerobic anoxygenic phototrophic bacteria found in seawater include species of *Erythrobacter*, *Roseibium*, *Roseivivax*, *Roseobacter*, *Roseovarius* and *Rubrimonas*. Fresh water is the habitat of other aerobic anoxygenic phototrophic bacteria including species of *Sandaracinobacter*, *Erythromonas*, *Erythromicrobium*, *Roseococcus*, *Porphyrobacter* and *Acidiphilium*. These have the pheophytin–quinone-type reaction centre. These genera include not only aerobic phototrophs but also species unable to synthesize BCHL *a*. It is not clear if the ability to synthesize the photosynthetic pigment was lost or transferred through lateral gene transfer during their evolution. Aerobic anoxygenic photosynthesis is also known in species of *Bradyrhizobium*, syntrophically growing on the stems of tropical legume plants. Although these are obligate aerobes with high carotenoid and low BCHL contents, they are closely related to purple photosynthetic bacteria in several aspects.

Some of the aerobic anoxygenic phototrophic bacteria divide in an unusual manner. Budding in addition to binary division occurs in *Porphyrobacter neustonensis* and *Erythromonas ursincola* (Section 6.14.1.4). Ternary fission and branching are exhibited by *Erythromicrobium ramosum* and *Erythromicrobium hydrolyticum* (Section 6.14.1.3).

11.2 | Photosynthetic pigments

Photosynthetic pigments include chlorophylls, carotenoids and phycobiliproteins in plants and cyanobacteria: phycobiliproteins are not found in photosynthetic bacteria (Table 11.3). Bacteria use bacteriochlorophylls in place of chlorophyll, and bacteriopheophytin to replace pheophytin.

Table 11.3. | *Photosynthetic pigments*

Pigment	Cyanobacteria	Purple bacteria	Green bacteria	Heliobacteria	Aerobic anoxygenic phototrophic bacteria
Reaction centre pigment	CHL *a*	BCHL *a, b*	BCHL *a*	BCHL *g*	BCHL *a*[a]
Antenna pigment	phycobiliprotein CHL *a*	BCHL *a* or *b*	BCHL *c, d* or *e*	BCHL *g*	BCHL *a*
Main carotenoid	dicyclic	aliphatic	aryl	aliphatic	dicyclic, aliphatic[b]

[a] BCHL *a* with Zn^{2+} or Mg^{2+}.
[b] Various carotenoids depending on the strain.

(a)

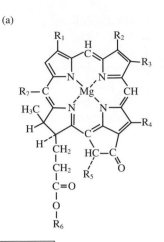

(b)

(c)

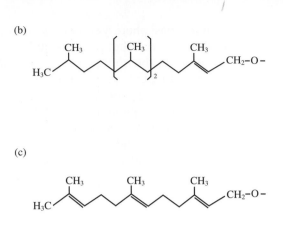

Figure 11.1 The structure of chlorophyll (see Table 11.4 for the side chains R1–R7).

(a) chlorophyll; (b) phytyl group; (c) farnesyl group.

11.2.1 Chlorophylls

Chlorophylls have a general structure of four pyrrole derivatives with covalently bound Mg^{2+} (Figure 11.1). They form a complex with proteins embedded in the membranes of the photosynthetic apparatus. There are several structurally different chlorophylls that have different side chains from the pyrrole rings (Table 11.4). These include chlorophyll *a*, and bacteriochlorophyll (BCHL) *a*, *b*, *c*, *d*, *e*, and *g*. Bacteriochlorophyll *a* with Zn^{2+} in place of Mg^{2+} occurs in aerobic anoxygenic phototrophic bacteria.

As listed in Table 11.3, cyanobacteria possess chlorophyll *a* as do plants. Green bacteria have reaction centres with BCHL *a* and antenna molecules with BCHL *c*, *d* or *e*. In contrast, purple bacterial reaction centres and antenna molecules contain BCHL *a* or *b*. Aerobic anoxygenic phototrophic bacteria contain BCHL *a* with Zn^{2+} or Mg^{2+} in both the reaction centre and antenna molecules. Cyanobacteria with CHL *a* absorb relatively short wavelength light while light of wavelength over 700 nm is absorbed by photosynthetic bacteria containing BCHL *a* or *b*. The purple bacteria absorb near infrared light of wavelength around 800 nm (see Figure 11.4).

11.2.2 Carotenoids

In addition to chlorophylls, photosynthetic organisms have carotenoids. They absorb light over wavelengths of 400–600 nm and transfer the energy to chlorophylls, and also protect biological materials from photooxidation caused by reactive oxygen derivatives. The common carotenoid in cyanobacteria is *β*-carotene. Over 30 different carotenoids are known in purple bacteria and in aerobic anoxygenic phototrophic bacteria, and spirilloxanthin (without a bezene ring in its structure) is the most common among them. The typical carotenoid of green bacteria is isorenieratene (Figure 11.2). The major carotenoid in heliobacteria is neurosporene that is similar in structure to spirilloxanthin.

Table 11.4. *The structure and maximum absorbance of chlorophylls*

Side chain[a]	CHL a	BCHL					
		a	b	c	d	e	g
R1	$-CH=CH_2$	$-CO-CH_3$	$-CO-CH_3$	$-CHOH-CH_3$	$-CHOH-CH_3$	$-CHOH-CH_3$	$-CH=CH_2$
R2	$-CH_3$	$-CH_3$	$-CH_3$	$-CH_3$	$-CH_3$	$-CHO$	$-CH_3$
R3	$-C_2H_5$	$-C_2H_5$	$=CH-CH_3$	$-C_2H_5$	$-C_2H_5$	$-C_2H_5$	$=CH-CH_3$
R4	$-CH_3$	$-CH_3$	$-CH_3$	$-C_2H_5$	$-C_2H_5$	$-C_2H_5$	$-CH_3$
R5	$-CO-OCH_3$	$-CO-OCH_3$	$-CO-OCH_3$	$-H$	$-H$	$-H$	$-CO-OCH_3$
R6	phytyl	phytyl	phytyl	farnesyl	farnesyl	farnesyl	geranyl-geranyl
R7	$-H$	$-H$	$-H$	$-CH_3$	$-H$	$-CH_3$	$-H$
Maximum absorbance (nm)	680–685	850–910	1020–1035	745–760	725–745	715–725	788, 670

CHL, chlorophyll; BCHL, bacteriochlorophyll.
[a] The positions of side chains are shown in Figure 11.1.

(a) β-carotene

(b) isorenieratene

(c) spirilloxanthin

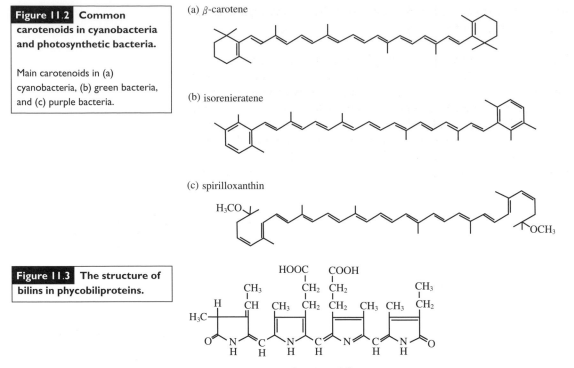

phycocyanobilin

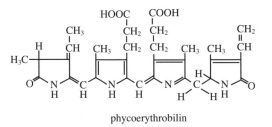

phycoerythrobilin

11.2.3 Phycobiliproteins

Phycobiliproteins are soluble proteins containing bilin, which has a structure of four pyrroles in a linear form and is found in photosynthetic eukaryotes and cyanobacteria (Figure 11.3). Cyanobacteria with a blue-green colour contain allophycocyanin and phycocyanin, while phycoerythrin is the phycobiliprotein in red-coloured cyanobacteria. This protein is not found in photosynthetic bacteria.

11.2.4 Pheophytin

This pigment has the structure of chlorophyll but without Mg^{2+} or Zn^{2+}. Cyanobacteria and photosynthetic eukaryotes possess pheophytin, while bacteriopheophytin is found in photosynthetic bacteria with the pheophytin–quinone-type reaction centre serving as an electron carrier. This pigment is not found in green sulfur bacteria and heliobacteria that have the iron–sulfur-type reaction centre.

Before its function was established, pheophytin was regarded as a chlorophyll degradation product.

11.2.5 Absorption spectra of photosynthetic cells

Each photosynthetic organism has specific photosynthetic pigments in different ratios, and absorbs light of a specific wavelength depending on their pigments (Figure 11.4). Cyanobacteria with CHL *a*, carotenoids and phycobiliproteins absorb light at wavelengths shorter than 700 nm. On the other hand, photosynthetic bacteria absorb light at wavelengths shorter than 600 nm with carotenoids and the BCHLs absorb light at wavelengths above 700 nm. Green bacterial cells

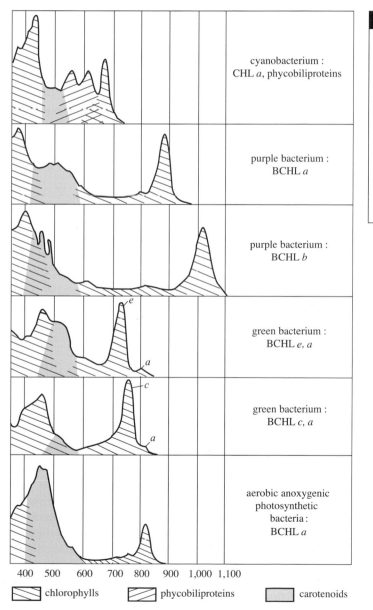

cyanobacterium :
CHL *a*, phycobiliproteins

purple bacterium :
BCHL *a*

purple bacterium :
BCHL *b*

green bacterium :
BCHL *e, a*

green bacterium :
BCHL *c, a*

aerobic anoxygenic
photosynthetic
bacteria :
BCHL *a*

400 500 600 700 800 900 1,000 1,100

chlorophylls   phycobiliproteins   carotenoids

Figure 11.4 **Absorption spectra of photosynthetic organisms.**

The photosynthetic pigments possessed determine the absorption spectra of photosynthetic cells. Cyanobacteria absorb light at wavelengths shorter than 700 nm with CHL *a*. Light of longer wavelengths is absorbed by photosynthetic bacteria that possess BCHLs.

absorb light at wavelengths between 700 and 800 nm, and also below 600 nm, using BCHL *a*, *b* and *e* and carotenoids in their antenna molecules. The heliobacteria contain BCHL *g* that has maximum light absorption at 788 nm, and shows a similar absorption spectrum to green bacteria. BCHL *a* and *b* absorb long wavelength light over 800 nm in purple bacteria. Aerobic anoxygenic phototrophic bacteria absorb light at 450 nm with carotenoids and at 750 nm with BCHL *a*.

11.3 | Photosynthetic apparatus

Photosynthetic organisms utilize light energy to reduce $NADP^+$ and to synthesize ATP through the proton motive force. This energy transduction is facilitated by the photosynthetic pigments and electron carriers arranged in the photosynthetic apparatus. Separate photosynthetic structures are not found in the heliobacteria and aerobic anoxygenic phototrophic bacteria. Their reaction centres and antenna molecules are located at the cytoplasmic membrane. The reaction centre is the key component for the primary events in the photochemical conversion of light into biological energy. Coupling to secondary electron donors and acceptors allows the electrons and accompanying protons to be transferred to other components of the photosynthetic apparatus synthesizing ATP or reducing $NADP^+$.

Photosynthetic reaction centres can be classed in two categories based on the nature of the electron acceptors. Those of the purple bacteria, filamentous anoxygenic phototrophic bacteria, and photosystem II of cyanobacteria belong to the pheophytin–quinone type, while the iron–sulfur type are found in green sulfur bacteria, heliobacteria, and photosystem I of cyanobacteria. Pheophytin (or bacteriopheophytin) and iron–sulfur centres participate in the electron transfer reactions of each type of reaction centre.

11.3.1 Thylakoids of cyanobacteria

Cyanobacteria have thylakoids with the photosynthetic pigments arranged similarly to the chloroplasts of eukaryotic cells. The thylakoid has a bilayer membrane structure consisting of galactosyl diglyceride containing one or two galactose molecules in place of the phosphate of phospholipids (Figure 11.5). Thylakoids convert light energy into biological energy in the cytoplasm of the cyanobacteria.

Chlorophyll *a* and proteins form the reaction centre on the thylakoid membrane. A small fraction of chlorophyll *a* is found in the reaction centre, and the majority forms antenna molecules known as the phycobilisome with phycobiliproteins and carotenoids. In addition to chlorophyll *a*, reaction centres have various electron carriers that convert the light energy into a proton motive force. Photosystem I contains [Fe-S] proteins and quinones, and photosystem II contains pheophytin and quinones. The phycobilisomes occupy the cytoplasmic side of the reaction centre (Figure 11.5).

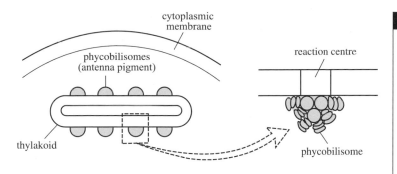

11.3.2   Green bacteria

Green bacteria have a photosynthetic apparatus called the chlorosome on their cytoplasmic membrane. Chlorosomes contain antenna molecules, and their baseplates are bound to the reaction centre that is a part of the cytoplasmic membrane (Figure 11.6). The chlorosome has the structure of a galactosyl diglyceride monolayer membrane filled with rod-shaped antenna molecules. Bacteriochlorophyll (BCHL) c, d or e constitute the antenna molecules together with carotenoids. The baseplate contains BCHL a that transfers photons to the reaction centre. The reaction centre contains BCHL a and the electron transport chains. In addition to BCHLs, [Fe-S] proteins are found in the reaction centres of green sulfur bacteria, and bacteriopheophytin (BPHE) in those of filamentous anoxygenic phototrophic bacteria.

The obligate anaerobe *Chlorobium tepidum* thrives in anaerobic aquatic environments where sulfide is available with very dim light. To capture light efficiently this bacterium contains about 200–250 chlorosomes per cell with more than 200×10^3 BCHL c molecules per chlorosome. Thus, this bacterium contains up to 50×10^6 BCHL c molecules per cell.

11.3.3   Purple bacteria

The purple bacteria have a less developed photosynthetic structure than cyanobacteria and the green bacteria. An intracellular membrane structure contains antenna molecules and reaction centres (Figure 11.7). This is a phospholipid bilayer membrane continuous with the cytoplasmic membrane. The shape of the intracellular membrane structure varies depending on the strain.

11.3.4   Heliobacteria and aerobic anoxygenic phototrophic bacteria

These organisms are the least developed in terms of the photosynthetic apparatus and lack differentiated structures such as chlorosomes or intracytoplasmic membranes. The antenna molecules and reaction centres reside within the cytoplasmic membrane.

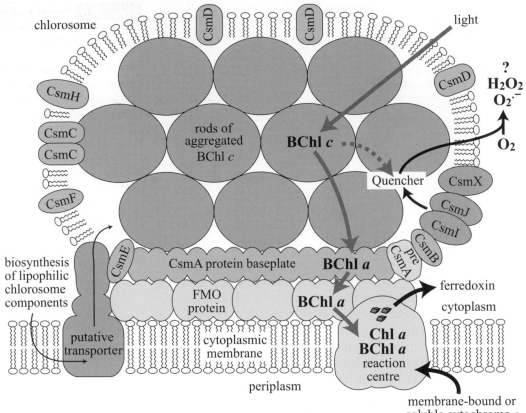

Figure 11.6 Chlorosome bound to the cytoplasmic membrane in green bacteria.

(Modified from *Arch. Microbiol.* **182**:265–276, 2004)

The monolayered chlorosome contains antenna molecules, and its baseplate is bound to the reaction centre that is a part of the cytoplasmic membrane.

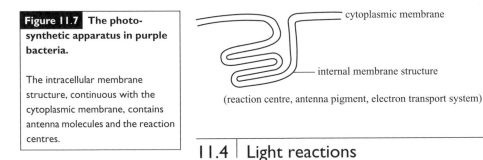

Figure 11.7 The photosynthetic apparatus in purple bacteria.

The intracellular membrane structure, continuous with the cytoplasmic membrane, contains antenna molecules and the reaction centres.

cytoplasmic membrane

internal membrane structure

(reaction centre, antenna pigment, electron transport system)

11.4 | Light reactions

Photosynthesis is a process utilizing light energy for biosynthesis. It can be divided into light reactions that convert light energy into biological energy and dark reactions that utilize the biological energy in biosynthesis.

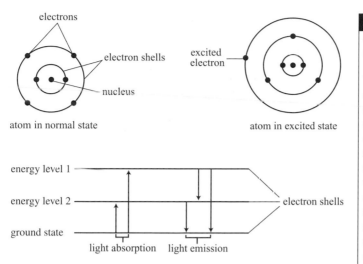

Figure 11.8 **Photon exciting an electron of a light-absorbing pigment molecule.**

The excitation takes place when the energy of the photon is similar to the energy needed to excite the electron. The excited electron returns to the original unexcited state coupled to one of three reactions. These reactions are (1) transfer of energy to the adjacent molecule exciting its electron (resonance transfer), (2) oxidation of the excited pigment transferring an electron to a second compound or (3) emission of fluorescent light.

11.4.1 Properties of light

Light is a form of electromagnetic radiation which travels in rhythmic waves transported in discrete particle units called photons. The number of photons per unit time is the intensity of the light. The energy carried by the photons is related to the frequency of the wave. The higher the frequency, the higher the energy carried.

When a photon is radiated to a surface, it may be reflected, transmitted, or absorbed. Different pigments absorb photons of different wavelengths depending on the nature of the absorbing pigment. When a photon is absorbed by a pigment, the energy of light is converted to kinetic energy in the form of an excited electron. When a molecule (or atom) absorbs a photon an electron is boosted to a higher energy level by transferring an electron from the normal shell to the outer shell (Figure 11.8).

11.4.2 Excitation of antenna molecules and resonance transfer

The excited electron is very unstable and returns to the original unexcited state coupled to one of three reactions. These reactions are (1) transfer of energy to the adjacent molecule exciting its electron (resonance transfer), (2) reduction of a second compound (oxidation, photo-induced charge separation) or (3) emission of fluorescent light.

The antenna molecules are excited on absorbing photons. The energy gained by the pigments through such excitation is referred to as the exciton. The exciton is transferred from antenna molecules to the reaction centre through resonance transfer, and this reaction takes about 0.1 picosecond. This energy is referred to as resonance energy. The electron transport reaction at the reaction centre is initiated by the oxidation with the consumption of resonance energy.

Resonance transfer takes place from molecules of a higher exciton to those with a lower exciton. For this reason, photosynthetic pigments are arranged in such a way in the phycobilisome of the cyanobacterial thylakoid and in the bacterial photosynthetic apparatus to facilitate such resonance transfer.

11.4.3 Electron transport

When the resonance energy excites the reaction centre chlorophyll, its redox potential becomes very low, enough to transfer electrons to a lower potential electron carrier. The oxidized chlorophylls are reduced again, either oxidizing externally supplied electron donor(s) in a process known as non-cyclic electron transport or by taking the original electrons through the cyclic electron transport system.

11.4.3.1 Photosystem I and II in cyanobacteria

As in photosynthetic eukaryotes, cyanobacteria have photosystem II to supply reducing power, oxidizing water through non-cyclic electron transport, as well as photosystem I to generate a proton motive force through cyclic electron transport. Chlorophyll *a* serves as the photosynthetic pigment in the reaction centres of both photosystems. The reaction centre complex of photosystem I has maximum absorption at a wavelength of 700 nm while that of photosystem II is at 680 nm. These are referred to as RCI or P700, and RCII or P680, respectively. The differences in the maximum absorption of chlorophyll *a* are due to the proteins forming the complex.

When excited by photons, the redox potential of RCI (P700) decreases to $-1.0 \, \mathrm{V}$ from $+0.5 \, \mathrm{V}$. The excited P700 reduces the primary acceptor (A$_0$, CHL *a*). Electrons are transferred from A$_0$ to ferredoxin through phylloquinone and several [Fe-S] centres. The electrons from the reduced ferredoxin are transferred back to P700 through the cytochrome *bf* complex and plastocyanin in a process known as cyclic electron transport (Figure 11.9). The free energy changes in cyclic electron transport are conserved as a proton motive force transporting protons into the thylakoid.

Alternatively, electrons are transferred from the reduced ferredoxin to NADP$^+$. Photosystem II replaces the electrons used to reduce NADP$^+$ in photosystem I. When RCII absorbs light, the redox potential decreases from $+1.0 \, \mathrm{V}$ to $-0.8 \, \mathrm{V}$. Electrons move from the excited RCII to pheophytin before being transferred to photosystem I via various electron carriers (Figure 11.9).

The P680 (RCII) of photosystem II has a redox potential of $+1.0 \, \mathrm{V}$ which is higher than that of O_2/H_2O. Oxidized P680 is reduced, oxidizing water to molecular oxygen. This reaction is catalyzed by a manganoprotein. Since O_2 is evolved, this is referred to as oxygenic photosynthesis.

11.4.3.2 Green sulfur bacteria

Green sulfur bacteria utilize light energy through anoxygenic cyclic electron transport, similar to that of cyanobacterial photosystem I, with a P840 iron–sulfur-type reaction centre. Bacteriochlorophyll *a* (BCHL *a*) is the photosynthetic pigment of P840 (Figure 11.10). Antenna molecules on the chlorosome transfer the exciton to the

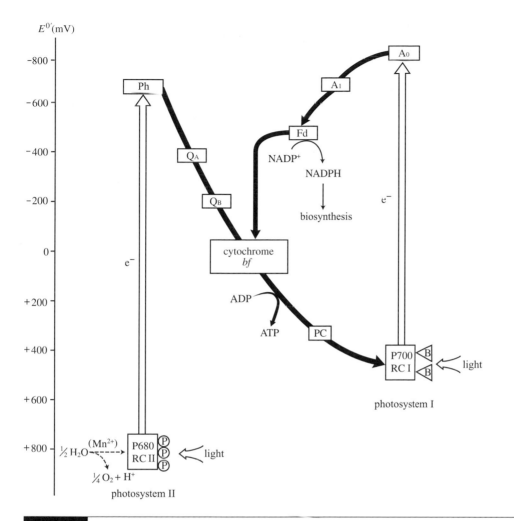

$E^{0'}$(mV)

Figure 11.9 **Photosynthetic electron transport chains of cyanobacteria.**

Cyanobacteria have two different reaction centres, RCI or P700, and RCII or P680. These consist of chlorophyll *a* and a protein complex. When RCI absorbs photons, chlorophyll *a* is excited and transfers electrons to the primary acceptor (A_0, CHL *a*). Electrons are transferred from A_0 to ferredoxin through phylloquinone (A_1) and several [Fe-S] centres. The electrons from the reduced ferredoxin are transferred back to P700 through the cytochrome *bf* complex and plastocyanin (PC) in a process known as cyclic electron transport. The free energy changes in cyclic electron transport are conserved as a proton motive force transporting protons into the thylakoid. $NADP^+$ is reduced, taking electrons from the ferredoxin (Fd). To replace the electrons channelled to NADPH, RCII oxidizes water reducing pheophytin (Ph) and transferring the electrons to the cyclic electron transport chain. Electron transport at RCII is referred to as non-cyclic electron transport.

Ph, pheophytin; Q_A and Q_B, plastoquinone A and B, respectively; PC, plastocyanin; A_0, primary electron acceptor (CHL *a*); A_1, phylloquinone; Fd, ferredoxin.

reaction centre on the cytoplasmic membrane. BCHL *a* mediates exciton transfer through the baseplate.

The normal state P840 has a redox potential of $+0.3$ V, and this decreases to lower than -1.0 V, low enough to reduce the primary acceptor (A_0, BCHL *a*) when excited by photons. Electrons either flow

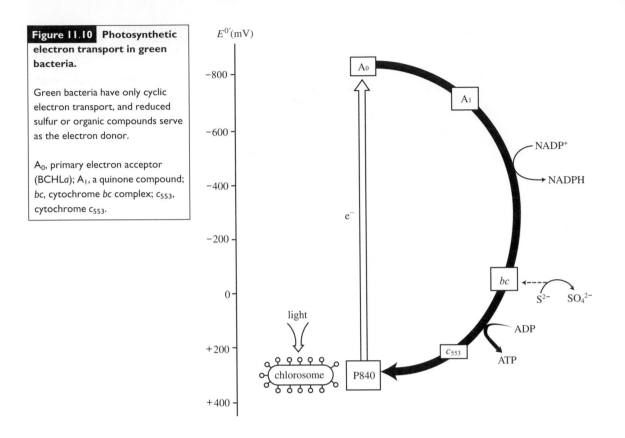

Figure 11.10 **Photosynthetic electron transport in green bacteria.**

Green bacteria have only cyclic electron transport, and reduced sulfur or organic compounds serve as the electron donor.

A_0, primary electron acceptor (BCHLa); A_1, a quinone compound; bc, cytochrome bc complex; c_{553}, cytochrome c_{553}.

back to P840 generating a proton motive force or are transferred to NAD(P)$^+$. Electrons that are consumed to reduce NAD(P)$^+$ are replaced by oxidizing electron donors such as sulfide. These organisms cannot grow chemoorganotrophically under aerobic conditions. On the other hand, the other members of the green bacteria, the filamentous anoxygenic phototrophic bacteria, can grow chemoorganotrophically under aerobic conditions. The latter have a different photosynthetic electron transport system from that of the green sulfur bacteria which is similar to that of the purple bacteria (see Figure 11.11), with a pheophytin–quinone-type reaction centre.

Heliobacteria have P788 as the reaction centre, which is different from the green sulfur bacteria, although the photosynthetic mechanisms are similar.

11.4.3.3 Purple bacteria

Purple bacteria have a reaction centre with maximum absorption at 870 nm, which is excited by photons with a decrease in redox potential. The excited P870 reduces bacteriopheophytin to begin the cyclic electron transport for the generation of a proton motive force. NAD(P)$^+$ is reduced through reverse electron transport oxidizing quinol. Sulfur and organic compounds are used as electron donors reducing cytochrome c_2 (Figure 11.11). Purple sulfur bacteria cannot grow under aerobic conditions, and purple non-sulfur bacteria grow chemoorganotrophically under aerobic conditions.

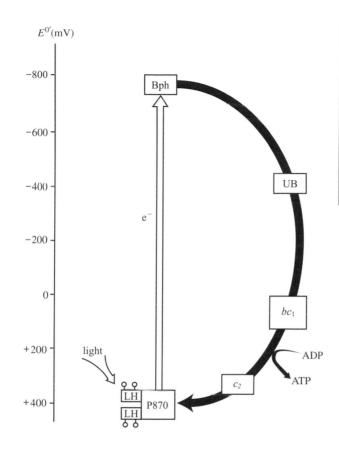

$E^{0'}$(mV)

Figure 11.11 **Light reactions in purple bacteria.**

Purple bacteria utilize light energy to generate a proton motive force which is used to synthesize ATP or reduce NAD(P)$^+$ through reverse electron transport.

Bph, bacteriopheophytin; UB, ubiquinone; bc_1, cytochrome bc_1 complex; c_2, cytochrome c_2.

11.4.3.4 Aerobic anoxygenic photosynthetic bacteria

These bacteria synthesize bacteriochlorophyll a and carotenoids only under aerobic conditions. They do not use light energy under anaerobic conditions. They have a pheophytin–quinone-type reaction centre. The cytoplasmic membrane accommodates the reaction centres and antenna molecules. They have a similar cyclic electron transport system as the purple bacteria, as depicted in Figure 11.11, but NAD(P)H is supplied from the metabolism of organic compounds.

11.5 | Carbon metabolism in phototrophs

According to their carbon sources, phototrophs are classified into photoorganotrophs and photolithotrophs. With a few exceptions, cyanobacteria are photolithotrophs, and the green bacteria and purple bacteria can grow photolithotrophically fixing CO_2 as well as photoorganotrophically (Table 11.2). Heliobacteria and aerobic anoxygenic photosynthetic bacteria do not fix CO_2.

11.5.1 CO_2 fixation

The Calvin cycle is the most common CO_2 fixing mechanism in phototrophs. The key enzymes of the Calvin cycle, phosphoribulokinase and ribulose-1,5-bisphosphate carboxylase, are not found in green sulfur bacteria and in one of the filamentous anoxygenic phototrophic bacteria, *Chloroflexus aurantiacus*. Green sulfur bacteria fix CO_2 into acetyl-CoA through the reductive TCA cycle (Section 10.8.2), and a less common 3-hydroxypropionate pathway is employed by *Chloroflexus aurantiacus* (Section 10.8.4). CO_2 is fixed through the Calvin cycle in cyanobacteria, purple bacteria and filamentous anoxygenic phototrophic bacteria (Table 11.2).

11.5.2 Carbon metabolism in photoorganotrophs

Most photosynthetic bacteria use simple organic compounds as their carbon sources and electron donors with ATP and NAD(P)H generated from the light reactions.

11.5.2.1 Purple bacteria, heliobacteria and aerobic anoxygenic photosynthetic bacteria

As photoorganotrophs, purple non-sulfur bacteria preferentially use organic compounds as their carbon source. Sugars are metabolized through the EMP or ED pathway depending on the organism. CO_2 from glycolysis is fixed under photosynthetic conditions (Figure 11.12). Under dark conditions, purple non-sulfur bacteria can grow with or without molecular oxygen.

Acetate is converted to acetyl-CoA before being metabolized through the TCA cycle and glyoxylate cycle, or by pyruvate:ferredoxin oxidoreductase to pyruvate. CO_2 is fixed to consume the excess reducing equivalents generated from the metabolism of organic compounds under photosynthetic conditions. For this reason, purple bacteria require CO_2 for photoorganotrophic growth on compounds more reduced than acetate. Butyrate metabolism is a good example of this (Figure 11.13).

Figure 11.12

Photoorganotrophic metabolism in purple bacteria.

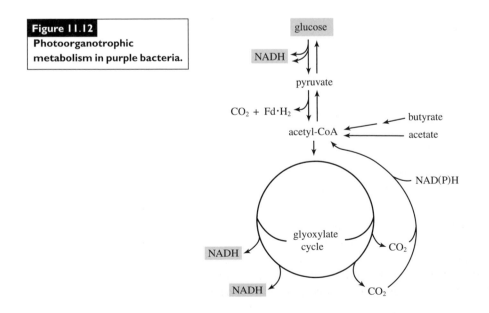

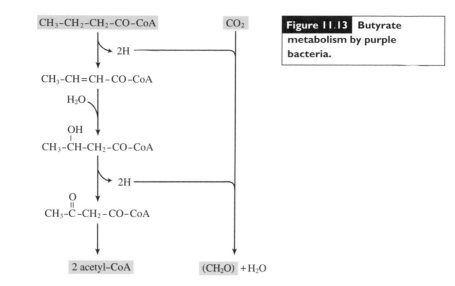

Figure 11.13 Butyrate metabolism by purple bacteria.

Heliobacteria and aerobic anoxygenic photosynthetic bacteria supplement their energy requirements through light reactions while growing as chemoorganotrophs.

11.5.2.2 Green sulfur bacteria

Green sulfur bacteria obtain energy required for their growth from the light reactions, and can use simple organic carbon sources such as acetate, but not as the electron donor. Acetate is assimilated only when CO_2 and sulfur compounds are available as electron donors. On the other hand, filamentous anoxygenic phototrophic bacteria can use organic electron donors in phototrophic metabolism and can grow chemoorganotrophically like purple non-sulfur bacteria.

11.5.2.3 Cyanobacteria

The majority of cyanobacteria grow photolithotrophically, and few grow photoorganotrophically. Obligately photolithotrophic cyanobacteria cannot use glucose, though they metabolize glycogen as storage materials with glycolytic enzymes. It is likely that they do not have sugar transport systems. The photoorganotrophic cyanobacteria use glucose but cannot use other organic compounds because they are devoid of a functional TCA cycle. They metabolize glucose through the oxidative HMP cycle (Figure 11.14) as in *Thiobacillus novellus* (Section 4.3.2). The cyanobacteria cannot grow under dark conditions.

11.6 | Photophosphorylation in halophilic archaea

Halophilic archaea, including species of *Halobacterium*, can grow at a NaCl concentration of over 2.5 M. They swim away from the light when enough electron donors and O_2 are available. With a limited O_2 supply, they move towards the light. They generate a proton motive force transporting H^+ and Cl^- across the membrane

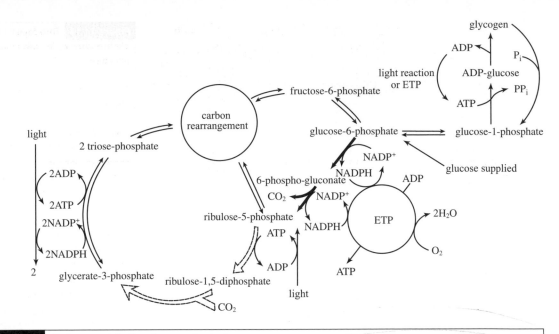

Figure 11.14 Carbon metabolism in cyanobacteria.

Photoorganotrophic cyanobacteria metabolize glucose through the oxidative HMP cycle (thick solid lines and carbon rearrangement reactions) and fix CO_2 through the Calvin cycle (double lines). Some enzymes participate in both metabolic pathways.

Figure 11.15 The structure of retinal bound to rhodopsins in species of *Halobacterium*.

(Gottschalk, G. 1986, *Bacterial Metabolism*, 2nd edn., Figure 9.20. Springer, New York)

Rhodopsin has the structure of a Schiff's base between the chromophore retinal and a lysine residue of the apoprotein. The Schiff's base exports H^+ with light energy.

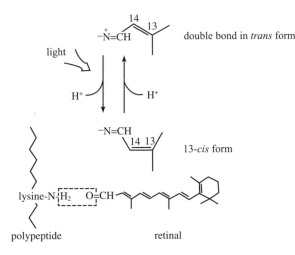

using light energy in a process known as photophosphorylation. A group of proteins known as rhodopsins facilitate motility and photophosphorylation.

Sensory rhodopsin II is synthesized for movement away from the light when the dissolved O_2 (DO) concentration is high, and light attracts halophilic archaea under a limited O_2 supply. Sensory rhodopsin I is synthesized with bacteriorhodopsin and halorhodopsin under low DO conditions. With light energy, bacteriorhodopsin exports H^+ and halorhodopsin imports Cl^- to generate a proton motive force. These rhodopsins are purple proteins bound with retinal (Figure 11.15).

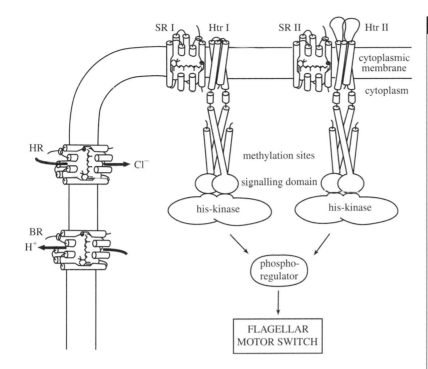

Figure 11.16 **Rhodopsins and their functions in halophilic archaea.**

(*Mol. Microbiol.* **28**:1051–1058, 1998)

Halophilic archaea have sensory rhodopsin I and II to control phototaxis, and bacteriorhodopsin and halorhodopsin to generate a proton motive force utilizing light energy. Bacteriorhodopsin exports H^+, and halorhodopsin imports Cl^-, utilizing light energy. Sensory rhodopsins with a similar structure pass the information to the transducer proteins to control phototaxis. At low dissolved O_2 concentrations, the organisms move towards the light through the actions of sensory rhodopsin I and transducer I to use it. Bacteriorhodopsin and halorhodopsin convert light energy into a proton motive force. Sensory rhodopsin II is synthesized in order for them to move away from the light at high dissolved O_2 concentrations.

BR, bacteriorhodopsin; HR, halorhodopsin; SRI, sensory rhodopsin I; HtrI, transducer I; SRII, sensory rhodopsin II; HtrII, transducer II.

Rhodopsins have a similar structure, with retinal forming a Schiff's base with a lysine residue of the peptide. The Schiff's base releases H^+ when it receives light. Bacteriorhodopsin exports H^+, and the sensory rhodopsins pass the information to transducer proteins to control phototaxis (Figure 11.16).

FURTHER READING

Photosynthetic bacteria

Buchan, A., Gonzalez, J. M. & Moran, M. A. (2005). Overview of the marine *Roseobacter* lineage. *Applied and Environmental Microbiology* **71**, 5665–5677.

Fleischman, D. & Kramerb, D. (1998). Photosynthetic rhizobia. *Biochimica et Biophysica Acta - Bioenergetics* **1364**, 17–36.

Gupta, R. S., Mukhtar, T. & Singh, B. (1999). Evolutionary relationships among photosynthetic prokaryotes (*Heliobacterium chlorum*, *Chloroflexus aurantiacus*, cyanobacteria, *Chlorobium tepidum* and proteobacteria): implications regarding the origin of photosynthesis. *Molecular Microbiology* **32**, 893–906.

Hiraishi, A. & Shimada, K. (2001). Aerobic anoxygenic photosynthetic bacteria with zinc-bacteriochlorophyll. *Journal of General and Applied Microbiology* **47**, 161–180.

Imhoff, J. F. (2001). True marine and halophilic anoxygenic phototrophic bacteria. *Archive of Microbiology* **176**, 243–254.

Lehto, K. M., Lehto, H. J. & Kanervo, E. A. (2006). Suitability of different photosynthetic organisms for an extraterrestrial biological life support system. *Research in Microbiology* **157**, 69–76.

Morgan-Kiss, R. M., Priscu, J. C., Pocock, T., Gudynaite-Savitch, L. & Huner, N. P. A. (2006). Adaptation and acclimation of photosynthetic microorganisms to permanently cold environments. *Microbiology and Molecular Biology Reviews* **70**, 222–252.

Yurkov, V. V. & Beatty, J. T. (1998). Aerobic anoxygenic phototrophic bacteria. *Microbiology and Molecular Biology Reviews* **62**, 695–724.

Zhang, C. C., Laurent, S., Sakr, S., Peng & Bedu, S. (2006). Heterocyst differentiation and pattern formation in cyanobacteria: a chorus of signals. *Molecular Microbiology* **59**, 367–375.

Photosynthetic apparatus and pigments

Barber, J. (2002). Photosystem II: a multisubunit membrane protein that oxidises water. *Current Opinion in Structural Biology* **12**, 523–530.

Elsen, S., Swem, L. R., Swem, D. L. & Bauer, C. E. (2004). RegB/RegA, a highly conserved redox-responding global two-component regulatory system. *Microbiology and Molecular Biology Reviews* **68**, 263–279.

Elsen, S., Jaubert, M., Pignol, D. & Giraud, E. (2005). PpsR: a multifaceted regulator of photosynthesis gene expression in purple bacteria. *Molecular Microbiology* **57**, 17–26.

Frigaard, N.-U. & Bryant, D. A. (2004). Seeing green bacteria in a new light: genomics-enabled studies of the photosynthetic apparatus in green sulfur bacteria and filamentous anoxygenic phototrophic bacteria. *Archives of Microbiology* **182**, 265–276.

Gregor, J. & Klug, G. (1999). Regulation of bacterial photosynthesis genes by oxygen and light. *FEMS Microbiology Letters* **179**, 1–9.

Jones, M. R., Fyfe, P. K., Roszak, A. W., Isaacs, N. W. & Cogdell, R. J. (2002). Protein-lipid interactions in the purple bacterial reaction centre. *Microbiology and Molecular Biology Reviews* **1565**, 206–214.

Kovacs, A. T., Rakhely, G. & Kovacs, K. L. (2005). The PpsR regulator family. *Research in Microbiology* **156**, 619–625.

MacColl, R. (2004). Allophycocyanin and energy transfer. *Biochimica et Biophysica Acta – Bioenergetics* **1657**, 73–81.

Oh, J. I. & Kaplan, S. (2001). Generalized approach to the regulation and integration of gene expression. *Molecular Microbiology* **39**, 1116–1123.

Samsonoff, W. A. & MacColl, R. (2001). Biliproteins and phycobilisomes from cyanobacteria and red algae at the extremes of habitat. *Archives of Microbiology* **176**, 400–405.

Umeno, D., Tobias, A. V. & Arnold, F. H. (2005). Diversifying carotenoid biosynthetic pathways by directed evolution. *Microbiology and Molecular Biology Reviews* **69**, 51–78.

Light reactions

Berry, S. & Rumberg, B. (2001). Kinetic modeling of the photosynthetic electron transport chain. *Bioelectrochemistry* **53**, 35–53.

Bryant, D. A. & Frigaard, N. U. (2006). Prokaryotic photosynthesis and phototrophy illuminated. *Trends in Microbiology* **14**, 488–496.

Iverson, T. M. (2006). Evolution and unique bioenergetic mechanisms in oxygenic photosynthesis. *Current Opinion in Chemical Biology* **10**, 91–100.

Krauss, N. (2003). Mechanisms for photosystems I and II. *Current Opinion in Chemical Biology* **7**, 540–550.

Oprian, D. D. (2003). Phototaxis, chemotaxis and the missing link. *Trends in Biochemical Sciences* **28**, 167–169.

Roegner, M., Boekema, E. J. & Barber, J. (1996). How does photosystem 2 split water? The structural basis of efficient energy conversion. *Trends in Biochemical Sciences* **21**, 44–49.

Vredenberg, W. J. (1997). Electrogenesis in the photosynthetic membrane: fields, fact and features. *Bioelectrochemistry and Bioenergetics* **44**, 1–11.

Carbon metabolism

Berg, I. A., Keppen, O. I., Krasil'nikova, E. N., Ugol'kova, N. V. & Ivanovsky, R. N. (2005). Carbon metabolism of filamentous anoxygenic phototrophic bacteria of the family Oscillochloridaceae. *Microbiology-Moscow* **74**, 258–264.

Garcia-Fernandez, J. M. & Diez, J. (2004). Adaptive mechanisms of nitrogen and carbon assimilatory pathways in the marine cyanobacteria *Prochlorococcus*. *Research in Microbiology* **155**, 795–802.

Hartman, F. C. & Harpel, M. R. (1994). Structure, function, regulation, and assembly of D-ribulose-1,5-bisphosphate carboxylase/oxygenase. *Annual Review of Biochemistry* **63**, 197–234.

Photophosphorylation

Neutze, R., Pebay-Peyroula, E., Edman, K., Royant, A., Navarro, J. & Landau, E. M. (2002). Bacteriorhodopsin: a high-resolution structural view of vectorial proton transport. *Biochimica et Biophysica Acta – Biomembranes* **1565**, 144–167.

Schertler, G. F. (2005). Structure of rhodopsin and the metarhodopsin I photo-intermediate. *Current Opinion in Structural Biology* **15**, 408–415.

Spudich, J. L. (2006). The multitalented microbial sensory rhodopsins. *Trends in Microbiology* **14**, 480–487.

12

Metabolic regulation

Life processes transform materials available from the environment into cell components. Organic materials are converted to carbon skeletons for monomer and polymer synthesis, as well as being used to supply energy. Microbes synthesize monomers in the proportions needed for growth. This is possible through the regulation of the reactions of anabolism and catabolism. With a few exceptions, microbial ecosystems are oligotrophic with a limited availability of nutrients, the raw materials used for biosynthesis. Furthermore, nutrients are not usually found in balanced concentrations while the organisms have to compete with each other for available nutrients.

Unlike animals and plants, unicellular microbial cells are more directly coupled to their environment, which changes continuously. Many of these changes are stressful so organisms have evolved to cope with this situation. They regulate their metabolism to adapt to the ever-changing environment.

Since almost all biological reactions are catalyzed by enzymes, metabolism is regulated by controlling the synthesis of enzymes and their activity (Table 12.1). Metabolic regulation through the dynamic interactions between DNA or RNA and the regulatory apparatus employed determine major characteristics of organisms. In this chapter, different mechanisms of metabolic regulation are discussed in terms of enzyme synthesis through transcription and translation and enzyme activity modulation.

12.1 | Mechanisms regulating enzyme synthesis

The rate of biological reactions catalyzed by enzymes is determined by the concentration and activity of the enzymes. Various mechanisms regulating the synthesis of individual enzymes are discussed here before multigene regulation is considered.

Table 12.1. | *Regulatory mechanisms that control the synthesis and activity of enzymes*

	Mechanism
Enzyme synthesis	
Transcription	promoter structure and sigma (σ) factor
	activator – positive control
	repressor – negative control
	termination – antitermination
Translation	attenuation
	autogenous translational repression
	mRNA stability
Enzyme activity	feedback inhibition
	feedforward activation
	chemical modification
	physical modification
	degradation

12.1.1 Regulation of transcription by promoter structure and sigma (σ) factor activity

RNA polymerase synthesizes messenger RNA (mRNA) that is needed for protein synthesis according to the DNA template. RNA synthesis is initiated with the recognition of the promoter by the σ-factor of the RNA polymerase (Section 6.11). Several σ-factors have been found in all the bacteria tested. They are grouped in two main families: the $\sigma-70$ and $\sigma-54$ families. The latter is composed exclusively of σ^{54}, while the $\sigma-70$ family includes all others. The $\sigma-70$ family is further divided into four groups (Table 12.2). The group 1 σ^D is essential, and responsible for the transcription of most genes in exponentially growing cells. A group 2 σ-factor, σ^S, participates in the expression of stationary phase proteins. Group 3 σ-factors include σ^F, responsible for flagella synthesis, and σ^H, the heat shock response σ-factor in *Escherichia coli*. Sigma factors involved in spore formation in *Bacillus subtilis* belong to group 3. σ^E is a member of group 4 σ-factors. This group includes the extracytoplasmic function σ-factor subfamily, the largest and most divergent group.

Each σ-factor recognizes a different promoter. The promoter region in *Escherichia coli* consists of six bases, each at -35 and -10 bases upstream of the transcription start site. Genes for proteins needed in large quantities have strong promoters with a high affinity for the σ-factors. For example, the housekeeping σ^D (σ^{70}) recognizes the promoter region consisting of TTGACA at -35 and TATAAT at -10 most efficiently but is much less effective for those with substituted base(s) in *Escherichia coli*. It should be emphasized that promoter activity is regulated by other transcription factors such as a

Table 12.2. | *RNA polymerase σ-factors in* Escherichia coli

σ-factor	Gene	Function	Promoter consensus sequence		Anti-σ factor
			−35	−10	
$\sigma-70$ family					
Group 1					
σ^D (σ^{70})	rpoD	housekeeping	TTGACA	TATAAT	AsiA [a]
Group 2					
σ^S (σ^{38})	rpoS, katF	stringent response general stress response	–	CTATACT	RssB
Group 3					
σ^F (σ^{28})	rpoF, fliA	chemotaxis	TAAA	GCCGATAA	FlgM
σ^H (σ^{32})	rpoH, htpR	heat shock protein	CTTGAAA	CCCATnT	DnaK
Group 4					
σ^E (σ^{24})	rpoE	stress response	GAACTT	TCTRA	RseA
$\sigma-54$ family					
σ^N (σ^{54})	rpoN, ntrA	nitrogen metabolism	TGGCAC	TTGCW	

R: A or G; n: A, T, G, or C; W: A or T.

[a] Bacteriophage T4.

repressor and an activator as well as others that interact with DNA around the promoters, as discussed later.

While the essential housekeeping σ^D functions in exponentially growing cells, other σ-factors are activated according to the growth conditions, including σ^E, σ^F, σ^H, σ^N and σ^S. The promoters recognized by σ^S (σ^{38}) do not have a −35 region but have a longer −10 region known as the extended –10 region.

Sigma factors with similar functions have different names in different organisms. The housekeeping σ-factor is called σ^D in Gram-negative bacteria and σ^A in Gram-positive bacteria. HrdB is the housekeeping σ-factor in species of the genus *Streptomyces*. Consequently, σ-factors with the same name have different functions in different bacteria: σ^D is the housekeeping σ-factor in *Escherichia coli* but is the σ-factor for flagella formation in *Bacillus subtilis*. In a related terminological problem, the extracytoplasmic function σ-factor, σ^E, should not be regarded as functioning extracytoplasmically. Instead, σ^E participates in the expression of genes needed to repair denatured proteins of extracytoplasmic location.

Since a σ-factor recognizes multiple promoters throughout the chromosome, a specific σ-factor participates in the transcription of functionally unrelated genes. Regulation by σ-factor–promoter interaction is a type of global regulation system discussed later (Section 12.2).

Proteins known as anti-sigma (anti-σ) factors bind σ-factors and inhibit their activity. When *Escherichia coli* is infected with bacteriophage T4, a phage-originating anti-σ factor, AsiA, inhibits the activity of the bacterial housekeeping σ^D to produce more phage proteins. An anti-σ factor for σ^E, RseA, is a membrane protein. Under normal growth conditions σ^E binds RseA. When σ^E is required, either an anti-anti-σ factor binds RseA releasing σ^E or RseA is degraded to release σ^E. In the case of σ^F responsible for flagellin synthesis, the anti-σ factor, FlgM, is exported into the environment through the flagellin export mechanism when more flagellin is needed (Section 12.2.11). The stationary phase σ-factor is inactivated by the anti-σ factor RssB, before being hydrolyzed by an ATP-dependent protease (ClpXP) when the σ^S is needed (Section 12.2.1).

Promoter activity is also regulated by the superhelix DNA structure and various general DNA-binding proteins such as H-NS, Fis, and StpA in addition to the specific activators and repressors.

12.1.2 Induction of enzymes

12.1.2.1 Inducible and constitutive enzymes

When *Escherichia coli* is transferred from a glucose medium to a lactose medium, the bacterium synthesizes β-galactosidase to hydrolyze the lactose into glucose and galactose (Figure 12.1). Enzymes synthesized in the presence of substrates are referred to as inducible enzymes and the substrate is termed the inducer. Constitutive enzymes are those enzymes that are produced under all growth conditions. Inducible enzymes are generally those used in the catabolism of carbohydrates such as polysaccharides (cellulose, starch, etc.), oligosaccharides (lactose, trehalose, raffinose, etc.) and minor sugars (arabinose and rhamnose), and aromatic compounds.

When a single inducer induces more than two enzymes, they are produced either simultaneously or sequentially. The former is referred to as coordinate induction, and the latter as sequential induction (Figure 12.2). Genes of coordinate induction are in the same operon, and genes from separate operons are induced sequentially.

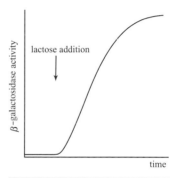

Figure 12.1 Induction of β-galactosidase by lactose in *Escherichia coli*.

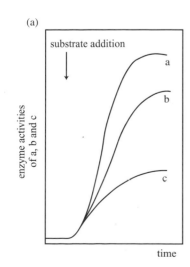

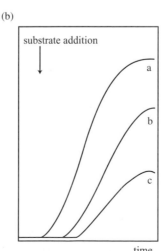

Figure 12.2 Coordinate induction and sequential induction of multiple enzymes by a single inducer.

(Gottschalk, G. 1986, *Bacterial Metabolism*, 2nd edn., Figure 7.2. Springer, New York)

Enzymes from genes of the same operon are induced simultaneously in coordinate induction (a) and genes from different operons are induced sequentially (b). The product of the first enzyme reaction is the inducer of the second enzyme in sequential induction.

12.1.2.2 Enzyme induction

Enzyme induction is regulated at the level of transcription. Lactose induces the production of β-galactosidase, permease and transacetylase. Their structural genes form an operon (*lac* operon) with a promoter and operator (Figure 12.3a). The regulatory gene (*lacI*) next to the 5′ end of the operon is expressed constitutively with its own promotor. In the absence of the inducer, the repressor protein binds the operator region of the *lac* operon, inhibiting RNA polymerase from binding the promoter region. Consequently, the structural genes are not transcribed (Figure 12.3b1). On the other hand, when the inducer is available, it binds the repressor protein, removing it from the operator

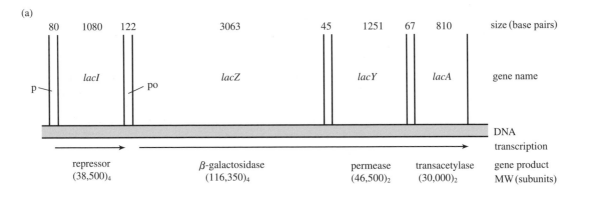

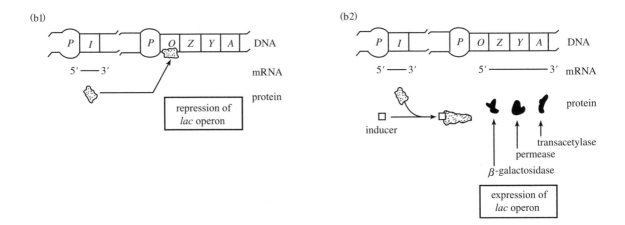

Figure 12.3 Induction mechanism of the *lac* operon in *Escherichia coli*.

(Gottschalk, G. 1986, *Bacterial Metabolism*, 2nd edn., Figure 7.4. Springer, New York)

The *lac* operon consists of a promoter, operator and structural genes, and a repressor protein is produced from the *lacI* gene that has its own promoter (a). In the absence of inducer, the repressor protein binds the operator region of the operons, preventing transcription of the structural genes (b1). The inducer forms a complex with the repressor protein. The complex cannot bind the operator region, and the structural genes are transcribed (b2). The structural genes are transcribed as constitutive enzymes in the *lacI* mutant. The regulation by repressor proteins is referred to as negative control.

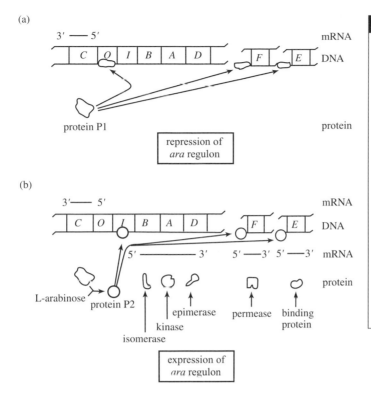

(a)

protein P1

repression of
ara regulon

(b)

L-arabinose
protein P2

epimerase
kinase
isomerase

permease   binding
protein

expression of
ara regulon

Figure 12.4 **Induction of enzymes of arabinose metabolism in *Escherichia coli*.**

(Gottschalk, G. 1986, *Bacterial Metabolism*, 2nd edn., Figure 7.5. Springer, New York)

The regulator protein AraC is in the protein P1 form to bind the operator region, inhibiting transcription of the structural genes in the absence of arabinose (a). When the inducer arabinose binds, protein P1 is converted to protein P2. Protein P2 activates transcription of the structural genes. *araC* mutants cannot use arabinose. The term positive control is used to describe metabolic regulation by activators.

region (Figure 12.3b2). The repressor protein is not produced in a *lacI* mutant. This mutant transcribes the structural genes as constitutive enzymes in the absence of the inducer. In this sense, the regulation by repressor proteins is referred to as negative control.

12.1.2.3 Positive and negative control
As stated above, the *lac* operon is regulated by a negative control mechanism. Activator proteins are involved in the regulation of catabolic genes for arabinose, rhamnose, maltose and others. Genes for arabinose catabolism consist of *araA*, *B*, *C*, *D*, *E*, and *F*; *araC* is a regulatory gene encoding an activator protein. *araC* mutants are unable to use arabinose, since an AraC complex with the inducer activates the transcription of the structural genes (Figure 12.4). *araA*, *B*, *C* and *D* form an operon, and *araE* and *araF* occupy other parts of the chromosome. The term regulon is used to define genes of the same metabolism controlled by the same effectors scattered around the chromosome, such as *ara* genes. Regulation by an activator, as in the *ara* regulon, is referred to as positive control.

12.1.3 Catabolite repression
When *Escherichia coli* or *Bacillus subtilis* is cultivated in a medium containing glucose and lactose, they grow in a distinct two-phase pattern as shown in Figure 12.5. This is called diauxic growth or diauxie. This is due to the fact that the readily utilizable glucose and its metabolites repress the utilization of lactose. The term catabolite repression

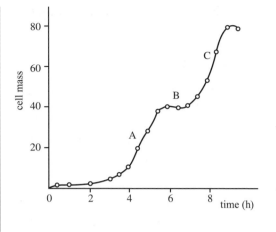

Figure 12.5 **Diauxic growth of** *Escherichia coli* **on glucose and lactose.**

When *Escherichia coli* is cultivated on a mixture of glucose and lactose, the bacterium grows on glucose, repressing lactose utilization (A) at the beginning. When glucose is exhausted the bacterium grows again on lactose (C) after a lag period (B). This is referred to as diauxic growth or diauxie. Diauxic growth is the result of a regulatory mechanism known as catabolite repression.

is used to describe this regulatory mechanism, which regulates many catabolic genes and operons in bacteria.

Gram-negative bacteria have two separate catabolite repression mechanisms, one involving a cAMP–CRP (cAMP receptor protein) complex and the other a catabolite repressor/activator (Cra) protein. In Gram-positive bacteria, the catabolite control protein A (CcpA) has a similar function.

12.1.3.1 Carbon catabolite repression by the cAMP–CRP complex

The primary carbon catabolite repression (CCR) in Gram-negative bacteria is related to the intracellular cyclic AMP (cAMP) concentration. When the readily utilizable substrate is exhausted, the cAMP concentration increases, and this cyclic nucleotide forms a complex with the cAMP receptor protein (CRP or catabolite activator protein, CAP). This complex controls many operons (Section 12.2).

Glucose is transported through the phosphotransferase (PT) system (group translocation mechanism) in many bacteria. When the glucose concentration is low, enzyme II_A of the PT system remains phosphorylated (Section 3.5). The phosphorylated enzyme II_A activates the activity of adenylate cyclase, a cytoplasmic membrane enzyme. This enzyme converts ATP to cAMP:

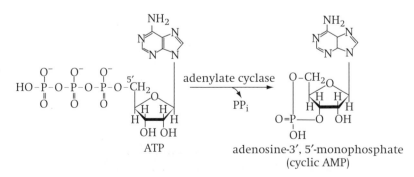

A cytoplasmic enzyme, phosphodiesterase, hydrolyzes cAMP, but the activity is very low:

$$\text{cAMP} + H_2O \xrightarrow{\text{phosphodiesterase}} \text{AMP}$$

When adenylate cyclase activity is low due to the low level of phosphorylated enzyme II_A with active glucose transport, the cAMP concentration is kept low, and CRP cannot form the complex. Under the opposite conditions, the rate of cAMP formation is higher than that of its hydrolysis facilitating cAMP–CRP complex formation. This complex activates the transcription of many operons including the *lac* operon.

When lactose is provided with a low level of glucose, lactose binds the repressor protein as shown in Figure 12.3 and the cAMP–CRP complex binds the CRP site of the promoter region, activating transcription of the structural genes (Figure 12.6). Studies with CRP and adenylate cyclase mutants showed that the cAMP–CRP complex regulates the expression of over 200 proteins. Carbon catabolite repression (CCR) by the cAMP–CRP complex is known only in Gram-negative bacteria. This regulatory mechanism is an example of global regulation (Section 12.2).

12.1.3.2 Catabolite repressor/activator

In addition to the cAMP–CRP complex, Cra (catabolite repressor/ activator) protein functions as a catabolite repressor and activator

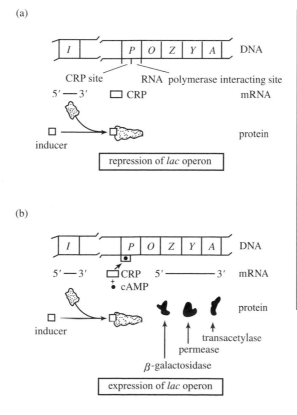

(a)

CRP site   RNA polymerase interacting site

5'—3'   □ CRP

inducer

repression of *lac* operon

(b)

5'—3'   □CRP   5'———3'   mRNA
    • cAMP

inducer

transacetylase
permease
β-galactosidase

expression of *lac* operon

Figure 12.6 **Activation of the** *lac* **operon by the inducer and the cAMP–CRP complex.**

(Gottschalk, G. 1986, *Bacterial Metabolism*, 2nd edn., Figure 7.7. Springer, New York)

Even when the repressor protein is inactivated by the inducer, RNA polymerase cannot bind the promoter to initiate transcription (a). When the cAMP–CRP complex binds the CRP site of the promoter region, the enzyme binds the promoter region to initiate transcription (b). The repression by the readily utilizable substrate or its metabolites is referred to as carbon catabolite repression.

Figure 12.7 cAMP–CRP independent catabolite repression by Cra protein in *Escherichia coli.*

(*J. Bacteriol.* **178**:3411–3417, 1996)

Cra (catabolite repressor/ activator) protein functions as a catabolite repressor and activator for enzymes related to sugar metabolism. When a non-carbohydrate substrate is used, Cra protein binds operator regions of the genes for PEP carboxykinase (*pckA*) and pyruvate kinase (*pykF*), activating the former and repressing the latter (a). When glucose is used as the substrate (b), the Cra protein is inactivated through complex formation with the catabolite and loses its function as an activator or repressor.

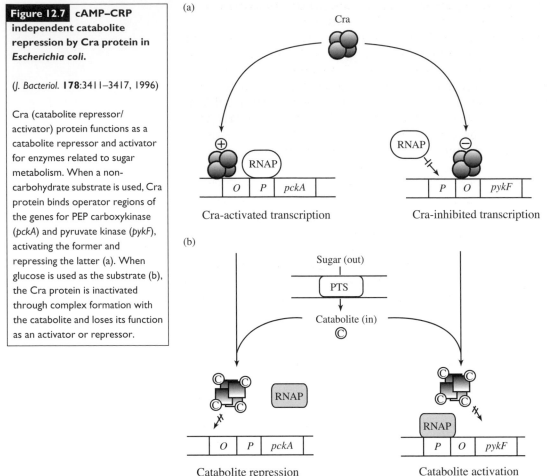

(a)

Cra-activated transcription

Cra-inhibited transcription

(b)

Catabolite repression (Deactivation)

Catabolite activation (Derepression)

in enteric bacteria including *Escherichia coli.* Cra-negative mutants synthesize more enzymes of glycolysis and the phosphotranferase (PT) system and fewer enzymes that are involved in gluconeogenesis, TCA cycle, glyoxylate cycle and electron transport than the wild-type strain. In addition, the mutants cannot utilize non-carbohydrate substrates including acetate, ethanol, pyruvate, alanine, citrate and malate. With a limited supply of readily utilizable glucose, the Cra protein positively regulates transcription of enzymes needed for the utilization of other substrates and represses the genes for glycolytic enzymes (Figure 12.7). When acetate or ethanol is used as the substrate, Cra protein activates the expression of genes for the enzymes of gluconeogenesis (fructose-1,6-diphosphatase) and anaplerotic sequence (PEP carboxykinase), and represses the transcription of genes for glycolytic enzymes (phosphofructokinase). Genes activated by the Cra protein are those that have a Cra-binding site upstream of the promoter region (left-hand side of Figure 12.7), and the genes are repressed where the Cra-binding site overlaps the promoter region or occupies it downstream (right-hand side of Figure 12.7).

12.1.3.3 Carbon catabolite repression in Gram-positive bacteria with a low G + C content

The cAMP–CRP complex is not known in Gram-positive bacteria, and these bacteria have a different carbon catabolite repression (CCR) mechanism. The HPr protein of the phosphotransferase system in Gram-positive bacteria with a low G + C content is phosphorylated at the histidine-15 residue by being coupled to the conversion of PEP to pyruvate as in Gram-negative bacteria. HPr in these bacteria has a second phosphorylation site at serine-46 in addition to histidine-15. PEP phosphorylates the histidine-15 for the group translocation while HPr(ser) kinase (PstK) phosphorylates the serine-46 consuming ATP. When glucose is available, the cell maintains high levels of glucose metabolites (fructose-1,6-diphosphate) and non-phosphorylated HPr. Fructose-1,6-diphosphate activates HPr(ser) kinase to phosphorylate the serine residue of the non-phosphorylated HPr protein. This phosphorylated HPr protein [HPr(Ser-P)] forms a complex with CcpA (catabolite control protein A). The CcpA–HPr(Ser-P) complex represses the expression of genes for enzymes of gluconate, glucitol and mannitol metabolism, and activates others through binding to the *cre* (catabolite responsive element) that is located either upstream or within their promoters (Figure 12.8). ATP-dependent HPr(ser) phosphorylation is not known in Gram-negative bacteria.

The CcpA–HPr(Ser-P) complex is an important regulator in *Bacillus subtilis*. More than 28 genes or operons have been experimentally demonstrated to be directly regulated by this complex. Moreover, recent global gene expression analyses combined with analyses of *cre* occurrences in the chromosome sequence suggest that approximately 10% of the genes might be directly regulated by this regulator in *Bacillus subtilis*.

Control by the CcpA–HPr(Ser-P) complex is known in low G + C content Gram-positive bacteria including the genera *Bacillus*, *Staphylococcus*, *Streptococcus*, *Lactococcus*, *Enterococcus*, *Mycoplasma*, *Clostridium*, and *Listeria*. This complex regulates many genes analogous to the cAMP–CRP complex in Gram-negative bacteria. In addition to catabolite repression, HPr(Ser-P) is responsible for the repression of gene expression through the inducer exclusion/expulsion mechanism. When glucose is present, the expression of proteins for lactose utilization is repressed (inducer exclusion), and lactose is exported (inducer expulsion). HPr(Ser-P) facilitates this inducer exclusion/expulsion.

When the glucose supply is limited, the expression of some hydrolases, such as levanase and β-glucosidase, are activated in *Bacillus subtilis*. The CcpA–HPr(Ser-P) complex is not involved in this regulation (upper part of Figure 12.8). With a limited supply of readily utilizable substrate, HPr is phosphorylated at histidine-15. HPr(His-P) phosphorylates the transcription regulatory protein (LevR) or transcription antiterminator (LicT). Phosphorylated LevR and LicT activate the expression of the hydrolases. This regulatory mechanism is referred to as a terminator/antiterminator mechanism (Section 12.1.6).

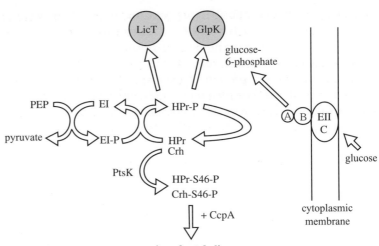

Figure 12.8 Carbon catabolite repression in *Bacillus subtilis*.

(*Curr. Opin. Microbiol.* **2**:195–201, 1999)

Catabolite repression by cAMP–CRP is not known in Gram-positive bacteria. HPr of the Gram-positive bacterial phosphotransferase has two sites for phosphorylation: 15-histidine and 46-serine. While 15-histidine phosphorylation is phosphoenolpyruvate (PEP)-dependent for sugar group translocation, 46-serine phosphorylation is ATP-dependent, catalyzed by the HPr kinase activated by fructose-1,6-diphosphate. HPr(Ser-P) forms a complex with catabolite control protein A (CcpA). The CcpA–HPr(Ser-P) complex binds a catabolite responsive element (*cre*) of various catabolic operons to repress or activate their transcription. When readily utilizable substrates are not present, the concentration of HPr(His-P) increases to activate the genes for alternative carbon sources, either directly (glycerol kinase, GlpK) or through a terminator/antiterminator mechanism (LicT, Section 12.1.6).

LicT, transcription antiterminator; GlpK, glycerol kinase; PtsK, HPr kinase; Crh, catabolite repression HPr.

A difference between CRP-dependent catabolite repression found in Gram-negative bacteria and CcpA-dependent catabolite repression found in Gram-positive bacteria is the strictness of the coupling between the PTS (phosphotransferase system), transport and regulatory functions. In the CRP-dependent mechanism, PTS enzyme II_A activates the primary sensor adenylate cyclase, while HPr(ser) kinase, the primary sensor in the CcpA-dependent mechanism, is activated by glycolytic intermediates. A search for HPr(ser) kinase gene homologues showed that this gene is widespread not only in Gram-positive bacteria but also in Gram-negative bacteria and mycoplasmas.

Metabolic control by the effectors of catabolite repression, cAMP–CRP and CcpA–HPr(Ser-P) complexes, and the Cra protein are examples of global regulation mechanisms (Section 12.2).

Similar regulators to CcpA have been identified in *Bacillus subtilis*. Crh (catabolite repression HPr) protein has a similar function in catabolite repression in *Bacillus subtilis*. Crh has a similar structure to HPr, but does not participate in group translocation. There is evidence to suggest that Crh is involved in the repression of catabolic genes by non-sugar electron donors. CcpB is the regulator for catabolite repression of gluconate (*gnt*) and xylose (*xyl*) operons by glucose, mannitol, and sucrose. CcpC represses transcription of genes that encode enzymes of the TCA cycle.

As stated earlier (Section 4.1.4.3), CggR (central glycolytic gene regulator protein) represses the genes encoding enzymes catalyzing the glycolytic steps from glyceraldehyde-3-phosphate to PEP in *Bacillus subtilis* when the bacterium is in a gluconeogenic state, while fructose-1,6-diphosphate inhibits CggR activity to activate glycolysis.

12.1.4 Repression and attenuation by final metabolic products

Under any given conditions, microbes reproduce themselves in the most efficient way and have to synthesize cell constituents in the right proportions. For this reason, anabolism should be regulated. Just as the enzymes of catabolism are induced by inducers, the final products repress the anabolic enzymes. When *Escherichia coli* grows on a glucose–mineral salts medium, glucose flux is tightly controlled to produce the cell constituents in the right proportions. If a monomer of a cell constituent is available, e.g. an amino acid, the bacterium shuts down production of this amino acid. Repression and attenuation by the final anabolic product(s) is the regulatory mechanism involved. Regulation by repression depends on repressor proteins, while proteins are not directly involved in the attenuation mechanism.

12.1.4.1 Repression

The expression of genes for the enzymes of amino acid biosynthesis is regulated through repression or attenuation in *Escherichia coli*. Figure 12.9 shows the repression of enzymes for tryptophan

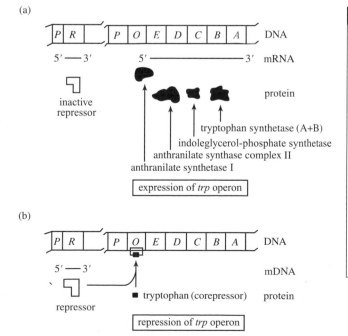

(a)

inactive repressor

protein

tryptophan synthetase (A+B)
indoleglycerol-phosphate synthetase
anthranilate synthase complex II
anthranilate synthetase I

expression of *trp* operon

(b)

repressor

tryptophan (corepressor)   protein

repression of *trp* operon

Figure 12.9 Transcriptional regulation of the tryptophan operon through repression.

(Gottschalk, G. 1986, *Bacterial Metabolism*, 2nd edn., Figure 7.8. Springer, New York)

When the intracellular tryptophan concentration is low, the repressor protein is inactive and the operon is transcribed (a). When the tryptophan concentration increases, it binds to the repressor protein to make the active form. The active repressor binds the operator region of the operon to inhibit transcription (b).

biosynthesis. A repressor protein is synthesized from a separate operon. When the tryptophan concentration is low, the repressor protein is inactive, and the *trp* operon is transcribed. When the tryptophan concentration increases, the amino acid binds the repressor protein to make the repressor active. In this case tryptophan is referred to as a corepressor. The active repressor binds to the operator region of the operon to repress its transcription.

12.1.4.2 Attenuation

In bacteria, gene expression is commonly regulated at the level of the initiation of transcription in response to external signals. Due to the low stability of most bacterial mRNA (Section 12.1.9), this provides an efficient way to control the synthesis of any given protein. Most frequently, transcription initiation is tightly regulated by σ-factors, repressors or activators, as discussed above. In several operons, however, transcription is constitutively initiated, but elongation can only proceed under specific conditions. Transcription elongation is regulated in response to external signals by the nascent mRNA with or without the aid of other effectors. In both cases the mRNA forms alternative hairpin structures known as a terminator and an antiterminator. For convenience, the discussion here is on the attenuation process where mRNA senses the physiological conditions directly without the aid of other effector molecules. In some cases the formation of the terminator/antiterminator is aided by small proteins, metabolites, tRNAs and antisense RNAs as discussed later (Section 12.1.6).

The repressor protein for the histidine operon has not been identified in *Escherichia coli*. The histidine operon is regulated through a mechanism known as attenuation. Control by attenuation is possible only in prokaryotic cells where mRNA is translated while it is being synthesized.

The tryptophan operon is under dual control, primarily through repression and attenuation. Even when the *trp* operon is fully derepressed, about 90% of the RNA polymerase transcribing the operon does not reach the structural genes due to the attenuation mechanism. About 160 base pairs, known as the leader sequence, occupy a region between the promoter and the start of the structural genes of the *trp* operon. The transcript of the leader sequence functions as the regulator (Figure 12.10).

As stated above, attenuation is known only in prokaryotes, where translation starts while mRNA is synthesized. The transcript of the leader sequence (leader mRNA) contains a relatively high proportion of codons for the amino acid whose biosynthesis is encoded by the operon, and four regions that can make hairpin (stem-and-loop) structures (Figure 12.11).

When tryptophan is available at adequate levels to meet cellular requirements, the ribosomes move to the end of the second sequence of the four hairpin-making regions on the mRNA during its translation. At this point the third and the fourth regions form a hairpin

(a)

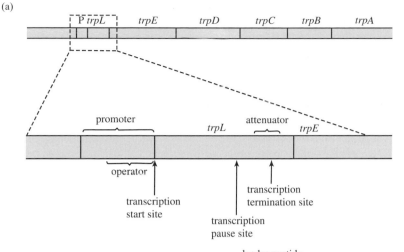

leader peptide

Met Lys Ala Ile Phe Val Leu Lys Gly Trp Trp Arg Thr Ser

(b)

```
 1                        20                      40                      60
pppAAGUUCACGUAAAAAGGGUAUCGACAAUGAAAGCAAUUUUCGUACUGAAAGGUUGGUGGCGCACUUCC
           80                      100                     120
UGAAACGGGCAGUGUAUUCACCAUGCGUAAAGCAAUCAGAUACCCAGCCCGCCUAAUGAGCGGGCUUUU
              140                     160                     180
       UUUUGAACAAAAUUAGAGAAUAACAAUGCAAACACAAAAACCGACUCUCGAACUGCU
```

Met Gln Thr Gln Lys Pro Thr Leu Glu Leu Leu

trpE polypeptide

structure. This hairpin removes the RNA polymerase from the DNA before it reaches the start of the structural genes. This hairpin is referred to as an attenuator or terminator. However, when the tryptophan level is low, the tryptophanyl-tRNA becomes limiting and the ribosome translating the leader sequence stalls at a trytophan codon before it reaches the second hairpin-forming region. At this point the leader mRNA forms a hairpin structure between the second and the third region, preventing the formation of the terminator between the third and the fourth segments, and the transcription continues to the structural genes. This hairpin between the second and the third regions is referred to as an antiterminator.

When the overall amino acid availability is low, the ribosome cannot start translation, and the leader mRNA adopts a conformation with two hairpin structures between the first and the second region, and the third and the fourth region (terminator).

The rate of transcription elongation in vivo is about 40 to 50 nucleotides/second. The decision between termination and antitermination is therefore made within a very short timeframe. Termination should be prevented by sequestration of the third region of the hairpin-forming

Figure 12.10 The structure of the *trp* operon with the leader sequence for transcriptional control through attenuation.

The distance is about 160 base pairs between the promoter and the start of the structural genes in the *trp* operon. This segment is the leader sequence (*trpL*) that controls the transcription through attenuation. The transcript from the leader sequence has four regions that can make hairpin structures (see Figure 12.11).

(a) *trp* operon; (b) base sequence of the leader sequence.

(a) tryptophan in adequate levels

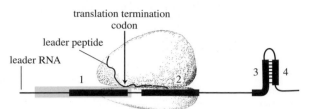

Hairpin formation between sequences 3 and 4 to function as a terminator.

(b) tryptophan in limited conditions

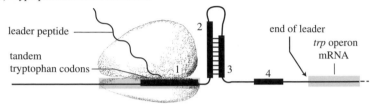

Hairpin formation between sequences 2 and 3 when the ribosome stops at sequence 1 due to lack of activated tryptophan.

(c) all amino acids in low concentration

When amino acids are in low concentration, translation is not initiated to form a hairpin between sequences 1 and 2, and 3 and 4.

Figure 12.11 **Transcriptional control of the *trp* operon through attenuation.**

The leader sequence of the messenger RNA transcribed from the *trp* operon contains codons (UGG) for tryptophan and four segments (thick lines) that can form hairpin structures. The hairpin structure formed between the 3rd and 4th segment functions as a terminator separating transcribing RNA polymerase from DNA before it reaches the structural gene. (a) With adequate tryptophan levels to meet cell requirements, ribosomes move to the end of the second hairpin-forming region that contains the tryptophan codons. At this point the 3rd and 4th segments form the terminator hairpin to remove RNA polymerase from the DNA before it reaches the start of the structural gene. The structural genes are not transcribed. (b) When the tryptophan concentration is low, the ribosomes stop at the tryptophan codon within the leader sequence before they reach the second segment of the hairpin-forming regions. At this point the 2nd and the 3rd segments form a hairpin structure preventing the formation of the terminator hairpin structure between the 3rd and the 4th segments. The transcription continues through the structural genes. The hairpin structure between the 2nd and the 3rd segments is referred to as the antiterminator. (c) Under general starvation conditions, translation cannot proceed and hairpin structures are formed between the 1st and 2nd, and the 3rd and 4th segments: the latter one functions as the terminator.

---

sequences by pairing with the second region to form the antiterminator element before synthesis of the fourth region. To synchronize translation with transcription of the leader sequence, transcription pauses at a transcription pause site within the leader sequence.

The synthesis of many amino acids and pyridine nucleotides is controlled by this mechanism alone or by a dual mechanism involving repression. Attenuation is not known in archaea and eukaryotes.

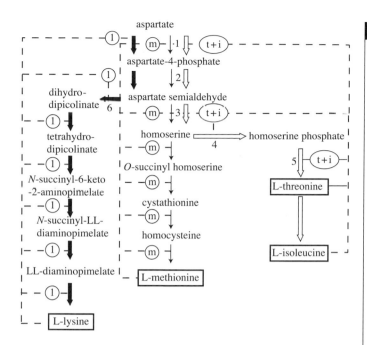

Figure 12.12 Transcription of genes for the synthesis of aspartate series amino acids is regulated through a dual repression–attenuation mechanism.

(Gottschalk, G. 1986, *Bacterial Metabolism*, 2nd edn., Figure 7.10. Springer, New York)

Aspartate series amino acids include lysine, methionine, threonine and isoleucine. Isoleucine is synthesized through threonine. Aspartate kinase, the enzyme catalyzing the first reaction, occurs as three isoenzymes. The final products control their transcription separately. Lysine and methionine regulate transcription of the specific isoenzymes through repression, while the third isoenzyme is regulated by threonine and isoleucine through attenuation. This regulation is referred to as divalent attenuation. Since other reactions branch off from aspartate semialdehyde and homoserine, the enzymes catalyzing the first reaction of each branch are subject to a similar regulatory control.

1, aspartate kinase; 2, aspartate semialdehyde dehydrogenase; 3, homoserine dehydrogenase; 4, homoserine kinase; 5, threonine synthase; 6, dihydrodipicolinate cyclohydrolase.

Attenuation is a regulatory mechanism in that leader mRNAs directly sense physiological signals by binding an effector molecule without the involvement of regulatory proteins or ribosomes. This is an example of 'riboswitch'. The term riboswitch is defined as structured domains that usually reside in the non-coding regions of mRNAs, where they bind metabolites and control gene expression (Section 12.1.6).

12.1.5 Regulation of gene expression by multiple end products

In an anabolic process where more than two products are formed from a common precursor, an excess of one product should not interfere with the synthesis of the other product. To ensure this, the first enzyme of the common pathway has one of the following properties:

(1) Multiple enzymes (isoenzymes) catalyze the same reaction, but their activities and synthesis are regulated by individual products.
(2) If a single enzyme catalyzes the reaction, gene expression and its activity are regulated collectively by the products.

The expression of genes for isoenzymes and their activities are regulated separately by individual products, though they catalyze the same reaction. Aspartate is the precursor for the synthesis of lysine, methionine, threonine and isoleucine (see Figures 6.13 and 6.15). The first reaction of this anabolic metabolism is catalyzed by aspartate kinase (Figure 12.12). This is an example of an isoenzyme. The activity of an isoenzyme as well as its expression is controlled by its own final product.

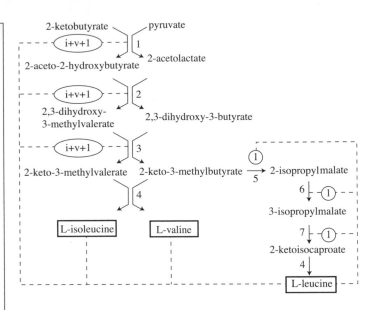

One of the three aspartate kinase isoenzymes is regulated collectively by threonine and isoleucine. Expression of the gene encoding this enzyme is regulated through an attenuation mechanism involving threonine and isoleucine. This regulatory process is referred to as divalent attenuation. Figure 12.12 shows some other examples of divalent attenuation. They include one of two homoserine dehydrogenases, homoserine kinase and threonine synthase. Their gene expression and activity are regulated collectively by threonine and isoleucine (Figure 12.12).

Pyruvate is the precursor for the synthesis of valine and leucine, and 2-ketobutyrate the precursor for isoleucine. The same enzymes catalyze each of the first four reactions synthesizing valine, leucine and isoleucine (see Figure 6.15). Expression of their genes is regulated through an attenuation mechanism mediated collectively by these three amino acids (Figure 12.13). Multivalent attenuation is the term used to describe attenuation systems mediated by more than two effectors.

12.1.6 Termination and antitermination

Since the attenuation process (Section 12.1.4.2) was elucidated in amino acid synthesis by *Escherichia coli*, similar transcription regulatory mechanisms have been discovered in several systems. These include terminator/antiterminator formation aided by small proteins, metabolites, tRNAs and antisense RNAs. These regulatory processes are referred to as 'riboswitch' mechanisms.

12.1.6.1 Termination and antitermination aided by protein

The best characterized example of a small protein involved in this mechanism is TRAP (*trp* RNA binding attenuation protein) in *Bacillus subtilis*. When tryptophan is available at adequate levels to meet

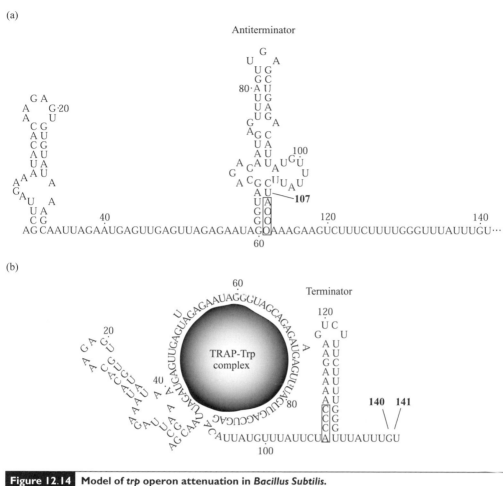

Figure 12.14 Model of *trp* operon attenuation in *Bacillus Subtilis*.

(Curr. Opin. Microbiol. **7**:132–139, 2004)

(a) In tryptophan-limiting conditions, TRAP is not activated and does not bind to the nascent transcript. Eventually, RNA polymerase overcomes the pause and resumes transcription. In this case, formation of the antiterminator prevents formation of the terminator, resulting in transcription readthrough into the *trp* structural genes. (b) In tryptophan-excess conditions, TRAP is activated and binds to the nascent mRNAs soon after they are synthesized. Pausing during the transcription of the leader sequence allows additional time for TRAP to bind. TRAP binding releases paused RNA polymerase and transcription resumes. Bound TRAP prevents formation of the antiterminator, thereby allowing formation of the terminator. Boxed nucleotides represent the overlap between the antiterminator and terminator structures.

cellular requirements, the amino acid binds TRAP to activate the protein. The activated TRAP binds to the leader mRNAs soon after they are synthesized, preventing the formation of antiterminator. RNA polymerase pauses at the pause site allowing additional time for the activated TRAP to bind. TRAP binding releases paused RNA polymerase and transcription resumes to the end of the leader sequence, allowing formation of the terminator. The terminator removes RNA polymerase from the DNA, terminating the transcription before the enzyme reaches the structural genes (Figure 12.14b). On the other hand, under tryptophan-limiting

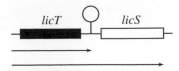

Figure 12.15 The structure of the β-glucanase (*lic*) operon, and its control by a terminator/antiterminator mechanism in *Bacillus subtilis*.

(*Mol. Microbiol.* **23**:413–421, 1997)

The β-glucanase operon consists of the transcription antiterminator gene (*licT*) and the β-glucanase structural gene (*licS*). These are transcribed as a single transcript. A sequence to form a hairpin occupies the middle of the mRNA between these two genes. The hairpin structure is a terminator as in the attenuation mechanism. When the supply of readily utilizable substrate (glucose) is low, the HPr protein of the PTS system is in the phosphorylated form, which phosphorylates the LicT protein (transcription antiterminator) to activate it. The activated LicT binds to the terminator region of the transcript, and *licS* is transcribed. With enough glucose, the LicT protein is inactive and the terminator prevents the transcription of *licS*.

conditions TRAP is not activated without the amino acid and does not bind to the nascent transcript, the leader mRNA. Eventually, RNA polymerase overcomes the pause and resumes transcription. In this case, the antiterminator is formed preventing formation of the terminator, resulting in transcription readthrough into the *trp* structural genes (Figure 12.14a). Since the mRNA region forming the terminator hairpin includes the ribosome-binding Shine–Dalgarno sequence, the binding of activated TRAP sequesters this sequence, blocking translation of the mRNAs that have escaped the termination.

The expression of the β-glucanase (*lic*) operon in *Bacillus subtilis* is another example of a protein-aided termination/antitermination process. In the presence of readily utilizable carbon sources such as glucose, enzymes for the utilization of β-glucan such as lichenan and salicin are not expressed in this bacterium. This regulation is due to the terminator, a hairpin structure formed within the mRNA. The β-glucanase operon consists of *licT* (transcription antiterminator, LicT) and *licS* (β-glucanase) in *Bacillus subtilis* (Figure 12.15). The antiterminator is activated by phosphorylated HPr (see Figure 12.8). When glucose is available, phosphorylated HPr(His-P) participates in glucose transport, and the HPr(His-P) concentration is low. In this case the antiterminator (LicT) is not activated, and the terminator is formed within the β-glucanase operon transcript inducing a premature transcription termination. In the absence of glucose, HPr(His-P) activates LicT, and the activated LicT prevents formation of the terminator. Consequently *licS* (β-glucanase) is transcribed. The antiterminator-binding site within the mRNA is referred to as the ribonucleic antiterminator (RAT).

Termination/antitermination is responsible for the regulation of 6-phospho-β-glucosidase in *Escherichia coli*, and operons for the utilization of sucrose, glycerol and histidine in addition to β-glucan in *Bacillus subtilis*. The RNA sequence of the terminator hairpin structure is encoded in the palindromic sequence of the DNA. This regulatory mechanism is known in genes of catabolism as well as anabolism.

12.1.6.2 Termination and antitermination aided by tRNA

In *Bacillus subtilis*, transcription of aminoacyl-tRNA synthetases and enzymes for the biosynthesis of certain amino acids are regulated by yet another mechanism known as tRNA-dependent transcription antitermination. Uncharged tRNA functions as the effector. The leader region of tyrosyl-tRNA synthetase mRNA contains a codon for tyrosine and a binding sequence for the 3′ acceptor end of uncharged tRNA^{Tyr}. Under tyrosine-excess conditions with a low uncharged tRNA^{Tyr} (effector) concentration, the anticodon of charged tyrosyl-tRNA^{Tyr} binds the tyrosine codon of the leader mRNA, but the acceptor end of the tyrosyl-tRNA^{Tyr} does not participate in mRNA binding, preventing antitermination formation (left side, Figure 12.16a). Under these conditions tyrosyl-tRNA synthetase is not expressed. When tyrosine is limiting, the effector, uncharged tRNA^{Tyr}, binds the leader mRNA through the acceptor end as well as the anticodon,

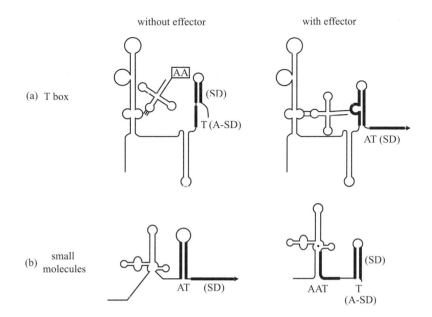

without effector   with effector

(a) T box

(b) small molecules

Figure 12.16   **RNA elements that directly bind effector molecules in tRNA-dependent transcription antitermination (a) and the termination and antitermination mechanism aided by metabolites (b).**

(*Curr. Opin. Microbiol.* **7**:126–131, 2004)

(a) For regulation by tRNA (T-box system), the uncharged tRNA (shown in cloverleaf form) serves as the effector. In the absence of the effector (or in the presence of charged tRNA, indicated by the boxed AA), the leader RNA folds into the terminator form indicated as T (left-hand side). In the presence of the effector (uncharged tRNA), the antiterminator (AT) is stabilized, preventing formation of the terminator hairpin and allowing transcription of the downstream coding regions indicated by the arrow (right-hand side). This arrangement can also be used to regulate gene expression at the level of translation initiation, in which case the 'terminator' helix (A-SD) sequesters the ribosome-binding Shine–Dalgarno sequence (SD) of the downstream gene. (b) For the termination and antitermination mechanism aided by metabolites, the antiterminator structure forms in the absence of effector (left-hand side), allowing expression of the downstream coding regions (arrow); addition of the effector (•, right-hand side) promotes stabilization of an anti-antiterminator element (AAT) that sequesters sequences required for the formation of the antiterminator (AT), permitting formation of the less stable terminator helix (T). As described above, the terminator hairpin can be replaced with a structure that occludes the Shine–Dalgarno sequence (SD) of the downstream gene to permit regulation at the level of translation initiation. Sequences involved in the formation of terminator as well as antiterminator are shown as bold lines.

leading to the formation and stabilization of the antiterminator to prevent transcription termination. In this case the terminator is not formed, and the enzyme is expressed. The uncharged tRNA binding site of the leader mRNA is referred to as the T-box, and the mechanism as the T-box transcription termination control system. As in the TRAP-mediated attenuation mechanism, translation is blocked since the ribosome-binding Shine–Dalgarno sequence participates in formation of the terminator hairpin.

The T-box transcription termination control system is widely used for the control of gene expression in Gram-positive bacteria, but rare in Gram-negative organisms. Genomic data analyses reveal high conservation of primary sequence and structural elements of the system. The T-box system regulates a variety of amino-acid-related genes.

A different mechanism known as the S-box system regulates genes involved in methionine metabolism. While both systems involve gene regulation at the level of premature termination of transcription, the molecular mechanisms employed are very different. In the T-box system, expression is induced by stabilization of an antiterminator structure in the leader by interaction with the uncharged tRNA, while an anti-antiterminator stabilized by an unknown factor is involved in the S-box system during growth under methionine-rich conditions, allowing formation of the terminator.

12.1.6.3 Termination and antitermination aided by metabolites

Genes for riboflavin synthesis are regulated through a similar mechanism as described above in *Bacillus subtilis*. A conserved leader sequence occupies a region upstream of these genes designated as the RFN element. When riboflavin is in excess, the metabolite binds the RFN element forming a terminator to inhibit transcription as well as translation in a similar manner to that depicted in Figure 12.16b.

Various metabolites are involved in direct interactions with the leader mRNA of the transcripts of enzymes, synthesizing them to regulate gene expression and/or translation. These include cobalamin in *Escherichia coli* (B_{12}-box), thiamine in *Rhizobium etli* (THI-box), guanine in *Bacillus subtilis* (G-box) and lysine in *B. subtilis* (L-box). Some of them regulate translation while others regulate transcription depending on the system and the bacterium.

It is worth mentioning that the replication of several plasmids in Gram-positive bacteria is regulated by antisense RNA-mediated transcription attenuation in a similar mechanism. Antisense RNA is discussed later (Section 12.1.9.4).

12.1.7 Two-component systems with sensor-regulator proteins

Microbes have mechanisms that monitor changes in environmental conditions so that they can regulate their metabolism accordingly. A membrane protein senses the changes and transfers the signal into the cytoplasm. The membrane sensor protein is autophosphorylated at a histidine residue exposed to the cytoplasm consuming ATP. The phosphate group is transferred to an aspartate residue of the soluble regulatory protein. The phosphorylated regulatory protein modulates transcription of the related genes. This is referred to as a two-component system. Two-component systems are involved in the regulation of more than one operon, and this is discussed in detail later (Figures 12.23 and 12.24, Section 12.2).

12.1.8 Autogenous regulation

Some proteins are able to control the transcription or translation of their own genes. These include proline dehydrogenase controlling its

(a) When proline is used as substrate

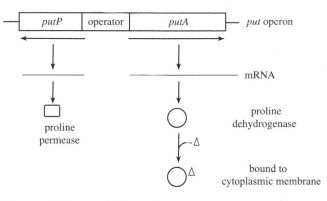

mRNA

proline permease

proline dehydrogenase

bound to cytoplasmic membrane

(b) When proline is not available as substrate

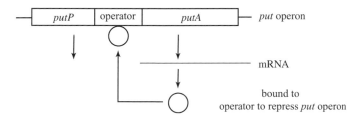

mRNA

bound to operator to repress *put* operon

transcription (Figure 12.17) and ribosomal proteins regulating their own translation.

A ribosome consists of three ribosomal RNAs (rRNA) and up to 55 ribosomal proteins (r-proteins). Elaborate regulations including one at the translational level ensure synthesis of these proteins in the correct proportions. The r-proteins bind rRNAs. When a r-protein is present in excess to bind the rRNAs, this binds its own mRNA to inhibit the initiation of translation. This process is referred to as autogenous regulation. The synthesis of many proteins associated with DNA and RNA is regulated through this regulatory mechanism.

Translation of other proteins is regulated in a similar manner. The S-layer constitutes the cell surface of many prokaryotes, and the S-layer protein is one of the most abundant single proteins (Section 2.3.3). This protein is encoded by a single gene and its mRNA is stable with a half-life of 10–22 minutes producing the protein efficiently. The mRNA has an untranslated region (5′UTR) of 30–358 bases that forms hairpin structures to protect the coding region from ribonuclease (Section 12.1.9.1). When excess S-layer protein is produced, this protein binds the 5′ UTR of its own mRNA inhibiting translation. These are examples of negative regulation of translation by a protein. Apo-aconitase binds 3′ UTR of its own mRNA to stabilize it in *Escherichia coli*. This is positive regulation.

When enteric bacteria, including *Escherichia coli*, use proline as a sole carbon source, the transcription of the *put* operon is regulated through autogenous regulation. Proline dehydrogenase and proline permease comprise the *put* operon. When proline is available as an energy source, proline dehydrogenase binds to the cytoplasmic membrane catalyzing a two-step oxidation of proline to glutamate (Figure 7.16, Section 7.5.6). When the substrate concentration is low, proline dehydrogenase binds the operator region of the *put* operon inhibiting the transcription (Figure 12.17). This is another example of autogenous regulation. When the enzyme concentration is higher than that of the binding site on the membrane, the free enzyme binds DNA inhibiting transcription initiation even when proline is available as an energy source.

12.1.9 Post-transcriptional regulation of gene expression

Gene expression is regulated not only at the transcriptional level as discussed above but also at the post-transcriptional level, and these mechanisms are integrated and coordinated to implement the information in the genome according to cellular needs. Post-transcriptional levels of regulation such as transcript turnover and translational control are an integral part of gene expression and approach the complexity and importance of transcriptional control. Post-transcriptional control is mediated by various combinations of RNA-binding proteins (RBPs) and small non-coding RNAs (sRNAs) that determine the fate of the tagged transcripts. Without regulation at the post-transcriptional level, regulation at the transcriptional level would be meaningless. Autogenous regulation of ribosomal protein synthesis (discussed above) is an example of post-transcriptional regulation.

12.1.9.1 RNA stability

Generally the half-life of a bacterial mRNA is in the range of several minutes and much shorter than those of rRNA and tRNA. The latter are referred to as stable RNAs. The stability of mRNA is determined by its intrinsic structure as well as by the proteins and RNAs that bind it, and related to its translation efficiency. The regulation at the transcription process becomes more meaningful with rapid mRNA turnover. *Escherichia coli* has both endo- and exo-type RNases. Endonuclease RNase E is the main enzyme degrading mRNA. This enzyme recognizes and hydrolyzes adenine- and uracil-rich sequences in the RNA.

A photosynthetic bacterium, *Rhodobacter capsulatus*, produces antenna proteins and reaction centre proteins from the *puf* operon. The photosynthetic apparatus consists of a reaction centre with about 15 antennae in this bacterium. A single transcript is produced from the *puf* operon, and 15-fold more antenna proteins are produced than the reaction centre proteins from this single transcript. This is possible because the mRNA for the antenna proteins is much more stable than that of the reaction centre proteins (Figure 12.18).

The transcript has an untranslated region of about 500 bases (5′-untranslated region, 5′ UTR) that forms three hairpin structures,

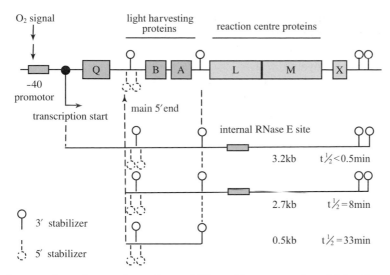

Figure 12.18 The stability of mRNA of *puf* genes in photosynthetic *Rhodobacter capsulatus*.

(*FEMS Microbiol. Rev.* **23**:353–370, 1999)

The photosynthetic structure contains about 15 antenna molecules per reaction centre. Proteins of the antenna molecule and the reaction centre are encoded by a single *puf* operon that is transcribed into one transcript. Proteins of the antenna molecules are produced 15 times greater than the reaction centre proteins. The difference in the efficiency of translation is due to differences in the stability of their mRNAs. The hairpin in the transcript prevents RNA degradation by endonuclease RNase E. The region with the base sequence for the endonuclease RNase E reaction is more prone to degradation.

and two sites for endonuclease RNase E, one each within the 5'-UTR and reaction centre proteins encoding region. Additionally, sequences between the coding regions and the 3'-end of this transcript can form hairpin structures. These structural features determine the stability of different translational units within the transcript.

The entire transcription has a half-life of about 30 seconds, but this increases to 8 minutes after RNase E cleaves a part of 5' UTR. When the transcript is processed to the translational unit for only the antenna proteins, the half-life is longer than 30 minutes (Figure 12.18).

12.1.9.2 mRNA structure and translational efficiency

When the growth temperature is shifted from 30 °C to 42 °C, *Escherichia coli* produces many proteins known as heat shock proteins (HSPs, Figure 12.26, Section 12.2.6) including sigma H (σ^H). Most of the HSPs are expressed through activation of transcription, and the expression of σ^H is regulated at the translational level. σ^H mRNA forms hairpin structures at physiological temperatures preventing translation. At an elevated temperature, the structure melts, allowing ribosomes to bind the Shine–Dalgarno sequence to initiate translation (Section 12.2.6).

12.1.9.3 Modulation of translation and stability of mRNA by protein

Many proteins are known to modulate the stability of mRNAs and their translation efficiency. Examples of these proteins include CsrA (carbon storage regulator) protein in *Escherichia coli* and RsmA (repressor of stationary phase metabolites) protein in *Erwinia carotovora*.

During the transition from exponential growth into stationary phase, non-sporulating bacteria, including *Escherichia coli*, readjust their physiological status from one that allows robust growth and metabolism to one that provides for greater stress resistance and an enhanced ability to scavenge remaining substrate from the medium. These phenotypic changes result from various global regulatory

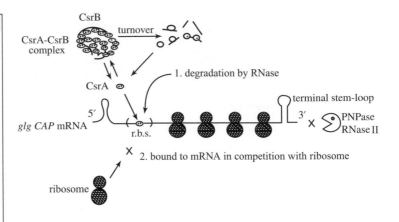

Figure 12.19 **The CsrA/CsrB system controls the translation and stability of mRNA encoding enzymes of glycogen metabolism.**

(*Mol. Microbiol.* **29**:1321–1330, 1998)

CrsA (carbon storage regulator) protein binds the ribosome binding site of mRNAs encoding ADP-glucose synthetase (GlgC), glycogen synthase (GlgA) and glycogen phosphorylase (GlgP), inhibiting their translation. CraA-bound mRNAs are prone to degradation by RNase. A small RNA (sRNA), CsrB binds CsrA protein in the ratio of 1 to 18 to inhibit its function. Another sRNA, CsrC, has a similar function.

mechanisms including σ-factor (Section 12.1.1), stringent response (Section 12.2.1), and the Csr system. Glycogen biosynthesis in *Escherichia coli* occurs primarily during the transition from the exponential to the stationary phase of growth when a nutrient other than carbon is limiting. The CsrA protein represses the translation of mRNAs for glycogen synthesis and for gluconeogenesis and activates genes for glycolysis and utilization of acetate.

Glycogen biosynthesis is catalyzed by ADP-glucose pyrophosphorylase (*glgC*), glycogen synthase (*glgA*) and glycogen phosphorylase (*glgP*) encoded in the *glgCAP* operon. Transcription of the *glgCAP* operon is activated by the cAMP–CRP complex (Section 12.1.3.1) and through the stringent response mediated by ppGpp (Section 12.2.1). CsrA protein modulates the translation and stability of the *glgCAP* transcript by binding to the ribosome-binding site (Shine–Dalgarno sequence, Section 6.12.2.2) of the target mRNAs, including the *glgCAP* transcript, inhibiting translation and promoting rapid degradation of the transcript by RNase II (Figure 12.19). A small non-coding RNA (sRNA), CsrB, antagonizes CsrA activity by sequestering this protein. Approximately 18 molecules of CsrA bind a molecule of CsrB. Another sRNA, CsrC, has a similar function. Over 50 sRNAs are known in *Escherichia coli* and are discussed later (Section 12.1.9.4).

The Csr system represses genes for gluconeogenesis and glycogen biosynthesis, and activates those for glycolysis, cell motility and acetate metabolism. In addition, CsrA represses extracellular polysaccharide production that promotes biofilm formation. A similar protein, RsmA, in *Erwinia carotovora* destabilizes mRNA for pectinase and represses its translation. This enzyme is a virulence factor in plants, and its transcription is regulated by quorum sensing (Section 12.2.8). An sRNA, AepH (exoenzyme regulatory protein) antagonizes RsmA activity.

Escherichia coli produces various cold shock proteins (CSPs) including CspA at low temperature. CspA is another example of a protein modulating translation. This protein prevents mRNAs from forming hairpin structures at low temperature facilitating translation by ribosomes. This will be discussed later (Section 12.2.7).

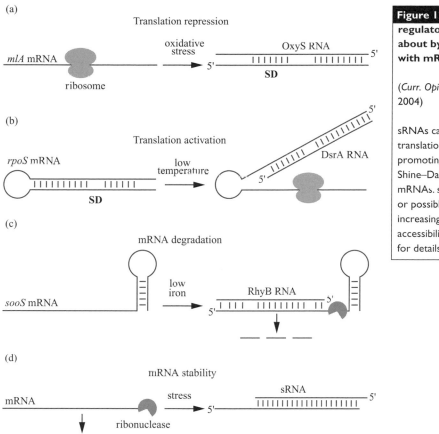

Figure 12.20 Different regulatory outcomes brought about by sRNA base pairing with mRNAs.

(*Curr. Opin. Microbiol.* **7**:140–144, 2004)

sRNAs can repress or activate translation by blocking or promoting ribosome binding to the Shine–Dalgarno sequence (SD) of mRNAs. sRNAs can also destabilize or possibly stabilize mRNAs by increasing or decreasing accessibility to RNase E (see text for details).

12.1.9.4 Modulation of translation and stability of mRNA by small RNA and small RNA–protein complex: riboregulation

More than 50 small non-coding RNAs have been identified in *Escherichia coli*. Their sizes range between 40 and 400 nucleotides. They are referred to as small RNAs (sRNAs) or non-coding RNAs (ncRNAs). They have been found in archaea, bacteria and eukaryota. Their functions are not fully understood yet, but some of them modulate translation and stability of mRNA. Their activities comprise three general mechanisms. They can (1) act as a member of RNA–protein complexes such as 4.5S RNA of the signal recognition particle (SRP, Section 3.8.1.1), (2) mimic the structure of other nucleic acids such as CsrB (Figure 12.19) and (3) base pair with other RNAs. Many of the regulatory sRNAs, such as the DsrA, MicF, OxyS, Spot42 and RyhB RNAs, act by base pairing to activate or repress translation, or to destabilize mRNAs (Figure 12.20).

When an *Escherichia coli* culture is shifted from 37 °C to 25 °C an sRNA, DrsA, concentration increases more than 25 times. This sRNA forms a base pair with the 5′ UTR region of *rpoS* mRNA, which encodes the stationary phase sigma factor σ^S, and leads to its

Table 12.3. | *Base pairing sRNAs of known function in* Escherichia coli

sRNA	Numbers of nucleotides	Target transcript(s)	Effects(s)
MicC	109	OmpC	translation repression
DicF	53	FtsZ	translation repression
RprA	105	RpoS	translation activation
DsrA	85	RpoS	translation activation
		Hns, RbsD	translation repression
MicF	93	OmpF	translation repression
GcvB	204	OppA, DppA	translation repression
RyhB	90	SodB	translation repression
			mRNA degradation
		SdhCDAB	mRNA degradation
Spot42	109	GalETKM	translation repression
OxyS	109	FhlA, RpoS	translation repression

OmpC, outer membrane protein; FtsZ, GTPase involved in cell division; RpoS, σ^S; Hns, histone-like DNA-binding protein; RbsD, D-ribose high-affinity transport system; OmpF, outer membrane protein F; OppA, oligopeptide transport, periplasmic binding protein; DppA, dipeptide binding protein; SodB, iron superoxide dismutase; SdhCDAB, succinate dehydrogenase; GalETKM, galactose operon; FhlA, transcriptional activator for formate hydrogen-lyase operon.
Source: Current Opinion in Microbiology, **7**,140–144, 2004.

increased translation. The DsrA sRNA promotes translation by preventing the formation of an inhibitory secondary structure that normally occludes the ribosome-binding site of the *rpoS* transcript. This sRNA down-regulates the translation of the DNA-binding protein H-NS, promoting degradation of its mRNA. At low temperature, *Escherichia coli* produces many proteins. They are referred to as cold-shock proteins (CSP), and are discussed later (Section 12.2.7).

Similar sRNAs are known in *Escherichia coli*, including MicF and OxyS, all induced under stress conditions (Table 12.3). Expression of the OxyS sRNA is strongly induced under oxidative stress imposed by hydrogen peroxide and other substances, and forms a base pair with the *fhlA* mRNA and represses its translation. FhlA is the transcriptional activator for the formate hydrogen-lyase operon that is an anaerobic enzyme (Section 8.6.4). The MicF sRNA blocks translation of the OmpF porin by base pairing with the *ompF* mRNA under high osmotic pressure. Another sRNA, MicC, regulates the other major porin, OmpC, under complementary conditions. These sRNAs presumably help the cell respond to environmental conditions beyond those that are sensed by the phosphorelay system that regulates both the *ompF* and *ompC* genes (Section 12.2.9). Changing the ratio of OmpF to OmpC in the cell envelope modulates the entry of small molecules into the cell (Section 2.3.3).

Spot42 is an sRNA with interesting function. The galactose operon, *galETKM* encodes UDP-galactose epimerase, galactose-1-phosphate uridylyltransferase, galactokinase and aldose-1-epimerase.

When *Escherichia coli* uses galactose as the substrate, this bacterium needs these enzyme activities. UDP-galactose epimerase (GalE) has a second role in the synthesis of UDP-galactose, a building block for the cell wall and capsule. When cells are growing on glucose, GalE is needed but not the others, including galactokinase (GalK). The expression of Spot42 is under the negative regulation of the CRP–cAMP complex, leading to higher levels of sRNA synthesis when cells are growing on glucose and lower levels when cells are growing on galactose. Spot42 pairs with, and negatively regulates, translation of GalK without perturbing GalE translation. This sRNA is unique in its role in regulating polarity within an operon.

The majority of sRNAs in *Escherichia coli* require a protein named Hfq for their function. Hfq was identified as an *Escherichia coli* host factor (also known as HF-I) required for initiation of plus-strand synthesis by the replicase of the Q RNA bacteriophage. This protein functions as an RNA chaperone preventing formation of secondary structures in sRNAs for efficient base pairing with the target mRNAs. Hfq is widely distributed in bacteria, but not found in some, including cyanobacteria, actinomycetes and species of *Chlamydia* and *Deinococcus*.

The functions of sRNAs requiring Hfq protein differ depending on their characteristics and on the binding site on the target mRNA as other sRNAs. Some repress translation by base pairing with the ribosome-binding Shine–Dalgarno (SD) sequence (Figure 12.20a), and the others activate by preventing secondary structures involving the SD sequence (Figure 12.20b). Some mRNAs become susceptible to RNase E (Figure 12.20c) and the others are stabilized (Figure 12.20d) through base pairing with sRNAs.

While most regulatory sRNAs participate in post-transcriptional regulation, 6S RNA acts on the transcription process. This sRNA is made in increased amounts when an *Escherichia coli* culture enters stationary phase to inhibit RNA polymerase containing the housekeeping σ^D (σ^{70}), but to activate that containing σ^S (σ^{38}).

The sRNAs discussed above are *trans*-acting, regulating transcripts from other operons. They are global regulators. A few of the regulatory sRNAs functioning by base pairing are encoded on the opposite strand of the DNA from which the target transcript is encoded. They are *cis*-acting sRNAs. An example is glutamine synthetase (GlnA) in *Clostridium acetobutylicum*. When the ammonia supply is limited, *glnA* is transcribed and an sRNA regulates translation (Figure 12.21). Classically, the *cis*-acting sRNA was referred to as 'antisense RNA', but this term is used as a synonym of sRNA.

12.2 | Global regulation: responses to environmental stress

In the previous section, various regulatory mechanisms of gene expression were discussed. Though some processes regulate a single operon

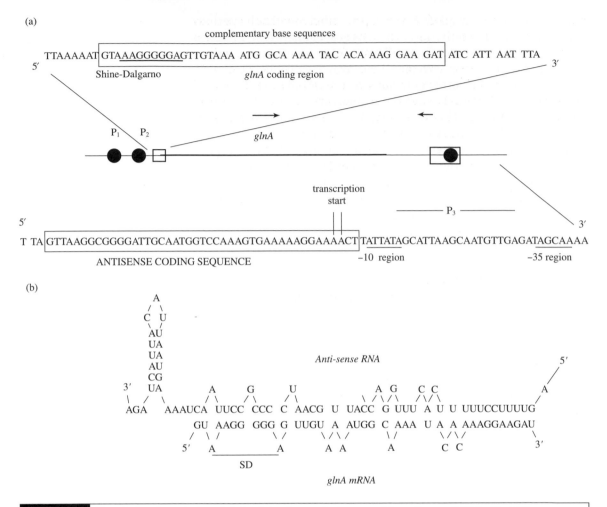

(a)

complementary base sequences

5′ TTAAAAAT GTAAAGGGGGAGTTGTAAA ATG GCA AAA TAC ACA AAG GAA GAT ATC ATT AAT TTA 3′

Shine-Dalgarno   *glnA* coding region

P₁   P₂

glnA

transcription start

P₃

5′   3′

T TA GTTAAGGCGGGGATTGCAATGGTCCAAAGTGAAAAAGGAAAACT TATTATAGCATTAAGCAATGTTGAGATAGCAAAA

ANTISENSE CODING SEQUENCE   −10 region   −35 region

(b)

Anti-sense RNA

glnA mRNA

Figure 12.21 **Translational regulation of *glnA* (glutamine synthetase gene) by antisense RNA in *Clostridium acetobutylicum*.**

(*J. Bacteriol.* **174**:7642–7647, 1992)

Some small RNAs are *cis*-acting regulators. They are encoded on the opposite strand of the DNA from which the target transcript is encoded. In this fermentative bacterium, *glaA* (glutamine synthetase gene) is transcribed under ammonia-limited conditions and the translation is regulated by an sRNA. Classically this sRNA was referred to as antisense RNA. Under conditions that repress *glaA* expression, the number of antisense RNA exceeds the *glaA* transcript by 1.6-fold, while fivefold more *glaA* transcript than the antisense RNA occurs under the opposite conditions.

---

or regulon, most of them regulate more than one operon or regulon. For example, the cAMP–CRP complex regulates enzymes of more than 200 metabolic processes. These regulatory processes are referred to as global regulation, a multigene system or pleiotropic control.

Genes modulated by global regulation are encoded in operons and regulons scattered throughout the chromosome, the expression of which is regulated by a common effector. They are involved in many metabolic processes. In this sense they are different from a regulon that is a collection of genes involved in a single metabolic process such as nitrogen fixation (*nif*) or arabinose catabolism (*ara*). The term modulon is used to define a collection of operons and regulons regulated by a common effector. Those regulated by the cAMP–CRP complex are referred as *crp* modulons. As the *lac* operon is regulated by the cAMP–CRP complex as well as by the *lac* repressor, each operon and regulon of a modulon is regulated by the common activator or repressor and by effectors specific for each of them.

Stresses such as heat shock and oxidative stress result in regulation of many operons and regulons. It is not known if this regulation is modulated by a common effector. In this case, the term stimulon is used to define a collection of modulons, regulons and operons regulated by a common environmental stimulation. Various global regulations are known in prokaryotic organisms, especially in bacteria (Table 12.4).

12.2.1 Stringent response

At the beginning of the stationary phase, most bacteria, including *Escherichia coli*, synthesize at least 50 new proteins, change their cell shape and size, and exhibit increased resistance against stresses such as oxidative stress, near-UV irradiation, potentially lethal heat shocks, hyperosmolarity, acidic pH, ethanol, and probably others yet to be identified. These phenotypic changes are a part of their survival strategy (Section 13.3). Complex regulatory networks known as the stringent response are involved in these phenotypic changes. Under nutrient-rich conditions, *Escherichia coli* grows at a high growth rate synthesizing proteins at high speed. When a culture is transferred from nutrient-rich conditions to nutrient-poor conditions, the growth rate is reduced with less protein synthesis. Two different mechanisms are known in the stringent response. They are the ribosomal pathway caused by a limitation in the amino acid supply and the carbon starvation pathway under carbon and energy source limited conditions. In both pathways, nutritional stresses result in the production of guanosine-3′-diphosphate-5′-triphosphate (guanosine pentaphosphate, pppGpp) or guanosine-3′-diphosphate-5′-diphosphate (guanosine tetraphosphate, ppGpp).

When the bacterial cell faces amino acid limitation where the amino acid supply cannot meet the demand for peptide synthesis by the ribosomes, uncharged tRNA enters the peptide-synthesizing ribosome. This activates the ribosome-associated (p)ppGpp synthetase I to produce (p)ppGpp transferring PP_i from ATP to GTP (GDP). This ribosome-associated enzyme is referred to as the RelA protein. RelA is derived from the fact that a mutant of this protein maintains high-rate peptide synthesis (relaxed mutant) even under amino-acid-limited conditions. This RelA-mediated stringent response is referred as the ribosomal pathway.

In the carbon starvation pathway, (p)ppGpp is produced by a RelA independent mechanism under carbon and energy source limited conditions by ppGpp synthetase II. Yet another enzyme, SpoT,

Table 12.4. | *Global regulation in bacteria*

Environment	Regulation	Bacteria	Control gene	Controlled gene(s)	Mechanism
Substrate availability					
Carbon limitation	catabolite repression	enteric bacteria	crp (activator)	lactose, maltose catabolism	expression by cAMP–CRP complex (see Figure 12.6)
Amino acid and energy limitation	stringent response	enteric bacteria	relA, spoT katF (rpoS)	enzymes in stationary phase	expression of σ^S by ppGpp
Substrate limitation	spore	Bacillus spp.	spoOA (activator) spoOF (modulator)	spore related σ and enzymes	alternative σ-factors
Nutrient gradient	chemotaxis	enteric bacteria	rpoF (F)	flagellar and other motility-related proteins	alternative σ-factor
Ammonia limitation	Ntr system	enteric bacteria	glnB, D, G, L	glutamine synthetase	activation by phosphorylated protein (see Figure 12.22)
Ammonia limitation	Nif system	nitrogen fixers	ntrA	nif regulon	alternative σ-factor (see Figure 6.9)
Phosphate limitation	Pho system	enteric bacteria	phoB, R, U	alkaline phosphatase	two-component system (see Figure 12.23)
Electron transport chain					
Aerobic	Arc system	E. coli	arcA, B	aerobic enzymes	two-component system (see Figure 12.24)
Anaerobic	anaerobic metabolism	E. coli	fnr	anaerobic enzymes	activation by reduced FNR (see Figure 12.25)
Deteriorating environment					
UV	SOS response	E. coli	lexA, recA	DNA repair system	repressed by LexA
Heat	heat shock	E. coli	rpoH	molecular chaperone	alternative σ-factors
Low temperature	cold shock	E. coli	CSP	many proteins	RNA chaperone (see Figure 12.29)
Cell density					
High cell density	quorum sensing	many bacteria	luxI, luxR	luxC, D, A, B, E	activated by LuxR–autoinducer complex (see Figure 12.30)

either synthesizes or hydrolyzes ppGpp to maintain a proper concentration of this regulatory nucleotide according to environmental conditions.

When the ppGpp concentration increases, RpoS (σ^S) synthesis is activated, and the synthesis of stable tRNA and rRNA and the activity of housekeeping σ^D are inhibited. ppGpp activates the expression of σ^S at the level of translation not transcription. The exact mechanism of this translational activation is not fully understood, but directly and indirectly involves various proteins and sRNAs. σ^S activates the transcription of genes of stationary phase proteins that facilitate the phenotypic changes. (p)ppGpp is not known in those archaea that have been examined.

σ^S and σ^S-dependent genes are induced not only in stationary phase but also under various stress conditions. Therefore, σ^S is now seen as the master regulator of the general stress response that is triggered by many different stress signals. Its production is often accompanied by a reduction or cessation of growth, and provides the cells with the ability to survive against the actual stress as well as additional stresses not yet encountered (cross-protection). It should be mentioned that an individual stress signal activates a specific stress response which results in the induction of proteins that allow cells to cope with this specific stress situation only. While specific stress responses tend to eliminate the stress agent and/or to mediate repair of cellular damage that has already occurred, the general stress response renders cells broadly stress resistant in such a way that damage is avoided rather than needing to be repaired.

σ^S-controlled gene products generate changes in the cell envelope and overall morphology. Stressed *Escherichia coli* cells tend to become smaller and ovoid. Metabolism is also affected by σ^S-controlled genes, consistent with σ^S being important under conditions where cells switch from a metabolism directed toward maximal growth to a maintenance metabolism. σ^S also controls genes mediating programmed cell death in stationary phase, which may increase the chances for survival for a bacterial population under extreme stress by sacrificing a fraction of the population in order to provide nutrients for the remaining surviving cells. A number of virulence genes in pathogenic enteric bacteria have been found to be under σ^S control, consistent with the notion that host organisms provide stressful environments for invading pathogens.

Gram-positive bacteria, including *Bacillus subtilis*, have σ^B with the function of σ^S but the amino acid sequences are different. In spore-forming bacteria, specific σ-factors are expressed in sequence to produce proteins for the transformation of vegetative cells to spores under nutrient-limited conditions (Section 13.3.1).

12.2.2 Response to ammonia limitation

Nitrogen is essential as an element of amino acids and nucleic acid bases. Ammonia is assimilated into organic nitrogenous compounds

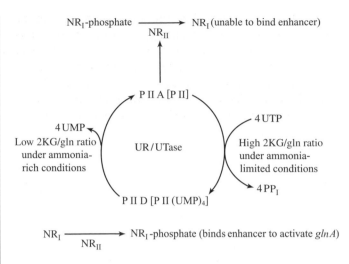

Figure 12.22 **Signal transduction cascade of ammonia availability regulating the transcription of *glnA* (glutamine synthetase gene).**

2-ketoglutarate/glutamine (2-KG/gln) ratio is dependent on the availability of ammonia. An enzyme, uridylylase/deuridylylase, of P II protein (GlnB) uridylylates P II protein in the P II A–NR$_{II}$ complex form to the P II D form, liberating NR$_{II}$ under high 2-KG/gln ratio conditions with limited ammonia supply. The free form of NR$_{II}$ phosphorylates NR$_I$ protein (GlnG, NtrC). The phosphorylated NR$_I$ binds the enhancer region of glutamine synthetase gene (*glnA*) activating transcription. P II A under ammonia-excess conditions adenylylates glutamine synthetase to a less active form, and P II D under the reverse conditions removes the adenylyl group from the enzyme to keep the activity high.

by the action of glutamate dehydrogenase and glutamine synthetase. Since the affinity for ammonia is low, glutamate dehydrogenase has low activity under low ammonia concentrations. Under these conditions, ammonia is assimilated by glutamine synthetase that has a higher affinity for ammonia than glutamate dehydrogenase (Section 6.2.3). Ammonia inactivates the expression of the *nif* regulon in nitrogen-fixing organisms (Section 6.2.1). These regulations are mediated by an alternative σ-factor affecting many regulons. In addition, individual regulons are under complex and separate regulatory mechanisms.

Escherichia coli transcribes the glutamine synthetase gene (*glnA*) at a low level to provide glutamine as the amine donor under ammonia-rich conditions recognized by the housekeeping σ^D (σ^{70}), but the same gene is transcribed with the aid of σ^N (σ^{54}) under ammonia-limited conditions. Transcription of *glnA* aided by σ^N requires the binding of phosphorylated NR$_I$ protein (NtrC). GlnG and NtrC are the same protein. NtrC is used in Chapter 6 (p. 135). On the enhancer region, upstream of the gene. NR$_I$ protein is phosphorylated through a complex cascade of reactions to transduce the signal of low ammonia supply to control transcription of the *glnA* gene (Figure 12.22).

The availability of ammonia determines the 2-ketoglutarate/glutamine (2KG/Gln) ratio. When the ammonia supply is limited, the 2KG/Gln ratio becomes high. This 2KG/Gln ratio regulates the activity of an enzyme uridylylase/deuridylylase of the P II protein (GlnB). When the ammonia supply is limited, this enzyme uridylylates P II protein transforming the P II A form to the P II D form [P II(UMP)$_4$] consuming 4 UTP. P II A forms a complex with another protein, NR$_{II}$, when the ammonia concentration is high, and the NR$_{II}$ protein is liberated when P II A is uridylylated to P II D with a limited ammonia supply. The P II A–NR$_{II}$ complex has a phosphatase activity on phosphorylated NR$_I$, while free NR$_{II}$ has kinase activity on NR$_I$. As a consequence of these reactions, NR$_I$ protein is phosphorylated to activate the transcription of the *glnA* gene binding the enhancer region when the ammonia supply is limited.

P II A/P II D regulates not only the transcription of glutamine synthetase but also the activity of this enzyme. P II A under ammonia-rich conditions adenylylates glutamine synthetase to a less active form, and P II D under the reverse conditions removes the adenylyl group from the enzyme to keep the activity high (Figure 12.36, Section 12.3.2). P II or similar proteins are identified in bacteria, archaea and eukaryota. They are believed to function as transducers of a variety of signals.

The transcription of the *nif* regulon is regulated in a similar manner. Under ammonia-limited conditions, σ^{54} (σ^N, NtrA) initiates the transcription of this regulon with the binding of an enhancer-binding protein, NR_1 (NtrC) to the enhancer region as discussed earlier (Section 6.2.1.5). Nitrogenase activity is also regulated by the availability of ammonia.

In addition to *gln* and *nif* operons, σ^N participates in the regulation of many other operons and regulons including amino acid transporters such as arginine (*argT*), histidine (*hisJQMP*) and glutamine (*glnHPQ*), amino acid metabolism such as arginine (*astCADBE*), carbon metabolism such as acetoacetate (*atoDAEB*) and propionate (*prpBCDE*), and hydrogenase (*hycABCDEFGHI*, *hydN-hycF* and *hypABCDE-fhlA*). Each of these has its own regulator.

The low G+C Gram positive bacterium *Bacillus subtilis* does not have an anabolic glutamate dehydrogenase and glutamine synthetase (GS) is the only enzyme used for ammonium assimilation in this bacterium. The expression of this enzyme (*gltAB* operon) is regulated by GlnR (transcriptional regulator) and TnrA (transcriptional regulator). When organic nitrogen compounds are in excess, GlnR and TnrA repress the expression of the *gltAB* operon. When the organic nitrogen supply is limited, TnrA activates various genes for nitrogen metabolism including asparagine degradation, urease, nitrate and nitrite assimilation, and ammonia uptake.

12.2.3 Response to phosphate limitation: the *pho* system

Phosphate is abundant in nature, but it is not always available to microbes due to its low solubility. Under phosphate-limited conditions, enteric bacteria synthesize around 100 new proteins including alkaline phosphatase (PhoA). The expression of genes for these proteins is regulated through the *pho* system. The *pho* system is a two-component system involving the sensor proteins, PhoU and PhoR, and a modulator protein, PhoB. Since a signal is transferred from a sensor protein to a modulator protein in a two-component system, this regulation can also be referred to as a signal-transducing system. In a two-component system, the histidine residue of the membrane-bound sensor protein is phosphorylated before the phosphate group is transferred to the aspartate residue of the soluble modulator protein.

When the phosphate supply is limited, the inorganic phosphate specific transporter (PST) senses it and passes this information

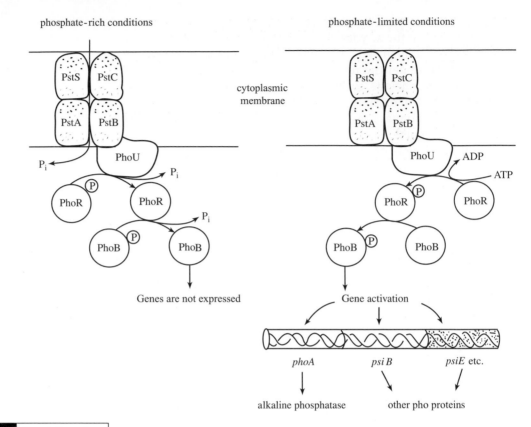

phosphate-rich conditions

phosphate-limited conditions

Figure 12.23 Regulation of the *pho* system under phosphate-limited conditions.

The inorganic phosphate specific transporter (PST) senses a limited phosphate supply and passes this information to PhoR through PhoU. PhoR is phosphorylated consuming ATP and transfers phosphate to a soluble regulator protein, PhoB. The phosphorylated PhoB activates the transcription of many operons. When phosphate is available, PhoB is dephosphorylated to an inactive form. This is an example of a two-component system. PhoU and PhoR are the sensor proteins and PhoB is the modulator protein.

to PhoR through PhoU. PhoR is phosphorylated consuming ATP. Phosphorylated PhoR transfers phosphate to PhoB. Phosphorylated PhoB is an activator (Figure 12.23). PST consists of PstS, PstC, PstA and PstB. PhoU and PhoR are membrane proteins, and PhoB is a soluble protein.

12.2.4 Regulation by molecular oxygen in facultative anaerobes

Facultative anaerobes including *Escherichia coli* metabolize glucose through the central metabolic pathway to generate ATP, NADPH and the carbon skeletons required for biosynthesis. Under anaerobic conditions, they operate an incomplete TCA fork (Section 5.4.1) instead of the TCA cycle. This metabolic shift means that activities of TCA cycle enzymes should be regulated (Table 12.5). In addition to the TCA cycle enzymes, transcription of other genes is regulated when the culture becomes anaerobic. Pyruvate is oxidized by pyruvate dehydrogenase under aerobic conditions, while pyruvate:formate lyase catalyzes a similar reaction under anaerobic conditions (Section 8.6.4). Genes for other anaerobic enzymes are activated under anaerobic conditions including fumarate reductase, nitrate reductase and hydrogenase in facultative anaerobes. The products of *arc* (aerobic respiration control) and *fnr* (fumarate nitrate reductase) mediate this regulation.

Table 12.5. *Activities of TCA cycle enzymes in* Escherichia coli *growing under aerobic and anaerobic conditions*

Enzyme	Activity (U/mg protein)	
	Aerobic growth	Anaerobic growth
Citrate synthase	51.5	10.5
Aconitase	317	16.1
Isocitrate dehydrogenase	1416	138
2-ketoglutarate dehydrogenase	17.4	0

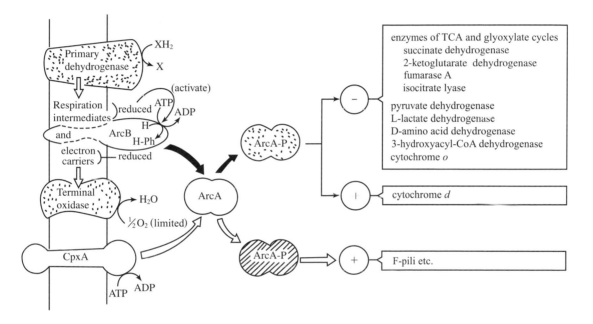

Figure 12.24 **Regulation by the *arc* system.**

Under oxygen-limited conditions, electron transport chain components are reduced, and with this signal a membrane sensor protein, ArcB, is autophosphorylated consuming ATP. Phosphate is transferred from the histidine residue of ArcB to the aspartate residue of the modulator protein, ArcA. The active ArcA-P modulates the expression of many operons including those for aerobic respiration and F-pili.

12.2.4.1 *arc* system

When the oxygen supply is limited, electrons cannot be transferred to oxygen and coenzyme Q of the electron transport system is predominantly in the reduced form. This signal is sensed by a membrane protein, ArcB, that is autophosphorylated consuming ATP. This phosphate is transferred to a soluble protein, ArcA. The phosphorylated ArcA inhibits the transcription of enzymes of aerobic respiration and activates the transcription of the cytochrome *d* gene. This has a higher affinity for oxygen than cytochrome *o* (Figure 12.24). The ArcB/ArcA system modulates the expression of at least 30 regulons including cytochrome oxidase, enzymes of the TCA and glyoxylate

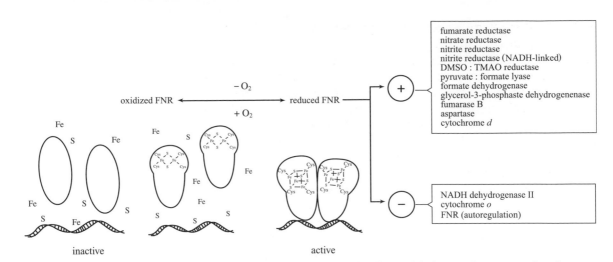

Figure 12.25 Modulation of gene expression by the FNR protein in *Escherichia coli*.

(*Ann. Rev. Microbiol.* **53**:495–523, 1999)

FNR is produced independent of oxygen availability. FNR with a [4Fe-4S] structure forms a homodimer that binds DNA and modulates the expression of many genes under anaerobic conditions. Under aerobic conditions [4Fe-4S] decomposes to [2Fe-2S] and decomposes further with extended exposure to molecular oxygen. FNR with the [2Fe-2S] structure does not bind DNA.

cycles, and fatty acid metabolism. This is another example of a two-component system. The ArcB/ArcA system also activates the gene for the F-pilus.

12.2.4.2 *fnr* system

A *fnr* (fumarate and nitrate reductase) mutant of *Escherichia coli* cannot use nitrate or fumarate as electron acceptors. The product of the *fnr* gene (FNR) modulates the expression of many operons. FNR is produced independent of oxygen availability. Under anaerobic conditions, the FNR protein with a [4Fe-4S] structure forms a homodimer that can bind DNA, but under aerobic conditions the [4Fe-4S] structure decomposes to a [2Fe-2S] structure that cannot form the dimer and does not bind DNA. When FNR with the [2Fe-2S] structure is exposed to oxygen, [2Fe-2S] is further decomposed. These decomposition processes are reversible (Figure 12.25).

Under anaerobic conditions, nitrate activates genes for enzymes needed for denitrification, and inhibits the expression of genes necessary to utilize fumarate, dimethylsulfoxide (DMSO) and trimethylamine-N-oxide (TMAO) as electron acceptors. Through these regulations, enteric bacteria can use electron acceptors of a higher redox potential preferentially over those of lower potential to conserve maximum energy under given conditions. For this regulation, each operon and regulon of a modulon is under the control of separate effector(s). Enteric bacteria, including *Escherichia coli*, can grow fermentatively under anaerobic conditions without electron acceptors (Section 8.6).

The availability of nitrate is sensed and this information is used to modulate nitrate metabolism in a two-component system similar to the *arc* system. When nitrate is available as an electron acceptor, a membrane sensor protein, NarX, is autophosphorylated and transfers the phosphate group to a soluble modulator protein, NarL. Phosphorylated NarL activates genes for denitrification and inhibits genes for enzymes of alternative electron acceptors such as fumarate and DMSO.

12.2.4.3 RegB/RegA system in purple non-sulfur photosynthetic bacteria

The purple non-sulfur bacteria grow photosynthetically under anaerobic conditions, and many of them grow chemoorganotrophically under aerobic conditions (Section 11.1.2). Their photosynthetic apparatus is synthesized only under anaerobic conditions. A two-component RgeB/RegA system modulates this regulation in *Rhodobacter capsulatus* and *Rhodobacter sphaeroides*. This system regulates not only photosynthesis but also numerous energy-generating and energy-utilizing processes such as carbon fixation, nitrogen fixation, hydrogen utilization, aerobic and anaerobic respiration, denitrification, electron transport, and aerotaxis. These are referred to as the RegB/RegA operons. As in the *arc* system, the membrane-bound sensor, RegB, senses the redox status, and is autophosphorylated at a histidine residue under oxygen-limited conditions. Phosphorylated RegB transfers phosphate to response regulator RegA at an aspartate residue. Both phosphorylated and unphosphorylated forms of the response regulator RegA are capable of activating or repressing a variety of genes in the regulon. Highly conserved homologues of RegB and RegA have been found in a wide number of photosynthetic and non-photosynthetic bacteria, with evidence suggesting that RegB/RegA plays a fundamental role in the transcription of redox-regulated genes in many bacterial species.

12.2.5 Oxidative stress

During aerobic electron transport reactions, oxygen is reduced not only to water but also to reactive oxygen species (ROS) including superoxide ($O_2^{\cdot-}$) (please refer to #1, Chapter 8 above, p. 253), hydrogen peroxide (H_2O_2) and the hydroxyl radical ($OH^{\cdot}$), which can damage cellular components including proteins, DNA, RNA and lipids. Under normal growth conditions, various enzymes are active to maintain ROS below toxic level. When an anaerobic culture is exposed to air or to deteriorating oxidants, the rate of ROS generation becomes higher than that of degeneration, leading to a deterioration of cellular function usually termed oxidative stress. To counter oxidative stress, organisms activate numerous mechanisms that detoxify ROS and repair cellular damage. When H_2O_2 is challenged bacteria produce over 30 new proteins. Similarly, new proteins are synthesized in bacteria in response to $O_2^{\cdot-}$ (please refer to #1, Chapter 8 above, p. 253). This regulation is referred to as the oxidative stress response, and protects the cell from other deteriorative oxidants such as hypochlorous acid (HOCl), nitric oxide ($NO^{\cdot}$) as well as others.

In *Escherichia coli*, an H_2O_2-sensing regulator, OxyR (hydrogen peroxide-inducible genes activator), senses H_2O_2 and activates related operons. OxyR has two cysteine residues that can form a disulfide bond when oxidized. OxyR with a mid-potential of around −185mV is in the reduced form in the cytoplasm that has a mid-potential of around −280mV under normal growth conditions. OxyR is oxidized forming an intramolecular disulfide bond when the intracellular H_2O_2 increases with the increase in the cytoplasmic redox potential. The oxidized OxyR binds and activates many operons including *katG* (catalase), *ahpCF* (alkylhydroperoxide reductase), *gorA* (glutathione reductase) and *grxA* (glutaredoxin 1). When the H_2O_2 concentration reaches a normal level, the oxidized OxyR is reduced by glutathione and glutaredoxin 1. Regulators, PerR and OhrR, with a similar function are known in *Bacillus subtilis*.

The SoxRS system regulates many operons according to the cytoplasmic $O_2^{\cdot-}$ (please refer to #1, Chapter 8 above, p. 253) concentration. SoxR is a redox-sensitive transcriptional activator and SoxS is a regulatory protein. A constitutive protein, SoxR, with 2 [2Fe-2S] clusters is oxidized to activate the transcription of *soxS*. SoxS in turn activates the *sox* operon to increase the SoxS concentration further. SoxS activates transcription of many oxidative stress response proteins including manganese superoxide dismutase (*sodA*), glucose-6-phosphate dehydrogenase (*zwf*), ferredoxin reductase (*fpr*) and endonuclease IV (*nfo*). SodA detoxifies $O_2^{\cdot-}$ (please refer to #1, Chapter 8 above, p. 253), Zwf and Fpr provide reducing power for SodA, and Nfo repairs damaged DNA.

12.2.6 Heat shock response

When a culture of *Escherichia coli* is transferred from the physiological temperature of 37 °C to 42 °C, a large number of proteins are synthesized. This change to cope with the elevated temperature is referred to as the heat shock response (HSR), and the proteins produced under this condition are called heat shock proteins (HSP). On the increase in temperature, a heat shock σ-factor (RpoH, σ^H or σ^{32}) is synthesized, and genes for HSP are transcribed by the RNA polymerase with σ^H (Figure 12.26).

HSP prevent heat denaturation of cell constituents and repair the denatured constituents. The expression of σ^H is regulated at the translational level. At a normal growth temperature, the *rpoH* transcript (mRNA) is not translated due to a hairpin formed within the molecule preventing ribosome binding. The hairpin structure is melted at an elevated temperature allowing ribosomes to translate the mRNA (Section 12.1.9.2).

HSP increase sharply to a maximum within 4–5 minutes after the temperature increase before they decrease to a new steady state concentration (Figure 12.27). HSP regulate the translation and the activity of σ^H that activates their transcription. This is an example of feedback regulation.

HSP are synthesized at a lower level at physiological temperature. They have a crucial role in all forms of life. They participate in protein folding into the correct forms (Section 6.12.3), and keep the nascent peptide in a translocation-compatible state during its translocation through a membrane (Section 3.8.1). For these reasons HSP are referred to as molecular chaperones. They are grouped as chaperones and chaperonins according to their functions (Table 12.6). They have homologous amino acid sequences not only between prokaryotes but also between prokaryotes and eukaryotes. Some of them that require energy for their activity consume ATP with their ATPase activity.

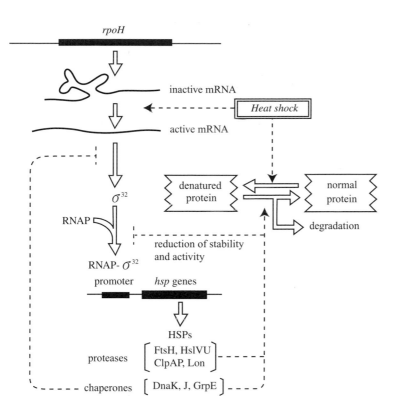

rpoH

inactive mRNA

Heat shock

active mRNA

σ^{32}

RNAP

denatured protein

normal protein

degradation

RNAP- σ^{32}

reduction of stability and activity

promoter   *hsp* genes

HSPs

FtsH, HslVU
ClpAP, Lon

proteases

chaperones   DnaK, J, GrpE

Figure 12.26 **Transcriptional regulation of heat shock protein genes in *Escherichia coli* at elevated temperature.**

(*Curr. Opin. Microbiol.* **2**:153–158, 1999)

Under physiological temperature conditions, the *rpoH* transcript forms a hairpin structure preventing ribosome binding, while the hairpin structure is melted at an elevated temperature allowing ribosomes to translate the mRNA. The heat shock protein (HSP) genes are recognized by RNA polymerase with σ^H and transcribed. HSPs prevent denaturation of cellular constituents, and repair or degrade the denatured proteins. When concentration of HSPs reaches a certain level, they separate σ^H from RNA polymerase and degrade it to prevent overproduction of HSPs. HSPs interact with σ^H mRNA inhibiting its translation. Through these mechanisms the concentration of σ^H and HSPs decreases to a stable level after a sharp increase after the temperature increase (see Figure 12.27).

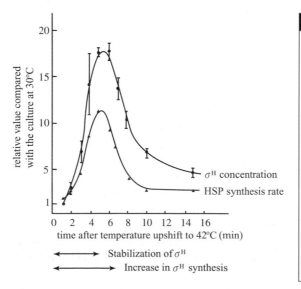

relative value compared with the culture at 30°C

time after temperature upshift to 42°C (min)

σ^H concentration

HSP synthesis rate

Stabilization of σ^H

Increase in σ^H synthesis

Figure 12.27 **Transcriptional regulation of HSPs in *Escherichia coli*.**

HSPs are synthesized at a low level at physiological temperatures for protein folding and translocation through the membrane. When their concentration reaches a certain level after a temperature up-shift, HSPs inhibit the translation and activity of σ^H, preventing the overproduction of HSPs. Through these mechanisms, the concentration of HSPs is maintained at an elevated level after a sharp increase with the temperature up-shift.

Table 12.6. *Bacterial molecular chaperones and their equivalents in eukaryotes*

Family	Protein	Function	Bacteria	Eukaryote
Chaperone				
Hsp70	Hsp70	Preventing the aggregation of unfolded peptide	DnaK	Hsp72
		Involved in protein translocation		Hsp73
		Regulating heat shock response		Hsp70
		With ATPase activity		BiP
	Hsp40	Co-chaperone regulating DnaK activity	DnaJ	Hsp40
	GrpE	Co-chaperone modulating ATP/ADP binding to DnaK		GrpE
Hsp90		General chaperone	HtpG	Hsp90
Hsp33		Heat and oxidative shock response	Hsp33	
Chaperonin				
Hsp60	Chaperonin 60	ATP-dependent protein folding	GroEL	Hsp60
	Chaperonin 10	Co-chaperonin to chaperonin 60	GroES	Hsp10
TRiC		Similar to GroEL	TriC (TF55)	TriC (TCP1)

Escherichia coli has an extracytoplasmic function sigma factor (σ^E) that has a similar function to σ^H. This σ-factor is activated when periplasmic proteins are denatured by heat or chemicals such as ethanol. Mutants defective in σ^E have not been isolated, suggesting that this σ-factor is an essential protein. σ^E recognizes at least ten genes in this bacterium but their functions have not been fully characterized. The activity of σ^E is regulated by anti-σ factor RseA, a cytoplasmic membrane protein. Under normal conditions, the periplasmic protein anti-anti-σ factor RseB binds the periplasmic side of RseA, increasing its affinity for σ^E at the cytoplasmic side. Denatured periplasmic protein binds RseB protein removing it from RseA. RseA without RseB at the periplasmic side loses its anti-σ factor activity releasing σ^E (Figure 12.28).

12.2.7 Cold shock response

When an exponentially growing culture of *Escherichia coli* is shifted from 37 °C to 15 °C, growth resumes after an acclimation phase characterized by the transient dramatic induction of a group of proteins known as cold induced proteins (CIP) with various functions. This regulation is referred to as the cold shock response (CSR). Under low temperature conditions, general protein synthesis is severely inhibited due to the formation of hairpin structures within mRNAs and the double stranded DNA structure becomes more stable, hindering transcription and replication. The membrane does not function properly due to reduced fluidity, and the reduced flexibility of proteins causes improper function. CIP facilitate the cell to cope with these problems under low temperature conditions. Among CIP, a group of low molecular weight proteins are involved in the

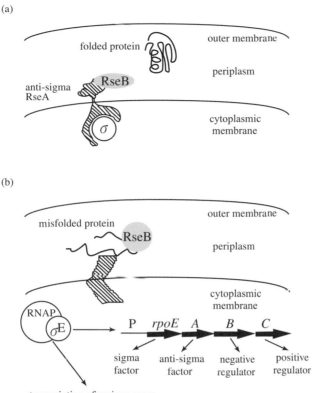

(a)

folded protein

outer membrane

periplasm

anti-sigma
RseA

RseB

cytoplasmic
membrane

σ

(b)

misfolded protein

outer membrane

RseB

periplasm

cytoplasmic
membrane

RNAP

σE

P   *rpoE*   A   B   C

sigma
factor

anti-sigma
factor

negative
regulator

positive
regulator

transcription of various genes

Figure 12.28 **Regulation of extracytoplasmic function sigma factor σ^E activity by anti-σ factor and anti-anti-σ factor.**

(*Mol. Microbiol.* **28**:1059–1066, 1998)

σ^E is activated when periplasmic proteins are denatured by heat or solvents. Under normal conditions, this σ-factor binds the cytoplasmic membrane-bound anti-σ factor, RseA, that forms a complex with RseB anti-anti-σ factor at the periplasmic side (a). When denatured periplasmic protein interacts with RseB, the latter is liberated from RseA. RseA without RseB at the periplasmic side loses its anti-σ factor activity and σ^E becomes active (b).

regulation of CIP expression. These are referred to as cold shock proteins (CSPs). In most of the literature, CSP is used interchangeably with CIP to describe all the proteins induced at low temperature.

CIP are grouped into two classes. Class I CIP are those produced only under low temperature conditions or that increase amounts more than ten times those at normal growth temperatures: those that increase less than ten times are termed class II CIP. The latter include RecA (recombinase A), initiation factor IF-2, DNA-binding protein H-NS and RNase R that are essential for replication, transcription and translation. Class I CIP include CspA (cold shock protein A), CspB, CdsA (cysteine sulfinate desulfinase), NusA (N utilization substance protein A that participates in the termination and antitermination process), PNP (polynucleotide phosphorylase) and RbfA (ribosome-binding factor A).

The induction of CSP is due to increased translation as well as transcription. The CSP genes including *cspA* contain an AT-rich UP region upstream of the promoter region that increases transcription at low temperature. In addition, transcripts of *csps* including *cspA* contain unique sequences that enhance their translation and stability at low temperature. These are the cold shock (CS) box of about 25 bases in the 5′-untranslated region (5′UTR) of about 150 bases and the downstream box (DB) in the coding region downstream of the ribosome-binding (RB) site (Figure 12.29(1)). CSPs have a high affinity

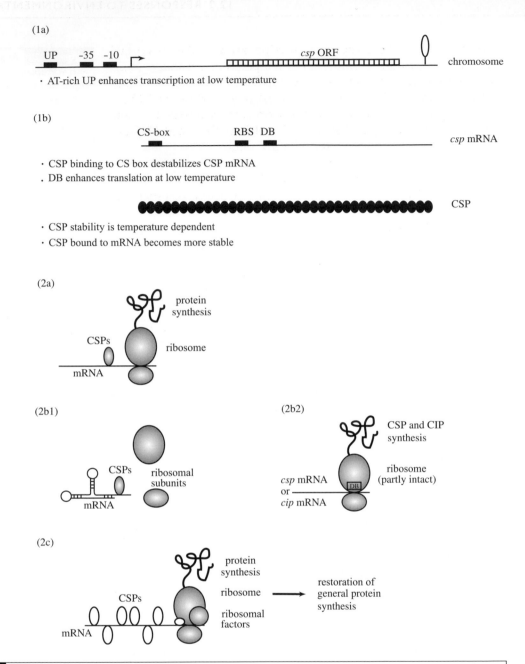

Figure 12.29 Structure of *cspA* (1), and translation of cold induced protein (CIP) and cold shock protein (CSP) and the function of CSP in the restoration of translational efficiency (2).

(System. Appl. Microbiol. **23**:165–173, 2000)

Upstream of the *cspA* promoter is an AT-rich UP region (1a) that enhances transcription of the gene. The *cspA* transcript contains the cold shock (CS) box in the 5′-untranslated region (5′UTR) and downstream box (DB) in the coding region (1b). The 5′UTR forms a secondary structure at low temperature that stabilizes the transcript. At normal growth temperature, the transcript is unstable without the secondary structure. DB increases translation with a high affinity for ribosomes at low temperature. At normal growth temperature, mRNAs are translated (2a), but at low temperature they are not translated due to the secondary structure formed in mRNAs and due to disruption in ribosomal structure (2b1). Transcripts of *csps* are translated with the increased affinity of the DB for ribosomes at low temperature (2b2). CSP bind mRNAs destabilizing the secondary structure. The CSP bound mRNA are transcribed normally at low temperature (2c). CSP binding mRNA are referred to as RNA chaperones.

to bind to the CS box. The *csp* transcripts are unstable at normal growth temperature, but stable at low temperature. Secondary structures in the unusually long 5′UTR are responsible for their stability at low temperature. When CSP are in excess they bind to the CS box, decreasing the secondary structure. The transcripts become unstable when CSP bind them. The DB has a high affinity for the 16S RNA of the ribosome increasing translation at low temperature.

At low temperature, the transcription of genes with a UP region is enhanced, and the *csp* transcripts are stabilized through the secondary structure at the CS box. The stabilized transcript is efficiently translated through the high affinity of the DB with the ribosome. CSP bind mRNAs reducing the secondary structure and restoring the translational efficiency at low temperature (Figure 12.29(2)). For this reason CSP are referred to as RNA chaperones. CSP genes have not been identified in the complete genome sequences of some bacteria pathogenic to warm-blooded animals including *Helicobacter pylori*, *Campylobacter jejuni* and *Mycobacterium genitalium*.

To maintain membrane fluidity, more unsaturated fatty acids are incorporated into the membrane phospholipids at low temperature. Unsaturated fatty acids are produced by two different mechanisms depending on the organisms. These are an anaerobic route only found in some bacteria and an aerobic route employed by both eukaryotes and prokaryotes (Section 6.6.1.3).

The cytoplasmic membrane of *Escherichia coli* consists of phospholipids containing three fatty acids. They are a saturated fatty acid, palmitic (hexadecanoic) acid, and two unsaturated fatty acids, palmitoleic (*cis*-9-hexadecenoic) acid and *cis*-vaccenic (*cis*-11-octadecenoic) acid. Lower growth temperatures result in an increase in the amount of the diunsaturated fatty acid, *cis*-vaccenic acid, in the membrane with a decrease in palmitic acid. This allows the organism to regulate membrane fluidity to optimize its function at various growth temperatures. This change is due to the increased activity of an enzyme involved in the synthesis of *cis*-vaccenic acid, 3-ketoacyl-acyl-ACP synthase (FabF). This enzyme is expressed constitutively and is stable at low temperature. Molecular oxygen is not involved in this pathway, which is referred to as the anaerobic route (Section 6.6.1.3).

In contrast, the increase in unsaturated fatty acid at low temperature involves transcriptional regulation in *Bacillus subtilis*. With the temperature downshift, this Gram-positive bacterium synthesizes a membrane-bound desaturase (Des) that uses existing phospholipids as substrates to introduce a *cis*- double bond at the fifth position of the fatty acyl chain consuming O_2. This transcriptional regulation is a two-component system consisting of a membrane-associated kinase, DesK, and a transcriptional regulator, DesR, which stringently controls the transcription of the *des* gene, coding for the desaturase. DesK senses the decrease in membrane fluidity, and autophosphorylates. This reaction is referred to as the aerobic route of unsaturated fatty acid synthesis (Section 6.6.1.3).

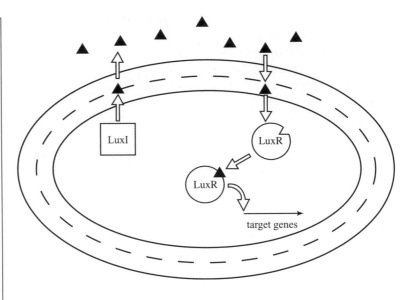

Figure 12.30 Metabolic regulation by quorum sensing in Gram-negative bacteria.

(*Curr. Opin. Microbiol.* **2**:582–587, 1999)

An autoinducer (black triangle) produced by autoinducer synthase (LuxI) diffuses through the membrane. When its concentration reaches a threshold level the autoinducer binds the transcriptional regulator (LuxR) activating it. The activated LuxR activates the expression of operons related to many physiological functions depending on the bacterium.

12.2.8 Quorum sensing

Luminescent bacteria emit light only when their cell concentration reaches a certain level (Section 5.9.1). This phenomenon is due to the fact that the genes for luminescence are expressed in response to a small molecular-weight signal compound secreted by the bacteria when it reaches a threshold concentration. This regulation is referred to as quorum sensing. Quorum sensing is known in many bacteria and has various physiological functions in, for example, motility, secondary metabolite production, conjugation, virulence and biofilm formation in addition to luminescence. This property is probably important for their survival at high cell densities. *N*-acyl homoserine lactones (AHL) are the small signal molecules involved in quorum sensing in Gram-negative bacteria, while Gram-positive bacteria use small peptides as the signal molecules in their quorum sensing. These small signal molecules are referred as autoinducers or pheromones. Autoinducers function not only in their producing organisms but also in other bacteria. For this reason, quorum sensing is referred to as cell-cell communication. In some cases more than two quorum sensing systems are known in a single bacterium.

The quorum sensing system in Gram-negative bacteria consists of an autoinducer synthase (LuxI) and transcriptional regulator (LuxR) as shown in Figure 12.30. The autoinducer produced by LuxI diffuses out of the cell. When the equilibrated concentration of the autoinducer reaches the threshold level, LuxI binds LuxR activating it as a regulator to activate a number of genes.

A number of bacteria have been identified which are capable of enzymic inactivation of AHL. These include Gram-positive bacteria including *Bacillus thuringiensis*, *Bacillus cereus*, *Bacillus mycoides* and *Arthrobacter* sp. as well as Gram-negative bacteria such as *Agrobacterium*

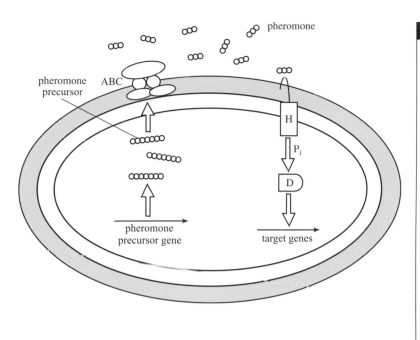

Figure 12.31 Quorum sensing in Gram-positive bacteria.

(*Curr. Opin. Microbiol.* **2**:582–587, 1999)

Peptides or modified peptides are used as the autoinducer (pheromone) of quorum sensing in Gram-positive bacteria. They are not permeable to the membrane and excreted through the ATP-binding cassette mechanism. When their concentration reaches the threshold level, they bind the membrane-bound sensor/kinase protein (H). A histidine residue of H is autophosphorylated when it binds the autoinducer peptide before the phosphate is transferred to the aspartate residue of the cytoplasmic regulator protein (D). The phosphorylated D modulates the expression of various genes.

tumefaciens, *Pseudomonas aeruginosa* and *Comamonas testosterone* among others. ALH-inactivating enzymes have been observed not only in the laboratory but also in the soil. These enzymes have been studied as a means to control the virulence of animal and plant pathogens.

Gram-positive bacteria use post-translationally modified peptides as their autoinducer. These autoinducer peptides are not permeable to the cytoplasmic membrane and are excreted through the ATP-binding cassette (ABC, Section 3.8.1.3) mechanism. When their extracellular concentration reaches the threshold level, they bind the membrane-bound sensor/kinase protein. The sensor protein is autophosphorylated consuming ATP when it binds the autoinducer peptide before transferring phosphate to the cytoplasmic regulator protein (Figure 12.31). The phosphorylated regulatory protein modulates various genes. This sensor/regulator is a typical example of a two-component system.

A-factor is the autoinducer in *Streptomyces griseus*. When its concentration builds up to the threshold level, a two-component system senses and regulates genes to produce aerial mycelium, leading to the formation of spores and secondary metabolites including streptomycin.

12.2.9 Response to changes in osmotic pressure
Gram-negative bacteria including *Escherichia coli* have a special class of outer membrane proteins (OMP) called porins forming water-filled channels through which low molecular weight hydrophilic solutes

Figure 12.32 **The EnvZ/ OmpR two-component system regulating the expression of the outer membrane proteins OmpC and OmpF.**

When the environmental osmotic pressure increases, the membrane-bound sensor protein EnvZ is autophosphorylated, consuming ATP at its histidine residue before transferring phosphate to the cytoplasmic modulator protein OmpR. The phosphorylated OmpR activates the expression of OmpC, a porin with a small pore, and represses OmpF that is another porin expressed under normal osmotic pressure.

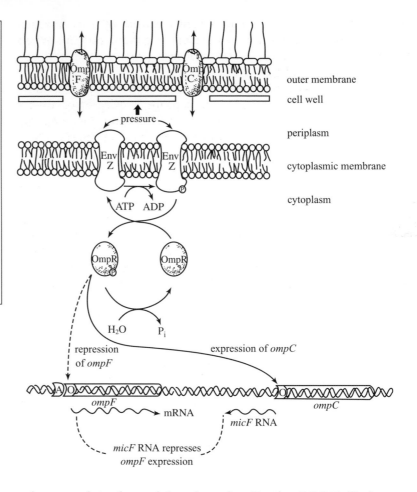

gain access into the periplasmic region (Section 2.3.3.3). Under normal osmotic pressure, outer membrane protein F (OmpF) is the major porin in *Escherichia coli*. Under increased osmotic pressure, a smaller pore-size OmpC replaces OmpF. A membrane-bound sensor protein, EnvZ, senses the increase in osmotic pressure to be autophosphorylated at its histidine residue. Subsequently the phosphate is transferred to an aspartate residue of the cytoplasmic modulator protein, OmpR. The phosphorylated OmpR activates the the expression of *ompF* and represses *ompC* (Figure 12.32). The EnvZ/OmpR two-component system modulates various genes in addition to *omp* genes. OmpC and OmpF are also subject to post-transcriptional regulation by the small regulatory RNA molecules *micC* and *micF*, respectively (Section 12.1.9.4).

12.2.10 Other two-component systems

Several two-component systems have been discussed including the *pho* system under phosphate-limited conditions, the *arc* system related to the availability of O_2, quorum sensing in Gram-positive bacteria, the EnvZ/OmpR system associated with changes in osmotic pressure, and NarC/NarL and NarQ/NarP related to denitrification

(Section 9.1.3). Two-component systems consist of a membrane-bound sensor protein and a cytoplasmic regulator protein. When the signal recognition domain of the sensor protein at the periplasmic side senses the signal, the autokinase domain of the protein at the cytoplasmic side consumes ATP to phosphorylate a histidine residue within the protein. This phosphate is transferred to the aspartate residue of the regulator protein. This phosphorylated regulator protein modulates the expression of the target genes.

Analyses of bacterial whole genomes show many putative two-component systems with base sequences of sensors/regulators. The function of many of them is not yet known. A few two-component systems are known that have not received detailed study. Enzymes of transport and metabolism of 4-carbon dicarboxylates such as succinate and fumarate are under the control of a two-component system consisting of a sensor protein, DcuS, and a regulator protein, DcuR in *Escherichia coli*. Another example of this kind is the utilization of organic sulfur compounds such as sulfonated detergents under sulfate-limited conditions in *Escherichia coli*, *Pseudomonas putida* and *Staphylococcus aureus*. This system is believed to be similar to the *pho* system.

The sporulation process in *Bacillus subtilis* is initiated with the phosphorylation of KinA or KinB consuming ATP in a similar way as a two-component system. The phosphate is transferred to Spo0 A through Spo0 F and Spo0 B. This is referred to as phosphorelay (Section 13.3.1). Histidine residues of KinB (or KinA) and Spo0 B are phosphorylated as the sensor protein, and aspartate residues of Spo0 F and Spo0 A are the phosphate receptor.

12.2.11 Chemotaxis

Motile bacteria synthesize proteins for chemotaxis including flagella formation when the substrate concentration becomes low. Bacterial flagella consist of a basal body, hook and filament (Section 2.3.1). The proteins forming the hook and filament are transported through the basal body. For this reason, the proteins of the basal body are synthesized before those forming the hook and filament.

The synthesis and function of the flagellar and chemotaxis system requires the expression of more than 50 genes which are scattered among at least 17 operons that constitute the coordinately regulated flagellar regulon. In *Escherichia coli* and *Salmonella typhimurium*, the flagella and chemotaxis regulons are expressed in three separate steps by transcriptional as well as post-transcriptional mechanisms according to substrate concentration and to the morphological development of the flagellar structure itself. With a limited supply of substrate, the cAMP–CRP complex activates the expression of the two early genes in the *flhDC* operon. The FlhD and FlhC proteins are transcriptional activators that activate the middle gene operons. The middle gene operons include *fliA* that encodes σ^{28} (σ^F). RNA polymerase with σ^{28} transcribes the late genes that encode proteins of the hook and filament (Figure 12.33). A similar alternative σ-factor, σ^D, is known in *Bacillus subtilis*.

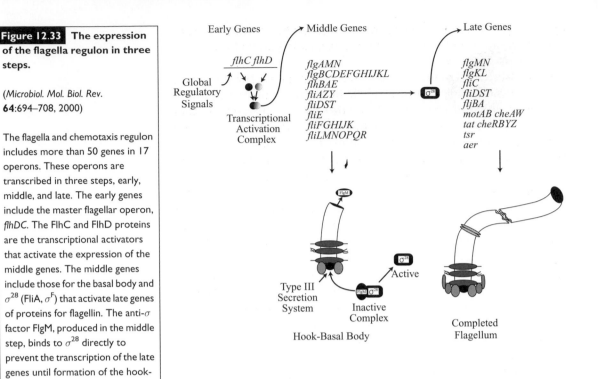

Figure 12.33 The expression of the flagella regulon in three steps.

(*Microbiol. Mol. Biol. Rev.* **64**:694–708, 2000)

The flagella and chemotaxis regulon includes more than 50 genes in 17 operons. These operons are transcribed in three steps, early, middle, and late. The early genes include the master flagellar operon, *flhDC*. The FlhC and FlhD proteins are the transcriptional activators that activate the expression of the middle genes. The middle genes include those for the basal body and σ^{28} (FliA, σ^{F}) that activate late genes of proteins for flagellin. The anti-σ factor FlgM, produced in the middle step, binds to σ^{28} directly to prevent the transcription of the late genes until formation of the hook-basal body is completed. Once the hook-basal body is completed, FlgM is secreted from the cell to free σ^{28}.

The anti-σ factor, FlgM, binds to σ^{28} directly to prevent the transcription of the late genes until the formation of the hook-basal body is completed. Once the hook-basal body is completed, FlgM is secreted from the cell through the flagellin export machinery to free σ^{28}. In addition to the transcriptional regulation, post-transcriptional control further ensures the efficiency of flagellar assembly. The translation of the anti-σ factor and hook protein are regulated. When the flagella and chemotaxis regulons are expressed, motility is regulated by the gradient of attractant or repellant through the mode of flagellar rotation. A cell typically travels in a three-dimensional random 'walk'. Intervals in which the cells swim in gently curved paths (runs) alternate with briefer periods of chaotic motion (tumbles) that randomly reorientate the next run. Cells run when the left-handed flagellar filaments rotate counterclockwise and coalesce into a bundle to propel the cell. Cells tumble when the flagella turn clockwise and disrupt the bundle. Cells extend runs (suppress tumbles) when they head up concentration gradients of attractants or down gradients of repellents.

Three classes of proteins are essential for chemotaxis. They are transmembrane receptors, cytoplasmic signalling components, and enzymes for adaptive methylation. Using these proteins a cell senses the spatial gradient and compares the instantaneous concentration of a compound. The transmembrane receptors consist of an amino-terminal transmembrane helix (TM1), a periplasmic ligand interaction domain, a second transmembrane helix (TM2), and a large cytoplasmic signalling and adaptation domain. The cytoplasmic domain contains five methylatable glutamate residues, and the receptors are therefore also called methyl-accepting chemotaxis proteins (MCP).

The attractant concentration is sensed by the receptor protein, and the signal is transduced to flagella by four proteins to determine the direction of flagella rotation. They are CheA, the histidine protein kinase; CheY, the response regulator; CheW, the receptor-coupling factor; and CheZ, an enhancer of CheY-P dephosphorylation. When the receptor protein is free from attractant the autophosphorylation activity of CheA is stimulated, which in turn increases phosphotransfer from CheA-P to CheY. CheY-P binds to the flagellar motor-switch complex to cause clockwise rotation. CheZ prevents accumulation of CheY-P by accelerating the decay of its intrinsically unstable aspartyl-phosphate residue. Under steady-state conditions, CheY-P is maintained at a level that generates the random walk. When an attractant (or an attractant–protein complex) binds the periplasmic side of the receptor protein, this protein undergoes a conformational change that suppresses CheA activity. Levels of CheY-P fall, and cells tumble less often. Thus, cells increase their run lengths as they enter areas of higher attractant concentration (Figure 12.34).

This response does not explain, however, how cells sense continually increasing attractant concentrations. To accomplish that, adaptation is necessary. Two enzymes, the methyltransferase CheR and the methylesterase CheB, are necessary for adaptive methylation. CheR is a constitutive enzyme that uses S-adenosylmethionine to methylate glutamate residues in the receptor protein. CheB is a target for phosphotransfer from CheA, and CheB-P removes methyl groups from the MCP. In steady state, methyl addition by CheR balances methyl removal by CheB-P to achieve an intermediate level of receptor methylation (0.5 to 1 methyl group per subunit) that maintains run–tumble behaviour. When an attractant binds a receptor and inhibits CheA activity, CheB-P levels fall, although more slowly than CheY-P levels, since CheB-P is not a substrate for CheZ. Increased methylation restores the ability of the receptor to stimulate CheA. Even after basal levels of CheY-P and CheB-P are regained, however, an attractant-bound receptor remains overmethylated, because its properties as a substrate for CheB-P are altered. Through these reactions the cell moves towards the attractant gradient (Figure 12.34).

12.2.12 Adaptive mutation

When an alkaline phosphatase mutant of *Escherichia coli* is cultivated under phosphate-limited conditions, the reversion rate is 15–20-fold higher than when cultivated under phosphate-excess conditions. Under phosphate-limited conditions, many genes including alkaline phosphatase (*phoA*) are expressed (Section 12.2.3). During the transcription a part of the gene is in an unpaired state (single strand) causing supercoiling of adjacent parts and changes in the DNA secondary stem-loop structures. These DNA structures are vulnerable to

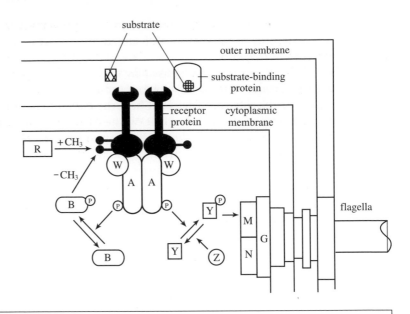

Figure 12.34 **Signal transducing pathway of chemotaxis to swim towards the attractant gradient.**

(*J. Bacteriol.* **180**:1009–1022, 1998)

The dimeric membrane-spanning chemoreceptors (paired black wrench-like proteins) form a complex with CheA and CheW polypeptides and stimulate the autokinase activity of CheA. CheA-P can transfer the phosphate to CheY. CheY-P interacts with FliM in the motor-switch complex to induce clockwise (CW) flagellar rotation, causing tumble. The decay of CheY-P is accelerated by CheZ. CheR is a constitutive methyltransferase that methylates certain glutamate residues in the cytoplasmic domains of the receptors. CheB is a methylesterase that is activated by phosphotransfer from CheA-P. CheB-P removes methyl groups from the receptors. When an attractant (cross-hatched square) binds directly or as a complex with substrate-binding proteins to the periplasmic domains of the receptors, this protein undergoes a conformational change that suppresses CheA activity. The level of CheY-P falls, and the flagella rotates counter clockwise (CCW). Thus, cells tumble less often and increase their run lengths as they enter areas of higher attractant concentration. In the adaptation pathway, reduced CheA activity decreases the CheB-P level, although more slowly than the CheY-P level. As methylesterase activity declines, the receptors become more highly methylated. Increased methylation counteracts the attractant-dependent inhibition of CheA activity. As CheA activity rises, the intracellular CheY-P concentration returns to its pre-stimulus value, and the flagellar motor resumes its pre-stimulus CW-to-CCW switching ratio.

Abbreviations: A, CheA; W, CheW; Y, CheY; Z, CheZ; R, CheR; B, CheB; G, FliG; M, FliM; N, FliN; p, phosphate; CH$_3$, methyl group (shown as lollipop-like forms on the cytoplasmic domains of the receptors).

mutation. The increased reversion rate is due to an increased mutation rate in the derepression of the *phoA* gene under phosphate-limited conditions.

As discussed previously, bacteria have mechanisms to derepress genes according to various stresses. These stresses are a selection pressure for beneficial mutations. Cells starved after they use up their usual carbon source in the presence of a carbon source they cannot use may produce mutants that can use it. The most common mechanism involved is gene derepression resulting in the constitutive production of a previously inducible enzyme or an enzyme with altered substrate specificity. Many examples exist, such as the use of altrose-galactoside via β-galactosidase, β-glycerol phosphate via alkaline phosphatase and xylitol via ribitol dehydrogenase. The metabolic steps required to metabolize a new related substrate are similar to those for existing pathways. Therefore, relatively minor changes to a duplicate copy of the existing gene may be required for recruitment to a new function. Other frequently observed mutations in response to carbon source starvation confer increased permeability for the limiting metabolite. These mutation and selection processes occurring under stress conditions are referred to as adaptive mutations.

12.3 | Regulation through modulation of enzyme activity: fine regulation

Metabolic regulation through the transcription and translation processes that take time are inadequate for microbes to cope with a rapidly changing environment. Furthermore, if the enzyme activity is stable, regulation through its synthesis would not be very efficient. For efficient metabolic regulation, enzyme activities are modulated in addition to their synthesis. Enzyme activities are regulated through various mechanisms including feedback inhibition, feedforward (precursor) activation, and physical and chemical modification of the enzyme proteins.

12.3.1 Feedback inhibition and feedforward activation

Amino acids are synthesized through a series of reactions. When an amino acid is available and sufficient to meet the needs of growth, the enzyme catalyzing the first reaction is inhibited. This process is referred to as feedback inhibition. For example, threonine dehydratase is inhibited when isoleucine is available (Section 6.4.1), and AMP and GMP inhibit amidophosphoribosyl transferase (Section 6.5.3).

In the case of a single precursor used for multiple product formation, isoenzymes are regulated separately according to the concentration of each product as in gene expression regulation through repression (Section 12.1.5). Aspartate is used to synthesize lysine, methionine and isoleucine through threonine (Figure 12.35). The first reaction is catalyzed by three separate isoenzymes. One is regulated by lysine, the other by both threonine and isoleucine. The third isoenzyme is not regulated, but methionine inhibits homoserine acyltransferase, the first enzyme specifically involved in its synthesis. When methionine is needed with threonine and lysine in excess, the third aspartate kinase isoenzyme is active and the other isoenzymes are inhibited by lysine and threonine. Enzymes using intermediates as common substrates are regulated similarly. These are dihydrodipicolinate synthase regulated by lysine, homoserine dehydrogenase regulated by threonine and homoserine acyltransferase regulated by methionine.

Figure 12.35 **Feedback inhibition of enzyme activities in the synthesis of aspartate family amino acids.**

(Gottschalk, G. 1986, *Bacterial Metabolism*, 2nd edn., Figure 7.12. Springer, New York)

Aspartate is phosphorylated by three isoenzymes of aspartate kinase (1). One of them is inhibited by lysine, and the other collectively by threonine and isoleucine. The third one is not regulated. The first enzymes specifically involved in the synthesis of each amino acid are under feedback inhibition control by the final product. These are dihydrodipicolinate synthase (2) regulated by lysine, homoserine dehydrogenase (3) regulated by threonine, and homoserine acyltransferase regulated by methionine (4). Isoleucine inhibits threonine dehydratase (5).

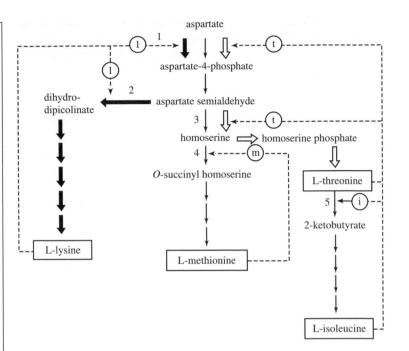

In some cases, a single enzyme catalyzes the first reaction that leads to the synthesis of multiple products. For example, *Escherichia coli* has isoenzymes of 3-deoxy-D-arabinoheptulosonate (DAHP) synthase catalyzing the first reaction of aromatic amino acid synthesis (Figure 6.18, Section 6.4.4), but a single enzyme catalyzes the reaction in *Ralstonia eutropha* (*Alcaligenes eutrophus*). In the latter case, each product inhibits the enzyme, and the inhibition is stronger in the presence of all the products than with just a single product. This regulatory mechanism is referred to as cumulative inhibition.

Enzymes controlled through feedback inhibition have a unique property. These enzymes have an inhibitor-binding site different from the active site. When the inhibitor binds the enzyme, the structure is changed to an inactive form. The inhibitor-binding site is referred to as an allosteric site, and the enzyme as an allosteric enzyme.

When fructose-1,6-diphosphate (FDP) is accumulated, this EMP pathway intermediate activates pyruvate kinase (Section 4.1.4.2). This enzyme activity modulation is referred to as feedforward activation or precursor activation. Another example of this regulatory mechanism is found in lactic acid bacteria, where lactate dehydrogenase is activated by FDP.

12.3.2 Enzyme activity modulation through structural changes

Inhibitor binding causes structural changes in allosteric enzymes. The activity of many enzymes is modulated through structural modification. Some are modified chemically and others physically.

Chemical modification includes phosphorylation, acetylation, methylation and adenylylation. Changes in ionic strength, pH and other factors cause physical modification in enzyme proteins, modulating their activities.

12.3.2.1 Phosphorylation

When acetate is used as the substrate, *Escherichia coli* metabolizes acetyl-CoA not only through the TCA cycle for ATP synthesis but also through the glyoxylate cycle to supply carbon skeletons (Section 5.3.2, Figure 5.4). Isocitrate is the common substrate for the dehydrogenase of the TCA cycle and lyase of the glyoxylate cycle. Since isocitrate dehydrogenase has a much higher affinity for the substrate than the other enzyme, its activity is regulated according to cellular need. When the adenylate energy charge is low and carbon skeletons are adequate, the bacterium needs to use the TCA cycle. Under these conditions isocitrate dehydrogenase is dephosphorylated to the active form. Under the reverse conditions a kinase phosphorylates this enzyme to an inactive form (Section 5.3.2.1, Figure 5.5). Many enzyme activities are modulated by a phosphorylation mechanism.

12.3.2.2 Adenylylation

Ammonia is used in the synthesis of glutamate and glutamine in the reactions catalyzed by glutamate dehydrogenase and glutamine synthetase, respectively. Since glutamate dehydrogenase has a low affinity for ammonia, this enzyme is active only at high ammonia concentrations. When the ammonia concentration is low, glutamate is produced through glutamine catalyzed by glutamine synthetase and glutamate synthase consuming ATP (Section 6.2.3):

$$\text{2-ketoglutarate} + NH_3 + NADPH + H^+ \xrightarrow{\text{glutamate dehydrogenase}} \text{glutamate} + NADP^+$$

$$\text{glutamate} + NH_3 + ATP \xrightarrow{\text{glutamine synthetase}} \text{glutamine} + ADP + P_i$$

$$\text{glutamine} + \text{2-ketoglutarate} + NADPH + H^+ \xrightarrow{\text{glutamate synthase}} \text{2 glutamate} + NADP^+$$

To prevent ATP consumption under ammonia-rich conditions, the glutamine synthetase gene is expressed at a low level (Section 12.2.2), and its activity is inhibited. Activity is modulated through adenylylation of the enzyme protein according to the 2-ketoglutarate/glutamine ratio that is determined by the ammonia concentration. Regulation of glutamine synthetase by adenylylation/ deadenylylation involves adenylyltransferase (ATase, also known as glutamine synthetase adenylyltransferase/removase encoded by

Figure 12.36 **Regulation of glutamine synthetase activity in *Escherichia coli* through adenylylation and deadenylylation of the enzyme protein according to the ratio of 2-ketoglutarate/glutamine determined by ammonia availability.**

(Gottschalk, G. 1986, *Bacterial Metabolism*, 2nd edn., Figure 7.17. Springer, New York)

When glutamine accumulates under ammonia-rich conditions, uridylyltransferase/uridylyl removing enzyme (UTase/UR, encoded by *glnD*) removes UMP from PIID [PII(UMP)$_4$] converting to PIIA. PIIA directs adenylyltransferase (ATase) adenylylate glutamine synthetase to a less active form. When 2-ketoglutarate is accumulated due to the limited supply of ammonia, PIIA is uridylylated to PIID that activates ATase to remove AMP from less active adenylylated glutamine synthetase to make the enzyme highly active. PIIA/PIID regulates the expression of the enzyme (Figure 12.22).

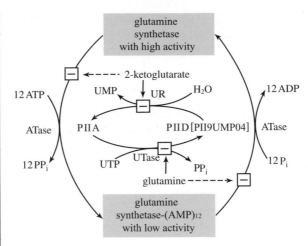

glnE), PII protein (encoded by *glnB*) and uridylyltransferase (UTase)/ uridylyl removing (UR) enzyme (encoded by *glnD*). UTase/UR senses the ammonia concentration uridylylating the PII protein under ammonia-limited conditions with the accumulation of 2-ketoglutarate and removing UMP when glutamine is accumulated under ammonia-rich conditions. PIID-(UMP)$_4$ formed under ammonia-limited conditions activates ATase to remove AMP from glutamine synthetase converting the enzyme to the active form. Under ammonia-rich conditions UR removes UMP converting PIID-(UMP)$_4$ to PIIA. PIIA activates ATase adenylylating the enzyme into a less active form (Figure 12.36).

It should be mentioned that glutamine synthetase activity is regulated not only by ammonia availability but also by the end-products of glutamine metabolism, ADP and AMP and other nucleotides through cumulative feedback inhibition (Section 6.2.3). Enzyme activity is inhibited even under ammonia-limited conditions when the culture is not actively growing.

12.3.2.3 Acetylation

The purple non-sulfur bacterium *Rhodopseudomonas gelatinosa* grows photoorganotrophically on citrate. Citrate lyase cleaves citrate to oxaloacetate and acetate. When this organism grows on carbon sources other than citrate under similar conditions, citrate synthase condenses oxaloacetate and acetyl-CoA into citrate.

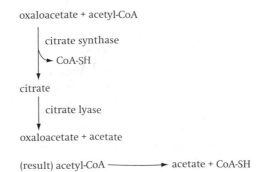

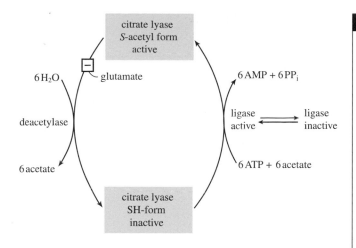

Figure 12.37 **Citrate lyase activity is regulated according to the cellular glutamate concentration through acetylation and deacetylation in *Rhodopseudomonas gelatinosa*.**

(Gottschalk, G. 1986, *Bacterial Metabolism*, 2nd edn., Figure 12.37. Springer, New York)

When citrate is used as a carbon and energy source, citrate lyase cleaves citrate to oxaloacetate and acetate in *Rhodopseudomonas gelatinosa*. Citrate synthase is active and synthesizes glutamate (2-ketoglutarate) through an incomplete TCA fork when carbon sources other than citrate are used. A futile cycle is possible when citrate lyase and citrate synthase are active at the same time. To avoid this futile cycle, the activity of citrate lyase is regulated. When glutamate is accumulated, a ligase acetylates citrate lyase to an active form. Deacetylase deacetylates this enzyme to an inactive form when citrate synthase activity is needed due to a low glutamate concentration.

As shown above, when citrate synthase and citrate lyase are active at the same time, acetyl-CoA is cleaved to acetate, wasting energy in the form of a high-energy acyl-CoA bond. To avoid this futile cycle, citrate lyase activity is regulated according to the concentration of glutamate. When a metabolite of citrate metabolism, i.e. glutamate, is accumulated, citrate lyase is acetylated to an active form. Citrate lyase is deacetylated to an inactive form under the reverse conditions when citrate synthase activity is needed (Figure 12.37).

12.3.2.4   Other chemical modifications

Table 12.7 summarizes enzyme activity modulation through chemical modification of the enzyme protein. As shown in the table, some proteins are modified by other means than those mentioned above. When nitrogenase activity is inhibited through the ammonia switch, this enzyme is modified with ADP-ribose from NAD^+ (Section 6.2.1.5). P II protein (GlnB) participating in the response to ammonia-limitation is uridylylated (Figure 12.36). The substrate-receptor protein involved in chemotaxis is methylated for its regulatory function (Section 12.2.11).

12.3.2.5   Regulation through physical modification and dissociation/association

Enzyme activity can be modulated not only through chemical modification but also through physical modification. Enzymes modifiable through this mechanism for activity regulation consist of multiple subunits. For example, clostridial lactate dehydrogenase is a homo-tetramer. It is active in the tetramer form, but loses its activity when it is dissociated into dimers. Lactate dehydrogenase is activated by fructose-1,6-diphosphate to accelerate carbohydrate metabolism through the EMP pathway increasing the rate of NADH oxidation to NAD^+. Under the reverse conditions, clostridia metabolize pyruvate to acetate and butyrate to synthesize more ATP, which inhibits lactate dehydrogenase.

Table 12.7. *Chemical modification of enzyme proteins for activity regulation*

Modified with	Source	Example
Phosphate	ATP	isocitrate dehydrogenase
Acetyl group	Acetate + ATP	citrate lyase
Adenylyl group	ATP	glutamine synthetase
Uridylyl group	UTP	PII (regulatory protein in nitrogen metabolism)
ADP-ribose	NAD^+	nitrogenase
Methyl group	SAM	substrate receptor protein in chemotaxis

Glutamate dehydrogenase activity is dependent on Mn^{2+}. This enzyme is a hexameric protein and the cation keeps this enzyme in the active taut form. Without the cation, the enzyme becomes the inactive relaxed form. Many proteins are degraded when they are not needed. Degradation is another form of metabolic regulation.

12.4 | Metabolic regulation and growth

As discussed in Chapter 6, growth conditions determine the growth rate and cell yield in a given organism. This is due to the differences in energy conservation efficiency and maintenance energy when growing on different carbon sources. This difference is the result of elaborate metabolic regulation.

12.4.1 Regulation in central metabolism

For the most efficient growth under given conditions, catabolism and anabolism are coordinately regulated not only through the expression of the genes but also through the control of enzyme activities. Figure 12.38 shows how the central metabolic pathways are regulated through the control of enzyme activities by inhibitors and activators. When the adenylate energy charge (EC, Section 5.6.2) is low with the accumulation of AMP and ADP, catabolic enzymes are activated, while enzymes of gluconeogenesis and glycogen synthesis are inhibited under similar conditions. Reduced pyridine nucleotides inhibit the TCA cycle. Some metabolic intermediates participate in the regulation, such as fructose-1,6-diphosphate, dihydroxyacetone phosphate, phosphoenolpyruvate, acetyl-CoA and 2-ketoglutarate.

12.4.2 Regulatory network

Control of gene expression is a fundamental process and it pervades most biological processes, from cell proliferation and differentiation to development. As discussed above, gene expression is regulated at transcriptional and post-transcriptional levels, and enzyme activity is

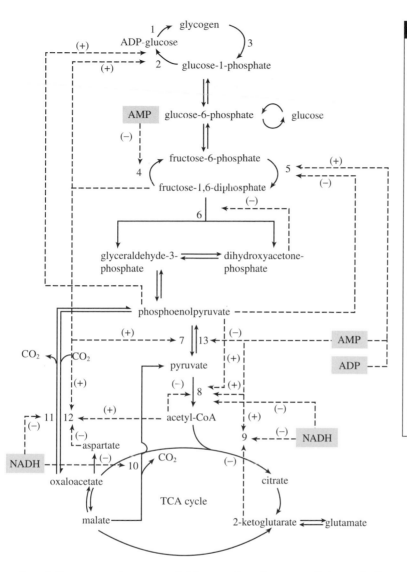

Figure 12.38 **Control of central metabolic pathways in bacteria.**

Detailed metabolic control is organism specific. Generally, the adenylate energy charge and the oxidized/reduced ratio of the pyridine nucleodies are the most important effectors, while some metabolic intermediates such as fructose-1,6-diphosphate and phosphoenolpyruvate are involved in the regulation as effectors.

1, glycogen synthase; 2, ADP-glucose pyrophosphorylase; 3, phosphorylase; 4, fructose-1,6-diphosphatase; 5, phosphofructokinase; 6, fructose-1,6-diphosphate aldolase; 7, pyruvate kinase; 8, pyruvate dehydrogenase; 9, citrate synthase; 10, malate enzyme; 11, PEP carboxykinase; 12, PEP carboxylase; 13, PEP synthetase. (+), activation; (−), inhibition.

regulated for the optimum activities under given conditions. Cells need to integrate intrinsic and environmental information and coordinate multiple regulatory mechanisms of gene expression to properly express their biological functions.

In this sense, it is interesting to note that nitrogen anabolism is regulated according to carbon catabolism. The CcpA–HPr(Ser-P) complex is an important regulator in *Bacillus subtilis*, regulating approximately 10% of the genes in this bacterium (Section 12.1.3.3). A *ccpA* mutant of this low G+C Gram-positive bacterium cannot grow on minimal media containing glucose and ammonia. *Bacillus subtilis* does not contain an anabolic glutamate dehydrogenase and glutamine synthetase (GS) is the only enzyme for ammonium assimilation. The expression of this enzyme (*gltAB* operon) is regulated not only by GlnR and TnrA (Section 12.2.2), but also by glucose or other glycolytically catabolized carbon sources. The latter regulation is mediated through

CcpA. This is an example of global gene regulation as an orchestrated system. It is believed that all catabolic and anabolic processes are regulated coordinately, as exemplified above, for efficient growth or survival in harsh environments.

12.4.3 Growth rate and regulation

The growth rate of microorganisms is dependent on the quality and quantity of their substrate. In a complex medium they grow fast, and need higher enzyme activities. To meet these needs, genes for proteins are activated for higher transcription and more ribosomes are needed to translate the increased transcripts (mRNA). When nutrients are depleted, ribosomal protein (r-proteins) synthesis is inhibited through autogenous regulation, and stationary phase proteins are synthesized through the stringent response. Under starvation conditions, RNAs and proteins are oxidized to supply energy needed for survival. This is referred to as endogenous respiration.

12.5 | Secondary metabolites

Some microorganisms produce compounds not directly related to their growth when their growth is impaired or during the cell differentiation such as sporulation. These compounds are referred to as secondary metabolites. Some of them are industrially important (Table 12.8). Antibiotics are the best known secondary metabolites.

A small molecular weight peptide known as A-factor induces the production of streptomycin in *Streptomyces griseus* and *Streptomyces bikiniensis*. Mutants of A-factor do not have the ability to produce the antibiotic compound, and the addition of A-factor to the culture restores this ability. A-factor is an autoinducer of quorum sensing. When its extracellular concentration reaches a certain level, the signal is transduced through a two-component system (Section 12.2.10). Many Grampositive bacteria produce peptide antibiotics known as bacteriocins. Nisin, a bacteriocin produced by *Lactococcus lactis*, is an autoinducer for quorum sensing in this bacterium.

Table 12.8. *Secondary metabolites produced during cell differentiation*

Secondary metabolite	Producer	Cellular event
Actinomycin D	*Streptomyces antibioticus*	sporulation
Amylase	*Bacillus subtilis*	sporulation
Butanol	*Clostridium acetobutylicum*	sporulation
Pamamycin	*Streptomyces alboniger*	aerial mycelium formation
Tyrocidin	*Bacillus brevis*	sporulation

12.6 | Metabolic regulation and the fermentation industry

'Fermentation' can be defined as 'anaerobic microbial growth through substrate-level phosphorylation under dark conditions' (Chapter 8). This term is also used to describe processes producing useful materials to a large industrial scale using microorganisms. Microorganisms have elaborate regulatory mechanisms for efficient growth but not for the production of specific materials. To improve fermentation efficiency, industry has developed and uses various mutants defective in regulatory mechanisms.

12.6.1 Fermentative production of antibiotics

When Fleming discovered penicillin in 1929, the fungus *Penicillium notatum* produced 1.2 mg/l penicillin. At present, the fermentation industry uses strains producing over 50 g/l penicillin. These industrial strains are mutants derived from the wild-type strains but with altered regulatory mechanisms. Industrial strains of other antibiotic producers have also been developed through extensive mutation programmes.

12.6.2 Fermentative amino acid production

Bacterial strains have been widely used to produce various amino acids. Coryneform bacteria of the genera *Brevibacterium* and *Corynebacterium* are the most commonly used industrial strains in amino acid production. These bacteria excrete amino acids into the surrounding environment when the membrane becomes more permeable to amino acids under biotin-limited conditions. Amino acid production is tightly controlled through various mechanisms, as discussed earlier (Section 12.1). The industrial strains used to produce amino acids are mutant strains with defects in regulation. These are selected based on their resistance properties to analogues (Table 12.9). When an amino acid analogue is added to a culture, the wild-type strain cannot grow since the analogue inhibits expression of the genes for production of the amino acid and the analogue cannot be used for biosynthesis. Auxotrophic mutants are used for the fermentative production of the intermediates. Guanine auxotrophs are used to produce adenine and hypoxanthine. Lately, whole genome sequences have been determined in some industrially important microorganisms and industrial strains have been developed through molecular biology approaches. These may aim to (1) increase enzyme activities, (2) relieve regulatory mechanisms and (3) improve membrane permeability. Such approaches are referred to as metabolic engineering.

Table 12.9. *Bacterial strains producing amino acids*

Amino acid	Strain	Genetic trait	Yield (g/l)
DL-alanine	*Microbacterium ammoniaphilum*	ArgHx[a]	60
L-arginine	*Brevibacterium flavum*	Gua[b], TA[a]	35
L-glutamate	*Corynebacterium glutamicum*	wild-type	>100
L-glutamine	*C. glutamicum*	wild-type	40
L-histidine	*B. flavum*	TA[a], SM[a], Eth[a], ABT[a]	10
L-isoleucine	*B. flavum*	OAHV[a], OMT[a]	15
L-leucine	*Brevibacterium lactofermentum*	Ile[b], Met[b], TA[a]	28
L-lysine	*B. flavum*	AEC[a]	57
L-methionine	*C. glutamicum*	Thr[b], Eth[a], MetHx[a]	2
L-phenylalanine	*C. glutamicum*	Tyr[b], PFP[a], PAP[a]	9
L-threonine	*B. flavum*	Met[b], AHV[a]	18
L-tryptophan	*C. glutamicum*	Phe[b], Tyr[b], 5MT[a]	12
L-valine	*Brevibacterium lactofermentum*	Phe[b], PFP[a], PAP[a], PAT[a], TyrHx[a], TA[a]	31

ABT, 2-aminobenzthiazole; AEC, *S*-(3-aminoethyl)-1-cysteine; AHV, 2-amino-3-hydroxdyvalerate; ArgHx, arginine hydroxamate; Eth, ethionine; Gua, guanine; Ile, isoleucine; Met, methionine; MetHx, methionine hydroxamate; 5MT, 5-methyltryptophan; OMT, *o*-methylthreonine; PAP, p-aminophenylalanine; PAT, p-aminotyrosine; PFP, p-fluorophenylalanine; Phe, phenylalanine; SM, selenomethionine; TA, 2-thiozolalanine; Thr, threonine; Tyr, tyrosine; TyrHx, tyrosine hydroxamate.
[a] Resistant to the analogue.
[b] Auxotrophic mutant.

FURTHER READING

Promoter and σ-factor

Ades, S. E. (2004). Control of the alternative sigma factor σ^E in *Escherichia coli*. *Current Opinion in Microbiology* **7**, 157–162.

Barnard, A., Wolfe, A. & Busby, S. (2004). Regulation at complex bacterial promoters: how bacteria use different promoter organizations to produce different regulatory outcomes. *Current Opinion in Microbiology* **7**, 102–108.

Bell, S. D. (2005). Archaeal transcriptional regulation – variation on a bacterial theme? *Trends in Microbiology* **13**, 262–265.

Borukhov, S. & Severinov, K. (2002). Role of the RNA polymerase sigma subunit in transcription initiation. *Research in Microbiology* **153**, 557–562.

Dove, S. L., Darst, S. A. & Hochschild, A. (2003). Region 4 of σ as a target for transcription regulation. *Molecular Microbiology* **48**, 863–874.

Geiduschek, E. P. & Ouhammouch, M. (2005). Archaeal transcription and its regulators. *Molecular Microbiology* **56**, 1397–1407.

Gourse, R. L., Ross, W. & Rutherford, S. T. (2006). General pathway for turning on promoters transcribed by RNA polymerases containing alternative σ factors. *Journal of Bacteriology* **188**, 4589–4591.

Gruber, T. M. & Gross, C. A. (2003). Multiple sigma subunits and the partitioning of bacterial transcription space. *Annual Review of Microbiology* **57**, 441–466.

Helmann, J. D. (2002). The extracytoplasmic function (ECF) sigma factors. *Advances in Microbial Physiology* **46**, 47–110.

Hinton, D. M. (2005). Molecular gymnastics: distortion of an RNA polymerase σ-factor. *Trends in Microbiology* **13**, 140–143.

Hughes, K. T. & Mathee, K. (1998). The anti-sigma factors. *Annual Review of Microbiology* **52**, 231–286.

Kazmierczak, M. J., Wiedmann, M. & Boor, K. J. (2005). Alternative sigma factors and their roles in bacterial virulence. *Microbiology and Molecular Biology Reviews* **69**, 527–543.

Enzyme induction – activation and repression

Geiduschek, E. P. & Ouhammouch, M. (2005). Archaeal transcription and its regulators. *Molecular Microbiology* **56**, 1397–1407.

Mullerhill, B. (1998). Some repressors of bacterial transcription. *Current Opinion in Microbiology* **1**, 145–151.

Pittard, J., Camakaris, H. & Yang, J. (2005). The TyrR regulon. *Molecular Microbiology* **55**, 16–26.

Rhodius, V. A. & Busby, S. J. W. (1998). Positive activation of gene expression. *Current Opinion in Microbiology* **1**, 152–159.

Rojo, F. (1999). Repression of transcription initiation in bacteria. *Journal of Bacteriology* **181**, 2987–2991.

Roy, S., Garges, S. & Adhya, S. (1998). Activation and repression of transcription by differential contact: two sides of a coin. *Journal of Biological Chemistry* **273**, 14059–14062.

Xu, H. & Hoover, T. R. (2001). Transcriptional regulation at a distance in bacteria. *Current Opinion in Microbiology* **4**, 138–144.

Attenuation

Alifano, P., Fani, R., Lio, P., Lazcano, A., Bazzicalupo, M., Carlomagno, M. S. & Bruni, C. B. (1996). Histidine biosynthetic pathway and genes: structure, regulation, and evolution. *Microbiological Reviews* **60**, 44.

Babitzke, P. (2004). Regulation of transcription attenuation and translation initiation by allosteric control of an RNA-binding protein: the *Bacillus subtilis* TRAP protein. *Current Opinion in Microbiology* **7**, 132–139.

Gollnick, P., Babitzke, P., Antson, A. & Yanofsky, C. (2005). Complexity in regulation of tryptophan biosynthesis in *Bacillus subtilis*. *Annual Review of Genetics* **39**, 47–68.

Mullerhill, B. (1998). Some repressors of bacterial transcription. *Current Opinion in Microbiology* **1**, 145–151.

Rojo, F. (1999). Repression of transcription initiation in bacteria. *Journal of Bacteriology* **181**, 2987–2991.

Yanofsky, C. (2004). The different roles of tryptophan transfer RNA in regulating trp operon expression in *E. coli* versus *B. subtilis*. *Trends in Genetics* **20**, 367–374.

Termination/antitermination

Amster-Choder, O. (2005). The *bgl* sensory system: a transmembrane signaling pathway controlling transcriptional antitermination. *Current Opinion in Microbiology* **8**, 127–134.

Condon, C., Squires, C. & Squires, C. L. (1995). Control of rRNA transcription in *Escherichia coli*. *Microbiological Reviews* **59**, 623.

Gollnick, P. & Babitzke, P. (2002). Transcription attenuation. *Biochimica et Biophysica Acta – Gene Structure and Expression* **1577**, 240–250.

Santangelo, T. J. & Roberts, J. W. (2002). RfaH, a bacterial transcription anti-terminator. *Molecular Cell* **9**, 698–700.

Shu, C. J. & Zhulin, I. B. (2002). ANTAR: an RNA-binding domain in transcription antitermination regulatory proteins. *Trends in Biochemical Sciences* **27**, 3–5.

Weisberg, R. A. & Gottesman, M. E. (1999). Processive antitermination. *Journal of Bacteriology* **181**, 359–367.

Two-component systems

Alexandre, G. & Zhulin, I. B. (2001). More than one way to sense chemicals [Review]. *Journal of Bacteriology* **183**, 4681–4686.

Backert, S. & Selbach, M. (2005). Tyrosine-phosphorylated bacterial effector proteins: the enemies within. *Trends in Microbiology* **13**, 476–484.

Beier, D. & Gross, R. (2006). Regulation of bacterial virulence by two-component systems. *Current Opinion in Microbiology* **9**, 143–152.

Buckler, D. R., Anand, G. S. & Stock, A. M. (2000). Response-regulator phosphorylation and activation: a two-way street? *Trends in Microbiology* **8**, 153–156.

Dunny, G. M. & Leonard, B. A. B. (1997). Cell-cell communication in Gram-positive bacteria. *Annual Review of Microbiology* **51**, 527–564.

Fabret, C., Feher, V. A. & Hoch, J. A. (1999). Two-component signal transduction in *Bacillus subtilis*: how one organism sees its world. *Journal of Bacteriology* **181**, 1975–1983.

Galperin, M. Y. (2004). Bacterial signal transduction network in a genomic perspective. *Environmental Microbiology* **6**, 552–567.

Hancock, L. & Perego, M. (2002). Two-component signal transduction in *Enterococcus faecalis*. *Journal of Bacteriology* **184**, 5819–5825.

Hoch, J. A. & Varughese, K. I. (2001). Keeping signals straight in phosphorelay signal transduction. *Journal of Bacteriology* **183**, 4941–4949.

Hoskisson, P. A. & Hutchings, M. I. (2006). MtrAB-LpqB: a conserved three-component system in actinobacteria? *Trends in Microbiology* **14**, 444–449.

Hutchings, M. I., Hoskisson, P. A., Chandra, G. & Chandra, G. (2004). Sensing and responding to diverse extracellular signals? Analysis of the sensor kinases and response regulators of *Streptomyces coelicolor* A3(2). *Microbiology-UK* **150**, 2795–2806.

Lux, R. & Shi, W. (2005). A novel bacterial signalling system with a combination of a Ser/Thr kinase cascade and a His/Asp two-component system. *Molecular Microbiology* **58**, 345–348.

Novick, R. P. (2003). Autoinduction and signal transduction in the regulation of staphylococcal virulence. *Molecular Microbiology* **48**, 1429–1449.

Ruiz, N. & Silhavy, T. J. (2005). Sensing external stress: watchdogs of the *Escherichia coli* cell envelope. *Current Opinion in Microbiology* **8**, 122–126.

Stock, A. M. & Guhaniyogi, J. (2006). A new perspective on response regulator activation. *Journal of Bacteriology* **188**, 7328–7330.

Stock, A. M., Robinson, V. L. & Goudreau, P. N. (2000). Two-component signal transduction. *Annual Review of Biochemistry* **69**, 183–215.

Varughese, K. (2002). Molecular recognition of bacterial phosphorelay proteins. *Current Opinion in Microbiology* **5**, 142–148.

West, A. H. & Stock, A. M. (2001). Histidine kinases and response regulator proteins in two-component signaling systems. *Trends in Biochemical Sciences* **26**, 369–376.

Autogenous control

Lindner, C., Hecker, M., Le Coq, D. & Deutscher, J. (2002). *Bacillus subtilis* mutant LicT antiterminators exhibiting enzyme I- and HPr-independent antitermination affect catabolite repression of the *bglPH* operon. *Journal of Bacteriology* **184**, 4819–4828.

Schneider, D. A., Ross, W. & Gourse, R. L. (2003). Control of rRNA expression in *Escherichia coli*. *Current Opinion in Microbiology* **6**, 151–156.

Post-transcriptional regulation

Agrawal, N., Dasaradhi, P. V. N., Mohmmed, A., Malhotra, P., Bhatnagar, R. K. & Mukherjee, S. K. (2003). RNA interference: biology, mechanism, and applications. *Microbiology and Molecular Biology Reviews* **67**, 657–685.

Altuvia, S. (2004). Regulatory small RNAs: the key to coordinating global regulatory circuits. *Journal of Bacteriology* **186**, 6679–6680.

Altuvia, S. & Wagner, E. G. (2000). Switching on and off with RNA. *Proceedings of the National Academy of Sciences, USA* **97**, 9824–9826.

Babitzke, P. (2004). Regulation of transcription attenuation and translation initiation by allosteric control of an RNA-binding protein: the *Bacillus subtilis* TRAP protein. *Current Opinion in Microbiology* **7**, 132–139.

Baulcombe, D. (2005). RNA silencing. *Trends in Biochemical Sciences* **30**, 290–293.

Boni, I. V. (2006). Diverse molecular mechanisms of translation initiation in prokaryotes. *Molecular Biology* **40**, 587–596.

Brantl, S. (2002). Antisense-RNA regulation and RNA interference. *Biochimica et Biophysica Acta – Gene Structure and Expression* **1575**, 15–25.

Condon, C. (2003). RNA processing and degradation in *Bacillus subtilis*. *Microbiology and Molecular Biology Reviews* **67**, 157–174.

Condon, C. (2006). Shutdown decay of mRNA. *Molecular Microbiology* **61**, 573–583.

Delihas, N. & Forst, S. (2001). MicF: an antisense RNA gene involved in response of *Escherichia coli* to global stress factors. *Journal of Molecular Biology* **313**, 1–12.

Denli, A. M. & Hannon, G. J. (2003). RNAi: an ever-growing puzzle. *Trends in Biochemical Sciences* **28**, 196–201.

Dennis, P. P., Omer, A. & Lowe, T. (2001). A guided tour: small RNA function in Archaea. *Molecular Microbiology* **40**, 509–519.

Eckstein, F. (2005). Small non-coding RNAs as magic bullets. *Trends in Biochemical Sciences* **30**, 445–452.

Filipowicz, W., Jaskiewicz, L., Kolb, F. A. & Pillai, R. S. (2005). Post-transcriptional gene silencing by siRNAs and miRNAs. *Current Opinion in Structural Biology* **15**, 331–341.

Franch, T. & Gerdes, K. (2000). U-turns and regulatory RNAs. *Current Opinion in Microbiology* **3**, 159–164.

Gelfand, M. S. (2006). Bacterial cis-regulatory RNA structures. *Molecular Biology* **40**, 541–550.

Gottesman, S. (2004). The small RNA regulators of *Escherichia coli*: roles and mechanisms. *Annual Review of Microbiology* **58**, 303–328.

Huttenhofer, A., Brosius, J. & Bachellerie, J. P. (2002). RNomics: identification and function of small, non-messenger RNAs. *Current Opinion in Chemical Biology* **6**, 835–843.

Kennell, D. (2002). Processing endoribonucleases and mRNA degradation in bacteria. *Journal of Bacteriology* **184**, 4645–4657.

Kushner, S. R. (2002). mRNA decay in *Escherichia coli* comes of age. *Journal of Bacteriology* **184**, 4658–4665.

Lease, R. A. & Belfort, M. (2000). Riboregulation by DsrA RNA: trans-actions for global economy. *Molecular Microbiology* **38**, 667–672.

Mata, J., Marguerat, S. & Bahler, J. (2005). Post-transcriptional control of gene expression: a genome-wide perspective. *Trends in Biochemical Sciences* **30**, 506–514.

Nair, V. & Zavolan, M. (2006). Virus-encoded microRNAs: novel regulators of gene expression. *Trends in Microbiology* **14**, 169–175.

Nogueira, T. & Springer, M. (2000). Post-transcriptional control by global regulators of gene expression in bacteria. *Current Opinion in Microbiology* **3**, 154–158.

Nudler, E. & Mironov, A. S. (2004). The riboswitch control of bacterial metabolism. *Trends in Biochemical Sciences* **29**, 11–17.

Rauhut, R. & Klug, G. (1999). mRNA degradation in bacteria. *FEMS Microbiology Reviews* **23**, 353–370.

Repoila, F., Majdalani, N. & Gottesman, S. (2003). Small non-coding RNAs, co-ordinators of adaptation processes in *Escherichia coli*: the RpoS paradigm. *Molecular Microbiology* **48**, 855–861.

Schlax, P. J. & Worhunsky, D. J. (2003). Translational repression mechanisms in prokaryotes. *Molecular Microbiology* **48**, 1157–1169.

Storz, G., Opdyke, J. A. & Zhang, A. (2004). Controlling mRNA stability and translation with small, noncoding RNAs. *Current Opinion in Microbiology* **7**, 140–144.

Takayama, K. & Kjelleberg, S. A. (2000). The role of RNA stability during bacterial stress responses and starvation. *Environmental Microbiology* **2**, 355–365.

Tang, G. (2005). siRNA and miRNA: an insight into RISCs. *Trends in Biochemical Sciences* **30**, 106–114.

Valentin-Hansen, P., Eriksen, M. & Udesen, C. (2004). The bacterial Sm-like protein Hfq: a key player in RNA transactions. *Molecular Microbiology* **51**, 1525–1533.

Wang, Y., Liu, C. L., Storey, J. D., Tibshirani, R. J., Herschlag, D. & Brown, P. O. (2002). Precision and functional specificity in mRNA decay. *Proceedings of the National Academy of Sciences, USA* **99**, 5860–5865.

Winkler, W. C. (2005). Riboswitches and the role of noncoding RNAs in bacterial metabolic control. *Current Opinion in Chemical Biology* **9**, 594–602.

Stringent response

Braeken, K., Moris, M., Daniels, R., Vanderleyden, J. & Michiels, J. (2006). New horizons for (p)ppGpp in bacterial and plant physiology. *Trends in Microbiology* **14**, 45–54.

Dennis, P. P., Ehrenberg, M. & Bremer, H. (2004). Control of rRNA synthesis in *Escherichia coli*: a systems biology approach. *Microbiology and Molecular Biology Reviews* **68**, 639–668.

Ferenci, T. (1999). Regulation by nutrient limitation. *Current Opinion in Microbiology* **2**, 208–213.

Ferenci, T. (2001). Hungry bacteria: definition and properties of a nutritional state. *Environmental Microbiology* **3**, 605–611.

Gralla, J. D. (2005). *Escherichia coli* ribosomal RNA transcription: regulatory roles for ppGpp, NTPs, architectural proteins and a polymerase-binding protein. *Molecular Microbiology* **55**, 973–977.

Hengge-Aronis, R. (1999). Interplay of global regulators and cell physiology in the general stress response of *Escherichia coli*. *Current Opinion in Microbiology* **2**, 148–152.

Hengge-Aronis, R. (2002). Signal transduction and regulatory mechanisms involved in control of the σ^S (RpoS) subunit of RNA polymerase. *Microbiology and Molecular Biology Reviews* **66**, 373–395.

Magnusson, L. U., Farewell, A. & Nystrom, T. (2005). ppGpp: a global regulator in *Escherichia coli*. *Trends in Microbiology* **13**, 236–242.

Spector, M. P. (1998). The starvation-stress response (SSR) of *Salmonella*. *Advances in Microbial Physiology* **40**, 233–279.

Venturi, V. (2003). Control of rpoS transcription in *Escherichia coli* and *Pseudomonas*: why so different? *Journal of Molecular Microbiology and Biotechnology* **49**, 1–9.

Nitrogen control

Arcondeguy, T., Jack, R. & Merrick, M. (2001). P-II signal transduction proteins, pivotal players in microbial nitrogen control. *Microbiology and Molecular Biology Reviews* **65**, 80–105.

Burkovski, A. (2003). Ammonium assimilation and nitrogen control in *Corynebacterium glutamicum* and its relatives: an example for new regulatory mechanisms in actinomycetes. *FEMS Microbiology Reviews* **27**, 617–628.

Charbit, A. (1996). Coordination of carbon and nitrogen metabolism. *Research in Microbiology* **147**, 513–518.

Commichau, F. M., Forchhammer, K. & Stulke, J. (2006). Regulatory links between carbon and nitrogen metabolism. *Current Opinion in Microbiology* **9**, 167–172.

Fisher, S. H. (1999). Regulation of nitrogen metabolism in *Bacillus subtilis*: vive la difference! *Molecular Microbiology* **32**, 223–232.

Forchhammer, K. (2004). Global carbon/nitrogen control by P_{II} signal transduction in cyanobacteria: from signals to targets. *FEMS Microbiology Reviews* **28**, 319–333.

Merrick, M. J. & Edwards, R. A. (1995). Nitrogen control in bacteria. *Microbiological Reviews* **59**, 604–622.

Ninfa, A. J. & Jiang, P. (2005). PII signal transduction proteins: sensors of α-ketoglutarate that regulate nitrogen metabolism. *Current Opinion in Microbiology* **8**, 168–173.

Reitzer, L. (2003). Nitrogen assimilation and global regulation in *Escherichia coli*. *Annual Review of Microbiology* **57**, 155–176.

Schwarz, R. & Forchhammer, K. (2005). Acclimation of unicellular cyanobacteria to macronutrient deficiency: emergence of a complex network of cellular responses. *Microbiology-UK* **151**, 2503–2514.

Silberbach, M. & Burkovski, A. (2006). Application of global analysis techniques to *Corynebacterium glutamicum*: new insights into nitrogen regulation. *Journal of Biotechnology* **126**, 101–110.

Quorum sensing

Atkinson, S., Sockett, R. E., Camara, M. & Williams, P. (2006). Quorum sensing and the lifestyle of *Yersinia*. *Current Issues in Molecular Biology* **8**, 1–10.

Bauer, W. & Robinson, J. (2002). Disruption of bacterial quorum sensing by other organisms. *Current Opinion in Biotechnology* **13**, 234–237.

Daniels, R., Vanderleyden, J. & Michiels, J. (2004). Quorum sensing and swarming migration in bacteria. *FEMS Microbiology Reviews* **28**, 261–289.

De Keersmaecker, S. C. J., Sonck, K. & Vanderleyden, J. (2006). Let LuxS speak up in AI-2 signaling. *Trends in Microbiology* **14**, 114–119.

Fuqua, C. (2006). The QscR quorum-sensing regulon of *Pseudomonas aeruginosa*: an orphan claims its identity. *Journal of Bacteriology* **188**, 3169–3171.

Gonzalez, J. E. & Marketon, M. M. (2003). Quorum sensing in nitrogen-fixing rhizobia. *Microbiology and Molecular Biology Reviews* **67**, 574–592.

Jacob, E. B., Becker, I., Shapira, Y. & Levine, H. (2004). Bacterial linguistic communication and social intelligence. *Trends in Microbiology* **12**, 366–372.

Keller, L. & Surette, M. G. (2006). Communication in bacteria: an ecological and evolutionary perspective. *Nature Reviews Microbiology* **4**, 249–258.

Kjelleberg, S. & Molin, S. (2002). Is there a role for quorum sensing signals in bacterial biofilms? *Current Opinion in Microbiology* **5**, 254–258.

Klose, K. E. (2006). Increased chatter: cyclic dipeptides as molecules of chemical communication in *Vibrio* spp. *Journal of Bacteriology* **188**, 2025–2026.

Miller, M. B. & Bassler, B. L. (2001). Quorum sensing in bacteria. *Annual Review of Microbiology* **55**, 165–199.

Pappas, K. M., Weingart, C. L. & Winans, S. C. (2004). Chemical communication in proteobacteria: biochemical and structural studies of signal synthases and receptors required for intercellular signalling. *Molecular Microbiology* **53**, 755–770.

Parsek, M. R. & Greenberg, E. P. (2005). Sociomicrobiology: the connections between quorum sensing and biofilms. *Trends in Microbiology* **13**, 27–33.

Rasmussen, T. B. & Givskov, M. (2006). Quorum sensing inhibitors: a bargain of effects. *Microbiology-UK* **152**, 895–904.

Reading, N. C. & Sperandio, V. (2006). Quorum sensing: the many languages of bacteria. *FEMS Microbiology Letters* **254**, 1–11.

Roche, D. M., Byers, J. T., Smith, D. S., Glansdorp, F. G., Spring, D. R. & Welch, M. (2004). Communications blackout? Do N-acylhomoserine-lactone-degrading enzymes have any role in quorum sensing? *Microbiology-UK* **150**, 2023–2028.

Sturme, M. H., Kleerebezem, M., Nakayama, J., Akkermans, A. D., Vaugha, E. E. & de Vos, W. M. (2002). Cell to cell communication by autoinducing peptides in Gram-positive bacteria. *Antonie van Leeuwenhoek* **81**, 233–243.

Suntharalingam, P. & Cvitkovitch, D. G. (2005). Quorum sensing in streptococcal biofilm formation. *Trends in Microbiology* **13**, 3–6.

Venturi, V. (2006). Regulation of quorum sensing in *Pseudomonas*. *FEMS Microbiology Reviews* **30**, 274–291.

Whitehead, N. A., Barnard, A. M. L., Slater, H., Simpson, N. J. L. & Salmond, G. P. C. (2001). Quorum-sensing in Gram-negative bacteria. *FEMS Microbiology Reviews* **25**, 365–404.

Withers, H., Swift, S. & Williams, P. (2001). Quorum sensing as an integral component of gene regulatory networks in Gram-negative bacteria. *Current Opinion in Microbiology* **4**, 186–193.

Zhang, L. H. & Dong, Y. H. (2004). Quorum sensing and signal interference: diverse implications. *Molecular Microbiology* **53**, 1563–1571.

FNR and arc systems

Bauer, C. E., Elsen, S. & Bird, T. H. (1999). Mechanisms for redox control of gene expression. *Annual Review of Microbiology* **53**, 495–523.

Beinert, H. & Kiley, P. (1996). Redox control of gene expression involving iron-sulfur proteins. Change of oxidation-state or assembly/disassembly of Fe-S clusters? *FEBS Letters* **382**, 218–219.

Elsen, S., Swem, L. R., Swem, D. L. & Bauer, C. E. (2004). RegB/RegA, a highly conserved redox-responding global two-component regulatory system. *Microbiology and Molecular Biology Reviews* **68**, 263–279.

Green, J. & Paget, M. S. (2004). Bacterial redox sensors. *Nature Reviews Microbiology* **2**, 954–966.

Imlay, J. A. (2006). Iron-sulphur clusters and the problem with oxygen. *Molecular Microbiology* **59**, 1073–1082.

Kiley, P. J. & Beinert, H. (1999). Oxygen sensing by the global regulator, FNR, the role of the iron sulfur cluster. *FEMS Microbiology Reviews* **22**, 341–352.

Kiley, P. J. & Beinert, H. (2003). The role of Fe-S proteins in sensing and regulation in bacteria. *Current Opinion in Microbiology* **6**, 181–185.

Nakano, M. M. & Zuber, P. (1998). Anaerobic growth of a 'strict aerobe' (*Bacillus subtilis*). *Annual Review of Microbiology* **52**, 165–190.

Sawers, G. (2001). A novel mechanism controls anaerobic and catabolite regulation of the *Escherichia coli tdc* operon. *Molecular Microbiology* **39**, 1285–1298.

Taylor, B. L., Zhulin, I. B. & Johnson, M. S. (1999). Aerotaxis and other energy-sensing behavior in bacteria. *Annual Review of Microbiology* **53**, 103–128.

Unden, G. (1998). Transcriptional regulation and energetics of alternative respiratory pathways in facultatively anaerobic bacteria. *Biochimica et Biophysica Acta – Bioenergetics* **1365**, 220–224.

Unden, G. & Schirawski, J. (1997). The oxygen-responsive transcriptional regulator FNR of *Escherichia coli*: the search for signals and reactions. *Molecular Microbiology* **25**, 205–210.

Pho system

Groisman, E. A. (2001). The pleiotropic two-component regulatory system PhoP-PhoQ. *Journal of Bacteriology* **183**, 1835–1842.

Hulett, F. M. (1996). The signal-transduction network for Pho regulation in *Bacillus subtilis*. *Molecular Microbiology* **19**, 933–939.

Lenburg, M. E. & Oshea, E. K. (1996). Signaling phosphate starvation. *Trends in Biochemical Sciences* **21**, 383–387.

Martin, J. F. (2004). Phosphate control of the biosynthesis of antibiotics and other secondary metabolites is mediated by the PhoR-PhoP system: an unfinished story. *Journal of Bacteriology* **186**, 5197–5201.

Vershinina, O. A. & Znamenskaya, L. V. (2002). The *pho* regulons of bacteria. *Microbiology-Moscow* **71**, 497–511.

Heat shock

Aertsen, A., Vanoirbeek, K., De Spiegeleer, P., Sermon, J., Hauben, K., Farewell, A., Nystrom, T. & Michiels, C. W. (2004). Heat shock protein-mediated resistance to high hydrostatic pressure in *Escherichia coli*. *Applied and Environmental Microbiology* **70**, 2660–2666.

Chhabra, S. R., He, Q., Huang, K. H., Gaucher, S. P., Alm, E. J., He, Z., Hadi, M. Z., Hazen, T. C., Wall, J. D., Zhou, J., Arkin, A. P. & Singh, A. K. (2006). Global analysis of heat shock response in *Desulfovibrio vulgaris* Hildenborough. *Journal of Bacteriology* **188**, 1817–1828.

Crapoulet, N., Barbry, P., Raoult, D. & Renesto, P. (2006). Global transcriptome analysis of *Tropheryma whipplei* in response to temperature stresses. *Journal of Bacteriology* **188**, 5228–5239.

Dubern, J. F., Lagendijk, E. L., Lugtenberg, B. J. J. & Bloemberg, G. V. (2005). The heat shock genes *dnaK, dnaJ*, and *grpE* are involved in regulation of putisolvin biosynthesis in *Pseudomonas putida* PCL1445. *Journal of Bacteriology* **187**, 5967–5976.

Engels, S., Schweitzer, J. E., Ludwig, C., Bott, M. & Schaffer, S. (2004). *clpC* and *clpP1P2* gene expression in *Corynebacterium glutamicum* is controlled by a regulatory network involving the transcriptional regulators ClgR and HspR as well as the ECF sigma factor σ^H. *Molecular Microbiology* **52**, 285–302.

Helmann, J. D., Wu, M. F. W., Kobel, P. A., Gamo, F. J., Wilson, M., Morshedi, M. M., Navre, M. & Paddon, C. (2001). Global transcriptional response of *Bacillus subtilis* to heat shock. *Journal of Bacteriology* **183**, 7318–7328.

Hillmann, F., Fischer, R. J. & Bahl, H. (2006). The rubrerythrin-like protein Hsp21 of *Clostridium acetobutylicum* is a general stress protein. *Archives of Microbiology* **185**, 270–276.

Kourennaia, O. V., Tsujikawa, L. & de Haseth, P. L. (2005). Mutational analysis of *Escherichia coli* heat shock transcription factor sigma 32 reveals similarities with sigma 70 in recognition of the −35 promoter element and differences in promoter DNA melting and −10 recognition. *Journal of Bacteriology* **187**, 6762–6769.

Laksanalamai, P., Maeder, D. L. & Robb, F. T. (2001). Regulation and mechanism of action of the small heat shock protein from the hyperthermophilic archaeon *Pyrococcus furiosus*. *Journal of Bacteriology* **183**, 5198–5202.

Musatovova, O., Dhandayuthapani, S. & Baseman, J. B. (2006). Transcriptional heat shock response in the smallest known self-replicating cell, *Mycoplasma genitalium*. *Journal of Bacteriology* **188**, 2845–2855.

Schmid, A. K., Howell, H. A., Battista, Jo. R., Peterson, S. N. & Lidstrom, M. E. (2005). HspR is a global negative regulator of heat shock gene expression in *Deinococcus radiodurans*. *Molecular Microbiology* **55**, 1579–1590.

Senn, M. M., Giachino, P., Homerova, D., Steinhuber, A., Strassner, J., Kormanec, J., Fluckiger, U., Berger-Bachi, B. & Bischoff, M. (2005). Molecular analysis and organization of the σ^B operon in *Staphylococcus aureus*. *Journal of Bacteriology* **187**, 8006–8019.

Servant, P. & Mazodier, P. (2001). Negative regulation of the heat shock response in *Streptomyces*. *Archives of Microbiology* **176**, 237–242.

Tachdjian, S. & Kelly, R. M. (2006). Dynamic metabolic adjustments and genome plasticity are implicated in the heat shock response of the extremely thermoacidophilic archaeon *Sulfolobus solfataricus*. *Journal of Bacteriology* **188**, 4553–4559.

Cold shock

Beckering, C. L., Steil, L., Weber, M. H. W., Volker, U. & Marahiel, M. A. (2002). Genomewide transcriptional analysis of the cold shock response in *Bacillus subtilis*. *Journal of Bacteriology* **184**, 6395–6402.

Beran, R. K. & Simons, R. W. (2001). Cold-temperature induction of *Escherichia coli* polynucleotide phosphorylase occurs by reversal of its autoregulation. *Molecular Microbiology* **39**, 112–125.

Cairrao, F., Cruz, A., Mori, H. & Arraiano, C. M. (2003). Cold shock induction of RNase R and its role in the maturation of the quality control mediator SsrA/tmRNA. *Molecular Microbiology* **50**, 1349–1360.

Cavicchioli, R., Thomas, T. & Curmi, P. M. G. (2000). Cold stress response in Archaea. *Extremophiles* **4**, 321–331.

Datta, P. P. & Bhadra, R. K. (2003). Cold shock response and major cold shock proteins of *Vibrio cholerae*. *Applied and Environmental Microbiology* **69**, 6361–6369.

Fang, L., Hou, Y. & Inouye, M. (1998). Role of the cold-box region in the 5′ untranslated region of the cspA mRNA in its transient expression at low temperature in *Escherichia coli*. *Journal of Bacteriology* **180**, 90–95.

Gao, H., Yang, Z. K., Wu, L., Thompson, D. K. & Zhou, J. (2006). Global transcriptome analysis of the cold shock response of *Shewanella oneidensis* MR-1 and mutational analysis of its classical cold shock proteins. *Journal of Bacteriology* **188**, 4560–4569.

Gerday, C., Aittaleb, M., Bentahir, M., Chessa, J. P., Claverie, P., Collins, T., D'Amico, S., Dumont, J., Garsoux, G., Georlette, D., Hoyoux, A., Lonhienne, T., Meuwis, M. A. & Feller, G. (2000). Cold-adapted enzymes: from fundamentals to biotechnology. *Trends in Biotechnology* **18**, 103–107.

Giangrossi, M., Exley, R. M., Le Hegarat, F. & Pon, C. L. (2001). Different *in vivo* localization of the *Escherichia coli* proteins CspD and CspA. *FEMS Microbiology Letters* **202**, 171–176.

Graumann, P. & Marahiel, M. A. (1996). Some like it cold: response of microorganisms to cold shock. *Archives of Microbiology* **166**, 293–300.

Graumann, P. L. & Marahiel, M. A. (1998). A superfamily of proteins that contain the cold-shock domain. *Trends in Biochemical Sciences* **23**, 286–290.

Hunger, K., Beckering, C. L., Wiegeshoff, F., Graumann, P. L. & Marahiel, M. A. (2006). Cold-induced putative DEAD box RNA helicases CshA and CshB are essential for cold adaptation and interact with cold shock protein B in *Bacillus subtilis*. *Journal of Bacteriology* **188**, 240–248.

Katzif, S., Lee, E. H., Law, A. B., Tzeng, Y. L. & Shafer, W. M. (2005). CspA regulates pigment production in *Staphylococcus aureus* through a SigB-dependent mechanism. *Journal of Bacteriology* **187**, 8181–8184.

Lopez, M. M., Yutani, K. & Makhatadze, G. I. (2001). Interactions of the cold shock protein CspB from *Bacillus subtilis* with single-stranded DNA: importance of the T base content and position within the template. *Journal of Biological Chemistry* **276**, 15511–15518.

Magg, C., Kubelka, J., Holtermann, G., Haas, E. & Schmid, F. X. (2006). Specificity of the initial collapse in the folding of the cold shock protein. *Journal of Molecular Biology* **360**, 1067–1080.

Martinez-Costa, O. H., Zalacain, M., Holmes, D. J. & Malpartida, F. (2003). The promoter of a cold-shock-like gene has pleiotropic effects on *Streptomyces* antibiotic biosynthesis. *FEMS Microbiology Letters* **220**, 215–221.

Phadtare, S. & Inouye, M. (2004). Genome-wide transcriptional analysis of the cold shock response in wild-type and cold-sensitive, quadruple-*csp*-deletion strains of *Escherichia coli*. *Journal of Bacteriology* **186**, 7007–7014.

Polissi, A., de Laurentis, W., Zangrossi, S., Briani, F., Longhi, V., Pesole, G. & Deho, G. (2003). Changes in *Escherichia coli* transcriptome during acclimatization at low temperature. *Research in Microbiology* **154**, 573–580.

Prud'homme-Genereux, A., Beran, R. K., Iost, I., Ramey, C. S., Mackie, G. A. & Simons, R. W. (2004). Physical and functional interactions among RNase E, polynucleotide phosphorylase and the cold-shock protein, CsdA: evidence for a 'cold shock degradosome'. *Molecular Microbiology* **54**, 1409–1421.

Sakamoto, T. & Murata, N. (2002). Regulation of the desaturation of fatty acids and its role in tolerance to cold and salt stress. *Current Opinion in Microbiology* **5**, 206–210.

Smirnova, G. V. & Zakirova, O. N. (2001). The role of antioxidant systems in the cold stress response of *Escherichia coli*. *Microbiology-Moscow* **70**, 45–50.

Weber, M. H. W., Klein, W., Muller, L., Niess, U. M. & Marahiel, M. A. (2001). Role of the *Bacillus subtilis* fatty acid desaturase in membrane adaptation during cold shock. *Molecular Microbiology* **39**, 1321–1329.

Wiegeshoff, F., Beckering, C. L., Debarbouille, M. & Marahiel, M. A. (2006). Sigma L is important for cold shock adaptation of *Bacillus subtilis*. *Journal of Bacteriology* **188**, 3130–3133.

Wouters, J. A., Rombouts, F. M., Kuipers, O. P., de Vos, W. M. & Abee, T. (2000). The role of cold-shock proteins in low-temperature adaptation of food-related bacteria. *Systematic and Applied Microbiology* **23**, 165–173.

Xia, B., Ke, H., Jiang, W. & Inouye, M. (2002). The Cold Box stem-loop proximal to the 5'-end of the *Escherichia coli* cspA gene stabilizes its mRNA at low temperature. *Journal of Biological Chemistry* **277**, 6005–6011.

Yamanaka, K. & Inouye, M. (2001). Selective mRNA degradation by polynucleotide phosphorylase in cold shock adaptation in *Escherichia coli*. *Journal of Bacteriology* **183**, 2808–2816.

Oxidative stress

Brioukhanov, A. L., Netrusov, A. I. & Eggen, R. I. L. (2006). The catalase and superoxide dismutase genes are transcriptionally up-regulated upon oxidative stress in the strictly anaerobic archaeon *Methanosarcina barkeri*. *Microbiology-UK* **152**, 1671–1677.

Gaudu, P., Dubrac, S. & Touati, D. (2000). Activation of SoxR by overproduction of desulfoferrodoxin: multiple ways to induce the *soxRS* regulon. *Journal of Bacteriology* **182**, 1761–1763.

Giro, M., Carrillo, N. & Krapp, A. R. (2006). Glucose-6-phosphate dehydrogenase and ferredoxin-NADP(H) reductase contribute to damage repair during the soxRS response of *Escherichia coli*. *Microbiology-UK* **152**, 1119–1128.

Gonzalez-Flecha, B. & Demple, B. (1999). Role for the *oxyS* gene in regulation of intracellular hydrogen peroxide in *Escherichia coli*. *Journal of Bacteriology* **181**, 3833–3836.

Hassett, D. J., Ma, J. F., Elkins, J. G., McDermott, T. R., Ochsner, U. A., West, S. E. H., Huang, C-T., Fredericks, J., Burnett, S., Stewart, P. S., McFeters, G., Passador, L. and Iglewski, B. H. (1999). Quorum sensing in *Pseudomonas aeruginosa* controls expression of catalase and superoxide dismutase genes and mediates biofilm susceptibility to hydrogen peroxide. *Molecular Microbiology* **34**, 1082–1093.

Imlay, J. A. (2003). Pathways of oxidative damage. *Annual Review of Microbiology* **57**, 395–418.

Manchado, M., Michan, C. & Pueyo, C. (2000). Hydrogen peroxide activates the SoxRS regulon *in vivo*. *Journal of Bacteriology* **182**, 6842–6844.

Ohara, N., Kikuchi, Y., Shoji, M., Naito, M. & Nakayama, K. (2006). Superoxide dismutase-encoding gene of the obligate anaerobe *Porphyromonas gingivalis* is regulated by the redox-sensing transcription activator OxyR. *Microbiology-UK* **152**, 955–966.

Rocha, E. R., Owens, G. & Smith, C. J. (2000). The redox-sensitive transcriptional activator OxyR regulates the peroxide response regulon in the obligate anaerobe *Bacteroides fragilis*. *Journal of Bacteriology* **182**, 5059–5069.

Toledano, M. B., Delaunay, A., Monceau, L. & Tacnet, F. (2004). Microbial H_2O_2 sensors as archetypical redox signaling modules. *Trends in Biochemical Sciences* **29**, 351–357.

Osmotic stress

Bohin, J. P. (2000). Osmoregulated periplasmic glucans in Proteobacteria. *FEMS Microbiology Letters* **186**, 11–19.

Kramer, R. & Morbach, S. (2004). BetP of *Corynebacterium glutamicum*, a transporter with three different functions: betaine transport, osmosensing, and osmoregulation. *Biochimica et Biophysica Acta – Bioenergetics* **1658**, 31–36.

Leonardo, M. R. & Forst, S. (1996). Re-examination of the role of the periplasmic domain of EnvZ in sensing of osmolarity signals in *Escherichia coli*. *Molecular Microbiology* **22**, 405–413.

Poolman, B., Spitzer, J. J. & Wood, J. M. (2004). Bacterial osmosensing: roles of membrane structure and electrostatics in lipid-protein and protein-protein interactions. *Biochimica et Biophysica Acta – Biomembranes* **1666**, 88–104.

Sleator, R. D. & Hill, C. (2002). Bacterial osmoadaptation: the role of osmolytes in bacterial stress and virulence. *FEMS Microbiology Reviews* **26**, 49–71.

Wood, J. M. (1999). Osmosensing by bacteria: signals and membrane-based sensors. *Microbiology and Molecular Biology Reviews* **63**, 230–262.

Chemotaxis

Aizawa, S., Harwood, C. S. & Kadner, R. J. (2000). Signaling components in bacterial locomotion and sensory reception. *Journal of Bacteriology* **182**, 1459–1471.

Alexandre, G., Greer-Phillips, S. & Zhulin, I. B. (2004). Ecological role of energy taxis in microorganisms. *FEMS Microbiology Reviews* **28**, 113–126.

Blair, D. F. (1995). How bacteria sense and swim. *Annual Review of Microbiology* **49**, 489–522.

Bren, A. & Eisenbach, M. (2000). How signals are heard during bacterial chemotaxis: protein-protein interactions in sensory signal propagation. *Journal of Bacteriology* **182**, 6865–6873.

Chilcott, G. S. & Hughes, K. T. (2000). Coupling of flagellar gene expression to flagellar assembly in *Salmonella enterica* serovar typhimurium and *Escherichia coli*. *Microbiology and Molecular Biology Reviews* **64**, 694–708.

Levit, M. N., Liu, Y. & Stock, J. B. (1998). Stimulus response coupling in bacterial chemotaxis: receptor dimers in signalling arrays. *Molecular Microbiology* **30**, 459–466.

Macnab, R. M. (1999). The bacterial flagellum: reversible rotary propeller and type III export apparatus. *Journal of Bacteriology* **181**, 7149–7153.

Manson, M. D., Armitage, J. P., Hoch, J. A. & Macnab, R. M. (1998). Bacterial locomotion and signal transduction. *Journal of Bacteriology* **180**, 1009–1022.

Mitchell, J. G. & Kogure, K. (2006). Bacterial motility: links to the environment and a driving force for microbial physics. *FEMS Microbiology Ecology* **55**, 3–16.

Parkinson, J. S. (2003). Bacterial chemotaxis: a new player in response regulator dephosphorylation. *Journal of Bacteriology* **185**, 1492–1494.

Parkinson, J. S., Ames, P. & Studdert, C. A. (2005). Collaborative signaling by bacterial chemoreceptors. *Current Opinion in Microbiology* **8**, 116–121.

Sourjik, V. (2004). Receptor clustering and signal processing in *E. coli* chemotaxis. *Trends in Microbiology* **12**, 569–576.

Szurmant, H. & Ordal, G. W. (2004). Diversity in chemotaxis mechanisms among the bacteria and archaea. *Microbiology and Molecular Biology Reviews* **68**, 301–319.

Taylor, B. L., Zhulin, I. B. & Johnson, M. S. (1999). Aerotaxis and other energy-sensing behavior in bacteria. *Annual Review of Microbiology* **53**, 103–128.

Adaptive mutation

Aertsen, A. & Michiels, C. W. (2005). Diversify or die: generation of diversity in response to stress. *Critical Reviews in Microbiology* **31**, 69–78.

Amzallag, G. N. (2004). Adaptive changes in bacteria: a consequence of non-linear transitions in chromosome topology? *Journal of Theoretical Biology* **229**, 361–369.

Dubnau, D. & Losick, R. (2006). Bistability in bacteria. *Molecular Microbiology* **61**, 564–572.

Foster, P. L. (1993). Adaptive mutation: the uses of adversity. *Annual Review of Microbiology* **47**, 467–504.

Foster, P. L. (2005). Stress responses and genetic variation in bacteria. *Mutation Research–Fundamental and Molecular Mechanisms of Mutagenesis* **569**, 3–11.

Goudreau, P. N. & Stock, A. M. (1998). Signal transduction in bacteria: molecular mechanisms of stimulus-response coupling. *Current Opinion in Microbiology* **1**, 160–169.

Igoshin, O. A., Price, C. W. & Savageau, M. A. (2006). Signalling network with a bistable hysteretic switch controls developmental activation of the sigmaF transcription factor in *Bacillus subtilis*. *Molecular Microbiology* **61**, 165–184.

Martin, B., Quentin, Y., Fichant, G. & Claverys, J. P. (2006). Independent evolution of competence regulatory cascades in streptococci? *Trends in Microbiology* **14**, 339–345.

Osorio, G. & Jerez, C. A. (1996). Adaptive response of the archaeon *Sulfolobus acidocaldarius* BC65 to phosphate starvation. *Microbiology-UK* **142**, 1531–1536.

Rosenberg, S. M. (1994). In pursuit of a molecular mechanism for adaptive mutation. *Genome* **37**, 893–899.

Sniegowski, P. D. & Lenski, R. E. (1995). Mutation and adaptation: the directed mutation controversy in evolutionary perspective. *Annual Review of Ecology and Systematics* **26**, 553–578.

Wright, B. E. (2004). Stress-directed adaptive mutations and evolution. *Molecular Microbiology* **52**, 643–650.

Enzyme activity modulation and metabolic flux

Arcondeguy, T., Jack, R. & Merrick, M. (2001). P-II signal transduction proteins, pivotal players in microbial nitrogen control. *Microbiology Molecular Biology Reviews* **65**, 80–105.

Commichau, F. M., Forchhammer, K. & Stülke, J. (2006). Regulatory links between carbon and nitrogen metabolism. *Current Opinion in Microbiology* **9**, 167–172.

Edwards, J. S., Covert, M. & Palsson, B. (2002). Metabolic modelling of microbes: the flux-balance approach. *Environmental Microbiology* **4**, 133–140.

Ehrmann, M. & Clausen, T. (2004). Proteolysis as a regulatory mechanism. *Annual Review of Genetics* **38**, 709–724.

El-Mansi, M., Cozzone, A. J., Shiloach, J. & Eikmanns, B. J. (2006). Control of carbon flux through enzymes of central and intermediary metabolism during growth of *Escherichia coli* on acetate. *Current Opinion in Microbiology* **9**, 173–179.

Forchhammer, K. (2004). Global carbon/nitrogen control by P_{II} signal transduction in cyanobacteria: from signals to targets. *FEMS Microbiology Reviews* **28**, 319–333.

Hengge, R. & Gourse, R. L. (2004). Cell regulation: tying together the cellular regulatory network. *Current Opinion in Microbiology* **7**, 99–101.

Herrero, A. (2004). New targets of the P_{II} signal transduction protein identified in cyanobacteria. *Molecular Microbiology* **52**, 1225–1228.

Jenal, U. & Hengge-Aronis, R. (2003). Regulation by proteolysis in bacterial cells. *Current Opinion in Microbiology* **6**, 163–172.

Ninfa, A. J. & Jiang, P. (2005). P_{II} signal transduction proteins: sensors of α-ketoglutarate that regulate nitrogen metabolism. *Current Opinion in Microbiology* **8**, 168–173.

Noirot, P. & Noirot-Gros, M. F. (2004). Protein interaction networks in bacteria. *Current Opinion in Microbiology* **7**, 505–512.

Russell, R. B., Alber, F., Aloy, P., Davis, F. P., Korkin, D., Pichaud, M., Topf, M. & Sali, A. (2004). A structural perspective on protein-protein interactions. *Current Opinion in Structural Biology* **14**, 313–324.

Energy, environment and microbial survival

As mentioned repeatedly in this book, the goal of life is preservation of the species through reproduction, but this requires energy. Although there are a few exceptional copiotrophic environments such as foodstuffs and animal guts, most ecosystems where microorganisms are found are oligotrophic. Those organisms that can utilize nutrients efficiently have a better chance of survival in such ecosystems. Further, many microbes synthesize reserve materials, when available nutrients are in excess, and utilize these under starvation conditions, while various resting cells are produced under conditions where growth is difficult. In this chapter, the main bacterial survival mechanisms are discussed in terms of reserve materials and resting cell types.

13.1 | Survival and energy

As discussed earlier, living microorganisms maintain a certain level of adenylate energy charge (EC) and proton motive force even under starvation conditions (Section 5.6.2). These forms of biological energy are needed for the basic metabolic processes necessary to survive such as transport and the turnover of macromolecules. Maintenance energy is the term used for this energy.

Under starvation conditions, cells utilize cellular components including reserve and non-essential materials for survival. This is referred to as endogenous metabolism. Almost all prokaryotes accumulate at least one type of reserve material under energy-rich conditions. During a period of starvation, the reserve material(s) are consumed through endogenous metabolism before the organism oxidizes other cellular constituents such as proteins and RNA that are not needed under the starvation conditions (Figure 13.1). When a population starves, some individuals die thereby providing an energy source for other members of the population in a mechanism known as programmed cell death (Section 13.3.5).

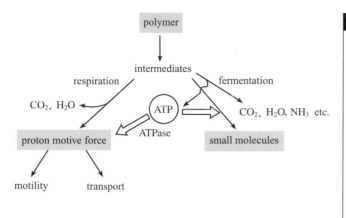

Figure 13.1 **Energy sources for survival under starvation conditions.**

(Dawes, E. A. 1986, *Microbial Energetics*, Figure 12.1. Blackie & Son, Glasgow)

When nutrients are unavailable in the environment, microbes utilize cellular constituents, including reserve materials and polymers that are not essential for survival. This is referred to as endogenous metabolism.

13.2 | Reserve materials in bacteria

RNA and proteins used for endogenous metabolism are not referred to as reserve materials since they have specific functions other than as substrates for endogenous metabolism. Reserve material can be defined as polymers synthesized when an energy source is supplied in excess, and used as a substrate for endogenous metabolism without any other cellular functions. Almost all prokaryotes accumulate at least one type of reserve material. It is likely that the ability to accumulate reserve materials is advantageous for survival in natural habitats.

Reserve materials in bacteria can be grouped into four categories according to their chemical nature. These are carbohydrates such as glycogen, lipids such as poly-β-hydroxybutyrate (PHB), polypeptides and polyphosphate.

13.2.1 Carbohydrate reserve materials: glycogen and trehalose

Glycogen is the most common carbohydrate reserve material in prokaryotes as in eukaryotes. Glycogen is a polysaccharide consisting of glucose with α-1,4-linkages as well as α-1,6-linkages. Many prokaryotes synthesize this polysaccharide when the energy source is in excess and when one or more essential elements are limiting, such as nitrogen.

Like all polysaccharides, glycogen is synthesized from an activated monomer. Glycogen synthase transfers glucose from ADP-glucose to the existing glycogen molecule forming an α-1,4-linkage (Figure 13.2, Section 6.9.1). ADP-glucose pyrophosphorylase is activated by fructose-6-phosphate, fructose-1,6-diphosphate, phosphoenolpyruvate and pyruvate, and repressed by AMP for regulation of glycogen synthesis.

Glycogen is utilized by the action of two enzymes, a debranching enzyme and glycogen phosphorylase. *Escherichia coli* has strong activities of these enzymes and uses glycogen at a high rate at the beginning of

Figure 13.2 Regulation of glycogen synthesis and hydrolysis.

PEP, phosphoenolpyruvate; FDP, fructose-1,6-diphosphate. +, activation; −, repression.

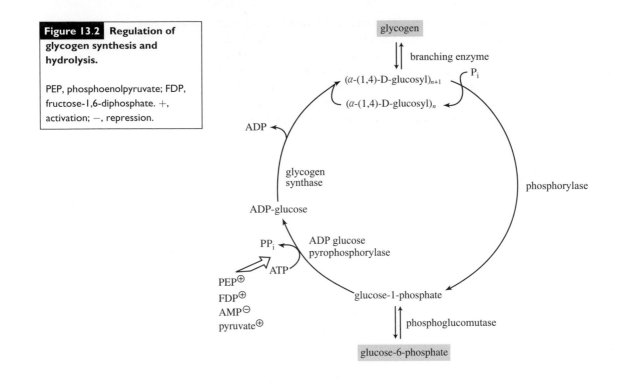

starvation; consequently this bacterium does not survive very long under starvation conditions. On the other hand, *Arthrobacter globiformis* stays viable for long periods of time under starvation conditions since this bacterium utilizes the reserve polysaccharide slowly with low activity of the debranching enzyme.

Some fungi synthesize trehalose as their reserve material. This disaccharide is a glucose dimer with an α-1,1-linkage. This disaccharide is not a reserve material in prokaryotes, but many bacteria can use trehalose as their sole carbon and energy source. Some bacteria, including *Desulfovibrio halophilus* and several purple sulfur and non-sulfur bacteria, synthesize trehalose under high osmotic pressure as a compatible solute, but this disaccharide is not regarded as a reserve material.

Trehalose-6-phosphate synthase condenses UDP-glucose and glucose-6-phosphate to trehalose-6-phosphate that is dephosphorylated to trehalose by trehalose-6-phosphatase. This disaccharide is metabolized in three different ways depending on the organism. Trehalase hydrolyzes it to two molecules of glucose, while trehalose phosphorylase cleaves it to glucose and glucose-1-phosphate. In other organisms, trehalose is phosphorylated to trehalose-6-phosphate by the action of trehalose kinase before being cleaved to glucose-6-phosphate and glucose by a hydrolase.

13.2.2   Lipid reserve materials

Poly-β-hydroxyalkanoates (PHAs), triacylglycerides (TAG), wax esters and hydrocarbons are synthesized as reserve materials in bacteria.

These lipophilic compounds are accumulated as inclusion bodies in the cytoplasm. Many proteins are associated with the inclusion bodies. These are mostly enzymes involved in the metabolism of the reserve material. PHAs are the most common reserve material in prokaryotes but are not found in eukaryotes. On the other hand, only a few prokaryotes have the property of accumulating triacylglycerides as a reserve material, a property that is widespread in eukaryotes. Wax esters are accumulated mostly in bacteria.

13.2.2.1 Poly-β-hydroxyalkanoate (PHA)

PHA is a reserve material accumulated in bacteria and archaea with a general structure:

$$HO-CH-CH_2-C-\left[-O-CH-CH_2-C-\right]-O-CH-CH_2-COOH$$

$$\begin{array}{ccccc} | & \| & | & \| & | \\ R & O & R & O & R \end{array}$$

The composition of the monomer differs depending on the strain but 3-hydroxybutyrate ($R = CH_3$) is the most abundant form. For this reason, PHA is usually called poly-β-hydroxybutyrate (PHB). Other monomers include 3-hydroxyalkanoate with carbon numbers of 5, 6, 7 and 8. Generally, the PHA in the inclusion body is a macromolecule with a molecular weight higher than 10^6 daltons. PHA is an ideal reserve material since this water-insoluble polymer does not increase the intracellular osmotic pressure and the energy content is higher than that of carbohydrates.

Under nitrogen- and oxygen-limited conditions, some bacteria including *Ralstonia eutropha* (*Alcaligenes eutrophus*) and *Azotobacter beijerinckii* accumulate up to 80% of their dry cell weight as PHA. An industrial process has been developed to produce PHA as a biodegradable plastic using a strain of *Ralstonia eutropha*.

PHB is synthesized through the polymerization of 3-hydroxybutyrate, which is a condensation product of acetyl-CoA in a series of reactions similar to the clostridial butyrate fermentation (Figure 13.3).

Azotobacter beijerinckii metabolizes carbohydrates though the ED pathway, and the resulting NADH is reoxidized through the electron transport system generating a proton motive force when the oxygen supply is not limited. When oxygen is limited, NADH is accumulated to inhibit enzymes reducing NAD^+ including glucose-6-phosphate dehydrogenase, citrate synthase and isocitrate dehydrogenase. When oxygen is limiting, acetyl-CoA is directed toward PHB synthesis reoxidizing NADH. For this reason, PHB is regarded as a reserve material and as an electron sink allowing the bacterium to continue glycolytic metabolism.

PHB depolymerase removes 3-hydroxybutyrate from PHB, and the product is oxidized to acetoacetate. Coenzyme A transferase activates acetoacetate to acetoacetyl-CoA consuming succinyl-CoA. Since acetoacetyl-CoA is involved both in synthesis and degradation, the enzymes related to this intermediate need to be strongly regulated

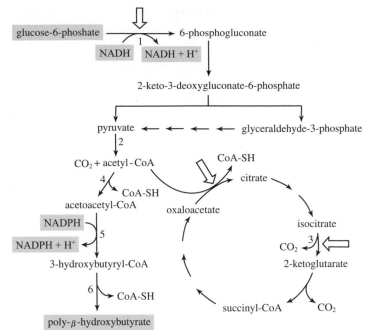

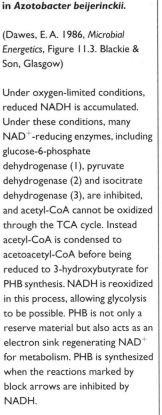

Figure 13.3 **Central metabolism and PHB synthesis in *Azotobacter beijerinckii*.**

(Dawes, E. A. 1986, *Microbial Energetics*, Figure 11.3. Blackie & Son, Glasgow)

Under oxygen-limited conditions, reduced NADH is accumulated. Under these conditions, many NAD$^+$-reducing enzymes, including glucose-6-phosphate dehydrogenase (1), pyruvate dehydrogenase (2) and isocitrate dehydrogenase (3), are inhibited, and acetyl-CoA cannot be oxidized through the TCA cycle. Instead acetyl-CoA is condensed to acetoacetyl-CoA before being reduced to 3-hydroxybutyrate for PHB synthesis. NADH is reoxidized in this process, allowing glycolysis to be possible. PHB is not only a reserve material but also acts as an electron sink regenerating NAD$^+$ for metabolism. PHB is synthesized when the reactions marked by block arrows are inhibited by NADH.

1, glucose-6-phosphate dehydrogenase; 2, pyruvate dehydrogenase; 3, isocitrate dehydrogenase; 4, acetyl-CoA acetyltransferase (thiolase); 5, 3-hydroxybutyryl-CoA dehydrogenase; 6, 3-hydroxybutyryl-CoA polymerase.

to avoid a futile cycle wasting a high energy bond in the form of succinyl-CoA (Figure 13.4).

13.2.2.2 Triacylglyceride (TAG)

TAG can serve as a reserve material in eukaryotes as well as in prokaryotes. Among prokaryotes, TAG is found mainly in the actinomycetes, including species of *Mycobacterium*, *Nocardia*, *Rhodococcus*, *Streptomyces*, *Micromonospora* and *Gordonia*. Among Gram-negative bacteria, species of *Acinetobacter* accumulate small amounts. A strain of *Rhodococcus opacus* accumulates TAG up to 87% of the cell dry weight.

TAG is synthesized from glycerol-3-phosphate and acyl-ACP. Phosphatidic acid is produced as in phospholipid synthesis (Section 6.6) before phosphatidate phosphatase removes the phosphate group. Finally the third acyl group is added to diacylglycerol from acyl-ACP by the action of diacylglycerol acyltransferase (DGAT). DGAT is believed to be associated with the cytoplasmic membrane. TAG is accumulated with an excess of carbon source but when nutrients other than the carbon source are limited as with PHA.

13.2.2.3 Wax ester and hydrocarbons

Wax ester (WE) is synthesized by various bacteria including species of *Acinetobacter*, *Moraxella*, *Micrococcus*, *Corynebacterium* and *Nocardia*. Under nitrogen-limited conditions, a strain of *Acinetobacter calcoaceticus* accumulates up to 25% of the cell dry weight. It is not known how WE is synthesized, but wax ester synthase is associated with the cytoplasmic membrane like DGAT.

Hydrocarbons are accumulated by various microorganisms, especially algae. These microbial hydrocarbons are either isoprenoids or

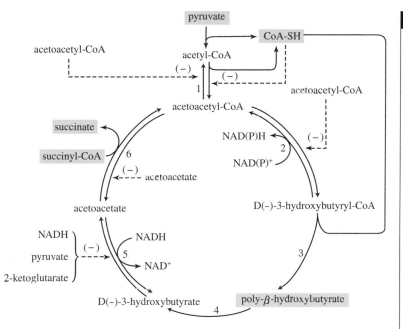

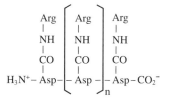

Figure 13.4 **PHB degradation and its regulation in _Azotobacter beijerinckii._**

(Dawes, E. A. 1986, _Microbial Energetics_, Figure 11.4. Blackie & Son, Glasgow)

PHB degradation is regulated not only by intermediates of central metabolism such as pyruvate and 2-ketoglutarate but also by NADH, since PHB functions as a reserve material as well as an electron sink. PHB is depolymerized to 3-hydroxybutyrate before being reduced to acetoacetate that is activated by coenzyme A transferase to acetoacetyl-CoA. Since acetoacetyl-CoA is an intermediate in degradation as well as in the synthesis of this reserve material, reactions involving this intermediate are tightly controlled to avoid a futile cycle which would waste a high energy bond in the form of succinyl-CoA.

1, acetoacetyl-CoA acetyltransferase; 2, 3-hydroxybutyryl-CoA dehydrogenase; 3, 3-hydroxybutyryl-CoA polymerase; 4, PHB depolymerase; 5, 3-hydroxybutyrate dehydrogenase; 6, coenzyme A transferase. – , inhibition.

alkanes. A strain of _Vibrio furnissii_ isolated from a sewage works accumulated lipid material extracellularly to 1.2 times its cell dry weight. Half of the extracellular lipid consisted of alkanes with a carbon number of 15–24.

13.2.3 Polypeptides as reserve materials

Cyanobacteria utilize not only the usual reserve materials such as glycogen, poly-β-hydroxyalkanoate (PHA) and polyphosphate, but also peptides. Many cyanobacteria accumulate cyanophycin as a reserve material for carbon and nitrogen. Cyanophycin has a structure composed of polyaspartate, each monomer of which is linked with a molecule of arginine (Figure 13.5). This peptide has a molecular weight of between 25,000 and 125,000 daltons. This peptide is synthesized under phosphate- or sulfate-limited conditions with nitrogen and light in excess. When the synthesis of protein and nucleic acids is inhibited, cyanophycin production is activated. Its synthesis is not inhibited by tetracycline, showing that cyanophycin is synthesized through a non-ribosomal mechanism.

When the nitrogen supply is limited, cyanophycin is degraded producing ammonia and carbon dioxide. It is not known if energy is conserved during oxidation of the amino acids. In heterocystous cyanobacteria, the heterocysts have a higher enzyme activity for cyanophycin synthesis than the vegetative cells.

Phycocyanin is a pigment peptide in the antenna molecule. When the nitrogen supply is limited, this peptide is also degraded as a nitrogen source. Cyanobacteria have a bluish-green colour under normal conditions but become yellowish-green under nitrogen-limited conditions because the blue coloured phycocyanin is degraded. It

Figure 13.5 **Structure of cyanophycin, a peptide reserve material in cyanobacteria.**

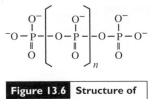

Figure 13.6 Structure of polyphosphate.

should be noted that phycocyanin is not a 'true' reserve material according to the definition given earlier.

13.2.4 Polyphosphate

It has been known for a long time that many bacteria have cytoplasmic granules that are stainable with basic dyes such as toluidine blue. These granules are composed of polyphosphate and are known in prokaryotes and eukaryotes. The number of phosphate residues ranges from two to over a million (Figure 13.6). Polyphosphate is consumed when the phosphate supply is limited, and in some organisms this can substitute for ATP in energy-requiring reactions. Polyphosphate has functions other than just as a phosphate and energy reserve material. These include regulation of the concentration of cytoplasmic cations because of its strong anionic properties, stabilization of the cytoplasmic membrane and regulation of gene expression and enzyme activity.

Polyphosphate is synthesized through the transfer of phosphate from ATP or from an intermediate of the glycolytic pathway, 1,3-diphosphoglycerate, to the existing polyphosphate. ATP-polyphosphate phosphotransferase (polyphosphate kinase) mediates the reaction with ATP. This reaction is reversible, and phosphate can be transferred not only to ADP but also to GDP, UDP and CDP.

$$(P)_n + ATP \xrightleftharpoons{\text{polyphosphate kinase}} (P)_{n+1} + ADP$$

The reaction with 1,3-diphosphoglycerate as the phosphate donor is mediated by 1,3-diphosphoglycerate-polyphosphate phosphotransferase:

1,3-diphosphoglycerate   3-phosphoglycerate

Polyphosphate synthesis is under elaborate regulation. In *Escherichia coli*, polyphosphate is synthesized during the stringent response (Section 12.2.1) with an increase in (p)ppGpp concentration. Polyphosphate is not synthesized in a PhoB mutant. PhoB is the regulator protein of the two-component *pho* system (Section 12.2.3). The free energy change in polyphosphate hydrolysis is -38 kJ/mol phosphate and this is bigger than that of ATP (-30.5 kJ/mol ATP) and smaller than that of 1,3-diphosphoglycerate (-49.4 kJ/mol 1,3-diphosphoglycerate). As mentioned above, polyphosphate hydrolysis is coupled to the synthesis of nucleotide triphosphate, including ATP, catalyzed by ATP-polyphosphate phosphotransferase (polyphosphate kinase). In some bacteria, including a strain of *Acinetobacter*, polyphosphate:AMP phosphotransferase converts AMP to ADP coupled to polyphosphate hydrolysis:

$$\text{(P)}_n + \text{AMP} \xrightarrow{\substack{\text{polyphosphate:} \\ \text{AMP phosphotransferase}}} \text{(P)}_{n-1} + \text{ADP}$$

Species of the genera *Propionibacterium* and *Micrococcus* possess polyphosphate glucokinase that phosphorylates glucose consuming polyphosphate:

$$\text{(P)}_n + \text{glucose} \xrightarrow{\text{polyphosphate glucokinase}} \text{(P)}_{n-1} + \text{glucose-6-phosphate}$$

Klebsiella pneumoniae (*Klebsiella aerogenes*) has a polyphosphatase that hydrolyzes polyphosphate without conserving energy. The function of this enzyme in this bacterium is not known.

$$\text{(P)}_n + \text{H}_2\text{O} \xrightarrow{\text{polyphosphatase}} \text{(P)}_{n-1} + \text{P}_i$$

Pyrophosphate (PP$_i$) is produced in many catabolic reactions from the hydrolysis of nucleoside triphosphate. The free energy change in the hydrolysis of PP$_i$ is -28.8 kJ/mol PP$_i$, which is smaller than that of ATP and polyphosphate. Many organisms have phosphatase activity that hydrolyzes PP$_i$ without conserving energy. In some bacteria, PP$_i$ functions like ATP. Phosphatase activity is low in species of *Desulfotomaculum*. These have an acetate:pyrophosphate kinase that phosphorylates acetate consuming PP$_i$ (Section 9.3.2). A purple non-sulfur bacterium, *Rhodospirillum rubrum*, has pyrophosphate phospho-hydrolase on the cytoplasmic membrane that synthesizes PP$_i$ using energy available from photosynthetic electron transport, and hydrolyzes it to build up a proton motive force.

Propionibacterium shermanii metabolizes glucose through the EMP pathway to pyruvate, which is then fermented to propionate (Section 8.7.1). In this bacterium, PP$_i$ is used as the substrate of pyrophosphate fructokinase (diphosphate – fructose-6-phosphate 1-phosphotransferase) and pyruvate phosphate dikinase.

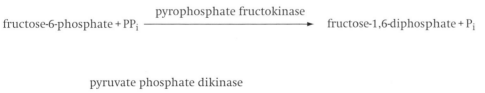

$$\text{fructose-6-phosphate} + \text{PP}_i \xrightarrow{\text{pyrophosphate fructokinase}} \text{fructose-1,6-diphosphate} + \text{P}_i$$

$$\text{PEP} + \text{AMP} + \text{PP}_i \xrightleftharpoons{\text{pyruvate phosphate dikinase}} \text{pyruvate} + \text{ATP} + \text{P}_i$$

13.3 | Resting cells

Many bacteria differentiate into resting cells when the growth environment becomes unfavourable. These include spores in low $G+C$ Gram-positive bacteria and actinomycetes, cysts, and viable but non-culturable cells. In addition to the formation of resting cells, some bacteria differentiate into other specialized cells including bacteroids in symbiotic nitrogen-fixing bacteria, heterocysts in cyanobacteria (Section 6.2.1.4), swarmer cells in *Caulobacter cereus*, and fruiting bodies in myxobacteria.

13.3.1 Sporulation in *Bacillus subtilis*

Some low $G + C$ Gram-positive bacteria, including species of *Bacillus* and *Clostridium*, form spores under nutrient-limited and other adverse growth conditions. Spores can maintain their viability for many years.

As in other cell differentiation processes, sporulation is the result of a complex regulated process which includes signal transduction from environmental and physiological factors. During the spore-forming process the cells cannot propagate. Sporulation is regulated through a series of phosphate transfers known as a phosphorelay. Phosphorylated Spo0 A (a two-component response regulator) plays a pivotal role in sporulation, activating the genes for sporulation and repressing the genes for vegetative growth (Figure 13.7). Spo0 A is phosphorylated through sporulation kinase A (KinA) or B (KinB), Spo0 F (a two-component response regulator) and Spo0 B (sporulation initiation phosphotransferase).

13.3.2 Cysts

In some bacteria such as *Azotobacter* and *Cytophaga*, resting cells known as cysts are formed. A cyst differs from a vegetative cell both in size and morphology. A cyst is resistant against desiccation and UV like a spore, but is not heat resistant. The outer wall of a cyst (exine) consists of protein and alginate. Genes for alginate production are transcribed by RNA polymerase with the extracytoplasmic function σ-factor, σ^E.

13.3.3 Viable but non-culturable (VBNC) cells

Many soil and marine bacteria are in a living state but fail to grow on routine bacteriological media. They are referred to as viable but non-culturable (VBNC) cells. Some authors claim that VBNC is not a proper term since some of these cells can be 'resuscitated' under proper conditions. Instead, the term 'yet-to-be-cultured' can be used. The VBNC state differs from the starvation survival state, which remains fully culturable even though the metabolic activity is very low, as in VBNC cells. Actively growing cells of various bacteria may enter the VBNC state under diverse environmental stresses including starvation, incubation outside the temperature range of growth, high osmotic pressure and exposure to white light. These stresses might be lethal if the cells did not enter this dormant state.

In addition to the failure to grow on routine bacteriological media, cells in the VBNC state differ from actively growing cells both morphologically and physiologically. VBNC cells are smaller and have increased cell wall cross-linking and extensively modified cytoplasmic membranes with altered fatty acid composition. The

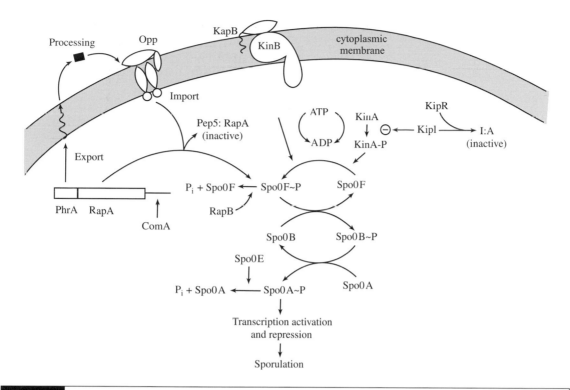

Figure 13.7 | **Sporulation in *Bacillus subtilis*.**

(*Trends Microbiol.* **6**:366–370, 1998, *Curr. Opin. Microbiol.* **1**:170–174, 1998)

KinA and KinB consume ATP to phosphorylate Spo0 F that transfers phosphate to Spo0 A through Spo0 B. This phosphate transfer is referred to as a phosphorelay. The phosphorylated Spo0 A represses the genes for vegetative cell growth and activates the genes for spore formation. The phosphorelay is under the control of four different steps. KapB regulates the activity of the membrane-bound KinB, and the cytoplasmic enzyme KinA is under the control of antikinase (KipI) and anti-antikinase (KipR). The phosphorylated Spo0 A is dephosphorylated by Spo0 E and dephosphorylation of Spo0 F-P is under the control of RapA and RapB. It is not fully understood how the signal is transduced to KapB, KipI, KipR, Spo0 E, RapA and RapB.

KapB, kinase-associated protein B; KinA, sporulation kinase A; KinB, sporulation kinase B; KipI, antikinase (KinA inhibitor); KipR, anti-antikinase (*kip* operon transcriptional regulator); RapA, response regulator aspartate phosphatase A; RapB, response regulator aspartate phosphatase B; PhrA, phosphatase (RapA) inhibitor; Spo0 A, two-component response regulator; Spo0 B, sporulation initiation phosphotransferase; Spo0 E, negative sporulation regulatory phosphatase; Spo0 F, two-component response regulator; ComA, two-component response regulator.

---

expression of starvation and cold shock proteins is increased, and the cells show reduced activities in nutrient uptake, respiration and the synthesis of macromolecules.

Several hypotheses have been proposed to explain why cells in the VBNC state cannot form colonies on routine media. High concentrations of nutrients may be toxic to cells in the VBNC state because of the formation of free radicals, although this is not known conclusively.

It is noteworthy that a protein produced by *Micrococcus luteus* at stationary phase promotes resuscitation of VBNC state cells of many high G + C Gram-positive bacteria including the producing organism. This protein is referred to as resuscitation-promoting factor (Rpf) or bacterial cytokine. Rpf is known in other bacteria including *Mycobacterium tuberculosis*, *Corynebacterium glutamicum* and *Streptomyces* spp.

13.3.4 Nanobacteria

In seawater and soil, cells with a size of less than 0.2 μm have been described for many years. These do not form colonies on routine solid media. They are referred to as ultramicrobacteria or nano-bacteria. It was believed that they are in a kind of VBNC state that developed under nutrient-limited conditions. However, a sea-water isolate, *Sphingomonas alaskensis*, has cells of this size while it grows. This bacterium is an oligotroph with a very small genome. This bacterium loses viability under nutrient-rich conditions, and is resistant to stresses including heat, hydrogen peroxide and ethanol.

13.3.5 Programmed cell death (PCD) in bacteria

Programmed cell death, or apoptosis, is a suicide process of active cells in multicellular eukaryotes. Cells no longer needed during a developmental process or those damaged by heat or other lethal agents are destroyed through proteolysis for the preservation of the tissue and/or the individual organism.

Although PCD has been regarded as a eukaryotic phenomenon, similar processes are known in some bacteria. Bacteroids and heterocysts (Section 6.2.1.4) do not divide or return to vegetative cells. When their function is no longer required, they are subject to autolysis like the parental cells of endospore formers. These are examples of PCD in bacteria that are similar to eukaryotic developmental PCD. Damaged cells are eliminated through autolysis as in a PCD mechanism. Damaged cells that have survived toxic stress induced by various compounds could consume nutrients but would produce few if any offspring which would be a burden for the whole population. PCD of damaged cells therefore benefits the entire population. Bacteria thus appear to be able to exhibit social behaviour. Many live in large, complex and organized communities such as biofilms, while fruiting bodies arise through the initiation of developmental processes that are directly analogous to multicellular eukaryotic organisms. Multicellularity in the bacterial world, in its various forms, might be more the norm than a deviation. With some of these known similarities to the developmental processes of eukaryotes, 'altruistic' behaviour in bacteria might be a predictable rather than an unexpected phenomena.

In *Staphylococcus aureus*, various genes have been identified that are related to the regulation of autolysis and murein hydrolase. The genes are believed to code for regulatory proteins involved in PCD. Actively metabolizing bacterial cells maintain a certain level of proton motive force (Section 5.7) with a relatively low pH microenvironment at the cell surface. When a cell is damaged with dissipation of the proton motive force, changes in the cell surface pH are sensed and transduced by a two-component system consisting of autolysin sensor kinase (LytS) and sensory transduction protein (LytR). It has been hypothesized that this signal represses the activity of antiholin-like protein A (LrgA) and B (LrgB), and activates holin-like protein A (CidA) and B (CidB). LrgA and LrgB inhibit murein hydrolase activity while the enzyme activity is activated by CidA and CidB for PCD. The homologues of *lrgAB* have been identified in many prokaryotes, including archaea and Gram-positive and Gram-negative bacteria, indicating that PCD is probably widespread in prokaryotes.

FURTHER READING

Survival and energy

Aertsen, A. & Michiels, C. (2004). Stress and how bacteria cope with death and survival. *Critical Reviews in Microbiology* **30**, 263–273.

Aertsen, A. & Michiels, C. W. (2005). Diversity or die: generation of diversity in response to stress. *Critical Reviews in Microbiology* **31**, 69–78.

Engelberg-Kulka, H. & Glaser, G. (1999). Addiction modules and programmed cell death and antideath in bacterial cultures. *Annual Review of Microbiology* **53**, 43–70.

Errington, J., Daniel, R. A. & Scheffers, D. J. (2003). Cytokinesis in bacteria. *Microbiology and Molecular Biology Reviews* **67**, 52–65.

Ferenci, T. (2001). Hungry bacteria: definition and properties of a nutritional state. *Environmental Microbiology* **3**, 605–611.

Lazazzera, B. A. (2000). Quorum sensing and starvation: signals for entry into stationary phase. *Current Opinion in Microbiology* **3**, 177–182.

Lewis, K. (2000). Programmed death in bacteria. *Microbiology and Molecular Biology Reviews* **64**, 503–514.

Matic, I., Taddei, F. & Radman, M. (2004). Survival versus maintenance of genetic stability: a conflict of priorities during stress. *Research in Microbiology* **155**, 337–341.

Morita, R. Y. (1999). Is H_2 the universal energy source for long-term survival? *Microbial Ecology* **38**, 307–320.

Mukamolova, G. V., Kaprelyants, A. S., Kell, D. B. & Young, M. (2003). Adoption of the transiently non-culturable state – a bacterial survival strategy? *Advances in Microbial Physiology* **47**, 65–129.

Nystroem, T. (1998). To be or not to be: the ultimate decision of the growth-arrested bacterial cell. *FEMS Microbiology Reviews* **21**, 283–290.

Nystrom, T. (2004). Growth versus maintenance: a trade-off dictated by RNA polymerase availability and sigma factor competition? *Molecular Microbiology* **54**, 855–862.

Peterson, C. N., Mandel, M. J. & Silhavy, T. J. (2005). *Escherichia coli* starvation diets: essential nutrients weigh in distinctly. *Journal of Bacteriology* **187**, 7549–7553.

Rice, K. C. & Bayles, K. W. (2003). Death's toolbox: examining the molecular components of bacterial programmed cell death. *Molecular Microbiology* **50**, 729–738.

Romling, U., Gomelsky, M. & Galperin, M. Y. (2005). C-di-GMP: the dawning of a novel bacterial signalling system. *Molecular Microbiology* **57**, 629–639.

Wai, S. N., Mizunoe, Y. & Yoshida, S. (1999). How *Vibrio cholerae* survive during starvation. *FEMS Microbiology Letters* **180**, 123–131.

Reserve materials

Aldor, I. S. & Keasling, J. D. (2003). Process design for microbial plastic factories: metabolic engineering of polyhydroxyalkanoates. *Current Opinion in Biotechnology* **14**, 475–483.

Alvarez, H. M. & Steinbuchel, A. (2002). Triacylglycerols in prokaryotic microorganisms. *Applied Microbiology and Biotechnology* **60**, 367–376.

Arguelles, J. C. (2000). Physiological roles of trehalose in bacteria and yeasts: a comparative analysis. *Archives of Microbiology* **174**, 217–224.

Brown, M. R. & Kornberg, A. (2004). Inorganic polyphosphate in the origin and survival of species. *Proceedings of the National Academy of Sciences, USA* **101**, 16085–16087.

Garcia-Contreras, R., Celis, H. & Romero, I. (2004). Importance of *Rhodospirillum rubrum* H$^+$-pyrophosphatase under low-energy conditions. *Journal of Bacteriology* **186**, 6651–6655.

Kornberg, A. (1995). Inorganic polyphosphate: towards making a forgotten polymer unforgettable. *Journal of Bacteriology* **177**, 491–496.

Kulaev, I. & Kulakovskaya, T. (2000). Polyphosphate and phosphate pump. *Annual Review of Microbiology* **54**, 709–734.

Kulaev, I., Vagabov, V. & Kulakovskaya, T. (1999). New aspects of inorganic polyphosphate metabolism and function. *Journal of Bioscience and Bioengineering* **88**, 111–129.

Ladygina, N., Dedyukhina, E. G. & Vainshtein, M. B. (2006). A review on microbial synthesis of hydrocarbons. *Process Biochemistry* **41**, 1001–1014.

Luengo, J. M., Garcia, B., Sandoval, A., Naharro, G. & Olivera, E. R. (2003). Bioplastics from microorganisms. *Current Opinion in Microbiology* **6**, 251–260.

Stubbe, J., Tian, J., He, A., Sinskey, A. J., Lawrence, A. G. & Liu, P. (2005). Nontemplate-dependent polymerization processes: polyhydroxyalkanoate synthases as a paradigm. *Annual Review of Biochemistry* **74**, 433–480.

Waltermann, M. & Steinbuchel, A. (2005). Neutral lipid bodies in prokaryotes: recent insights into structure, formation, and relationship to eukaryotic lipid depots. *Journal of Bacteriology* **187**, 3607–3619.

Resting cells

Barer, M. R. & Harwood, C. R. (1999). Bacterial viability and culturability. *Advances in Microbial Physiology* **41**, 93–137.

Cohen-Gonsaud, M., Keep, N. H., Davies, A. P., Ward, J., Henderson, B. & Labesse, G. (2004). Resuscitation-promoting factors possess a lysozyme-like domain. *Trends in Biochemical Sciences* **29**, 7–10.

Errington, J. (2001). Septation and chromosome segregation during sporulation in *Bacillus subtilis*. *Current Opinion in Microbiology* **4**, 660–666.

Hilbert, D. W. & Piggot, P. J. (2004). Compartmentalization of gene expression during *Bacillus subtilis* spore formation. *Microbiology and Molecular Biology Reviews* **68**, 234–262.

Hoch, J. A. (1998). Initiation of bacterial development. *Current Opinion in Microbiology* **1**, 170–174.

Kaprelyants, A. S., Gottschal, J. C. & Kell, D. B. (1993). Dormancy in non-sporulating bacteria. *FEMS Microbiology Reviews* **104**, 271–286.

Keep, N. H., Ward, J. M., Cohen-Gonsaud, M. & Henderson, B. (2006). Wake up! Peptidoglycan lysis and bacterial non-growth states. *Trends in Microbiology* **14**, 271–276.

Kell, D. B. & Young, M. (2000). Bacterial dormancy and culturability: the role of autocrine growth factors. *Current Opinion in Microbiology* **3**, 238–243.

Leadbetter, J. R. (2003). Cultivation of recalcitrant microbes: cells are alive, well and revealing their secrets in the 21st century laboratory. *Current Opinion in Microbiology* **6**, 274–281.

Moir, A. (2003). Bacterial spore germination and protein mobility. *Trends in Microbiology* **11**, 452–454.

Nystrom, T. (2004). Stationary-phase physiology. *Annual Review of Microbiology* **58**, 161–181.

Oliver, J. D. (2005). The viable but nonculturable state in bacteria. *Journal of Microbiology-Seoul* **43**, 93–100.

Paredes, C. J., Alsaker, K. V. & Papoutsakis, E. T. (2005). A comparative genomic view of clostridial sporulation and physiology. *Nature Reviews Microbiology* **3**, 969–978.

Setlow, P. (2003). Spore germination. *Current Opinion in Microbiology* **6**, 550–556.

Stephenson, K. & Hoch, J. A. (2002). Evolution of signalling in the sporulation phosphorelay. *Molecular Microbiology* **46**, 297–304.

Tyson, G. W. & Banfield, J. F. (2005). Cultivating the uncultivated: a community genomics perspective. *Trends in Microbiology* **13**, 411–415.

Vainshtein, M. B. & Kudryashova, E. B. (2000). Nanobacteria. *Microbiology-Moscow* **69**, 129–138.

Zhang, C. C., Laurent, S., Sakr, S., Peng & Bedu, S. (2006). Heterocyst differentiation and pattern formation in cyanobacteria: a chorus of signals. *Molecular Microbiology* **59**, 367–375.

Index